WATER POLICY ENTREPRENEURS

To our fathers, Gabe and Henk

Water Policy Entrepreneurs

A Research Companion to Water Transitions around the Globe

Edited by

Dave Huitema

Senior Researcher, Institute for Environmental Studies, Vrije Universiteit Amsterdam, the Netherlands

Sander Meijerink

Assistant Professor, Institute for Management Research, Radboud University Nijmegen, the Netherlands

Edward Elgar

Cheltenham, UK • Northampton, MA, USA

Published by
Edward Elgar Publishing Limited
The Lypiatts
15 Lansdown Road
Cheltenham
Glos GL50 2JA
UK

Edward Elgar Publishing, Inc.
William Pratt House
9 Dewey Court
Northampton
Massachusetts 01060
USA

A catalogue record for this book
is available from the British Library

Library of Congress Control Number: 2009936762

Mixed Sources
Product group from well-managed forests and other controlled sources
www.fsc.org Cert no. SA-COC-1565
© 1996 Forest Stewardship Council

ISBN 978 1 84844 331 0

Printed and bound in Great Britain by MPG Books Group, UK

Co-published by IWA Publishing, Alliance House, 12 Caxton Street, London SW1H 0QS, UK
Tel. +44 (0) 20 7654 5500, Fax +44 (0) 20 7654 5555
publications@iwap.co.uk
www.iwapublishing.com
ISBN 1843393158
ISBN13 9781843393153

Contents

Figures

Tables and boxes

Tables

Box

Maps

Contributors

Sezin Alp is a Master's student in the Department of Political Science and Public Administration at the Middle East Technical University, Turkey. She researches the neoliberal transformation of the water resources management policies in Turkey.

Argun Baskan is an assistant researcher in the Department of International Relations at the Middle East Technical University, Turkey. His work focuses on the international politics of energy resources.

Gert Becker is a researcher at the Institute for Environmental Studies (IVM) at the Vrije Universiteit Amsterdam. He tracks and examines new practices of adaptive management in the German part of the Rhine basin.

Anjali Bhat is a PhD candidate at the Centre for Development Research (ZEF) at the University of Bonn, Germany. Her doctoral research explores the impact of decentralization policies on integrated river basin management in Indonesia.

Dorri te Boekhorst is a junior researcher at Radboud University Nijmegen. Her research field is water management in the Yangtze River basin in the People's Republic of China.

Zsuzsanna Flachner is a researcher at the Research Institute for Soil Science and Agrochemistry, Hungarian Academy of Sciences. She also guest lectures at the Technical and Economic University, Budapest and organizes training for regional decision-makers on sustainable resource management. Her major research field is sustainability assessment and integrated river basin management.

Nuria Font is a professor of political science at the Autonomous University of Barcelona. Her research fields include environmental governance and European Union (EU) institutions and policy-making.

Victor Galaz is a researcher and research theme leader at the Stockholm Resilience Centre, Stockholm University. His research interests include resilience theory, governance theory, crises management, epidemics governance and applications of complexity theory to social systems.

Po Garden is the project director of Internews's Earth Journalism Network. At the time of contributing to this book, he was a researcher at the Unit for Social and Environmental Research (USER), Faculty of Social Science, Chiang Mai University, Chiang Mai, Thailand. His interests lie in climate change, science and decision-making, and local government.

Jaqui Goldin is the head of the African Water Issues Research Unit (AWIRU) at the University of Pretoria. She is also the director of Ecolink, an Environmental Education and Development Training Centre in Mpumalanga, South Africa. She investigates governance and the role of knowledge and trust between stakeholders in the water sector.

Joyeeta Gupta is a professor of climate change policy and law at the Institute for

Environmental Studies of the Vrije Universiteit Amsterdam and of water law and policy at UNESCO-IHE Institute for Water Education in Delft.

Sara Hughes is a doctoral candidate at the Bren School of Environmental Science and Management at the University of California, Santa Barbara. Her research focuses on institutional change and policy innovations, the science–policy interface and urban water infrastructure.

Dave Huitema is a senior researcher at the Institute for Environmental Studies (IVM) at the Vrije Universiteit Amsterdam. He coordinates the Research Cluster on Water Governance and Economics and investigates the adaptive governance of natural resources.

Helen Ingram is a professor emerita in planning, policy and design, and in political science, and a former Drew, Chace and Erin Warmington Chair in the School of Social Ecology, University of California, Irvine. Her research specializations are public policy, US–Mexico relations and environmental resource management.

Aysegül Kibaroglu is an associate professor of international relations at the Middle East Technical University, Turkey. Her areas of research include transboundary water politics, international law, political geography, environmental security and Turkish water policy.

Deusdedit Kibassa is a research scientist and a contributing researcher for the African Water Issues Research Unit (AWIRU). He participates in various research projects at the Institute of Resources Assessment (IRA) at the University of Dar es Salaam. His work has centred on water policy issues, energy and ecological interactions.

Louis Lebel is the founding and current director of the Unit for Social and Environmental Research (USER), Faculty of Social Science, Chiang Mai University, Thailand. He has ongoing research interest in ecology, public health, political science, social justice and environmental governance in the Southeast Asia region.

Lei Gang is programme manager of the WWF Yangtze Programme. Prior to joining WWF, he worked as technical chief of East Dongting Lake Nature Reserve. His major research field is water birds.

Raul Lejano is an associate professor of public policy at the University of California, Irvine. He is a theorist of collective action and studies innovative institutional designs for environmental governance.

Li Lifeng is the freshwater programme director at WWF International, and leads WWF's freshwater conservation in many large river basins. His research interests include river basin management, water and ecological footprints, and freshwater ecoregion conservation.

Piotr Matczak is a researcher at the Research Center for Agricultural and Forest Environment (Polish Academy of Sciences) and at the Center for Public Policy (Adam Mickiewicz University, Poznan, Poland). His work focuses on the social and economic background of environmental problems, particularly in relation to climate change, and on local governance.

Jennifer McKay is a professor of business law and director of the Centre for Comparative Water Policies and Laws at the University of South Australia, Adelaide. She considers the factors relevant to the adoption of sustainable development laws for major water user groups, such as water supply businesses and major agricultural and urban users.

Sander Meijerink is an assistant professor at the Institute for Management Research (IMR) at the Radboud University Nijmegen, the Netherlands. His research interests include policy continuity and change, water and spatial governance and climate adaptation.

Peter P. Mollinga is senior researcher at the Center for Development Research (ZEF) at the University of Bonn, Germany. His research fields are water governance and water politics, and inter- and transdisciplinary approaches to natural resources management. He consults on education and capacity-building for integrated water resources management in India and Indonesia.

Sakkarin Na Nan is a doctoral student at the Regional Center for Social Science and Sustainable Development at Chiang Mai University and a former assistant researcher at the Unit for Social and Environmental Research (USER), Chiang Mai University, Thailand.

Vishal Narain is an associate professor at the School of Public Policy and Governance at the Management Development Institute (MDI), India. He researches different aspects of water policy and institutions, water rights and local governance in water.

Per Olsson is a researcher and research theme leader at the Stockholm Resilience Centre, Stockholm University. His work on adaptive governance of dynamic land- and seascapes follows his primary research interest in linked social–ecological system dynamics and resilience.

Lena Partzsch is a postdoctoral researcher and project leader at the Helmholtz Centre for Environmental Research (UFZ) in Leipzig, Germany. As a graduate fellow of the Wuppertal Institute, her research focuses on new forms of governance with a special interest in environmental issues.

Toine Smits is a professor of sustainable water management at Radboud University Nijmegen, the Netherlands. His research fields are the sustainable management and functionality of river basins.

Joan Subirats is a professor of political science and director of the Institute of Government and Public Policy at the Autonomous University of Barcelona. His research fields include policy analysis, democratic innovation, public participation and local and regional government.

Nutthawat Subsin is an assistant researcher at Chiang Mai University and formerly a research assistant at the Unit for Social and Environmental Research (USER), Chiang Mai University, Thailand.

Anthony Richard Turton is a director of TouchStone Resources (Pty) Ltd. TouchStone seeks to grow economies in water- and energy-constrained circumstances by unlocking the potential of new water and new energy. He specializes in transboundary water issues and is a professional speaker.

Saskia E. Werners is a researcher at Wageningen University and Research Centre in the Netherlands. Her research on adaptation to climate change in managed river basins is firmly rooted in the global change community while branching out into institutional as well as biophysical aspects.

Margaret Wilder is an assistant professor of Latin American studies, public policy studies, and geography and development at the University of Arizona, Tucson, USA. She researches water governance, water and development, and the political ecology of water in Mexico and other developing countries.

Yu Xiubo is an associate professor at the Institute of Geographic Sciences and Natural Resources Research, Chinese Academy of Sciences. His research fields include integrated river basin management, ecosystem management and policy.

Zhang Chen is a programme manager at the World Wide Fund for Nature (WWF), Beijing Office. His work mainly covers wetland conservation in the Yangtze River basin.

Preface

The first plans for this book project were developed early in 2007, when we discovered our shared interest in water governance and policy science. In that year we started discussing possibilities for editing a volume on water policy transitions around the globe. The Amsterdam Conference on the Human Dimensions of Global Environmental Change in May 2007 offered a first opportunity to discuss our plans with others interested in continuity and change in water policies. There we met Per Olsson of Stockholm University, Saskia Werners of the Wageningen University and Research Centre, and Zsuzsanna Flachner of the Hungarian Academy of Sciences. We are happy that they belonged to the first group of scholars who committed to the book project. This contributed much to the quality of this book, but even more to the fun we would have on various other occasions when we met to review progress. At the conference we also established our first contact with Felicity Plester, Commissioning Editor of Edward Elgar Publishing. She was enthusiastic about our book proposal from the very beginning, providing a main reason for us to start working with Edward Elgar on publishing this book. Felicity was always there to answer our questions, allow flexibility over deadlines when necessary and to do an excellent job during the production process.

Another important milestone came with the acceptance of our panel proposal for the 4th ECPR General Conference, in Pisa, September 2007. In Italy, we met Sara Hughes of the University of California at Santa Barbara who at that time was a Fulbright scholar at the University of South Australia, Peter Mollinga of the University of Bonn, and Per Olsson. Our fruitful discussions happily led Sara and Peter to promise chapters on Australia and Indonesia respectively. Two months later, we presented a draft of the theoretical introduction to this book at the Conference on Adaptive and Integrated Water management (CAIWA), in Basel, Switzerland. Here we had the opportunity to introduce our project to Helen Ingram, professor emerita of planning, policy and design, and of political science at the University of California at Irvine. We had both met her before and admire her for her in-depth knowledge of water governance and policy science, her contagious enthusiasm and her social skills (something often lacking in established scholars). We were greatly honoured when Professor Ingram committed to write a chapter on water policy transitions in the United States of America (USA) and we have benefited greatly from her advice during the making of this book.

Because we received a lot of interest in our research questions, we decided to organize a two-day international symposium on the role of policy entrepreneurs in realizing water policy transitions. The symposium, organized in July 2008 at the Royal Netherlands Academy of Arts and Sciences (KNAW) and the Vrije Universiteit Amsterdam, was made possible through a conference subsidy of the Royal Netherlands Academy of Arts and Sciences, financial contributions by the EU NeWater (New Approaches to Adaptive Water Management under Uncertainty) project, the Dutch knowledge impulse programme Living with Water, and the research programme Governance and Places (GaP) of the Nijmegen School of Management, Radboud University Nijmegen. The first day of the symposium saw intense discussion by water management experts from

Australia, Asia, Africa, the US and Europe of the roles of policy entrepreneurs in water policy transitions. We are much indebted to Helen Ingram, Louis Lebel of Chiang Mai University and Anthony Turton of TouchStone Resources, for their guidance of the deliberations, and their help on defining key themes of transition management. We owe a debt of gratitude also to our other chapter authors for their attendance and input at the conference. On the second day of the symposium reviews of the draft chapters became the subject of the meeting. We are grateful to the reviewers, since their points and questions helped a great deal to sharpen the research questions and contributed much to the coherence of this volume. The social programme, which included a visit to the National Parc de Hoge Veluwe and the Kröller-Müller Museum, generated enjoyable and productive group dynamics. Saskia, thank you for your creative suggestions for the symposium programme and workshop in Amsterdam. Jan, thank you for your practical assistance during these days.

A first draft of the conclusions was presented at the Freude am Fluss final conference, 'Space for the River, Space for People?' in Nijmegen, the Netherlands, October 2008. Our thanks go to Thomas Birkland of North Carolina State University for his willingness to discuss our work there, and his valuable suggestions. Around this time, we also met Paul Sabatier of the University of California, Davis, who raised our awareness of some methodological pitfalls in the case comparison. Finally we are much indebted to Ron Wunderink of the Nijmegen School of Management for drawing most of the maps included in this volume and to Maria Gordon who did an excellent job in revising our English and that of some other chapter authors. The interesting thing is that she actually did much more than we asked her to do. She put her finger on inconsistencies and pushed us and the authors to explain our stories better. We are very grateful for her essential contribution to this book.

Dave Huitema
Sander Meijerink

PART I

INTRODUCTION

1 Transitions in water management: positioning this book

Dave Huitema and Sander Meijerink

1.1 Water management: a field in flux

Water is a resource that man cannot do without. The functions it fulfils for humans range from direct support of our biological systems and of agriculture, to serving as a repository for waste, as a medium for recreation and as a political and cultural symbol, to mention some of the most important. Given its indispensability to humans it is no surprise that managing water, taking care that users have a sufficient amount of the necessary quality at the right time, has preoccupied human societies since they formed.

Obviously, methods of water management have changed considerably since human beings began to intervene in the natural water cycle. The field of water management continues to be in flux. Climate change has brought predictions of an increase in extreme water events and of rising sea levels (see for example, Easterling et al., 2000; Vellinga and Van Verseveld, 2000; Cabanes et al., 2001; Gleick at al., 2001; Alley et al., 2005). Serious flaws in the traditional engineering approach to water management have become clear in the past few decades, including the massive social and ecological damage caused by dams (see for example World Commission on Dams, 2000; Gleick, 2003; Stone, 2008). There has been much poorly planned development in arid and semi-arid areas, which has increased demand for water and resulted in considerable, often unanticipated environmental effects (see for example Schlesinger et al., 1990; Genxu and Guodong, 1999; Turner et al., 2007). Today sees a worldwide drive for the privatization of services such as the provision of drinking water, but this push now finds increasing opposition (see for example Wolff and Hallstein, 2005).

Actors contribute to water problems and bear the burden of their consequences in largely asymmetrical ways, creating a significant collective action problem (Conca et al., 2006; Meinzen-Dick, 1997, 2007). Institutions able to overcome or address this are mainly in a formative stage. The interrelation of water problems with many other issues, including political tensions between countries and groups within countries, complicates the situation (Burchi and Spreij, 2003; Blomquist et al., 2005; Kemper et al., 2005). Water managers as a result find themselves pressed to act by protagonists with widely different values, worldviews and understanding of the problem at hand (Falkenmark et al., 2004).

Water issues can thus constitute yet another example of 'wicked problems' (Rittel and Webber, 1973). In seeking resolution water managers face relatively high levels of uncertainty about the consequences of their actions, since they deal with socio-ecological systems that exhibit, among other qualities, complexity, non-reducibility, spontaneity, variability and collectivity (Dryzek, 1987, pp. 28–33; Young et al., 2006). This makes it likely that water managers do not and will not completely understand the ecological systems in which they intervene. Yet they often have little room for error as much depends on them finding the right answers (Clark, 2002). One way out of this situation is to increase the capability to change, the adaptability, of water management. In explaining this concept we can refer to Young et al. (2006), who contrast adaptability with adaptation and adaptedness. Adaptation refers to the process of structural change in response to structural circumstances. Effective adaptation results in adaptedness, meaning that a certain dynamic structure is effective in dealing with its current external environment. Adaptability, then, is about the capacity of water management for change and to adapt to future variations in the environment. Some researchers suggest that adaptability is enhanced by an emphasis on 'soft solutions', such as flood retention areas, community-scale infrastructure, decentralized and open decision-making systems, and their like (Gleick, 2003; Tàbara and Pahl-Wostl, 2007). Whether or not these measures actually are of the type needed to advance adaptability is open to debate, but the fact is that most authors agree that the capability to change will be an important feature of any water management system, given the uncertain context in which water managers operate in these times.

This book aims to increase understanding of the way in which water management changes and how such change can be directed. The starting point of our journey is the fact that government, government regulations and government policies play an important role in water management. In simple societies, the individual water user or group of water users can alter the way in which water is managed. However this is not the case in modern and complex societies. Even in countries where private property rights over water are strong and relevant, such rights are only valid if supported by government and only as far as they have not been superseded or pre-empted by a publicly established system of rights and obligations. It follows that changes in water management will often be preceded by changes in government policy and regulations, or must receive government confirmation and sanction after the fact. Therefore, to understand changes in water management, we need to understand policy change and its opposite, policy stability. In this book, we are interested specifically in radical policy change, which we refer to as a transition. We are also interested in the extent to which and through what strategies individuals or organizations can manage change.

1.2 Policy change

The world watched as Barack Obama was elected in 2008 as President of the United States of America (USA). The dominant theme of his campaign was 'change', expressing an intention to reverse many of the policies of his predecessor. Considerations of a political nature aside, Obama's rhetoric on policy change is noteworthy because it indicates the importance of credible alternative policies – 'Change we can believe in' – and suggests that the achievement of change through a coordinated effort is possible – 'Yes we can'.

The case of the US elections involves the concept of policy change. Obama promised not only substantively new policies, he also promised a new approach to policies, one to hinge less on executive authority and more on consultation and collaboration. This stance recognizes two different types of possible change. Substantive change sees government start on a new course in addressing a policy issue, such as water management. Some governments, for example, have become less keen on building large dams for safety or power generation and some have installed policies that forbid the use of groundwater where it was previously freely accessible. Other changes are governance transitions, where alterations occur in the way in which decisions are reached and water organizations are set up. Examples include the introduction of new decision-making techniques, such as cost–benefit analysis, and new decision-making procedures, like stakeholder participation methods. Organizational change can be seen, for example, in the increasing number of river basin management organizations. In this volume we deal with both types of policy change, concentrating particularly on what we call transitions, described basically as radical change.

The US election rhetoric serves as a contrast with the type of discussions about change considered in this book. The elections were 'high politics', and as such their outcomes will provide a context for US water management in the years to come. Viewed from the subsystem of policy, the elections themselves were not the change, but rather reflected a changed political mood that potentially offered a window of opportunity to change water policies. We adopt this perspective as we focus on the water policy subsystem, analysing events external to the system as opportunities for or constraints on change.

It is not clear at the time of writing in April 2009 what an Obama government will mean in terms of water management. We propose that external events, from elections to disastrous events such as floods, are open to different interpretations in terms of policy. Their meaning viewed retrospectively for the water domain tends to be less obvious than is often assumed. Even if accounts of history portray the response to external events as 'natural' or 'unavoidable', such accounts often ignore the fact that other responses were possible and were considered (see Birkland, 1998). It becomes necessary, then, to examine the way external events were perceived and the responses they generated from those in a certain policy subsystem. In addition, change does not necessarily stem from an outside stimulus. It can also be triggered by events within the policy subsystem including, for instance, changed leadership, the emergence of new visions, the evaluation and monitoring of progress, and so on.

We need to be aware, too, that policy change in itself is often not directly connected to the situation 'on the ground'. This is partly a matter of implementation, partly a matter of the changed position of government. Policies may change on paper, but this does not necessarily produce change in the approach of those involved in implementation. In this book, we look at policy change on paper, but where possible also at the way in which such changed policies are implemented – or not.

In terms of the changed position of government, this book's emphasis on water policy rather than the more fashionable topic of 'water governance' is somewhat related to the personal interest of the authors. However it is also based on a substantive position in the debate on governance. Strictly speaking, governance is not a new term, although it has certainly grown in popularity in the 1990s and 2000s (Pierre and Peters, 2000, p. 2). The *Shorter Oxford English Dictionary* defines 'to govern' as to guide, direct or steer society. Political scientists used to treat governance as a synonym for government (Stoker, 1998, p. 17), but the majority now consider them to be analytically distinct terms. Bevir and Rhodes (2003, p. 45) recently defined governance as 'a change in the nature or meaning of government'. Governance and government are often regarded not as discrete entities, but as two poles on a continuum of different governing types (Finer, 1970). If the extreme form of government was the 'strong state' in the era of 'big government' (Pierre and Peters, 2000, p. 25), then the equally extreme form of governance is an essentially self-organizing and coordinating network of societal actors (Schout and Jordan, 2005).

In focusing on change and stability in government policy we do not necessarily deny the changed position of government in many countries as an example of governance. We do, however, contend that government continues to be a very important factor in the way water is governed. In this way we connect to Pierre and Peters (2000, p. 25) who are circumspect when they argue that government endures in the new era of 'governance', but that its form and function can vary in several important respects.

This book focuses mainly on policy change at the level of the nation state, complemented by two chapters on 'higher' levels – policy change at the global and the European Union (EU) level – to contextualize the analysis. Why such a focus? Governance experts, in disagreement about many things, do agree that the importance of the nation state has decreased considerably in past decades because of privatization, globalization and often also decentralization (for example Jordan et al., 2007). This is undeniably true, with no exception for water management, but the nation state still holds a lot of sway, if not by controlling – with other nation states – the actions of international organizations, then by controlling 'lower' levels of government or by sanctioning the outcomes of private governance. Governing in reality has become more of a multilevel game, and the empirical chapters in our book highlight this phenomenon. We reiterate, however, that the nation state does play an important role and thus warrants a good deal of attention.

1.3 The central questions and position of this book

This book analyses 15 cases of national water policy change. The additional discussion of policy changes at the global and EU levels provides some sense of the international context for the cases studied. Map 1.1 shows the countries in which the cases are located.

The authors were asked to consider the following questions about radical changes in water policy they had observed in their country in the past three or four decades:

1. Who were the main policy entrepreneurs in the cases studied, and what strategies have they used to realize change; and likewise, who were the main defenders of the status quo and what strategies have they used to block change?
2. How do institutions constrain or enable policy entrepreneurs' efforts to either realize or block change?
3. What are the implications of these findings for those who aim to direct change?

We have been trained as policy scientists, and our inspiration as editors for this book came from our academic discipline. In further positioning the book, we address several issues, including our own position vis-à-vis the state of play in the policy sciences and our relationship to the emerging literature more generally on transitions. In a later section we address the issues of country selection and methodology.

The issue of policy change has been studied extensively in the policy sciences, as our treatment in the next chapter shows in more detail. For now it is sufficient to note that the established theories on policy change cast much doubt on whether policy change can be deliberately effected, although for different reasons. A strong current of thought contends that policy change is a random process (Kingdon, 1995), one not much due to direction and planning. One stream of studies suggests that policy subsystems are resistant to change and that radical policy change only happens when the existing paradigm

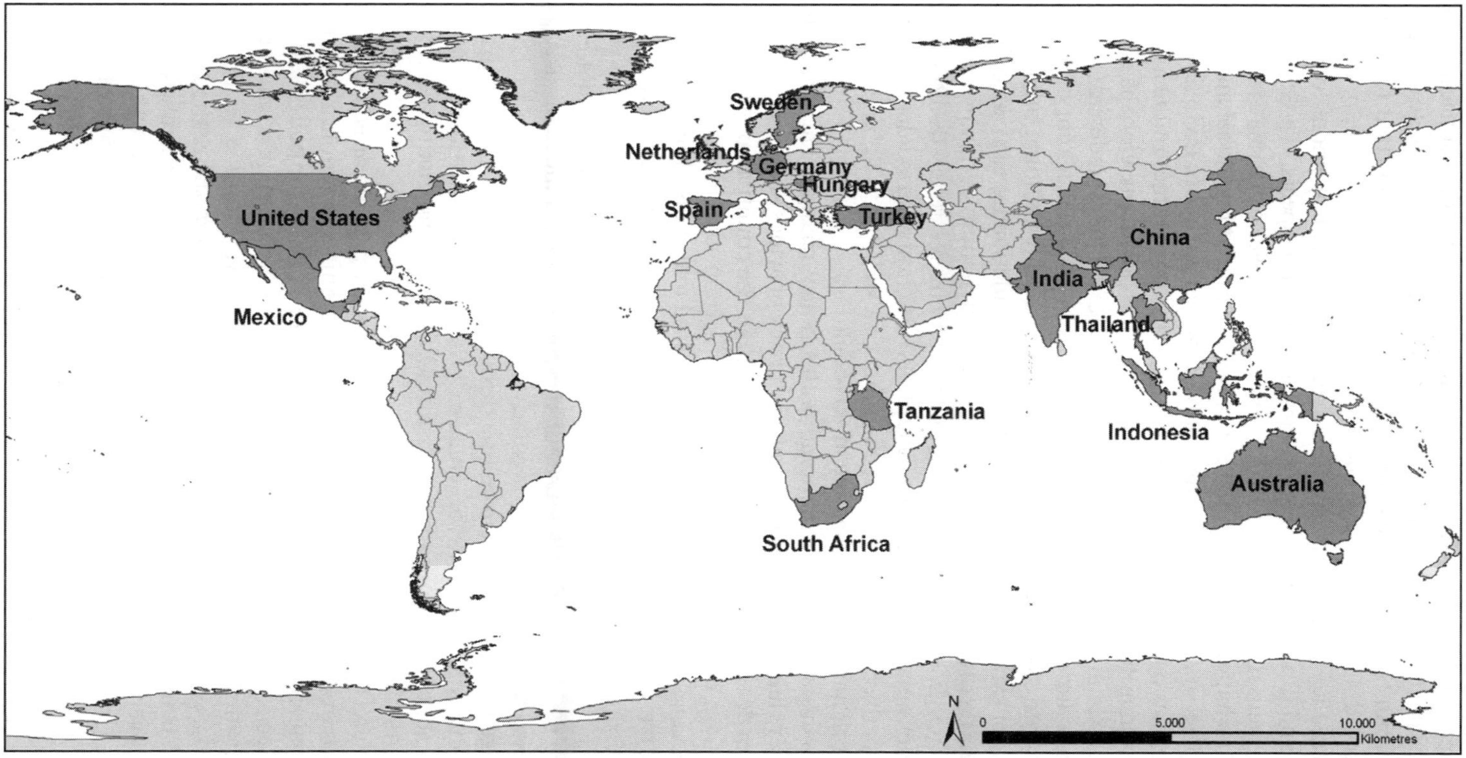

Map 1.1 Overview of countries as the locations of the cases presented

succumbs under external pressure and is 'punctuated' (see Baumgartner and Jones, 1991 for the foundation study of this stream). Yet another flow of studies (inspired by Sabatier and Jenkins-Smith, 1993) proposes that change is next to impossible without external shock events.

Given the need for adaptability and change, the emphasis in studies on resistance to change and the impossibility of planning change are worrying signs. The stakes are potentially high, particularly when the problem concerns ecological systems involving 'tipping points' beyond which their state fundamentally alters, resulting in serious consequences (see for example Lenton et al., 2008). Rapid and radical policy change may be key to the survival of societies as we know them today (see for example Duit and Galaz, 2008), but we need to distinguish between the normative and the positive. Even if it is normatively desirable that change can be managed, this might not be possible in reality. This book emanates from a more optimistic view than that taken, it seems, by the policy sciences in general. We can see quite clearly that policy change is usually not just a coincidence, policy paradigms do not simply break down on their own, and fundamental policy beliefs actually do change. It is also clear that some individuals and organizations are more successful than others in achieving policy change under similar circumstances. Indeed, more recent scholarship, such as work by Birkland (1997, 1998) and Olsson et al. (2006), indicates that policy change perhaps cannot be managed in the sense of being pre-planned and centrally controlled, but that it can at least be prepared for and 'steered' from point to point.

This gave us reason to revisit the various theoretical models mentioned above, and to find what they have to say about agency in policy change. The results of this exercise are explained in detail in the next chapter. Here it suffices to say that each of the models we examined does contain clues suggesting that certain individuals and organizations potentially effect policy change ('change agents') and about how they may achieve such an enterprise ('strategies'). The type of individual we are interested in is recognized in the policy sciences and other literatures as the change agent, usually one of a certain kind, such as a 'boundary worker', 'policy advocate' or 'visionary leader'. Subtle differences distinguish the meaning of such terms, but these are not the subject of discussion here, the various descriptions serving only to indicate the type of players that we have in mind. All such change agents will be referred to throughout this volume as 'policy entrepreneurs' (for example Kingdon, 1995). Our analysis in the next chapter produces a set of strategies that policy entrepreneurs may use, and which have been applied to the various cases in the book.

Because we use the term 'transition', we need to say a few words about the way in which our book relates to the burgeoning literature on socio-technical transitions and transition management (see for example Schot et al., 1994; Kemp et al., 2001; Rotmans et al., 2001; Geels, 2002; Smith et al., 2005; Kemp et al., 2007). A recent review of the transitions literature (Dewulf et al., 2008) suggests that the transition management literature has drawn a series of concepts and methods from several other theories, several of which emanate from the policy sciences (such as network governance and agenda-setting theories). The overview also indicates that the transition management literature, like this book, starts from the idea that radical change is necessary from an environmental perspective. In its approach to change, the transition management literature takes a perspective that reconciles planning and incrementalism in the sense that the best aspects of both approaches to change – the vision-providing aspect of planning and the uncertainty-handling aspect of incrementalism – are combined in what is labelled as 'goal-oriented modulation'.

Dewulf et al. (2008) criticize transition management theory on two grounds. First, they suggest that the transition management literature underestimates the complexities associated with goal-setting processes in goal-oriented transitions. Sustainability issues involve such a level of complexity and such a multitude of perspectives on problems and solutions, according to Dewulf et al., that the appropriateness and suitability of a transition goal ('a single outcome parameter') cannot be easily established, let alone allow a single person or organization (the 'transition manager') to set such a goal. We concur, which is one of the reasons why we speak here not just of change, but also about adaptability, necessary for further change if goals, means and results of change prove insufficient or unsatisfactory.

A second criticism developed by Dewulf et al. concerns the eclectic nature of transition management theory. Instead of striving for an integrated theory, Dewulf et al. (2008) suggest that multiple theories continue to be needed because they provide a set of conceptual tools to analyse situations and design interventions. Rather than merging theories, they advocate the distinctness of each theory and the exploration of zones where theories overlap or can inform each other. In the same vein, this book is intended as a contribution to the transition management literature from one discipline, the policy sciences. Within this discipline we look at various theories of policy change and check the extent to which they overlap in specifying a role for policy entrepreneurs in achieving policy change.

In this way we aim to contribute to the transition management literature, but take a somewhat different approach to that typical of the literature. As already indicated, we are less concerned than the transition management literature with the endpoint of transition processes. We are instead interested in a mechanism of change – policy entrepreneurs and their strategies – which has value regardless of the type of endpoint achieved by the change. Change will always be necessary and intermediate endpoints in policy development often need later adjustment ('adaptability'). In addition, by looking closely at policy change, our approach is much more limited than the one taken by the transition management literature so far, which essentially addresses a range of aspects, of which policy is only one. This book does not concern itself so much with socio-technical changes, although these obviously provide context for efforts to change policies. Many authors touch upon socio-technical regime factors, but our core interest is rather in the role of policy entrepreneurs across a range of such contexts, leaving a systematic exploration of context to others. We note also that the transition management literature has developed a scale that suggests that transitions can occur at three levels: the niche, the regime and the landscape (see for example Dewulf, 2008). Our approach is based on what can be called a 'jurisdictional scale', which distinguishes between various government levels (local, regional, national, transnational and global). Our principal interest lies in the national level and to some extent its interactions with lower and higher levels of government.

1.4 Methodology and case study selection

The selection of 15 case studies of national policy change in a range of countries (see Map 1.1) and of two cases of transnational policies requires attention to the method of selection, of analysis and, in our final chapter, of comparison.

In terms of case selection the plan to write this book originated from two panels on water policy transitions held at the Amsterdam Conference on the Human Dimensions of Global Environmental Change (May 2007) and the General Conference of the European Consortium for Political Research (ECPR) in Pisa (September 2007). A first version of the chapter on theory (Chapter 2) was presented at both events. In the course of organizing the panels, the first connections were made to other authors with an interest in water policy transitions. Further contacts were established at the Conference on Adaptive and Integrated Water Management (CAIWA) in Basel (December 2007). The positive response and the level of interest led to the organization of an open meeting on Transitions in Water Management in July 2008. As editors we actively sought contributions from diverse countries, aiming to include at least two contributions from Africa, Australasia, Europe and the Americas, and to feature stories based in world-leading countries such as China, India and the USA. This was largely accomplished although we regret not being able to include more case studies from Africa and South America.

No plan existed beforehand to include specific countries and the case study selection was not theory-led. This does not invalidate comparison of the cases presented. Hopkin (2002) holds in his treatment of state-of-the-art comparative international research that the 'most different systems approach' to international comparison is preferable. Such an approach works on the understanding that:

> [I]f a hypothesised relationship between two or more variables is replicated across a wide variety of different settings, then there are strong grounds for arguing that there is a causal link between the variables. This implies that attention should be shifted from the 'intersystemic' level, where variables such as the type of political regime are often examined, to the 'intrasystemic' level, in the hope of eliminating system-level variables (such as the political regime) from the inquiry and establishing generalisations across different settings. (Hopkin, 2002, p. 255)

We have indeed approached international comparison in this spirit. We present a great diversity of countries within the cases selected. If our hypothesized relationship between policy entrepreneurs and policy change mediated through policy entrepreneurs' strategies can be found across this diversity of settings, this will support our claim that policy entrepreneurs are key to policy change. We asked our authors to keep discussion of 'intersystemic' variables, such as the political regime, and of the national environmental situation, to a minimum. We are interested in the degree to which the 'intrasystemic' variable of institutions constrained or enabled policy entrepreneurs to bring about change. This singles out only one aspect of the many ways in which the countries are different, but holds relevance because it accounts for the possibility, for instance, that a policy entrepreneur operating in the autocratic policy system of China could be more constrained than a policy entrepreneur in the American system, which is relatively open.

Hopkin (2002) proposes that research techniques can vary according to the number of cases involved in a comparison, with quantitative strategies usually preferred for a large number of cases and qualitative strategies for a smaller quantity. Here, the number of cases is neither small nor large, but the authors have essentially followed a qualitative approach in an attempt to mine the richness of the policy change stories they have observed. Typically, the authors have used secondary analysis of the existing literature, documentation analysis and a set of interviews to assess and describe the dynamics in a particular country. In most cases, the authors have years of experience in analysing water management in the country concerned. Several measures have been taken to enhance comparability of the case studies. All authors were asked to read the chapter on theory (Chapter 2) as a basis for their analysis. This effectively meant that authors would identify policy entrepreneurs and their strategies and assess the effectiveness of these strategies in terms of policy change. The meeting in July 2008 clarified and improved the chapter on theory, and brought about an internal review process, which increased the coherence of this volume.

At the outset of the project we assumed that some of the policy entrepreneurs we would encounter operated at both national and international levels. Apart from a desire to contextualize our findings about policy change at the national level with a clear understanding of what is happening internationally, this was a major reason to include the international level in the book, represented by a chapter that focuses on the global water policy level and another centred on the EU level.

A few further remarks about the various country studies are in order. First, as the emphasis is on explaining change, most authors have analysed instances where transitions have been made – more or less – successfully. This approach largely precludes the analysis of failed transitions, which could be equally interesting from an analytical perspective. We found this a price worth paying as part of our research agenda (general question 3) inquired about lessons on how to direct change, which is where examples of the most successful changes can be very instructive. It should be clear, however, that policy processes tend to go in cycles, often bringing a second round for those who at first lost the battle over 'policy on paper.' Such players may attempt to influence the application of policy. We were minded initially to be satisfied with radical policy change on paper, but it became increasingly clear that opponents of change use the implementation process to block change on the ground. We have therefore reviewed implementation as part of our overall analysis.

Second, some remarks on the term 'case study'. Our collective interest in policy change at the national level since the 1960s and 1970s anticipated a relevant case in every country. Several authors, however, go 'below' the national level and analyse one or more (regional) situations that effectively constitute sub-case studies from the perspective of this book. For ease of reading, however, we use only the term 'case study'. We also take care, though, to identify the jurisdictional level of each transition featured in the book. The selection of such sub-case studies calls attention to the way national policy change in many cases is affected or exemplified by decision processes at the regional level. We asked the authors to base any selection of a sub-case study on its capacity to demonstrate either of these aspects and preferably as paradigmatic examples. Previous research indicated that this would entail cases characterized by high levels of media attention. Through our discussions at the various workshops, we came to realize that this approach leaves open a key question relating to the position of sub-case studies in the wider 'case study universe' in the various countries. It is difficult for most authors to answer this question, but we contend that limiting the topic to radical policy change and including mainly paradigmatic sub-case studies constrained the universe of instructive cases considerably, although probably less so in large countries than small ones.

1.5 Following chapter summaries

In Chapter 2 of this volume Meijerink and Huitema provide an overview of the way in which the policy science literature has explained policy change, focusing on the role of individuals, groups and organizations that advocate change. Although the policy sciences suggest that radical policy change is a rare phenomenon, a close reading of various influential theories of policy change suggests that policy entrepreneurs play a role in nudging policy in certain directions. Meijerink and Huitema skim these theories for descriptions of the way in which change can be effected, an effort that results in the identification of five strategies that policy entrepreneurs can apply: the development of new ideas; building coalitions and selling ideas; the recognition and exploitation of windows of opportunity; the use of multiple venues; and the orchestration and management of networks. These five strategies form the basic categorization applied to the featured cases, noting that they are not necessarily for use in any particular order and not necessarily all used in every change process.

In Chapter 3, Gupta sketches the way in which the law and policy regarding fresh water have evolved globally, thus describing the emerging field of global water governance. This chapter seeks to relate the context for many of the case studies of national change and the direction of such change as it interacts with what occurs at the global level. Such interactions are not necessarily 'top-down' in the sense that global developments steer national change. Rather, it seems that the global discourse on fresh water is heavily influenced by a limited number of rich countries (the 'global North') and that the poorer countries (the 'global South') – once they are accepted in the multitude of global fora – do not have much choice but to accept the institutional implications, including prescriptions such as water privatization. The acceptance of such ideas is often a condition for receiving international aid.

Gupta shows that a number of discourses flow at the global level, some of them water-specific. Every discourse has different proponents, but by comparison with some of the national case studies, it stands out that academics and other experts are prominent in shaping global discourses. Gupta distinguishes types of discourse as general-ideological, governance, environment-specific and water-specific. Of the water-specific discourses, many of the readers will recognize the following: the 'hydraulic mission', focusing on large-scale infrastructure; 'integrated water resource management', focusing on policy coordination and integration; and 'human rights to water', focusing on issues such as access to clean water. All of these discourses remain influential, propagated by different elements of the emerging and relatively fragmented global water governance system. The water-specific discourses interact with others, of which the general-ideological and the governance discourses figure most prominently. The general ideological discourse is one of neoliberalism, resulting in proposals for privatization and marketization (Ingram and Lejano speak of 'water as a commodity' in Chapter 4). In the discourse on governance, the term 'good governance' is very popular and ideas such as 'decentralization' and 'stakeholder participation' are advocated. As the rest of the book shows, many of these global ideas are enmeshed in discussions of water at the national level.

In Part II of this book, we follow some of the water transitions that have taken place in the US and Mexico. For Ingram and Lejano in Chapter 4 on a US transition, the concept of 'ways of knowing' takes centre stage. Ways of knowing are 'dynamic social sense-making systems' that are 'created through communication, discourse and relationships'. They distinguish, among others, ways of knowing based on economics, ecology, equity and sense of place. Each of these ways of knowing emphasizes different aspects of water, ranging from its status as a commodity to the importance of water for local identities. Most of these ways of knowing are also recognizable at the global level, as Gupta's chapter demonstrates. Ingram and Lejano suggest, on the basis of three case studies, that transitions occur when dominant ways of knowing water meet other ways of knowing, and that policy innovation is forged when a 'novel synthesis' is achieved. Ingram and Lejano not only address how policy change happens, but they also suggest a direction for such change. In their opinion, the challenge in times of climate change is to foster policy designs that incorporate or transcend different ways of knowing. This implies an imperative to build bridges across networks and for collective learning from problem-solving efforts, a theme also taken up by Olsson and Galaz in Chapter 17 based in Sweden.

In her case from Mexico, Wilder in Chapter 5 uses the concept of aperture as a perspective for the study of water policy change and to describe the breaking open of broad political and economic processes. Mexico is an interesting case because it has emerged, according to some, as a 'successful new globalizer'. The transitions discussed originate in Mexico's economic troubles in the 1980s. The reform of Mexican water law dates to 1992 and was in line with the ideas then advanced by the World Bank and Inter-American Development Bank, with the intention of achieving marketization, centralization and sustainability of the water management system.

Wilder focuses on the border state of Sonora and tracks the progress of these transitions by narrowing in on the functioning of irrigation districts and river basin councils. She contends that the water transition in Mexico cannot be regarded in isolation, but as part of a broader democratic transition working itself out in different parts of society, including water institutions. The chapter highlights the work of policy entrepreneurs, such as Julia Carabias Lillo who engaged in 'venue manipulation' by relocating water management in a different ministry, and Ulises Ruìz who proposed further legal amendments in 2003, remarking: 'Mexico's water policy appears stuck in a transitional phase.' The reason, according to Wilder, is that Mexico has seen alternating political parties take control since 2000, and that the political system operates in a way qualified by conflict and lack of consensus over the direction of Mexico's water policy, but radical policy change has occurred: the transition to a market logic and efficiency principles has been well consolidated, but the decentralization and sustainability policy proposals have become mired due to political fragmentation and the lack of strong political will. Decentralization presents a serious challenge to the status quo, and the galvanizing effect of the advocacy coalition pushing the change is fading as time passes since the focusing event that set the changes in motion. Policy entrepreneurs need time in such an environment, and run the risk of being replaced or ignored as administrations change.

In Part III, we look at transitions in several countries in Australasia. Te Boekhorst, Smits, Yu, Li, Lei and Zhang in Chapter 6 tell the story of the way China has moved towards integrated river basin management (IRBM). They relate the contribution made by the World Wide Fund for Nature (WWF) to China's transition. The WWF as a policy entrepreneur was also active in policy transitions in the Netherlands, Hungary and Germany (see Chapter 19 by Huitema and Meijerink, Chapter 14 by Werners et al., and Chapter 18 by Becker, respectively, in this volume).

In China, WWF launched two major programmes: the Yangtze: Partnership for a Living River and the WWF–HSBC Yangtze Programme. Both programmes encompassed a governance and a substantive transition, as they were developed in interactions with stakeholders and were geared toward restoring biodiversity. The WWF engaged in the strategies of development and selling of new ideas, coalition-building and venue manipulation. The willingness to consider alternative approaches to water management increased considerably when floods occurred in 1998. Pilot sites formed an important part of the strategy to use this window of opportunity, allowing the WWF to show on a small scale that its alternative approach was effective and viable. After much positive media attention for the pilot projects, the WWF used the success to start promoting IRBM at the highest political level in order to change the legal framework. A window of opportunity opened when the China Council for International Cooperation on Environment and Development (a high-level non-governmental advisory organization) launched an IRBM task force in which the WWF could engage directly with the Ministry of Water Resources and the Changjiang (Yangtze) Water Resources Commission (CWRC). Among the recommendations of the IRBM task force was the idea to found a Yangtze Forum, where issues pertaining to the river as a whole could be discussed. The consideration of these and other recommendations by the premier of the State Council was greatly expedited through the contacts with the Chinese bureaucracy.

In Chapter 7 on India, Narain traces the origins and effects of participatory irrigation management (PIM) on water management in several Indian states, such as Haryana, Maharashtra, Gujarat and Karnataka. His analysis suggests that the origin of PIM in the form of the founding of water user organizations can be found, at an abstract level, in several global discourses: on good governance (see also Gupta's Chapter 3 in this volume); on the analysis that institutional factors rather than technical factors are key in irrigation; and on public participation in water management. Ideas on PIM have strongly affected official policies in several Indian states as a result of the work of multilateral organizations such as the World Bank, certain non-governmental organizations (NGOs), and high-ranking public officials sympathetic to PIM.

Narain finds that NGOs especially have used pilot projects as a strategy to show that water user organizations might work, but that such NGOs lack the capacity to scale up such ideas. One reason for their ineffectiveness is the fact that they tend to exclude water management officials and farmers from their pilots, making these groups relatively passive receptors of the idea. Where the idea of PIM has been scaled up, this has been possible mainly through the help of high-ranking public officials. However these officials have not engaged with 'street-level bureaucrats' and the farmers themselves. This has had a major effect on the implementation of PIM in India. Official agreements are still needed and officials quote a lack of resources as reason to delay rehabilitation of the irrigation infrastructure before transfer of its management to water user groups. Several farmers, particularly those benefiting from the current situation, resist the increased role of water user associations and sometimes resort to threats and violence to undermine NGOs advocating PIM. In addition, the political support needed for successful implementation of PIM has sometimes waned as a result of changes in political leadership.

Lebel, Garden, Subsin and Na Nan in Chapter 8 on water management in Thailand focus on the Upper Ping basin in the north of Thailand. They analyse three transitions that have taken place since the late 1950s, referring to the transitions as 'wet to dry', 'farm to city' and 'good to service'. The first transition was aimed at increasing agricultural productivity, the second at assigning greater priority to urban interests in water management, and the third transition to shifting emphasis from consumable quantities of water to the service-rendering quality of water. Thailand's long experience 'lurching towards democratization and then retreating into centralized, elite rule' was an important context for all three transitions. The authors also point to the importance of the regional context. While periods of democratization have resulted in a well-organized opposition to dams, Thai investors and the government nowadays pursue their interest in water and energy projects in neighbouring countries under more authoritarian rule.

Each of the transitions described by Lebel et al. has its own origins, narrative, windows and arenas. Regarding the origins, the authors find that policy change at the national level in Thailand has often followed changes in actual practice on the ground, thus codifying practice rather than modifying it. Lebel et al. refer to various powerful narratives, such as the 'water is a natural right' and the 'price is right', which we also encounter in other countries. The narratives serve as pulling forces, directing change their way. Lebel et al. suggest that windows of opportunity are important, but that: 'not all windows are recognized beforehand, and sometimes actors only realize that they existed after they have been passed through. Some windows are not seen as opportunities for all the relevant actors, but as threats or looming crises – for some actors a window, for others a brick wall.' They distinguish between predictable windows, such as those depending on seasonal changes in the weather, and windows that become obvious only with hindsight. As for arenas, they find that in the Thai situation, where new institutions often overlay old ones, creating institutional density, complexity and redundancy, the result is alternative platforms for deliberation and policy influence at multiple levels. Consequently, negotiations over water are never over.

Bhat and Mollinga in Chapter 9 analyse the attempted change in Indonesian water policies, specifically policies concerning the operation and maintenance of the irrigation infrastructure. Attempts to reform the Irrigation Operation and Maintenance Policy (IOMP) in Indonesia were premised on the idea that the malfunctioning of the irrigation infrastructure was based not so much on failure at the construction stage, but in the sphere of maintenance. Governance of the infrastructure was to change to more participatory, intersectoral and integrated means. Proponents of these ideas included the World Bank, the Asian Development Bank, reform-minded middle-level bureaucrats, the National Planning Agency and several politicians. The Directorate General of Water Resources Development (DGWRD), however, saw the proposed reforms as a threat to its position because it would lose control over irrigation funds. Opponents of change managed to undermine consensus on the alternative approach, blocking reform through several bureaucratic regulations that kept irrigation projects effectively under their control.

Bhat and Mollinga show that the Asian financial crisis of 1997 and 1998 offered a window of opportunity for the reformers as the centralized top-down Suharto regime toppled. A new reform-minded government came to power, supported by the International Monetary Fund (IMF) and the World Bank and seeking to end nepotism and corruption. The proposed reforms became official government policy, and a new legislative and organizational framework for irrigation emerged, which took away many of the central government responsibilities in this domain. Despite support from the public and lower government levels, and despite a certain level of institutionalization in various provinces, the 'crisis reform' has since then largely been overtaken by 'non-crisis', 'politics as usual'. Opponents of the reform, supported by a new minister, have skilfully manipulated venues and thwarted the gist of the reforms, despite the strategic manoeuvering of proponents of the reform, who exist largely at the provincial level.

Hughes and McKay with Chapter 10 conclude the section on Australasian water transitions by analysing two innovations: the creation of environmental water allocations in the Murray River basin, and the regulation of water use by the forestry industry in the southeast of South Australia. In both cases they point to the importance of 'bureaucratic entrepreneurs', who used windows of opportunities in the problem stream, mainly related to a growing awareness of water scarcity, to move Australian water management towards a new phase where the environment is taken into account along with development interests. In both cases analysed, Hughes and McKay find that scientific communities and the knowledge they produce play a key role in developing new ideas for policies, and that the uncertainties surrounding scientific insights become important focus points in the debate on policy innovations. The potential for the 'politicization of science' and the 'scientification of policy' is great in such a context, which explains why both opponents and proponents could agree to the introduction of an independent environmental manager for the Murray River. Both proponents and opponents of policy innovations acted strategically by forming coalitions and orchestrating the network within which decisions were made about water. The forestry regulation case shows that the achievement of water transitions in Australia is a multilevel game, with discussions about the regulation of water use in a province always perceived from the perspective of interstate water transfers, and with innovations in one state being watched by other states and presented as a potential model.

Part IV of this book moves to Africa, where we present a story on water transitions in South Africa and the tale of water management in Tanzania. Turton describes in Chapter 11 how South African water management is changing from the paradigm of extraction to a paradigm of redistribution. This change is deeply embedded in the wider political context, particularly the end of the apartheid regime. Turton takes up the case of waste (radioactive and hazardous materials) related to gold and other forms of mining in South Africa. Whereas it is clear that the overall water regime has been changed significantly, with land ownership and the ownership of water now being separated, the issue of mining waste has not been addressed so well. Turton suggests that this is partly to do with the quality of the political leadership, which was indispensable, but not available everywhere. One consequence is that the apartheid legacy of mining waste still lingers.

Turton analyses how the 'liberation leaders', the mining industry itself, civil society and the scientific community with its funders, are struggling to change the current situation. He suggests that their attempts to create a transition have not gone further than idea-development and certain forms of networking. Complications arise over the liability of mining companies as they relocate away from South Africa, bringing a risk of collusion between the newly emerging political elite and the mining industry. The envisaged transition towards more sustainable forms of waste management remains in the balance.

Goldin and Kibassa in Chapter 12 describe the development of water policy in Tanzania in recent years, while highlighting the historical context that has led to a hybrid mix of formal institutional settings and customary traditions. Here too some of the international discourses that Gupta sketches in Chapter 3 can be seen at play, as change agents such as the World Bank and the International Monetary Fund collude with national and local elites, driving forward a water agenda that emphasizes expropriation of common property by the state and subsequent privatization. The reforms proposed in Tanzania are, on paper, in line with the ideals of integrated water resources management (IWRM). In reality, though, this is only partly true, as water users now need to pay, but do not have much of a say in the development of water policies.

Goldin and Kibassa show that national elites, which they refer to as the 'cunning state', stand to gain from privatization because the most important change agents in the Tanzanian context, the World Bank and IMF, make their financial contribution to the state coffers contingent upon adoption of these principles. In doing so, they ignore the customary institutions, such as the chiefs. Goldin and Kibassa show that they also ignore the democratic principles of IWRM through the organization of very selective participatory processes that serve to co-opt only local elites by promising them part of the gains created by privatization. Although Goldin and Kibassa thus describe a very successful transition, they find fault with the direction the transition has taken and suggest an alternative route that integrates and builds upon customary traditions and entails the social-adaptiveness embedded in these elements.

As the book moves to Europe, Partzsch kicks off Part V with Chapter 13 by comparing the way the European Water Framework Directive (WFD) came about with the development process of the EU Water Initiative. She asks whether the difference between the two policies in terms of water pricing and cost recovery – an issue also touched upon by Goldin and Kibassa – can be attributed to differences in advocacy coalitions involved in their production. She finds that the discussions on water pricing in the WFD were dominated by environmental concerns and that a successful coalition of the water industry and a broad range of NGOs managed to instil this piece of legislation with a philosophy of charging polluters. By contrast, water pricing in the Water Initiative context drew much criticism and came to be framed as charging users, especially poor users. This ruled out a large coalition of water industry and NGOs. Partzsch suggests that this has led 'to the failure of the initiative to begin a transition to water pricing and full cost recovery'. Based on the comparison of the two cases, Partzsch concludes that the difference in the composition of advocacy coalitions present in both processes explains the difference in outcomes, suggesting that advocacy coalitions are indeed key in explaining policy change.

Werners, Flachner and Matczak, in Chapter 14 on Hungary, describe how policy entrepreneurs such as the Bokartisz coalition, jointly with several innovative water experts such as Géza Molnár, scientists in the National Water Resources Research Center (VITUKI), municipalities and high-ranking national water officials, changed the management of the Tisza River. Water policy for this tributary of the Danube changed fundamentally from the year 2000 onwards. Whereas flood protection through hard engineering had previously dominated, flood plain rehabilitation and land use change became the prime focus of the new policy. The analysis of Werners et al. points to the importance of major floods and the cyanide spill in the year 2000. In themselves these events were not sufficient to change policy, but advocates of change successfully framed them as signs of fundamental problems with the dominant paradigm. The alternative approach gathered steam through a range of deliberate steps by local NGOs and municipalities (the Bokartisz coalition). These strategic steps included small pilot studies, the solicitation of scientific research to support the viability of the new approach, cementing local support by coalition-building and presentation of the plan to various fora.

Members of the coalition also tried to block an emerging plan that was based on more traditional ideas.

Their moment came when a dyke break and flood occurred in 2001, and elections brought a new government to power in 2002, one keen to distinguish itself from the previous administration. Water management issues were transferred to a new ministry (Environment and Water), creating opportunities for new linkages and bringing new and more open decision procedures. Accession to the EU brought a need to harmonize Hungarian water policies with EU policies that emphasize 'greener' water management (see also Chapter 13 by Partzsch and Chapter 17 by Olsson and Galaz in this volume). EU accession also implied that Hungary would gain access to EU funds that would help finance implementation of the new plans. Werners et al. conclude that a relatively complex constellation of factors and developments brought about policy change in the Tisza basin. They add, however, that up to April 2009 implementation of the new policies has been slow and incomplete.

Font and Subirats detail in Chapter 15 how Spanish water policy has changed in recent years. The traditional supply-based approach, with its heavily subsidized hydraulic projects, has been complemented by 'sustainable' policy innovations such as water markets and, to a limited extent, water demand restrictions. Font and Subirats determine that environmental NGOs, the scientific community, and the Socialist Party are especially responsible for this transition. The first signs of the transition became visible in the early 1990s when free-market economists and environmentalists (see also Partzsch, Chapter 13 in this volume) started questioning the existing paradigm, attacking its redistributive nature. The resulting debates formed the seed of a large-scale movement consisting of environmental NGOs, scientists such as Pedro Arrojo, and certain politicians from the Socialist Party. These groups formed a blocking coalition against the 2001 National Water Plan. In achieving their goals, they especially used the supranational venues offered by the EU (Parliament, Commission, Court of Justice) to challenge the National Water Plan, undermining potential EU financial contributions to its implementation and inviting regulatory scrutiny.

The outcome of the 2004 elections, which brought the Socialists to power, and the appointment of a minister with strong environmental credentials, opened a window for the coalition to advocate the alternative vision. The implementation of the EU's WFD appears structurally to have opened up the water sector to environmental interests, although the long-term effects remain to be seen.

The water transitions in Turkey described in Chapter 16 by Kibaroglu, Baskan and Alp relate to hydropower and irrigation water. Although these are clearly two different domains, they have in common the fact that the era of state provision has ended, replaced with an emphasis on private sector delivery. In the case of hydropower, private involvement has taken the shape of build–operate–transfer and build–own-and-operate schemes, with large private companies becoming responsible for hydropower generation and delivery. In the case of irrigation systems, privatization has taken the shape of devolution of responsibilities to water associations. These transitions did not originate within Turkey, but rather from water management ideas promoted by the World Bank (see also Chapter 3 by Gupta, Chapter 5 by Wilder and Chapter 12 by Goldin and Kibassa, in this volume).

A window of opportunity came with the economic and political crises of the late 1970s and early 1980s, which brought high-level politicians like Turgut Ozal to the fore. Well connected to the World Bank and supported by a bureaucracy that believed in neoliberal reform, such politicians pushed the transition through. Opposition came mainly from professional organizations, such as the Union of Chambers of Turkish Engineers and Architects, and to some extent from public officials in several government agencies and some academics. The opponents used a range of strategies to try and block the transitions, notably bringing about a set of court cases, starting critical societal debate and commenting on government documents. They tried to frame the transition as against the public interest and as a surrender of critical national infrastructure to foreign powers. The World Bank heralded the Turkish practice of establishing water associations as a success story. Kibaroglu et al., however, observe many shortcomings in the model, notably a capture of these organizations by local groups. This is in line with observations made by Goldin and Kibassa on the transition seen in Tanzania.

In Chapter 17 on water policy transition in Sweden, Olsson and Galaz find support for the notion that transitions in water management need to enhance the adaptive capacity of institutions, that is, to increase their ability to deal with uncertainty and abrupt change. They sketch the emerging characteristics of adaptive institutions and indicate that such institutions should allow for interactions across organizational levels, experimentation, new policies, and flexibility. As most current institutional set-ups do not meet these criteria, they use the Kristianstads Vattenrike Biosphere Reserve in southern Sweden as an example of how transitions towards adaptive institutions can be achieved. Olsson and Galaz suggest that the related transition hinged on three elements: changing peoples' perceptions and mental models by reframing the problems (water as an opportunity rather than water as an obstacle); the development of new institutional structures (particularly a 'bridging organization'); and finally the development of structural and ad hoc learning networks.

After detailing the secrets of this success story, they focus on the introduction of the EU's WFD, which they describe as the biggest ever transition in Swedish water management. An assessment from the perspective of adaptive capacity unfortunately suggests – at least in the Swedish implementation of the WFD – that the Directive does little to enhance adaptiveness because the directive does little to promote the learning process deemed necessary across government levels by Olsson and Galaz.

Becker, in his account of German water policies in Chapter 18, focuses on flood measures in four different states (*Länder*). The transitions that have occurred there, with slight variations between the states, move from what Becker labels 'a traditional, technocratic safety discourse' to 'an ecological risk perspective'. As in Hungary, Spain and the Netherlands, the transition specifically entails a replacement of hard engineering solutions by soft approaches that give more space for water. Important drivers of these transitions were the report of the International Commission for Research on Floods of the Rhine River, and the subsequent 1982 decision by France and Germany to attain flood safety standards by creating additional retention space. As Germany is a federal country, and water management tasks rest largely with the states, the implementation of this agreement fell largely upon the states. In each case, the idea of creating retention space was a breach with the past approach. The state bureaucracies trying to sell the idea needed to engage in coalition building.

Although WWF Germany and its Auen Institute were important in underpinning the new approach, Becker observes that the German environmental NGOs more generally were not very involved in water management issues. The policy entrepreneurs were thus mostly officials who had to operate with variable political and societal support. Flood events in 1993 and 1995 helped to prime the political scene in the various states, whereas the 2002 floods in the Elbe and Donau (Danube) rivers brought to light shortcomings in the various state approaches and interstate coordination. The 'ecological retention' and 'risk-based approach' pioneered at the state level were then scaled up to the federal level. Becker points to various problems that are currently surfacing in implementation of the new ideas.

Huitema and Meijerink detail in Chapter 19 how the transition towards 'greener' water management in the Netherlands is taking shape. The greening of water management started in the 1960s and 1970s and continues today. Initially, ambition was moderate as energy focused mainly on amelioration of the negative environmental impacts of the hard engineering approach to water management. In the 1980s and 1990s players raised their aim, as a more fundamental critique of the existing paradigm ensued, calling for the replacement of hard engineering by an agenda of 'living with water', with the attendant flood management strategies of river rehabilitation and water retention.

The two cases discussed, the Eastern Scheldt dam and the decision process on the large river policy, exhibit an interesting pattern. The main conclusion is that a persistent group of entrepreneurs (collective entrepreneurship) steered the transition along a relatively unpredictable path of natural, societal and political events. As in other chapters the importance of changed ideas in the scientific community, the selling of these ideas to relevant politicians with the power to muscle the new ideas through, the creation of coalitions of actors with diverging interests and the role of pilot projects are highlighted. Both case studies underpin the argument that floods and near floods provide windows of opportunity, highlighting problems with water management strategies. However, as the belief in traditional water management is deeply engrained in the Dutch psyche, such events do not necessarily help those who seek to advance alternative approaches. The large river case effectively shows how much ingenuity it takes for policy entrepreneurs to overcome standard societal reflexes in such situations.

Part VI of this book brings the book to a close with Chapter 20 by Meijerink and Huitema, a summary, findings and the conclusions drawn on the basis of the various cases presented. As noted above, the intention is for these to serve as a contribution to the transition management literature from the policy sciences and as a practical perspective on ways to bring about major change in water policy.

References

Alley, R.B., P.U. Clark, T. Huybrechts and I. Joughin (2005), 'Ice-sheet and see-level changes', *Science*, **310** (5747), 456–60.

Baumgartner, F.R. and B.D. Jones (1991), 'Agenda dynamics and policy subsystems', *Journal of Politics*, **53** (4), 1044–74.

Bevir, M. and R. Rhodes (2003), *Interpreting British Governance*, London: Routledge.

Birkland, T.A. (1997), *After Disaster: Agenda Setting, Public Policy, and Focusing Events*, Washington, DC: Georgetown University Press.

Birkland, T. (1998), 'Focusing events, mobilization, and agenda setting', *Journal of Public Policy*, **18** (1), 53–74.

Blomquist, W., A. Dinar and K. Kemper (2005), 'Comparison of institutional arrangements for river basin management in eight basins', World Bank Policy Research Working Paper 3636, June, Washington, DC: World Bank.

Burchi, S. and M. Spreij (2003), *Institutions for International Freshwater Management*, Paris: UNESCO.

Cabanes, C. A. Cazenave and C. Le Provost (2001), 'See level rise during past 40 years determined from satellite and in situ observations', *Science*, **294** (5543), 840–42.
Clark, M.J. (2002), 'Dealing with uncertainty: adaptive approaches to sustainable river management', *Aquatic Conservation: Marine and Freshwater Ecosystems*, **12**, 347–63.
Conca, K., F. Wu and C. Mei (2006), 'Global regime formation or complex institution building? The principled content of international river agreements', *International Studies Quarterly*, **50**, 263–85.
Dewulf, A., K. Termeer, R. Werkman and G. Breeman (2008), 'Steering system innovations: a theoretical exploration of transition management', paper presented to the 15th Annual Conference on Multi-Organizational Partnerships, Alliances and Networks (MOPAN), Boston, MA: 25–27 June.
Dryzek, J. (1987), *Ecological Rationality: Environment and Political Economy*, New York: Basil Blackwell.
Duit, A. and V. Galaz (2008), 'Governance and complexity: emerging issues for governance theory', *Governance*, **21** (3), 311–35.
Easterling, D.R., G.A. Meehl, C. Parmesan, S.A. Changnon, T.R. Karl and L.O. Mearns (2000), 'Climate extremes: observations, modelling, and impacts', *Science*, **289** (5487), 2068–74.
Falkenmark, M., L. Gottschalk, J. Lundquist and P. Wouters (2004), 'Towards integrated catchment management: increasing the dialogue between scientists, policy-makers and stakeholders', *Water Resources Development*, **20** (3), 297–309.
Finer, S. (1970), *Comparative Government*, Harmondsworth: Penguin.
Geels, F.W. (2002), 'Technological transitions as evolutionary reconfiguration processes: a multi-level perspective and a case-study', *Research Policy*, **31** (8/9), 1257–74.
Genxu, W. and C. Guodong (1999), 'Water resource development and its influence on the environment in arid areas of China: the case of the Hei River basin', *Journal of Arid Environments*, **43** (2), 121–31.
Gleick, P. (2003), 'Global freshwater resources: soft-path solutions for the 21st century', *Science*, **302** (5650), 1524–8.
Gleick, P., A. Sing and H. Shi (2001), *Threats to the World's Freshwater Resources*, Oakland, CA: Pacific Institute.
Hopkin, J. (2002), 'Comparative methods', in D. Marsh and G. Stoker (eds), *Theory and Methods in Political Science*, 2nd edn, Houndsmills: Palgrave Mcmillan, pp. 249–70.
Jordan, A., T. Rayner, D. Huitema, C. Haug, E. Massey, J. Stripple and F. Berkhout (2007), *Climate Policy and Governance: A Theoretical and Methodological Framework to Guide Research*, Amsterdam: IVM.
Kemp, R., A. Rip and J. Schot (2001), 'Constructing transition paths through the management of niches', in Raghu Garud and Peter Karnøe (eds), *Path Dependence and Creation*, Mahwah, NJ and London: Lawrence Erlbaum, pp. 269–99.
Kemp, R., D. Loorbach and J. Rotmans (2007), 'Transition management as a model for managing processes of co-evolution', *The International Journal of Sustainable Development and World Ecology*, **14**, 78–91.
Kemper, K., A. Dinar and W. Blomquist (eds) (2005), *Institutional and Policy Analysis of River Basin Management Decentralization: The Principle of Managing Water Resources at the Lowest Appropriate Level – When and Why Does It (Not) Work in Practice?*, Washington, DC: World Bank.
Kingdon, J.W. (1995), *Agendas, Alternatives and Public Policies*, 2nd edn, New York: HarperCollins.
Lenton, T.M., H. Held, E. Kriegler, J.W. Hall, W. Lucht, S. Rahmstorf and H.J. Schellnhuber (2008), 'Tipping elements in the Earth's climate system', in *Proceedings of the National Academy of Sciences*, **105** (6), Online Early Edition, 12 February, 1786–93.
Meinzen-Dick, R. (1997), 'Farmer participation in irrigation: 20 years of experience and lessons for the future', *Irrigation and Drainage Systems*, **11** (2), 103–18.
Meinzen-Dick, R. (2007), 'Beyond panaceas in water institutions', *Proceedings of the National Academy of Sciences*, **104** (39), 15200–205.
Olsson, P., L.H. Gunderson, S.R. Carpenter, P. Ryan, L. Lebel, C. Folke and C.S. Holling (2006), 'Shooting the rapids: navigating transitions to adaptive governance of social–ecological systems', *Ecology and Society*, **11** (1), http://www.ecologyandsociety.org/vol11/iss1/art18/.
Pierre, J. and B.G. Peters (2000), *Governance Politics and the State*, Basingstoke: Macmillan.
Rittel, H. and M. Webber (1973), 'Dilemmas in a general theory of planning', *Policy Sciences*, **4**, 155–69.
Rotmans, J., R. Kemp, and M. van Asselt (2001), 'More evolution than revolution: transition management in public policy', *Foresight*, **3** (1), 15–31.
Sabatier, P.A. and H.C. Jenkins-Smith (eds) (1993), *Policy Change and Learning: An Advocacy Coalition Approach*, Boulder, CO: Westview Press.
Schlesinger, W.H., J.F. Reynolds, G.L. Cunningham, L.F. Huenneke, W.M. Jarrell, R.A. Virginia and Walter G. Whitford (1990), 'Biological feedbacks in global desertification', *Science*, **247** (4946), 1043–8.
Schot, J.W., R.J.F. Hoogma and B. Elzen (1994), 'Strategies for shifting technological systems: the case of the automobile system', *Futures*, **26** (10), 1060–76.
Schout, A. and A. Jordan (2005), 'Coordinated European governance: self-organizing or centrally steered?' *Public Administration*, **83** (1), 201–20.

Smith, A., A. Stirling and F. Berkhout (2005), 'The governance of sustainable socio-technical transitions', *Research Policy*, **34** (10), 1491–1510.

Stoker, G. (1998), 'Governance as theory', *International Social Science Journal*, **155** (1), 17–28.

Stone, R. (2008), 'China's environmental challenges. Three Gorges dam: into the unknown', *Science*, **321** (5889), 628–32.

Tàbara, D. and C. Pahl-Wostl (2007), 'Sustainability learning in natural resource use and management', *Ecology and Society*, **12** (2), 3, available online at: http://www.ecologyandsociety.org/vol12/iss2/art3/.

Turner II, B.L., E.F. Lambin and A. Reenberg (2007), 'The emergence of land change science for global environmental change and sustainability', *Proceedings of the National Academy of Sciences*, **104** (52), 20666–71.

Vellinga, P. and W.J. van Verseveld (2000), *Climate Change and Extreme Weather*, Gland, Switzerland: WWF.

Wolff, G. and E. Hallstein (2005), *Beyond Privatization: Restructuring Water Systems to Improve Performance*, Oakland, CA: Pacific Institute.

World Commission on Dams (2000), *Dams and Development*, London: Earthscan.

Young, O.R., F. Berkhout, G.C. Gallopin, M.A. Janssen, E. Ostrom and S. van der Leeuw (2006), 'The globalization of socio-ecological systems: an agenda for scientific research', *Global Environmental Change*, **16** (3), 304–16.

2 Understanding and managing water policy transitions: a policy science perspective

Sander Meijerink and Dave Huitema

2.1 Introduction

In this chapter we lay a theoretical foundation for the case studies presented in the following chapters. The theoretical framework applied in this book is based on the policy science literature. Policy science is often defined as a subdiscipline of mainstream political science, one that covers 'the totality of public decision-making' and 'investigates the complex links between inchoate public demands and the detailed implementation of policy choices' (John, 1998, p. 3). The policy science literature is inspired by a large variety of approaches and theories, which are based on different ontological, epistemological and theoretical stands. Sets of theories that have informed policy sciences include (John, 1998): institutional approaches, group and network approaches, socio-economic approaches, rational choice theory and idea-based approaches. Because of our research interest in radical policy change and the strategies of policy entrepreneurs, we have included in our analysis a wide range of theories on policy continuity and change, such as the punctuated equilibrium framework (Baumgartner and Jones, 1991), the multiple streams framework (Kingdon, 1995), the advocacy coalition framework (Sabatier, 1993), discourse theory (Hajer, 1995) and policy network theory (for example Kickert et al., 1997). We have skimmed these theories for what they have to say on the roles and strategies of policy entrepreneurs.

In this chapter we first explore the concept of policy stability or continuity and the different ways in which it is defined and explained. We move on to discuss policy change and the conditions conducive to its occurrence. A review of concepts from the policy science literature is included to expand the description of both policy continuity and change, and to give a theory-based definition of transitions. The broader, underlying question addressed in this chapter concerns what the policy sciences tell us about possibilities for realizing change. This helps in turn to answer the question: what strategies can actors use to bring about policy change?

2.2 The conceptualization of policy stability

The policy science literature reveals many ways of conceptualizing, let alone explaining, policy stability and change. A distinction is made between on the one hand incremental and on the other hand radical, fundamental, deep or paradigmatic policy change (John, 1998; Huitema et al., 2006; Schlager, 2007). Below we explore how the various theories on policy continuity and change conceptualize and explain policy stability, and fundamental and incremental policy change, respectively.

Most theories argue that under 'normal' circumstances policies are quite stable and only develop or change incrementally. This is often called 'normal politics'. Policy domains tend to stability because they are captured by groups of actors who share an interest in maintaining the status quo and who resist attempts to change prevailing policies and policy programmes. In the political sciences these groups used to be called 'iron triangles'. Policy network theory sees actor groups as forming issue networks or policy communities (Rhodes and Marsh, 1990), while the advocacy coalition framework identifies these actors as members of a dominant umbrella group promoting a particular policy configuration or type (Sabatier, 1993; Sabatier and Weible, 2007). Whatever the group of actors is called, the members share a common understanding of a policy domain, the main policy problems and the desirability and feasibility of various policy options. The policy science literature has produced a rich diversity of conceptualizations of such policy ideas or views held in common. They may, for example, be described and analysed as a belief system (Sabatier, 1993), a discursive structure (Hajer, 1995, 2005; Fischer, 2003), an institutionalized policy image (Baumgartner and Jones, 1991, 2002; Jones and Baumgartner, 2005), a policy frame (Schön and Rein, 1994) or a policy paradigm (Hall, 1993). An example from the water policy domain: the institutionalized policy image of manifold benefits of taming and controlling water has characterized the dam-building era in many countries. Today, however, this image has been largely replaced by one of dams as detrimental to river ecosystems, communities and fisheries.

How, then, should we account for policy stability? In other words, why is it so difficult to bring about policy change, particularly in a world full of problems and in need of new solutions? A first and obvious explanation is that the group of actors in power possesses the means and resources to remain in power. This frequently brings the response of resistance, typically strong, to change. Resistance can stem from, for instance, the relatively simple desire to be left in peace to work within the policy paradigm or equilibrium of the moment. It can also arise as a result of ideology and convictions, and from vested interests, where for example careers can depend on policy programmes. Along with resistance to change, persistence in power often leads to issue legislation reflecting and supporting extant policy ideas, to the formulation of decision-making rules favourable to current policies and actors and to strategic manipulation of various decision-making fora. Policy scientists writing on policy stability and change strongly agree that resistance and strategic manoeuvering both influence the process of policy change.

Bachrach and Baratz (1970) point to various barriers to policy change. They first identify an ideological barrier based on the dominant pattern of values. A problem can be debated only to the extent to which it is recognized as a problem that needs to be solved. In this way, problem definitions need to be compatible with the dominant policy discourses, frames or beliefs. Second, institutional barriers, such as mandatory procedures and how public tasks and competences are organized, can determine the opportunity for a problem to be incorporated in the agenda-setting process and the way in which it is addressed. Third, the public decision-making process determines whether proposals to tackle a problem will be adopted, rejected or amended. A fourth barrier lies in implementation, where wrong interpretations of policy, partial execution or lapses may prevent successful problem-solving and discredit associated policy.

According to Baumgartner and Jones (2002; also True et al., 2007) policy monopolies can dampen pressure for change. This involves processes of negative feedback very similar to Kuhn's theory about scientific revolutions. Policy monopolies are able to accommodate pressure for change for a long time. Only sufficient and prolonged pressure may result in a fundamental rethinking of policy principles.

Policy stability is not the result only of strategic manoeuvering and defence of the status quo. Historical institutionalists argue that 'history matters' (North, 1990) and that the legacy of the past conditions the future (Gains et al., 2005). From this comes the notion of 'path-dependence', which refers to the idea that preceding steps in a particular direction help to ensure further movement in the same direction (Needham and Louw, 2006). Path-dependent developments are often explained by increasing returns (Gains et al., 2005). In an increasing returns process, Pierson (2000, p. 252) explains: 'The probability of further steps along the same path increases with each move down that path. This is because the relative benefits of the current activity compared with other possible options increase over time.' In other words, considerations of benefit and cost often lead to the decision to continue the same practice. Hence increasing returns processes are also described as self-reinforcing processes (ibid.). The costs of exit (that is, of choosing a different institutional path) rise with each step taken.

The usefulness of the concept of increasing returns in policy studies is the subject of debate. Kay (2005), for example, argues that increasing returns processes are sufficient, but not necessary for path-dependency. There are several other non-increasing returns mechanisms which may explain a path-dependent development of policies. Kay discusses various examples (2005), including: (1) that 'policies involve investments or disinvestments in administrative infrastructure; this transforms governmental capacity and the set of possible future policies that may be enacted'; and (2) that 'policies involve the establishment of formal or informal contracts with individuals, which are costly to change'. Kay takes up the high transaction costs of changing the type of contract between the parties involved. The transaction costs of reaching agreement on a contract type familiar to the parties are considerably lower than those for an unfamiliar type. Path-dependency can be induced, for example, when water managers with long experience of command-and-control-based strategies choose not to experiment with more interactive policy processes. Such mechanisms do not rely on an increasing returns process, but rather the potential for 'high future switching costs' of a policy (ibid.). Finally, in the water policy domain, physical infrastructure, such as dams or dykes, may also account for a path-dependent development of policies or 'lock-ins'. Once construction is in place, the land use has changed accordingly and investments, public and private, have been made. This tends to make water and spatial policies next to irreversible, although cases presented in this book, such as the implementation of integrated river basin management in China and Germany, show that infrastructure is not an insurmountable obstacle.

The epistemic communities (EC) framework (Haas, 1992) also addresses explicitly the question of policy persistence or stability. Viewed within this framework the ideas and policies put forward by epistemic communities, once adopted and institutionalized, are seen as easily gaining the status of orthodoxy (ibid). In EC theory the metaphor of developmental biology is used to understand policy evolution (Adler and Haas, 1992; Haas, 1992). This is based on the observation that 'evolutionary changes to structures, once in place, are largely irreversible and virtually determine the array of subsequent choices available to the species' (Adler and Haas, 1992, p. 372). This evolutionary model of path-dependency implies that the effects of epistemic involvement in policy change are not easily reversed, which can be significant as epistemic communities are often strongly related to and inform advocacy coalitions (Dudley and Richardson, 1996; Meijerink, 2005).

A path-dependent development of policies does not mean an absence of fresh dynamics. Changes in this case, however, are always marginal or incremental; they fit within the dominant policy paradigm, and do not affect the basic interests of the (network of) actors in power. In the advocacy coalition framework these changes are defined as changes to the secondary aspects of a policy belief system, which by definition is narrow in scope. Such changes may relate to detailed regulations, budget allocation within a specific programme, and so on, but do not affect the core of a policy programme. Cross-coalition learning means that a dominant coalition incorporates some elements of the belief system of opposing coalitions in its own. This may account for minor changes to the status quo (Jenkins-Smith and Sabatier, 1993). Hall (1993) would categorize such minor changes as changes in the fine-tuning of instruments or changes at the level of instruments at best.

2.3 The conceptualization of major policy change

What, then, about major, fundamental, radical, deep or paradigmatic policy change, change which entails a fundamental reorientation of policies, and which we equate with transitions? In the advocacy coalition framework such change is defined as a transformation of core policy beliefs. Sabatier and Jenkins-Smith have defined 11 components of such core beliefs: 'including the priority of different policy-related values, whose welfare counts, the relative authority of government and markets, the proper roles of the general public, elected officials, civil servants, and experts, and the relative seriousness and causes of policy problems in the subsystem as a whole' (Sabatier and Weible, 2007, p. 195). In this way major policy change covers not only substantive changes to policy, but also changes of the dominant governance paradigms, such as a coalition's reliance on the free market, on civil society or on government intervention for realizing its objectives. Hall (1993) defines paradigmatic policy change as a change at the level of overarching goals. True et al. (2007) speak about change as large-scale departures from the past, which include new issue definitions or policy images.

What is needed for fundamental policy change? And under what circumstances may we expect such change to occur? Drawing on the policy science literature a first answer is a set of new policy ideas challenging the status quo. Richardson (2000) finds another biological analogy in his expressive and revealing comparison between policy ideas and viruses, in the way new or competing policy ideas may be able to disrupt seemingly stable policy networks. Similarly in their article 'The politics of path dependency', Peters et al. (2005) argue that new ideas may influence the trajectory of policy development. As indicated above, such new policy ideas can be conceptualized as part of a new policy core belief (Sabatier, 1993), a new policy image (Baumgartner and Jones, 2002), a new policy discourse (Hajer, 1995) or as a new policy frame (Schön and Rein, 1994), and as generally entailing a different value orientation, new problem perceptions and new policy preferences.

Ideas and agency are strongly related. Both Richardson (2000) and Peters et al. (2005) argue that for ideas to be able to change the course of a policy, they must first be articulated and inserted into the political debate. Typical agents that produce and proliferate ideas are social movements (Benford and Snow, 2000), advocacy coalitions (Sabatier, 1993), epistemic communities (Haas, 1992) and shadow networks (Olsson et al., 2006). All these types of groups actively try to influence governmental decision-makers in their attempts to change existing policies and institutions. Just as a policy monopoly can be analysed as a group of actors sharing a particular worldview, so groups of actors sharing a different worldview can be analysed as policy enterprises when they challenge the status quo and try to make their worldview dominant.

Fundamental policy change can be a painstaking exercise for many reasons. Because of negative feedback, increasing returns and other mechanisms that keep a policy on a particular path, policy communities tend to be rather stable over time. In the view of historical institutionalists, radical policy change occurs in exceptional cases of major performance crises or external shocks, and entails a change of policy path (Knill and Lenschow, 2001, p. 193). Triggers of such 'discontinuous institutional change' include wars, revolutions, conquests and natural disasters (North, 1990).

The advocacy coalition framework highlights the importance of less drastic perturbations external to the governing policy subsystem (or external system events) as factors behind fundamental policy change. Such events range from changes in socio-economic conditions, changes in public opinion and changes in a governing coalition, to policy decisions and impacts from other subsystems (Sabatier, 1993; Sabatier and Weible, 2007). If such events take place, opposing coalitions may successfully challenge dominant policy beliefs.

The focus of the advocacy coalition framework on external subsystem events is inspired by Kingdon's agenda-setting theory (1995). John Kingdon can be credited with introducing to the policy sciences the concept of 'windows of opportunity' – a combination of favourable circumstances that creates an opening to influence the official policy agenda. This is tied to the unique contribution of agenda-setting theory of the identification of the separate processes of problem-setting, finding solutions and political legitimization. The often-assumed irrationality of the political process becomes much more comprehensible when viewed in these terms. Actors advocating policy change need to be aware of the dynamics of agenda-setting and to look out for and create windows of opportunity. Considering the sequence of the agenda-setting process, Kingdon thinks in terms of a problem stream (public attention to issues: the public agenda), a political stream (political attention to issues: the political agenda) and a policy stream (attention of policy-makers to policy options: the official agenda). A window of opportunity can arise in any one stream or when two or all three streams – problems, policies and politics – converge. A full convergence dramatically increases the probability of an item entering and rising on the official decision-making agenda. Such a window can open quite predictably or quite unpredictably.

No single stream is sufficient by itself to place an item firmly on the policy decision agenda. If one of the three elements is missing – if a problem has not been identified, is not sufficiently compelling or is not presented well; if support is not forthcoming from political players; or a solution in the form of a policy option is not available – then an issue's place on the decision agenda will be fleeting. The window may open for a short time, but if the three streams are not coupled quickly, the window closes. A subject can rise on the agenda abruptly, but remain there only briefly. Generally, an item rises due to the combined effect of several factors coming together at a given point in time, not to each in isolation.

Kingdon considers developments in the political stream as particularly powerful agenda-setting factors. Consensus builds in the political stream through bargaining more than by persuasion. Proposals are evaluated in terms of their political support and opposition, and only partly against logical or analytical criteria. The policy stream sees a process of consideration situated more in the policy community, where ideas themselves are important and where rational reasoning and evidence of effectiveness is required. The problem stream features the activities of action groups, statistics and news providing information on the state of the world, and so on. Events that focus attention or shock can also raise awareness of issues (see also Birkland, 1997). As a consequence of activity in any one or two of the three streams, a window of opportunity may open, but often only for a limited period of time and without a policy transition unless there is convergence with and sufficient consensus across the remaining stream(s).

To give but a few examples from Dutch water policy of policy windows: the storm surge of 1953 and the extremely high water levels in the main Dutch rivers in 1993 and 1995 clearly served as focusing events (Huitema and Meijerink, Chapter 19 in this volume). It is clear, too, that these events had an impact on policy-making. It is less evident, however, to what extent these events contributed to fundamental policy change: that is, whether they caused the selection of a new policy path, and served as a factor in a critical policy juncture, or whether they resulted merely in an acceleration of policies on an existing path. After all, plans for total enclosure of the estuaries in the southwestern Netherlands had already been developed before 1953. In fact the Delta Plan policies put in place following the flood are very similar to the policies of the 1930s aimed at the enclosure of the Zuiderzee and construction of the Afsluitdijk (Huitema and Kuks, 2004). In similar vein, the government's first reaction after the 1993 and 1995 river floods was to issue emergency legislation to speed up the implementation of the already planned projects for strengthening dykes. It is also true, though, that the floods were increasingly perceived as a sign of the failure of the existing policy paradigm of fighting water with technical means only. As a consequence the alternative paradigm of giving water more space gained importance.

As an example of a political window – one that opens as a result of attention from political players – several studies point to the importance of the Dutch government led by Prime Minister Den Uyl in 1973 for its comprehension of the issue and factors related to decision-making on the construction of the famous Eastern Scheldt storm surge barrier. A coalition of environmentalists was allowed access to decision-making once Den Uyl's cabinet came to power and was therefore better able to sell their ideas. This example, just like the interesting example of Thatcherism as used by both Peters et al. (2005) and Richardson (2000), supports Kingdon's consideration as above that the political stream may be particularly influential.

Although punctuated equilibrium theory (another originating in the field of biological evolution) acknowledges the importance of windows of opportunity, it describes an additional mechanism of policy change. Punctuated equilibrium posits the notion that fundamental policy change occurs rarely and rapidly in brief spurts that establish new policy paths that branch off from ones followed up to the point of transition. The case of California water policy examined by Ingram and Fraser (2006) provides an illustration of analysis using this theory. Policy stability is explained by punctuated equilibrium as deriving from negative feedback processes, while it attributes policy change to positive feedback mechanisms. Positive feedback occurs when changes, even modest ones, cause the effects of a policy to be amplified (True et al., 2007). This is also called the 'bandwagon effect'. Whether or not positive feedback occurs depends on the interplay between policy image and policy venue, where the strength of the former can ensure take-up of an issue at the latter, and vice versa, the one reinforcing the other. Most policy issues are ambiguous, allowing them to be portrayed in different ways. It is exactly for this reason that different and competing policy images often arise.

Baumgartner and Jones (1991) give the example of nuclear power. Nuclear power was initially associated with economic progress and technical expertise. When opponents raised images of danger and environmental degradation, helped by incidents such as those on Three Mile Island and at Chernobyl, the nuclear power industry began to collapse (ibid.). Now, in the face of climate change, a competing image is emerging, portraying nuclear energy as more sustainable than energy from fossil fuels. New images attract new supporters. This is where the opportunity structure – the configuration of venues and processes that delineate potential paths for ideas to reach the political agenda – of modern societies plays an important role as issues move from political subsystems to the macro political agenda, opening opportunities at one or more venues for major policy change.

Modern Western societies are characterized by complex opportunity structures. Policy issues are debated in many places including, for example, various levels of government, fora of scientists and legislatures. As Richardson (2000) argues, these venues constitute an interest group resource that can be exploited in the form of 'venue shopping', a term coined by Baumgartner and Jones (1991). The impact of the European Union on water policy developments offers an interesting case study. The European level sees various venues that are relevant to decision-making on water management issues as discussed, for example, by Kaika (2003) in her analysis of how the environmental lobby managed to exploit changing decision-making rules.

In addition to new ideas, windows of opportunity and opportunity structures, the course of interactions and negotiations between opposing coalitions can be relevant to understanding policy change (Sabatier et al., 2005; Sabatier and Weible, 2007). The latest contributions to the advocacy coalition framework owe much to the theory of alternative dispute resolution. The basic argument is that in some cases fundamental policy change may be based on negotiated agreements (Sabatier and Weible, 2007). In this way, fundamental policy change does not have to be accompanied by a change of the dominant coalition, but may instead result from an effective negotiation process to end a policy stalemate (Sabatier et al., 2005). One of the negotiation strategies that may account for such change is that of issue linkage, where different issue areas are linked to each other for strategic rather than substantive reasons (Fischhendler et al., 2004; Meijerink, 2008). In these situations, parties involved in the negotiations may be willing to accept major policy change even when they do not support such a change because they are compensated in another issue area. These observations indicate the relevance of the theories of alternative dispute resolution (and also negotiation analysis) to understanding and managing policy transitions.

Lowry (2006) adds still another account of policy change. In his international comparative case study of dam politics, which uses a policy science perspective, he develops the concept of focusing projects. Basically he argues that next to focusing events, focusing projects often play an important role in processes of policy change. Each case study discussed by Lowry reveals a central controversial project around which a solid image of failure starts to build, demonstrating the bankruptcy of the policy paradigm, triggering the debate between members of different coalitions and producing positive feedback that magnifies the effects of both the debate and the policy image. Thus the entry of and debate over new ideas in the political arena can form a major source of policy change and are crucial to understanding it. Ideas, however, need an agent, and an agent in turn needs to apply various strategies as described in the next section to get their ideas accepted.

In spite of any ambition for each theoretical framework to offer explanations for policy stability and/or policy change, the developers of most of the theories treated here recognize the importance of serendipity and coincidence in understanding the policy process (Schlager, 2007). While the frameworks emphasize the importance of variables, such as external system events, shock events or governmental turnover, it is acknowledged that these events for the most part cannot be predicted. Moreover it is understood that these events as such do not account for policy change, but must be exploited by policy entrepreneurs or policy opponents.

In this section we have described various generic mechanisms of policy change that are discussed in the policy science literature and that may be helpful in understanding transitions. To understand specific cases of transition, however, these insights should be supplemented with knowledge about the institutional context, since it is an important intermediate variable in most of the frameworks discussed. Theories of venue change and venue shopping, for example, assume the presence of multiple venues as part of the opportunity structure. Yet whereas such opportunity structures are present in most Western democracies, they are not as prevalent in developing countries, where institutions can include very constrained decision-making procedures.

Fundamental changes are not necessarily discontinuous or the result of a sudden shift, but can evolve out of a process of continuous incremental change. Kay (2005) proposes representing policy as 'a vector in policy space', that is, it has a velocity and a direction. Moments of crisis, then, may cause a change of velocity and/or a change of direction. Building on the vector notion, he argues that a seemingly small change of direction, 'may turn out in retrospect to have been a critical juncture' (Kay, 2005, p. 566). In addition, fundamental shifts that seem at first sight to have taken place within a short time horizon may have been preceded by a long period of more incremental change, such as the growth of a new coalition, a gradual change of national mood and so on (Meijerink, 2005). In-depth studies of policy processes show that it is often more difficult to distinguish between fundamental and incremental policy change than some accounts of policy change would suggest. These findings are interesting not only for improving the understanding of policy transitions, but also for learning how to manage them.

The policy sciences, then, can help in the crucial task of understanding the range and type of agency and processes in order fully to comprehend what occurs in a policy transition. From the knowledge of the workings of a transition come questions about how such understanding can be applied. What strategies can be used to purposefully alter policy practices and to bring about structural policy change? Or, from a deeper perspective, is it possible to develop a generic theory about successful strategies of transition management?

2.4 Strategies of transition management

2.4.1 Introduction

In this section we highlight strategies for instigating, realizing and implementing policy change. As Bachrach and Baratz (1970) show, change can be instigated both from within and from outside government. Within government the individuals seeking change may be politicians or bureaucrats. Outside government, they can be members of non-governmental organizations (NGOs), academics or individual citizens.

Working to achieve policy change obviously involves different challenges for different types of policy entrepreneurs. For instance, the political leader may have a political party and a bureaucracy to back him or her, which the employee of a small NGO probably does not. The more or less generic strategies listed below may or may not be at the disposal of any particular player. Whether or not various types of individuals seeking change have access to and are able to use a certain strategy is an empirical question finding answers in the cases analysed in this book. Although we do not suggest that the strategies below are necessarily used in chronological order, for ease of presentation they are listed as if for application in sequence. In fact we expect the strategies to be used in wildly differing orders and combinations to suit varying circumstances and types of policy entrepreneurs. Which strategies are used and combined, in which order, by whom and to what effect is, again, an empirical question for exploration in the following chapters.

2.4.2 The development of new ideas

Infrastructure and regulation in water management can be seen as an expression of ideas that guide their development, ideas referred to as above as a 'policy paradigm' and so on, and sometimes given an extra edge as in terms such as 'hegemonic policy discourse' and 'institutionalized policy monopoly'. To aim for policy change, then, requires the development of an alternative idea or approach for managing water, described in the literature as a new policy frame (Schön and Rein, 1994), a new policy image (Baumgartner and Jones, 2002), alternative system configurations (Olsson et al., 2006), an alternative policy path (Pierson, 2000, 2004), new long-term visions and transition agendas (Loorbach and Rotmans, 2004; Van der Brugge et al., 2005), a new policy theory or new storyline (Hajer, 1995).

There is a long-standing debate in the policy sciences on the relative importance of ideas as opposed to interests (for example Majone, 1992). Here we do not take a position. We are sympathetic to the notion that ideas shape interests, but also aware that (perceived) interests can be a motivating factor behind the entertainment and development of ideas. The link between interests and ideas may vary according to the forum as, for example, the discourse of scientists occurs in an environment that is less accepting of interest-based arguments (argumentation) than the forum of markets, parties and politicians that allows much more open expression of interests (brokering). We make the point, however, that policy change requires at least the germ of an idea to provide the direction in which the situation might change. The policy sciences suggest that more extreme visions of alternative futures develop among actors who are outside the government. Because they are free from governmental constraints they are also more likely to pursue major change than incremental change. One hypothesis of the advocacy coalition framework states that: 'Within a coalition, administrative agencies will usually advocate more moderate positions than their interest group allies' (Sabatier and Weible, 2007, p. 220). Roberts and King (1996) argue also that inside the government individual entrepreneurs pursuing incremental change are more common. These are hypotheses for testing in the empirical chapters of this book.

2.4.3 Build coalitions and sell ideas

There are obviously few actors who can manage policy change on their own. The degree to which collaboration is needed depends on various factors, including the extent of change sought and the institutional arrangements surrounding the decision process. However, collaboration appears to be necessary in any situation and by implication drives the building of coalitions. Such coalitions are referred to as 'discourse coalitions', 'advocacy coalitions' and 'shadow networks'. Coalition-building is often a delicate task as it entails sensitive issues such as differences of opinion and power asymmetries among actors. Various theories from the policy sciences propose different mechanisms through which coalitions are built, again raising questions for the following chapters of this book. Discourse analysts such as Hajer (1995) suggest that storylines or narratives, preferably with a certain ambiguity or multiple interpretability, are key in attracting new actors to new ways of understanding. This attraction is referred to by Hajer as 'affinity', a concept that stresses the attractiveness of a new vision in coalition-building. Benford and Snow (2000) think along similar 'ideational' lines as they speak of 'frame alignment' as the key factor in coalition-building. Folke et al. (2005) and Olsson et al. (2006) discuss the creation of 'shadow networks', describing the fact that there is more room outside official policy-making circles for open debate and ideas that may be somewhat unconventional. The role of academics and scientists connected to lower-ranking bureaucrats is important in such shadow networks, which can use boundary objects, such as books or models in common, to develop a shared alternative vision. Sabatier (1993) sees coalition-building as a way to 'pool' resources, observing that coalition-building efforts emphasize shared beliefs and explicit agreements on how to use the resources of the actors involved to achieve common goals.

2.4.4 Recognize and exploit windows of opportunity

In the previous section, we point to the relevance of John Kingdon's concept of a window of opportunity for understanding policy change (Kingdon, 1995). Such windows, however, need to be recognized and exploited – crucial tasks for policy entrepreneurs. These are typically the actors who recognize problem or political windows and work to open and expand them. They do so by linking solutions to problems and by efforts to get the resulting policy packages accepted by decision-makers, thus bringing about a convergence of the problem, policy and political streams. According to Kingdon such entrepreneurs can be found anywhere, but share some important characteristics: they often have a good reputation and good networking skills and are willing to spend resources, most notably time, in a change process. To be successful, policy entrepreneurs need to be both good advocates of new policy ideas and good policy brokers.

2.4.5 Recognize, exploit, create and/or manipulate the multiple venues in modern societies

In a world where democratic thought – expressed for instance in the separation of powers – is close to being the only system that forms public discourse on the functioning of government (and governance for that matter), and in a world where the Internet allows networking on a scale previously unseen, there are many 'venues' that can serve as a base for the instigation of policy change. As discussed above 'venue shopping' describes the strategic behaviour associated with the choice between the various possible places where an individual or group can try and effect change. When engaged in venue shopping, policy actors: 'try to alter the roster of participants who are involved in the issue by seeking out the most favorable venue for the consideration of their issues. In this process, both the institutional structures within which policies are made and the individual strategies of policy entrepreneurs play important roles' (Baumgartner and Jones, 1991, p. 1045). Within government, actors may decide to by-pass a decision process that offers unfavourable terms for representing their arguments, waiting instead for the next opportunity. Alternatively actors can try to exploit venues for the representation of arguments that were not originally intended for that forum, thereby attempting to change the nature of the venue. The strategies used by policy entrepreneurs range from exploitation of the advantage brought simply by the existence of multiple venues (Richardson, 2000), to the highly directed use of multiple venues at various levels of government (Baumgartner and Jones, 2002; Pralle, 2003) and of those outside government (media, research, and so on), to the manipulation of venues (their composition, decision-making rules, and so on), and the creation of new venues.

2.4.6 Orchestrate and manage networks

A relationship exists between coalition-building and network orchestration. We distinguish between them, however, on the basis that any coalition will be confronted with a much broader set of actors engaged in a certain policy domain, a set we refer to as a network. We demonstrate in our introduction how some see networks as spontaneous, self-organizing entities. From a state-centrist view, this is correct, because networks will exist without state intervention – although the state can actively alter the existence and operation of networks (for example Putnam, 1994). However a less state-centred view sees networks as far from self-organizing and spontaneous. In fact much effort goes into creating and maintaining a network. Policy networks range in nature; they can be relatively closely knit and rather well aligned in terms of collective views and actions (policy communities), but they can also be relatively ad hoc and short-lived (issue networks) (Rhodes and Marsh, 1990). Fundamental policy change is likely to require the alteration, manipulation, breaking open or breaking up of policy communities that have crystallized around a policy domain. Policy scientists have focused much attention on the development of network typologies and strategies for network orchestration and identify two such strategies (Kickert et al., 1997; Meier and O'Toole, 2001; Klijn, 1996, 2005): management of the interactions within the current network settings; or creation or change of the current network setting.

2.5 Conclusions

This chapter explores the concept of policy change and stability from a policy science perspective and brings us to the following observations. The policy sciences clearly offer a wide range of conceptualizations of policy stability and change and of the role of agents in bringing about change. While some of the theories discussed above essentially focus only on policy change or stability, we have come across several theories that address both phenomena in an interrelated though obviously different way. Such theories include several typologies of the degree to which policies change. The sophistication of such typologies varies, but they essentially make a distinction between relatively shallow (sometimes called instrumental) and deeper levels of policy change (often called paradigmatic). Here we are interested in deeper levels of policy change, which we equate with transitions. How to distinguish transitions from shallow levels of change is debatable, but we argue that such transitions should become visible in a change either in the substance of policies or in a governance paradigm. The more fundamental the change, the more it resembles what we would call a transition. Similarly, the greater the level of policy stability, the less the likelihood that we would identify a transition.

In this book, we highlight the role of individuals or organizations working to achieve a transition at the level of the nation state. Our discussion of theories from the policy sciences shows that transitions are related to the emergence of new ideas that challenge the existing paradigm. Such ideas must be carried by individuals or groups that need to prepare them for consideration in the public and policy arenas. Those, or the allies of those, who have developed an idea must then find a 'window' that facilitates its public launch. Additionally policy entrepreneurs should look for appropriate venues to air their ideas before various audiences and to undermine the substance, procedures and organizations that work from the 'old' paradigm. This requires the skills and strategies of policy entrepreneurs as captured in further chapters.

Of course policy entrepreneurs seeking policy change do not work in a void. We draw attention to the institutional arrangements in management, adjacent policy fields and the government of the country concerned. It has been observed, for instance, that in countries with an elitist or corporatist approach to policy-making, the likelihood of radical policy change is usually much smaller than in countries with open government systems, because the latter offer actors seeking change a better opportunity structure. But does this always hold true even when, as in the USA, a relatively open government includes many checks and balances? What of the notion, too, that countries with a corporatist approach can be capable of pragmatic radicalism? To provide a variety of scenarios, this book contains case studies from both 'democratic' and 'undemocratic' countries.

Regardless of the wider institutional context we propose that policy entrepreneurs employ a set of strategies that closely conforms to the way policy change is achieved according to the policy sciences. Analysis shows that actors seeking change develop new ideas, sell their ideas, create and/or use windows of opportunity, select and use venues and orchestrate or manage networks to achieve transitions. We do not yet have sufficient research to support ideas about the order and combinations of strategies, nor about the types of agents actually applying these strategies. This produces, then, the need for empirical analysis of various cases. Transitions in water policy, as they occur in various nations and exhibit a range of conditions, timescales, actors and results, present rich material for just such analysis. We hope the following case studies will help to fill the knowledge gaps relating to policy transitions, in addition to shedding light on the key elements of change in the increasingly important realm of water supply and usage.

References

Adler, E. and P.M. Haas (1992), 'Conclusion: epistemic communities, world order, and the creation of a reflective research program', *International Organization*, **46** (1), 367–90.

Bachrach, P. and M.S. Baratz (1970), *Power and Poverty: Theory and Practice*, New York: Oxford University Press.

Baumgartner, F.R. and B.D. Jones (1991), 'Agenda dynamics and policy subsystems', *Journal of Politics*, **53** (4), 1044–74.

Baumgartner, F.R. and B.D. Jones (eds) (2002), *Policy Dynamics*, Chicago, IL: University of Chicago Press.

Benford, R.D. and D.A. Snow (2000), 'Framing processes and social movements: an overview and assessment', *Annual Review of Sociology*, **26**, 611–39.

Birkland, T.A. (1997), *After Disaster: Agenda Setting, Public Policy, and Focusing Events*, Washington, DC: Georgetown University Press.

Dudley, G. and J. Richardson (1996), 'Why does policy change over time? Adversarial policy communities, alternative policy arenas, and British trunk roads policy 1945–95', *Journal of European Public Policy*, **3** (1), 63–83.

Fischer, F. (2003), *Reframing Public Policy: Discursive Politics and Deliberative Practices*, Oxford: Oxford University Press.

Fischhendler, I., E. Feitelsen and D. Eaton (2004), 'The short-term and long-term ramifications of linkages involving natural resources: the US–Mexico transboundary water case', *Environment and Planning C: Government and Policy*, **22** (5), 633–50.

Folke, C., T. Hahn, P. Olsson and J. Norberg (2005), 'Adaptive governance of social–ecological systems', *Annual Review of Environmental Resources*, **30**, 441–73.

Gains, F., P.C. John and G. Stoker (2005), 'Path dependency and the reform of English local government', *Public Administration*, **83** (1), 25–46.

Haas, P.M. (1992), 'Introduction: epistemic communities and international policy coordination', *International Organization*, **46** (1), 1–35.

Hajer, M.A. (1995), *The Politics of Environmental Discourse: Ecological Modernization and the Policy Process*, Oxford: Oxford University Press.

Hajer, M.A. (2005), 'Coalitions, practices and meaning in environmental politics: from acid rain to BSE', in D. Howarth and J. Torfing (eds), *Discourse Theory in European Politics: Identity, Policy and Governance*, Basingstoke, Hampshire: Palgrave Macmillan, pp. 297–315.

Hall, P.A. (1993), 'Policy paradigms, social learning, and the state', *Comparative Politics*, **25** (3), 275–96.

Huitema, D., W. Egas, S. Hanegraaff, S. Kuks and K. Lulofs (2006), 'Coupling strategies in water management: an exploration on the basis of literature from the policy sciences and related fields', Amsterdam (IVM).

Huitema, D. and S.M.M. Kuks (2004), 'Harboring water in a crowded European delta', in J.Th.A. Bressers and S.M.M. Kuks (eds), *Integrated Governance and Water Basin Management: Conditions for Regime Change towards Sustainability*, Dordrecht: Kluwer Academic Publishers, pp. 61–100.

Ingram, H. and L. Fraser (2006), 'Path dependency and adroit innovation: the case of California water', in R. Repetto (ed.), *Punctuated Equilibrium and the Dynamics of US Environmental Policy*, New Haven, CT, USA and London, UK: Yale University Press, pp. 78–109.

Jenkins-Smith, H.C. and P.A. Sabatier (1993), 'The dynamics of policy-oriented learning', in H.C. Jenkins-Smith and P.A. Sabatier (eds), *Policy Change and Learning: An Advocacy Coalition Approach*, Boulder, CO: Westview Press, pp. 41–56.

John, P. (1998), *Analysing Public Policy*, London: Cassell.

Jones, B. and F. Baumgartner (2005), *The Politics of Attention: How Government Prioritizes Problems*, Chicago, IL: Chicago University Press.

Kaika, M. (2003), 'The Water Framework Directive: a new Directive for a changing social, political and economic European framework', *European Planning Studies*, **11** (3), 299–316.

Kay, A. (2005), 'A critique of the use of path dependency in policy studies', *Public Administration*, **83** (3), 553–71.

Kickert, W.J.M., E.H. Klijn and J.F.M. Koppenjan (1997), *Managing Complex Networks: Strategies for the Public Sector*, London: Sage.

Kingdon, J.W. (1995), *Agendas, Alternatives and Public Policies*, 2nd edn, New York: HarperCollins.

Klijn, E.H. (1996), *Regels en sturing in netwerken. De invloed van netwerkregels op de herstructurering van naoorlogse wijken* (*Rules and Steering in Networks: The Influence of Network Rules on the Restructuring in Postwar Neighbourhoods*), Delft: Eburon.

Klijn, E.H. (2005), 'Netwerken als perspectief op beleid en uitvoering van beleid' ('A network perspective on policy and policy implementation'), *Beleidswetenschap* (*Policy Science*), **19** (4), 32–54.

Knill, C. and A. Lenschow (2001), 'Seek and ye shall find! Linking different perspectives on institutional change', *Comparative Political Studies*, **34** (2), 187–215.

Loorbach, D. and J. Rotmans (2004), 'Managing transitions for sustainable development', in A.J. Wieczorek and X. Olsthoorn (eds), *Industrial Transformation: Disciplinary Approaches towards Transformation Research*, Dordrecht: Kluwer.

Lowry, W. (2006), 'Potential focusing projects and policy change', *Policy Studies Journal*, **34** (3), 313–35.

Majone, G. (1992), *Evidence, Argument and Persuasion in the Policy Process*, New Haven, CT: Yale University Press.

Meier, K.J. and L.J. O'Toole (2001), 'Managerial strategies and behavior in networks: a model with evidence from US public education', *Journal of Public Administration and Theory*, **11** (3), 271–93.

Meijerink, S. (2005), 'Understanding policy stability and change: the interplay of advocacy coalitions and

epistemic communities, windows of opportunity, and Dutch coastal flooding policy 1945–2003', *Journal of European Public Policy*, **12** (6), 1060–77.
Meijerink, S. (2008), 'Explaining continuity and change in international policies: issue linkage, venue change, and learning on policies for the River Scheldt estuary 1967–2005', *Environment and Planning A*, **40** (4), 848–66.
Needham, B. and E. Louw (2006), 'Institutional economics and policies for changing land markets: the case of industrial estates in the Netherlands', *Journal of Property Research*, **23** (1), 75–90.
North, D. (1990), *Institutions, Institutional Change and Economic Performance*, Cambridge: Cambridge University Press.
Olsson, P., L.H. Gunderson, S.R. Carpenter, P. Ryan, L. Lebel, C. Folke and C.S. Holling (2006), 'Shooting the rapids: navigating transitions to adaptive governance of social–ecological systems', *Ecology and Society*, **11** (1), 18, www.ecologyandsociety.org/vol11/iss1/art18/.
Peters, B.G., J. Pierre and D.S. King (2005), 'The politics of path dependency: political conflict in historical institutionalism', *Journal of Politics*, **67** (4), 1275–1300.
Pierson, P. (2000), 'Increasing returns, path dependence, and the study of politics', *American Political Science Review*, **94** (2), 251–67.
Pierson, P. (2004), *Politics in Time: History, Institutions and Social Analysis*, Princeton, NJ: Princeton University Press.
Pralle, S.B. (2003), 'Venue shopping, political strategy, and policy change: the internationalization of Canadian forest advocacy', *Journal of Public Policy*, **23** (3), 233–60.
Putnam, R. (1994), *Making Democracy Work: Civic Traditions in Modern Italy*, Princeton, NJ: Princeton University Press.
Rhodes, R.A.W. and D. Marsh (1990), 'Policy networks: a British perspective', *Journal of Theoretical Politics*, **2** (3), 293–317.
Richardson, J. (2000), 'Government, interest groups and policy change', *Political Studies*, **48** (5), 1006–25.
Roberts, N.C. and P.J. King (1996), *Transforming Public Policy*, San Francisco, CA: Jossey-Bass.
Sabatier, P.A. (1993), 'Policy change over a decade or more', in P.A. Sabatier and H.C. Jenkins-Smith (eds), *Policy Change and Learning: An Advocacy Coalition Approach*, Boulder, CO: Westview Press, pp. 13–39.
Sabatier, P.A., W. Focht, M. Lubell, Z. Trachtenberg, A. Vedlitz and M. Matlock (eds.) (2005), *Swimming Upstream: Collaborative Approaches to Watershed Management*, Cambridge, MA: MIT Press.
Sabatier, P.A. and C.M. Weible (2007), 'The advocacy coalition framework: innovations and clarifications', in P.A. Sabatier (ed.), *Theories of the Policy Process*, Boulder, CO: Westview Press, pp. 189–220.
Schlager, E. (2007), 'A comparison of frameworks, theories, and models of policy processes', in P.A. Sabatier (ed.), *Theories of the Policy Process*, Boulder, CO: Westview Press, pp. 293–319.
Schön, D. and M. Rein (1994), *Frame Reflection*, New York: Basic Books.
True, J.L., D. Jones and F.R. Baumgartner (2007), 'Punctuated equilibrium theory: explaining stability and change in public policymaking', in P.A. Sabatier (ed.), *Theories of the Policy Process*, Boulder, CO: Westview Press, pp. 155–87.
Van der Brugge, R., J. Rotmans and D. Loorbach (2005), 'The transition in Dutch water management', *Regional Environmental Change*, **5** (4), 164–76.

3 Driving forces in global freshwater governance

Joyeeta Gupta

3.1 Introduction: the problem

Although water governance has a very long history going back some 5000 years, and transboundary water governance goes back some centuries, global water governance, to the extent that it is possible to talk about that, is of relatively recent origin (Pahl-Wostl et al., 2008; Schnurr, 2008). This can be attributed to a number of factors. First, water has for long been seen as a local to regional physical body and its management was confined to relevant local and regional administrative bodies. Second, the scale of water problems has been relatively small and focused and did not call for a global water governance process. Third, the interconnections between environmental, economic and water challenges were not quite so obvious.

However as social issues and those of global politics and economics changed dramatically in the twentieth century, the need for water governance at the global level was felt strongly in United Nations (UN) bodies, non-UN bodies and academic circles, but water is very distinct from other fields of governance. Thus while governance related to the depletion of the ozone layer is highly centralized, and governance of climate change matters is relatively centralized within the UN multilateral framework under which a number of different regional and international efforts tend to converge, water governance is highly diffuse and dispersed throughout the UN system and beyond it (Pahl-Wostl et al., 2008). It is necessary to understand the nature of the water governance field to be able to explore what has driven these different developments in different contexts.

Against this brief background this chapter explores the following research question: what are the contours of the global water governance agenda and what have been the strategic moments of change in the agenda? What discourses drive the water agenda in the different fora? What factors influence how policy entrepreneurs use discourses to promote and/or block transitions?

This chapter is based on a literature review, but the literature on global water governance as such is quite limited, and the driving forces and trends have been derived inductively from the available literature and could therefore perhaps best be seen as hypotheses for future testing rather than firm conclusions. It should be noted upfront that while there is considerable literature on international and global water law, there is much less available on the role and contribution of other social drivers in global water governance. This may reflect the fact that international water law has been on the UN agenda for at least a century while global water governance has a shorter pedigree. Although global water governance refers to fresh water as well as the oceans, the focus here is primarily on the global fresh water regime. This chapter discusses the contours of global water governance, the ideological discourses underlying different competing governance processes and the key factors influencing change at the global level, before drawing some conclusions.

There are clear links between the contours, the discourses and the actors. As a preview of how the chapter unfolds, the following section presents a chronological table of key global water events followed by an assessment of legal and policy discussions within the UN context and a discussion of the role of non-state actors and hybrid bodies. Section 3.3 clusters the discourses into four groups, namely general ideological discourses, general governance discourses, water-specific discourses and environment-specific discourses and relates them to the actors in global water governance. Section 3.4 focuses on specific factors – namely the role of science, the shift from legal to hybrid governance systems, social movements and official development assistance in shaping water governance.

3.2 The structure and drivers of global water governance

3.2.1 Introduction

Unlike most other global fields where governance is framed within the context of the UN and often as consistent with the agreements made within the UN, global water governance has developed differently in different fora. There have been discussions within the UN, discussions within or between non-state actors and between state and non-state actors – hybrid bodies. A description of these bodies and their contribution follows the chronological table (Table 3.1) of the key events influencing global freshwater governance over the last 50 years starting with the adoption of the Helsinki Rules on the Uses of the Waters of International Watercourses in 1966 (ILA, 1966). The text that follows, however, does not take a chronological perspective, but instead examines water governance in terms of the activities of the key actors.

3.2.2 Discussions within the UN

Water governance within the UN has been scattered through UN agencies. Within the United Nations, water was not seen as a sectoral issue, but rather as a part of many other existing challenges and, hence, UN agencies (for example Food and Agriculture Organization, World Health Organization), funds (for example Global Environment Facility) and programmes (for example United Nations Development Programme, United Nations Environment Programme) deal with water to the extent that water is an integral part of the activities of these bodies.

However, interstate water disputes have existed for over a century, and in 1970 the United Nations General Assembly requested the International Law Commission (ILC) to assess the state of water law and progressively develop the law. This mission was ultimately completed in 1997 when the General Assembly adopted the UN Convention on the Law of the Non-Navigable Uses of International Watercourses (UN Watercourses Law, 1997) that looks at common principles for regulating transboundary water issues (see McCaffrey, 2001). This law has not been ratified by a sufficient number of countries to enter into force, possibly because of the way in which the treaty was negotiated – within the ILC and not within an intergovernmental negotiating committee (Gupta and Dellapenna, 2009) – but the convention has been influential in shaping many regional agreements including the Southern African Revised Protocol on Shared Watercourses (Revised SADC Protocol, 2000).

On oceans the UN Convention on the Law of the Sea of 1984 and its protocols provides the legal framework and this is complemented by other conventions like those on whaling and ocean dumping. Furthermore, although a number of relevant global treaties

Table 3.1 Key events in global freshwater governance (1960–2008)

Year	Activity	Actor
1966	Helsinki Rules on the Uses of the Waters of International Watercourses	International Law Association
1972	Stockholm Declaration on the Human Environment	UN Conference on the Human Environment
1977	Mar del Plata Conference on water	UN Water Conference
1980s	International Drinking Water Supply and Sanitation decade	UN General Assembly
1992	Agenda 21 chapter on freshwater governance and ocean governance	United Nations Conference on Environment and Development
1992	Dublin Principles	International Conference on Water and the Environment
1996	World Water Council set up; holds conference every three years	Hybrid body set up partly following a resolution of the International Water Resources Association
1996	Global Water Partnership	Set up by World Bank, the United Nations Development Programme and the Swedish International Development Agency
1997	Law on the Non-Navigational Uses of International Watercourses	UN's International Law Commission
2000	Millennium Development Goals	UN General Assembly
2000	Report on Dams	World Commission on Dams
2002	General Comment on the Right to Water	UN Economic and Social Council
2005–15	UN Water Decade	UN General Assembly
2008	Appointment of expert to look into the human right to water and sanitation	UN Human Rights Council

were negotiated on environmental issues during the 30 years of its evolution such as on transboundary movement of hazardous wastes, climate change, biodiversity, desertification, and so on, the UN Watercourses Convention was not much influenced by this process (Dellapenna, 2000; Dellapenna and Gupta, 2008).

On the issue of a human right to water (and sanitation), progress has been relatively slow. Although this right is possibly implicit in earlier human rights declarations, it was the 2002 General Comment on the Right to Water (General Comment, 2002) adopted by 140 countries to the Covenant on Economic, Social and Cultural Rights that first articulated this right and indicates that there is some political momentum on the human right to water issue. In 2008, the UN Human Rights Council decided that it would be appropriate to appoint an expert with a three-year mandate to investigate the issue of a human right to water and sanitation.

Bilateral and multilateral agreements on water have been negotiated since the eighth century, and since 1820, 400 water-related agreements have been established (Oregon State University, 2002). The UN Economic Commission for Europe adopted legislation on Transboundary Watercourses and Lakes in 1992. The European Water Framework Directive (EU WFD, 2000) is the only supranational agreement on the issue (see Chapter

13 by Partzsch, this volume). Since 1990 there have been several agreements between countries including those in southern Africa (van der Zaag, 2009) and the Middle East (Sabel, 2009).

There have been relatively few international court cases on water issues. The prominent cases are those on the River Oder before the Permanent Court of International Justice (PCIJ) in 1929, the diversion of waters from the Maas before the PCIJ in 1937, the Lac Lannoux Tribunal of 1957, the Gut Dam case of 1968 and the Gabĉikovo Nagymaros case, which have all provided precedents that have shaped international water law. Judgments on the Helmand River (1872), the Kuskh River (1893) and the Zaramulla River (1945) focus on water quantity issues, while those on the Rhine (2004) and the ongoing pulp mills case between Uruguay and Argentina (2006) focus on water quality issues and indicate how legal developments may unfold in the future.

For historical reasons and because of the disciplinary and sectoral nature of water discussions, legal and policy developments at global level have not gone hand in hand, nor have the processes of water law and environmental law been synchronized. In the legal arena the discourses have tended to focus on equitable use of water and the 'no harm' principle, that limit sovereignty; the human right to water; and to a limited extent included environmental principles in water governance. The dominant discourse in Western water agreements has been optimal utilization and reduction of pollution (for example Hall, 2009; de Castro, 2009) and in Southern agreements the focus has been on sharing water resources.

Although, in general, policy developments precede legal articulation, in the water field UN policy processes within UN agencies have proceeded in parallel with legal developments in water governance (within the International Law Commission). On the policy front, following the UN Conference on the Human Environment in 1972, the Water Conference in 1977 at Mar Del Plata, and the UN Decade on Water and Sanitation in the 1980s, the most significant progress was made at the UN Conference on Environment and Development where two chapters focused on fresh water and oceans setting out an agenda for the global community (Agenda 21, 1992). In 2000 the Millennium Development Goals were adopted and aimed to reduce by 50 per cent the proportion of people without sustainable access to safe drinking water and basic sanitation.

In 2003 UN Water was set up as an official collaborative platform among 25 UN agencies to promote coordination in the development and implementation of water policy following the 2002 World Summit on Sustainable Development. It also publishes the World Water Assessment Report. In 2005 a 'Water for Life' decade was set in motion as part of the implementation of the Millennium Goals. The United Nations' Educational, Scientific and Cultural Organization (UNESCO) International Hydrological Programme produces regular publications and the United Nations Environment Programme (UNEP) and the Global Environment Facility (GEF) publish the Global International Waters Assessment (GIWA).

In the policy arena there has been a tendency to recognize the need to provide access to water to all humans, while also acknowledging that water may need to be priced in order to be effectively managed. Further, many argue that an integrated and coherent UN approach to water management is needed.

3.2.3 Non-state actors: scholars, non-governmental organizations (NGOs), industry

A highly influential group of actors in water governance worldwide has been the expert communities of legal scholars at the Institut de Droit International and the International Law Association. The International Law Association prepared its Helsinki Rules on the Uses of the Waters of International Rivers in 1966 (ILA, 1966), far in advance of the work of the UN International Law Commission. This document presented the state of the art of water law and was strongly influential, despite its non-legal status, in persuading countries (such as the riparian nations of the Mekong Basin) all over the world to adopt its key features (Bourne, 1996).

The Law Association completed its Berlin Rules in 2004 that form an up-to-date assessment of the latest legal discourses on water (international and national, surface and groundwater) governance, integrating all relevant international legal rules from international water, humanitarian, environmental and human rights law (Dellapenna and Gupta, 2008). The Rules incorporate the principles of public participation, integrated management, sustainable development, the precautionary approach and environmental impact assessments. In incorporating policy and scientific developments in a legal text, the Berlin Rules are ahead of their time and probably of international water law, but may be a good indicator of how water law will develop in the coming decades, to judge by the influence of the Helsinki Rules.

The community of legal scholars within the International Law Association and the International Law Commission has been most influential in water law because of its long history of engagement in the process and because the members of this community are also engaged in water policy formulation in their national contexts. The discourses espoused and promoted are equitable sharing of water resources and the no-harm principle. The Berlin Rules take a more comprehensive approach and promote the integration of national and international law, and water and environmental law.

As hydrological sciences developed in the twentieth century, water engineers and professionals formed a relatively new community that also engaged heavily in water governance issues, especially the notion of taming the waters in an effort to meet human needs and wants. Ideas promoted included technocratic engineering solutions (for example dams) to water management, the river basin management approach and, more recently, inter-basin transfers. These were promoted in the 1970s through professional communities, specific government agencies and development cooperation agencies leading to about 800 000 dams in 2008, of which more than 40 000 are large dams.

Water professionals have formed effective networks to deal with water issues. For example the International Water Association, the result of a merging of the International Water Supply Association and the International Water Quality Association, aims to create a network of professionals working in the area of water. They claim on their website to provide expert, global, innovative, diverse and future-focused policies.[1] Other professional organizations include the International Water Resources Association, and the International Network of Basin Organizations. Their publications promote integration discourses such as river basin management and integrated water resources management.

There is no clear epistemic community in the area of global water governance, but a number of journals – *Water Resources Research*, *Water Policy*, *Journal of Water Law*, the *Urban Water Journal*, and so on – encourage scientific papers on the subject. The

World Water Assessment Programme discusses water issues, but does not provide a systematic assessment of the state of the art of water science.

Multinational water corporations are dominant actors in water governance. Companies like Suez, Bechtel and Vivendi promote private sector interests in water governance often directly through engagement in (hybrid) organizations, or indirectly by influencing national governments to promote their interests. Some 10 per cent of formerly public water systems are now in private hands (Hall and Lobina, 2006). Many of these have actively supported private sector participation in water management and the notion that water should be seen as an economic commodity.

3.2.4 Hybrid bodies

Increasingly, multinationals are cooperating with governments and non-state actors to create new hybrid international commissions, meetings and councils, possibly to fill a perceived gap in water governance. The World Commission on Dams, set up by the World Bank and the International Union for the Conservation of Nature (IUCN) in 1997, concluded that large dams could have serious negative impacts on society (WCD, 2000). Although its report is not legally binding and many argue is not entirely legitimate since by including specific types of actors (environmental NGOs) it also excluded others (Scheumann, 2008; Dingwerth, 2004), it has been influential nonetheless. It has tried to integrate social and environmental concerns into the engineering and economic equation used by states to promote large dams. The Dublin Conference of 1992 was organized outside the purview of the United Nations and emphasized, on the one hand, the need for water to be seen as an economic good and, on the other hand, the need to take gender issues into account.

The World Water Council was established in 1996, representing a range of stakeholder categories among 300 members from 60 countries. The Council holds a World Water Forum every three years. Its mission is:

> to promote awareness, build political commitment and trigger action on critical water issues at all levels, including the highest decision-making level, to facilitate the efficient conservation, protection, development, planning, management and use of water in all its dimensions on an environmentally sustainable basis for the benefit of all life on earth. (http://www.worldwatercouncil.org/index.php?id=92, accessed 4 November 2008)

The Global Water Partnership is a partnership of government agencies, professional organizations and multilateral development agencies 'committed to the Dublin–Rio principles' (as its own website puts it), with the mission to help countries implement sustainable water resources management.

3.2.5 Inferences

On the basis of the above, the following inferences can be drawn. First, that water governance, unlike the agreements on addressing climate change and the depletion of the ozone layer, is highly scattered at global level and according to Pahl-Wostl et al. (2008) constitutes a mobius web system of governance. Many of the developments in the water field have taken place independently of each other. No single agency has a clear-cut mandate to push water governance issues further. Second, these competing developments have followed different trajectories. Water law has evolved gradually over the last few centuries through what has been in some ways a bottom-up process. Policy evolution, influenced by scientific ideas, has tended to take a top-down path and is of relatively recent origin. The hybrid bodies and the non-state actors reflect a more modern trend of diagonal interaction between different types of bodies.

Third, global water governance is punctuated by defining moments in terms of ideas. The hydraulic mission with its focus on harnessing the waters to irrigate lands was a key idea launched soon after the perceived success of modern dam building and irrigation works, and often found support expressed in infrastructural investment by aid agencies soon after the end of the Second World War. The 1966 Helsinki Rules, used by many countries, and the 1997 Watercourses Convention, presented the comprehensive framework on equity issues. The 1992 Dublin Declaration, with its important references to gender, presented water as an economic good, which led to many more activities encouraging private sector participation in water management. In 1992, the adoption of the Rio Declaration emphasized sustainable development and related environmental principles and these are being implemented by professional organizations and national governments. Implementation draws on instruments such as the 1992 UN Economic Commission for Europe (ECE) Convention, the European Union (EU)'s Water Framework Directive (EU WFD 2000), and the Berlin Rules of 2004. The 2002 General Comment on the Right to Water (General Comment, 2002) clearly articulated water as a human right and is currently being developed within the UN Human Rights Council. Finally, each of these processes has its own discursive framework, taken up in the next section.

3.3 Driving forces: the key discourses

3.3.1 Introduction: lack of a global problem definition

This section explores the driving forces moving and used by different actors promoting global governance on water. Although Agenda 21 (1992) attempted a problem definition for global water, there is no clear and commonly shared belief about water problems worldwide. Problems are very different in different parts of the world. In the global South a key issue affecting between 1 and 2 billion people is access to sanitation and drinking water. In the global North problems tend to be focused more on pollution, quality and standards. The links between climate change, biodiversity loss, deforestation and water have not yet been clearly framed as a common problem, which leads to a lack of real consensus on how water issues should be addressed. The idea that there should be a common comprehensive global agenda on water issues is in its infancy.

A key factor driving change has been the growing discourses that shape human behaviour. Four types of discourses appear to affect global water policy, as listed in Table 3.2: general ideological discourses, general governance discourses, water-specific discourses and environment-specific discourses. These are discussed below, moving from general to specific.

3.3.2 General ideological discourses driving change

General discourses include the discourse on neoliberalism and private sector management of water resources. The neoliberal economic model is dominant and promotes fiscal discipline, a movement away from general subsidies, tax reform, market-based interest

Table 3.2 Discourses, actors and fora in global water governance

Discourse	Dominant actor	Forum
General		
Liberalization and private sector participation	Economists; banks; aid agencies; private sector	Development bank policy; hybrid meetings and bodies (Dublin Conference, Global Water Partnership; World Water council, World Water Forums)
Governance		
Governance as opposed to government; good governance; decentralization	Policy analysis; banks; institutional economists	Development bank policy; policy of aid agencies
Stakeholder participation; subsidiarity	Policy analysis; legal scholars	Rio Declaration; UN Aarhus Convention
Water discourses		
Hydraulic mission: harness the power of water	Water engineers, investors	Development banks and aid agencies
Equitable sharing and no harm	Water lawyers	UN (Watercourses Convention; and many treaties) Non-UN (ILA Rules)
Water as a human right	Water lawyers and water NGOs and aid agencies	UN (ECOSOC); Human Rights Council
Indigenous peoples' rights	Indigenous peoples and their organizations	UN (ILO; international treaties)
IWRM	Hydrologists, water professionals and scholars, World Bank	Dublin Declaration; Agenda 21; GWP
Virtual water trade	Trade and water scientists	
Environmental discourses		
Sustainable development	All actors; since sustainable development is amenable to many definitions	UN, non-UN and hybrid fora
Principles of environmental protection	Environmentalists	Mostly UN fora

and exchange rates, trade liberalization, privatization of state enterprises, deregulation, property rights and individualism. This discourse returns to classical liberalism following the perceived failure of import substitution strategies encouraged in the 1970s. A core element of the neoliberal agenda is smaller government and a greater role for the private sector.

Influence of neoliberal economic discourse on the water sector began to come to the fore when water was first recognized as an economic good in the Dublin Declaration of 1992 and subsequently in the World Water Forum meetings. The issue was not settled in 1992 as the UN Conference on Environment and Development clearly stated that human needs should be given priority. The conflict between water as an economic good and as a basic necessity led the 2002 UN World Summit on Sustainable Development to be more cautious and to argue instead that states should 'employ the full range of policy instruments, including regulation, monitoring, voluntary measures, market and information-based tools, land use management and cost recovery of water services, without cost recovery objectives becoming a barrier to access to safe water by poor people' (WSSD, 2002, art. 4(26)b).

In the Western world the rise of neoliberalism led to various experiments with liberalizing the water sector and allowing private sector participation in water management (Barlow and Clarke, 2002; Dellapenna, 2009). In the South, the lack of government resources to invest in the water sector, combined with conditional loans and assistance and the argument that the private sector had the resources and skills to manage the sector, led to a process of privatization and deregulation (Petrella, 2001; Gleick et al., 2002; UNWWDR, 2003). While the record of water deregulation and privatization in the developed world is generally mixed, the developing world exhibits a number of broken contracts (Schouten and Schwartz, 2006; Tecco, 2008) and there are many stories of failure (Petrella, 2001; Shiva, 2002; Barlow and Clarke, 2002). Poor results have often stemmed from a hike in water prices following liberalization, putting water out of reach for the poorest people and even those of the lower middle class.

3.3.3 Governance discourses

General governance discourses include discussions on governance versus government, good governance, decentralization, subsidiarity and stakeholder participation in policy-making.

The governance versus government discourse originated in the 1980s and called for a movement from hierarchical to non-hierarchical government involving less institutionalized processes where social actors and networks had a significant role to play (Hanf and Sharpf, 1978; Kooiman, 1997; Rhodes, 1997; Pierre, 2000; Hajer and Wagenaar, 2003; Blatter, 2003; Arts and van Tatenhove, 2005). This inclusionary process was meant to eliminate the elitist bias in government approaches and led to the development of three important concepts – network governance and management (Kickert et al., 1997; Koppenjan and Klijn, 2004) and deliberative policy-making (Hajer and Wagenaar, 2003; Fischer, 2003).

In recent decades, the good governance discourse has become very popular. Its principles include legitimacy, transparency, accountability, efficiency, effectiveness, responsiveness and, sometimes, the notion of subsidiarity. Good governance has become a major credo in development work and much of the water discourse also refers to this. For example the UN World Water Development Report (UNWWDR, 2003, p. 373) defines good water governance as an inclusive participatory, transparent, equitable, accountable, coherent, responsive, integrated and ethical form of governance.

Stakeholder participation in decision-making is a discourse in itself, although it is also seen as part of the governance and good governance discourses in the governance field, and has pervaded many sectoral debates including that over water in both developed and developing countries. It is seen as positive because it can potentially increase the legitimacy and effectiveness of policy (Hildering, 2004; UNWWDR, 2006), but may also lead to non-harmonious policy, externalization of extra territorial impacts and can reflect local power politics in a way that is not conducive to equity issues (Stiglitz, 2000; Lemon, 2001; Keeley and Scoones, 2003). It has been promoted by academic writing (van de Kerkhof, 2004), policy documents (for example World Water Vision, in Cosgrove and Rijsberman, 2000), and legal documents of NGOs (ILA, 2004) and of states (Aarhus Convention, 1998; Article 10 of the Rio Declaration).

Running parallel to the discourse on stakeholder participation, the discourse on decentralization is seen as a way to deepen democracy by bringing governance closer to the governed. This process is held to make governance more legitimate and effective through the incorporation of stakeholder knowledge and accountability to local communities (Rondinelli et al., 1984; UNWWDR, 2006). The discourse is still very popular although in many countries there is increasing recognition that water is a scarce and valuable resource and may need comprehensive, and possibly hierarchical, management at central government level. Such trends are visible in Israel (Laster and Livney, 2009), Australia (McCay and Marsden, 2009) and South Africa (Kidd, 2009).

3.3.4 Water-specific discourses

Introduction Water-specific discourses include the hydraulic mission of harnessing the power of water (Mollinga, 2008) and the equitable distribution and no harm discourses. In recent years, river basin management, catchment management, and integrated and adaptive water resource management have become influential. The growing discourse on water as a human right is also key.

The hydraulic mission The end of the Second World War, as societies began to harness water resources on a huge scale for human development, brought the so-called hydraulic mission (Allan, 2006). The associated discourse focused on the importance of enhancing supply and bringing water to areas without easy access. This idea was adopted in both developed and developing countries, and especially rain-dependent economies. The frequent underuse of engineering works led to pressure to engage in social engineering approaches to match the civil engineering approaches as a way to promote the mission (Mollinga, 2008). This was a highly dominant discourse and one marketed by water professionals and aid agencies.

The equitable use and no harm discourse: legal discourses Case law demonstrated the need to look at watercourses from the perspective of equity. Equity issues have increasingly been taken into account in watercourse treaties. The International Law Association identified 11 criteria in its Helsinki Rules (1966), which included taking into account the following elements relating to a basin: its geography; its hydrology; the climate affecting it; the past utilization of its waters; the economic and social needs of the basin states; the population dependent on the waters; the comparative costs of the alternative means of satisfying the economic and social needs of each basin state; the availability of other resources; the avoidance of unnecessary waste; the practicability of compensation of one state by another; and the degree to which the needs of a basin state can be satisfied without causing substantial injury to a co-basin state. These criteria are used to determine how water resources are to be shared between countries. The 1992 UNECE Convention also refers in Article 2.2 to reasonable and equitable utilization, but does not further elaborate on principles. However such elaboration was undertaken in the UN Watercourses Convention of 1997. The UN Convention further specified that it was up to parties to determine how weights should be assigned to specific criteria. The principles were derived mostly from existing state practice, legal precedents and court cases. The discourse on equitable use and no harm is closely linked to the idea that domestic activities should not lead to harm in other states and the two principles together serve to limit the sovereign rights of states to do as they please with the waters flowing through their territory. Equitable utilization forms the single theme of a discourse studied and supported by legal scholars (Fuentes, 1999; Tanzi and Arcari, 2001; McCaffrey, 2001; Benvenisti, 2003).

Discourse on river basin management, catchment councils and integrated water resource management Another key discourse arose around the need to manage rivers in terms of their hydrological unity – a social engineering idea. This led to the concept of river basin management and catchment-related bodies, which implies going beyond the administrative boundaries to deal with water within its hydrological boundaries. Hooper (2005) discusses the development of the international river basin management paradigm (IRBM), the need to manage rivers at river basin level and how best to achieve integrated water resources management (IWRM). IRBM is seen as a subset of IWRM. These related concepts call for integrating other related concerns into water management. Promoted by academics (for example Koudstaal et al., 1992) and political and policy declarations (for example Agenda 21, 1992; the Dublin Declaration, 1992), the discourse is seen as integrating different types of water (green and blue, surface and ground, and so on) and as requiring a strong governance framework in order to promote a synergetic and equitable approach to water policy (UNWWDR, 2006). Integration, which may be theoretically simple, is very complex in practice to address within government agencies both within the developed world (Rahman et al., 2003 on the EU Water Framework Directive; see also Chapter 13 by Partzsch, this volume) and the developing world. This discourse has been marketed not just by water professionals, but also by the World Bank (Hartje, 2008).

The human rights discourse A driving force behind the human rights to water discourse is the recognition that a substantial portion of the human population does not have access to water and sanitation, combined with the growing recognition of the importance of human rights in general.

The human right to water (and possibly sanitation) is seen as a corollary to basic social and economic rights that guarantee the ability of humans to live a life of dignity (McCaffrey, 1992; Smets, 2000; UNWWDR, 2003; Hildering, 2004). Although such a right implicitly derives from the existing human rights declarations, it was explicitly adopted in the 2002 General Comment on the Right to Water (General Comment, 2002). In addition, in March 2008 the Human Rights Council adopted a resolution by consensus to appoint a special rapporteur on the right to water and sanitation. The rapporteur has a three-year period to come up with concrete proposals. The recognition of such a right to water (sanitation services an additional possibility) would make it obligatory for states to provide such services. However it should be noted that mere recognition of such a right may not lead to actual change, as rights are not self-enforcing and humans must assert their rights in judicial systems. The adoption of the Millennium Development Goals goes halfway to recognizing the vital importance of the human need for water, since it only aims to halve the number of people without access to water and sanitation. Some of the aid agencies are focusing actively on this area (Kipping, 2008). The World Bank has traditionally avoided engaging in debates on human rights, and transnational corporations tend to provide at best lukewarm support for the idea (Sacher and Windfur, 2008).

Another discourse relates to the rights of indigenous people who have lost their 'ownership' rights to land and water over the centuries through exploitation and conquests in countries like the USA, Canada and Australia, as well as many developing countries. There are two elements of this discourse: the return of the right of indigenous people to ownership of land and water; and respect for the way they deal with nature and water (Nakashima and Chiba, 2006; Boelens, 2006). The efforts of indigenous peoples have been supported by the adoption of a number of international treaties (for example ILO Convention 169; Chapters 18 and 26 of Agenda 21; Articles 8j and 10c of the Convention on Biological Diversity; and the UN Draft Declaration on the Rights of Indigenous Peoples), but they have an uphill battle in claiming ownership of land and water resources in their home countries. The shift from communal to private was undertaken to allow for agricultural exploitation and industrialization. Gupta and Dellapenna (2009) show that this shift occurred across the board in many parts of the world to allow conquerors and powerful actors to consolidate their hold over natural resources and the factors of production.

The virtual water discourse In recent years the virtual water discourse has become increasingly important. It focuses on the notion that societies trade not only in water, but also in products that include water in their production and distribution process. An assessment of trade in virtual water reveals that water-poor areas are often exporting products rich in water, and water-rich areas are often importing products that embody water. The assessments help to enrich understanding of 'illogical' social decisions, but may not translate easily into policy ideas. Susanne Neubert in collaboration with Lena Horlemann (2008) submit that four sorts of objections are commonly made to this discourse, namely: that there are strategic issues that need to be taken into account; that the assumptions in the analysis are not valid; that the conditions for successful trade are limited; and that the consequences of virtual water trade may have major implementation challenges.

3.3.5 Environment-specific discourses

Environment-specific discourses include those on sustainable development and principles of environmental management. The merger of the environmental and developmental discourses led to the development of the concept of sustainable development. Launched politically in the report of the World Commission on Environment and Development (WCED, 1987), this concept has led to considerable research on the subject and has been adopted in a number of international laws (for example the United Nations Framework

Convention on Climate Change, the Biodiversity Convention, and so on) and political documents (for example the Rio Declaration, 1992; WSSD, 2002). Although there is confusion regarding the exact meaning of sustainable development and whether it is a process or a goal, whether or not it calls for trade-offs between social, economic and environmental objectives, and whether or not it calls for meeting the needs of present generations before meeting those of future generations, the concept inspires many of the environment- and development-related aspirations for the future.

More concretely, a large number of environmental principles have been adopted over the years, starting with the Stockholm Declaration of 1972 and its restriction of sovereignty to ensure no extra-territorial harm. A number of principles were adopted later, in the Rio Declaration of 1992, including: the precautionary principle; the polluter pays principle; the environmental impact assessment principle; the liability and compensation principle; the common, but differentiated responsibility principle; and many others. These principles are increasingly included in transboundary water agreements and have also been included in the Berlin Rules of 2004.

3.3.6 Inferences

Four types of discourses have been discussed here. I suggest that they sometimes reinforce each other, but more often compete in the global arena to provide the framework for global water governance. They compete in terms of their focus on market versus human rights approaches, integrated and harmonized versus stakeholder and decentralized approaches, and in terms of managing water resources to enhance development as well as taking sustainable development into account. Such competition may actually be a step towards synthesis, since compromise between the discourses is possible. This depends, however, on how actively different actors lobby for different discourses. The clash between the notions of water as a human right and water as an economic good in the current framing of issues can be witnessed in the Millennium Development Goals, which are based more on the notion of needs than rights. Furthermore the equitable utilization concept favoured in the legal sphere does not get much emphasis in the other spheres. Ultimately the notion of integrated water resource management, while in itself a very solid concept, is not easy or affordable to implement.

3.4 Factors of change

3.4.1 Introduction

This section addresses the question: what factors influence the adoption of discourses? It argues that there are five drivers: the science underlying the discourses and the actors pushing the science; the shift from regulatory approaches to hybrid approaches; social movements; the organizations for water management; and, specifically, the instrument of development cooperation as a dominant factor related to change.

3.4.2 The related science

Scientific research often provides tailored ammunition for policy change. Scientific results provide an independent influence on policy (epistemic influence on converging state policy theory; Haas, 1989), but scientific results that are used by lobby groups often have more weight in the process.

In terms of water science a number of issues need to be considered. First, water science is highly scattered among a number of different journals. Although this is not unique to the water field, it is clearly significant. Water science is not an integrated science and no global assessment of water science exists that could be taken into account in policy-making, as undertaken by the Intergovernmental Panel on Climate Change and the Millennium Ecosystem Assessment. Conca (2005) shows that epistemic communities are dispersed, and their efforts to define integrated water resource management have not been entirely successful.

Second, water science is affected by disciplinary and ideological biases. Economists tend to emphasize neoliberal issues and to concentrate on water as an economic good. Human rights lawyers tend to focus on equity issues and the right to water. Engineers tend to deal with scarcity by proposing technical solutions (for example dams, inter-basin transfers). A quick scan of the papers in Scopus[2] reveals that in the period 1998–2007 there were three times as many articles on water and ecology than on water and development, which may be taken to reflect a Northern bias in the science as 31 Northern countries produce 97.5 per cent of the most-cited papers. In the same period while 177 law articles focused on water and rights and 62 engineering articles included a discussion of rights, only two economics articles included such discussion. While 146 economics articles focused on water and markets, only 27 legal articles and 76 engineering articles did so. While 408 articles focusing on dams and reservoirs appeared in engineering articles, only 31 articles were published on these subjects in economics journals and only 14 in law journals (Gupta and van der Zaag, 2009). Although this quick overview merely indicates that disciplines tend to focus on issues of concern to those disciplines, it also points to the potential problems of bias in the scope and weight of scientific evidence reviewed in decision-making about water issues, and needs to be followed up through more detailed research.

Third, while some science may have a limited independent influence on policy processes through expert epistemic networks, especially within the more organized context of the UN Economic Commission of Europe and the EU, this is less the case at the global level.

Fourth, most of the time scientific results are promoted by actors that lobby for the implementation of those scientific results. Thus water as an economic good that necessitates private sector participation is an idea that has been advocated through aid agencies in the form of conditions on aid, but also by engineers and policy-makers. Water as a human right has been marketed more by non-state actors and legal scholars. While these ideas are marketed as so-called neutral science-based ideas, they often reflect underlying discourses and have impacts on decision-making. Agents choose the science based on their own interests, or when it matches their intuitive expectations or preferred existing policies (Rich, 1991).

Finally, at the international level, there is a major structural imbalance in scientific developments between rich and poor countries (Annan, 2003). This imbalance affects the way that science influences the policy-making discourse (cf. on climate changes, Gupta, 1997).

Current science is dominated by the global North and focuses primarily on issues of the North; where it focuses on issues of the South, it does not often take contextual aspects into account, presenting instead technological fixes or ready-made solutions.

'Ninety-five per cent of the new science in the world is created in the countries comprising only one-fifth of the world's population' (Annan, 2003, p. 1485), and much of this science does not focus on the needs of the developing world.

3.4.3 From structured approaches to hybrid meetings and social movements

At the global level, actors can use legal regulatory instruments or extra-legal suasive instruments. Economic instruments may be subsumed under the former or the latter. Within the UN system there is a progression between the adoption of ideas based on consensus in political declarations (so-called soft law) and the subsequent adoption in legally binding agreements (hard law), instruments based on the principles of the law of treaties. There are clear rules of procedure and the dominant actors are states. Non-state actors are observers. The process allows for consensus-forming, leading to legally binding decisions that are supported generally, but not always by non-compliance mechanisms. Many states and environmental and water groups support such a multilateral arrangement as a legitimate and fair means of addressing problems.

In the domain of water there has been a clear preference by states and non-state actors to choose hybrid fora through which ideas are launched. This occurred parallel to the former UN Secretary General's support for tripartite approaches and Type 2 mechanisms (that is, cooperative agreements between state and non-state actors) introduced at the World Summit on Sustainable Development at the insistence of the US government, which opposed the then current form of multilateral approaches. These hybrid fora do not have clearly established rules of procedure and all actors may have a role in these bodies. There are no clear procedures for taking decisions and such decisions can never reach the status of legally binding treaties, becoming at best contracts under private international law. The latter of course are items the private sector is very much at home with, hence its actors prefer to use hybrid sessions to pre-empt a formal rule-making process (as they have done in the past with the UN Code for Transnational Companies) and to create an atmosphere conducive to allowing private sector participation in the management of natural resources (Gupta, 2005). In the extra-legal sphere, global standard-setting and labelling by non-state actors are forms of the new legal pluralism that has become visible. While these meetings have limited impact (Gleick and Lane, 2005), they provide a venue for social movements to express themselves, and these social movements increasingly have a say in global water governance.

3.4.4 Social movements

With the rise of large dams and their impacts on human populations, a number of social movements have emerged worldwide that oppose major dams and support ecological management of water resources (Conca, 2005). Such movements include that against the Narmada Dam led by Medha Patkar. In recent decades such movements have cooperated globally, sometimes under the leadership of Rivers International. In 1997 the First International Meeting of People Affected by Dams was held in Curitiba, Brazil, followed by a second in November 2003, in Rasi Salai, Thailand. This movement argues against damming the world's rivers and calls for building better sustainable management systems for water. A second wave of social movements has opposed the privatization of water and lobbies for water justice.

3.4.5 Organizational frameworks

River basin management has in the twentieth century been institutionalized in all kinds of management bodies. A taxonomy of river basin organizations includes advisory committees (for example catchment coordinating committees), authorities (for example Tennessee Valley Authority, Niger Valley Authority), associations (for example Missouri River Basin Association), commissions (for example Great Lakes Commission, International Commission for the Protection of the Rhine), councils (for example Fraser Basin Council), corporations (for example Damodar Valley Corporation), tribunals (for example Valencia Water Court), trusts (Hawkesbury–Nepean Catchment Management Trust) and federations (Hooper, 2005). These bodies also implement the established discourses in the area of water management and in their early years mostly promoted dam construction and, more recently, integrated water resource management.

3.4.6 Aid as a driver of change

A key instrument of change in the North–South context is development cooperation. The motivations for aid range from altruism to self-interest. A good proportion of the resources have been linked to donor country ideologies and interests (Pronk, 2003).

Although development cooperation is often seen as having failed to promote economic growth in the developing countries (Boone, 1994, 1996; Dollar and Easterley, 1999; Rajan and Subramanian, 2005), it has nevertheless been a driving factor in changing policies and policy frameworks in the developing world. This can be seen in various chapters in this volume. Early irrigation schemes and infrastructural investments were supported by the banks and aid agencies. Subsequently, structural adjustment programmes were imposed through aid and lending agencies in the 1980s in most indebted countries. In recent years, the concept of good governance has been heavily marketed in developing countries (Santiso, 2001; Anders, 2005). While in the previous decades the emphasis was on 'donor conditionality' as an instrument for changing the behaviour of developing countries, it was seen as a failing instrument except in terms of changing policies on paper (Joint European NGO Report, 2006). This led to the new movement on 'country ownership' (Dollar and Easterley, 1999, p. 574), but although this was supposed to imply demand-driven ideas, the developing countries in effect wrote proposals that had to be aligned to the conditions of the bank and/or aid agencies, and often hired consultants to do this making a mockery of 'country ownership' (Anders, 2005).

Scheumann et al. (2008) discuss the role of development cooperation in water management and each of the chapters in the edited volume show how aid agencies and banks have often promoted both civil engineering works (irrigation works and dams) as well as social engineering ideas (catchment councils, good governance and so on). Often such ideas get manipulated by powerful local social actors to serve their own interests, and do not yield the anticipated results as they are not in line with local contexts and demand.

3.4.7 Inferences

This section has presented five possible factors influencing global governance. It argues that water science tends to be incomplete and biased in favour of the interests of the developed world and that the shift from hierarchical to hybrid systems of governance has implied a shift from legalistic approaches to talking shops of stakeholder participation. While social and environmental movements have tried to represent the concerns of the local people in river basins from local to global levels, the institutionalized framework for water management has led to the institutionalization of discourses in water management. The latter has been helped by aid agencies and banks that have also marketed these ideas.

3.5 Analysis and conclusions: what would a transition to good global water governance imply?

This chapter set out to understand the contours of the current system of global water governance. It demonstrates that water governance is spread through the UN system, the non-UN system and in hybrid collaborations between UN and non-UN actors. It shows that there is some degree of competition between these actors, and each promotes its own interests and discourses. Although there appears to be a common rhetoric – namely the rhetoric of sustainable management of water resources – sustainable development is a concept that anyone can subscribe to and which therefore has many adherents. This in itself is not a common defining agenda for the global water problem.

Discussion in this chapter makes the following points. First, that water governance at the global level is highly diffuse and spread out among different fora, involving strong competition between the activities undertaken by different bodies. Second, four obvious tensions surround discourses at the global level:

1. The tension between the needs discourse, supported by the private sector and the international banks, and the human rights discourse, supported by social movements and legal scholars. The needs discourse is subsumed by the neoliberal discourse.
2. The tension between the water law discourse (traditional positive international water lawyers who see international law as a reflection of state practice) and the environmental discourse (environmentalists, Berlin Rules), which implies that these two discourses have not merged at the global level, even though at the supranational, regional and national levels there are considerable efforts in this direction.
3. The tension between the large dams discourse (engineers, banks, and increasingly pushed as a greenhouse gas-friendly energy alternative) and the discourse that focuses on alternative approaches to water management (social movements).
4. The tension between integrated water management (water professionals, catchment bodies) and hierarchical centralized management of scarce strategic resources, and between tendencies to decentralize authority and engage stakeholders.

Neoliberal ideology has dominated, favouring the large dam approach and private sector participation in water management. The call for smaller government and a greater role for the private sector is compatible with social engineering approaches aiming to promote good governance and to establish river basin organizations. Economists, banks and development agencies tend to focus on neoliberal views and market them primarily through hybrid agencies, since the rules of procedure within the UN system are more rigid and call for formal democratic processes. The social practice model focuses more on human rights, eco- and society-friendly management of water resources, and equitable sharing of resources. Social movements support this approach. Ironically the call for more inclusive approaches from the social movements has led to greater stakeholder engagement, that has led the more dominant actors to promote their visions through their advocacy of hybrid organizations. However, questions arise over the legitimacy and effectiveness of organizations that do not use the existing legal machinery of the UN to move from soft policy to hard law (cf. Gleick and Lane, 2005). This practice, however, may be more consistent with the general developments in global governance, which tend to be hierarchical and pluralist in nature.

A main limiting factor in global water governance (as in other resource fields) is the absence of transnational science. Current Anglo-American science is very limited. There is an urgent need for transnational science that openly deals with values and assumptions and varies these to reflect the positions and views of different countries and peoples, that focuses on issues that concern people from other countries, and that understands existing governance structures and tries incrementally to build on these instead of trying to make poor copies of Western systems.

Returning, then, to the heading of this section. What would a transition to good global water governance imply? First, that a common definition exists of the global water problem. Second, that actors understand how to develop a path towards addressing the problem in a manner consistent with existing cultures and systems of administration. Third, that elements of the problem can be prioritized and simple, context-relevant and affordable solutions to each element can be sought out in working towards addressing them all. In this respect, I see a key contradiction as I review the current priorities of global water governance: are the goals of private sector participation and the Millennium Development Goals anything but irreconcilable? After all, providing water and sanitation to the poor is not likely to meet the profit motives of the private sector.

Acknowledgement

The author has worked on this chapter as part of the Netherlands Organization of Scientific Research VIDI project on International and Private Environmental Governance: Sustainable Development, Good Governance and the Rule of Law (contract number: 452-02-031).

Notes

1. See http://www.iwahq.org/uploads/iwa%20hq/website%20files/about%20iwa/IWA%20brand%20values.pdf, accessed 4 November 2008.
2. See for more information: http://www.scopus.com/scopus/home.url.

References

Agenda 21 (1992), *United Nations Conference on Environment and Development; Rio Declaration and Agenda 21*, Report on the UN Conference on Environment and Development, Rio de Janeiro, 3–14 June 1992, UN Doc. A/CONF.151/26/Rev.1 (Vols 1–3).

Allan, J.A. (2006), 'IWRM: the new sanctioned discourse?' in P.P. Mollinga, A. Dixit and K. Athukorala (eds), *IWRM in South Asia: Global Theory, Emerging Practice and Local Needs*, Water in South Asia Series, New Delhi: Sage, pp. 38–63.

Anders, G. (2005), 'Civil servants in Malawi: moonlighting, kinship and corruption in the shadow of good governance', PhD Manuscript, Law Faculty, Erasmus University.

Annan, K. (2003), 'A challenge to the world's scientists', *Science*, **299**, 1485.

Arts, B. and J. van Tatenhove (2005), 'Policy and power: a conceptual framework between the "old" and "new" policy idioms', *Policy Sciences*, **37** (3–4), 339–56.

Barlow, M. and T. Clarke (2002), *Blue Gold: The Battle Against Corporate Theft of the World's Water*, London: Earthscan.

Benvenisti, E. (2003), *Sharing Transboundary Resources: International Law and Optimal Resource Uses*, Cambridge: Cambridge University Press.

Blatter, J.K. (2003), 'Debordering the world of states: toward a multi-level system in Europe and a multi-polity system in North America? Insights from border regions', in N. Brenner, B. Jessop, M. Jones and G. MacLeod (eds), *State/Space: A Reader*, Oxford: Blackwell Publishing, pp. 185–207.

Boelens, R. (2006), 'Local rights and legal recognition: the struggle for indigenous water rights and the cultural politics of participation', in UNESCO (ed.), *Water and Indigenous Peoples*, Paris: UNESCO, pp. 46–61.

Boone, P. (1994), 'The impacts of foreign aid on savings and growth', London School of Economics, mimeo.

Boone, P. (1996), 'Politics and the effectiveness of foreign aid', *European Economic Review*, **40**, 289–329.

Bourne, C.B. (1996), 'The International Law Association's contribution to international water resources law', *Natural Resources Journal*, **36** (2), 155–216.

Conca, K. (2005), *Governing Water: Contentious Transnational Politics and Global Institution Building*, Cambridge, MA: MIT Press.

Cosgrove, W.J. and F.R. Rijsberman (2000), *World Water Vision: Making Water Everybody's Business*, London: Earthscan Publications.

de Castro, P.C. (2009), 'European Community water policy', in J. Dellapenna and J. Gupta (eds), *The Evolution of the Law and Politics of Water*, Dordrecht: Springer Verlag, pp. 227–43.

Dellapenna, J.W. (2000), 'The customary international law of transboundary fresh waters', *International Journal of Global Environmental Issues*, **1**, 264–305.

Dellapenna, J.W. (2009), 'The market alternative', in J. Dellapenna and J. Gupta (eds), *The Evolution of the Law and Politics of Water*, Dordrecht: Springer Verlag, pp. 373–88.

Dellapenna, J. and J. Gupta (2008), 'The evolving legal framework for global water governance', *Global Governance*, **14** (3), 437–53.

Dingwerth, K. (2004), 'The democratic legitimacy of public private rule making: what can we learn from the World Commission on Dams?', *Global Governance*, **11** (1), 65–83.

Dollar, D. and W. Easterley (1999), 'The search for the key: aid, investment and policies in Africa', *Journal of African Economies*, **8** (4), 546–77.

Dublin Declaration (1992), *Guiding Principles: The Dublin Statement on Water and Sustainable Development*, adopted on 31 January by the International Conference on Water and the Environment, Dublin.

EU WFD (2000), *EU Water Framework Directive*, EC Directive 2000/60/EC.

Fischer, F. (2003), *Reframing Public Policy: Discursive Politics and Deliberative Practices*, Oxford, UK and New York, USA: Oxford University Press.

Fuentes, X. (1999), 'Sustainable development and the equitable utilization of international watercourses', in J. Crawford and V. Lowe (eds), *The British Yearbook of International Law 1998*, Oxford: Clarendon Press, pp. 119–200.

General Comment (2002), *General Comment No. 15: The Right to Water, Substantive Issues Arising in the Implementation of the International Covenant on Economic, Social and Cultural Rights*, UN Economic and Social Council, UN Doc. E/C.12/2000/11 (26 November 2002).

Gleick, P.H. and J. Lane (2005), 'Large international water meetings: time for a reappraisal', *Water International*, **30**, 410–14.

Gleick, P., G. Wolff, G. Chalecki and R. Reyes (2002), 'The privatization of water and water systems', in P. Gleick et al. (eds), *The World's Water: The Biennial Report on Freshwater Resources 2002–2003*, Washington, DC: Island Press, pp. 33–56.

Gupta, J. (1997), *The Climate Change Convention and Developing Countries: From Conflict to Consensus?* Environment and Policy Series, Dordrecht: Kluwer Academic Publishers.

Gupta, J. (2005), 'Non-state actors in international governance and law: a challenge or blessing', *ILSA Journal of International and Comparative Law*, **11** (3), 497–517.

Gupta, J. and J.W. Dellapenna (2009), 'The challenges for the twenty-first century: a critical approach', in J. Dellapenna and J. Gupta (eds.), *The Evolution of the Law and Politics of Water*, Dordrecht: Springer Verlag, pp. 391–410.

Gupta, J. and P. van der Zaag (2009), 'The politics of water science: unresolved water problems and biased research agendas', *Global Environmental Politics*, **9** (2), 14–23.

Haas, P.M. (1989), 'Do regimes matter? Epistemic communities and Mediterranean pollution control', *International Organization*, **43** (3), 377–403.

Hajer, M.A. and H. Wagenaar (eds) (2003), *Deliberative Policy Analysis: Understanding Governance in the Network Society*, Cambridge: Cambridge University Press.

Hall, D. and E. Lobina (2006), 'Pipe dreams: the failure of private sector to invest in water services in developing countries', PSIRU report for Public Services International and World Development Movement, London.

Hall, N. (2009), 'The North American Great Lakes', in J. Dellapenna and J. Gupta (eds), *The Evolution of the Law and Politics of Water*, Dordrecht: Springer Verlag, pp. 281–98.

Hanf, K. and F.W. Scharpf (eds) (1978), *Interorganizational Policy Making*, London: Sage.

Hartje, V. (2008), 'The World Bank's water sector policy reforms', in W. Scheumann, S. Neubert and

M. Kipping (eds), *Water Politics and Development Cooperation: Local Power Plays and Global Governance*, Dordrecht: Springer Verlag, pp. 301–32.
Hildering, A. (2004), 'International law for sustainable development in water management', Amsterdam: PhD manuscript, Vrije Universiteit Amsterdam.
Hooper, B. (2005), *Integrated River Basin Management: Learning from International Experience*, London: International Water Association Publishing.
ILA (1966), 'The Helsinki Rules on the uses of the waters of international rivers', *Report of the Fifty-Second Conference (Helsinki)*, International Law Association, London.
ILA (2004), 'The Berlin Rules on water resources', *Report of the Seventy-First Conference (Berlin)*, London: International Law Association.
Joint European NGO Report (2006), 'EU aid: genuine leadership or misleading figures? An independent analysis of European Governments' aid levels', Brussels: Concord.
Keeley, J. and I. Scoones (2003), *Understanding Environmental Policy Processes: Cases from Africa*, London: Earthscan.
Kerkhof, M. van de (2004), *Debating Climate Change: A Study on Stakeholder Participation in an Integrated Assessment of Long-Term Climate Policy in the Netherlands*, Utrecht: Lemma Publishing.
Kickert, W.J.M., E.H. Klijn and J.F.M. Koppenjan (eds) (1997), *Managing Complex Networks: Strategies for the Public Sector*, London: Sage.
Kidd, M. (2009), 'South Africa: the development of water law', in J. Dellapenna and J. Gupta (eds), *The Evolution of the Law and Politics of Water*, Dordrecht: Springer Verlag, pp. 87–104.
Kipping, M. (2008), 'Challenges for German development cooperation in the water sector', in W. Scheumann, S. Neubert and M. Kipping (eds), *Water Politics and Development Cooperation: Local Power Plays and Global Governance*, Dordrecht: Springer, pp. 333–52.
Kooiman, J. (1997), 'Governance and governability: using complexity, dynamics and diversity', in J. Kooiman (ed.), *Modern Governance: New Government–Society Interactions*, London: Sage, pp. 35–50.
Koppenjan, J.F.M. and E.H. Klijn (2004), *Managing Uncertainties in Networks*, London: Routledge.
Koudstaal, R., F. Rijsberman and H. Savenije (1992), 'Water and sustainable development', *Natural Resources Forum*, November, 277–90.
Laster, R. and D. Livney (2009), 'Israel: the evolution of water law and policy', in J. Dellapenna and J. Gupta (eds), *The Evolution of the Law and Politics of Water*, Dordrecht: Springer Verlag, pp. 121–38.
Lemon, A. (2001), 'Gender and forestry: integrating local knowledge into environmental planning and management', *International Journal of Sustainable Development*, **4** (3), 265–81.
McCaffrey, S.C. (1992), 'A human right to water: domestic and international implications', *Georgetown International Environmental Law Review*, **5**, 1–23.
McCaffrey, S.C. (2001), *The Law of International Watercourses: Non-Navigational Uses*, Oxford: Oxford University Press.
McCay, J. and S. Marsden (2009), 'Australia: the problem of sustainability in water', in J. Dellapenna and J. Gupta (eds), *The Evolution of the Law and Politics of Water*, Dordrecht: Springer Verlag, pp. 175–88.
Mollinga, P. (2008), 'Water policy – water politics', in W. Scheumann, S. Neubert and M. Kipping (eds), *Water Politics and Development Cooperation: Local Power Plays and Global Governance*, Dordrecht: Springer, pp. 1–29.
Nakashima, D. and M. Chiba (2006), 'Introduction', in R. Boelens, D. Nakashima and M. Chiba (eds), *Water and Indigenous Peoples*, Paris: UNESCO, pp. 12–16.
Neubert, S. in collaboration with L. Horlemann (2008), 'Strategic virtual water trade: a critical anlysis of the debate', in W. Scheumann, N. Neubert and M. Kipping (eds), *Water Politics and Development Cooperation: Local Power Plays and Global Governance, Dordrecht: Springer Verlag, pp. 123–45.*
Oregon State University (2002), *Atlas of International Freshwater Agreements*, Nairobi: UNEP.
Pahl-Wostl, C., J. Gupta and D. Petry (2008), 'Governance and the global water system: a theoretical exploration', *Global Governance*, **14** (4), 419–36.
Petrella, R. (2001), *The Water Manifesto: Arguments for a World Water Contract*, London: Zed Books.
Pierre, J. (ed.) (2000), *Debating Governance*, Oxford: Oxford University Press.
Pronk, J.P. (2003), 'Aid as a catalyst: a rejoinder', *Development and Change*, **343** (3), 383–400.
Rahman, M.M., O. Varis and T. Kajander (2003), 'EU Water Framework Directives international principles concerning IWRM: the seven mismatches', paper prepared by the Helsinki University of Technology.
Rajan, R.G. and A. Subramanian (2005), 'Aid and growth: what does the cross-country evidence really show?', IMF Working Paper, WP/05/127.
Revised SADC Protocol (2000), *Revised Protocol on Shared Watercourse Systems of the South African Development Community*.
Rhodes, R.A.W. (1997), *Understanding Governance*, Buckingham: Open University Press.
Rich, R.F. (1991), 'Knowledge creation, diffusion and utilization: perspectives of the founding editor of *Knowledge*', *Knowledge: Creation, Diffusion and Innovation*, **12**, 319–37.

Rio Declaration (1992), *Rio Declaration and Agenda 21, Report on the UN Conference on Environment and Development*, Rio de Janeiro, 3–14 June 1992, UN doc. A/CONF.151/26/Rev.1 (Vols 1–3),
Rondinelli, D., J. Nethis and S. Chemma (1984), *Decentralization in Developing Countries: A Review of Current Experience*, Washington, DC: World Bank.
Sabel, R. (2009), 'The Jordan basin: evolution of the rules', in J. Dellapenna and J. Gupta (eds), *The Evolution of the Law and Politics of Water*, Dordrecht: Springer Verlag, pp. 263–80.
Sacher, D. and M. Windfur (2008), 'The debate on "water as a human right" and its implications for development assistance', in W. Scheumann, S. Neubert and M. Kipping (eds), *Water Politics and Development Cooperation: Local Power Plays and Global Governance*, Dordrecht: Springer, pp. 147–76.
Santiso, C. (2001), Good governance and aid effectiveness: the World Bank and conditionality, *Georgetown Public Policy Review*, **7** (1), 1–22.
Scheumann, W. (2008), 'How global norms for dams reach decision-makers', in W. Scheumann, S. Neubert and M. Kipping (eds), *Water Politics and Development Cooperation: Local Power Plays and Global Governance*, Dordrecht: Springer, pp. 55–80.
Schuemann, W., S. Neubert and M. Kipping (eds) (2008), *Water Politics and Development Cooperation: Local Power Plays and Global Governance*, Dordrecht: Springer Verlag.
Schnurr, M. (2008), 'Global water governance: managing complexity on a global scale', in W. Scheumann, S. Neubert and M. Kipping (eds), *Water Politics and Development Cooperation: Local Power Plays and Global Governance*, Dordrecht: Springer, pp. 107–20.
Schouten, M. and K. Schwartz (2006), 'Water as a political good: implications for investments', *International Environmental Agreements: Politics, Law and Economics*, **6** (4), 407–21.
Shiva, V. (2002), *Water Wars: Privatization, Pollution and Profit*, Cambridge, MA: South End Press.
Smets, H. (2000), 'The right to water as a human right', *Environmental Policy and Law*, **30** (5), 248–50.
Stiglitz, J. (2000), 'Vers un nouveau paradigme du developpement', *L'économie politique*, **5**, 6–39.
Tanzi, A. and M. Arcari (2001), *The United Nations Convention on the Law of International Watercourses*, London: Kluwer Law International.
Tecco, N. (2008), 'Financially sustainable investments in developing countries water sectors: what conditions could promote private sector involvement?' *INEA*, **8** (2), 129–42.
UN Watercourses Law (1997), *UN Convention on the Law of the Non-Navigational Uses of International Watercourses* (IWC) (1997), UN Doc. A/51/869; 37 ILM 719.
UNWWDR (2003), *Water for People, Water for Life. The United Nations World Water Development Report*, Barcelona: UNESCO and Berghahn Books.
UNWWDR (2006), *Water: A Shared Responsibility. The United Nations World Water Development Report 2*, Paris and New York: UNESCO and Berghahn Books.
van der Zaag, P. (2009), 'Southern Africa: evolving regional water law and politics', in J. Dellapenna and J. Gupta (eds), *The Evolution of the Law and Politics of Water*, Dordrecht: Springer Verlag, pp. 245–63.
WCED (World Commission on Environment and Development) (1987), *Our Common Future*, Oxford: Oxford University Press.
World Commission on Dams (WCD) (2000), *Dams and Development: A New Framework for Decision-making. The Report of the World Commission on Dams*, London: Earthscan Publishing.
WSSD (2002), *Johannesburg Plan of Implementation*, UN Doc. A Res/55/199.

PART II

THE AMERICAS

4 Transitions: transcending multiple ways of knowing water resources in the United States

Helen Ingram and Raul Lejano

4.1 Introduction

Water is a complex, multifaceted resource reflecting very different values and ways of knowing. The history of water as a resource reprised in this chapter reveals a period of stability during which the dominant way of knowing water, as an engine for economic development, encountered other ways of knowing resulting in transitions, or substantial policy changes. Today US water policy is highly contentious with at least four widely accepted ways of knowing water to which many subscribe. Transitions happen as policy innovations are forged in particular contexts. Transitions in such contexts depend upon networks being built between existing ways of knowing that allow for collaborative action. Such knowledge networks may be facilitated by, but are different from, many institutional changes embraced by supporters of integrated water resources management (IWRM).

The purpose of this chapter is to enrich the framework introduced in Part I of this book by adding a new theoretical perspective. We contend that transitions require particular strategies by policy entrepreneurs to forge links between people, agencies and groups who comprehend water through very different ways of knowing. In fact the multiplicity of ways of knowing about water resources stem from previous policy initiatives to introduce new perspectives into water management and policy. This chapter illustrates that policy science research coming from national study commissions, centres within universities and Washington-based think tanks has sometimes been substantial in changing policy frames (Kingdon, 1984) and fostering emergence and inclusion of other ways of knowing. Further, history reveals the important role of policy entrepreneurs, many of whom came from government, in bridging different ways of knowing. Today there are multiple and only partially overlapping ways of knowing associated with water resources. The challenge is to forge water policy designs that incorporate and even transcend different ways of knowing. Many contemporary institutional reforms, like IWRM, are not innovative and are unlikely to lead to meaningful change without bridges being built across multiple ways of knowing.

Contemporary water transitions that integrate divergent ways of knowing occur in different venues and at a variety of levels of government. They sometimes involve agencies and actors previously peripheral to water resource decisions. This chapter offers three examples of substantial change – one local, one regional and one national – although, as the cases illustrate, all levels of government, along with private as well as public actors, are involved in water transitions. The examples illustrate many of the influences set out by Meijerink and Huitema in the theoretical introduction to this book (Chapter 2), particularly the importance of ideas, networks and contexts, the sequencing of actions and events, and the roles of policy entrepreneurs. Cases illustrate strategies for managing interaction among knowledge networks that we believe to be a helpful theoretical addition fully consistent with the theories in Part I of this book.

4.2 Ways of knowing

The term 'ways of knowing' is relatively new (see Feldman et al., 2006), one that integrates many of the terms employed in Chapter 1 including policy frames, institutionalized policy image, policy networks and discourses, core beliefs, ideas and epistemic communities. Other researchers have coined related terms such as schema, perception, mental model, narrative, social construction, scene, script or perspective (Dewulf et al., 2005). Given the plethora of closely related terms, the introduction of yet another construct demands careful definition and justification.

Ways of knowing (WoK) are dynamic social sense-making systems (including both ideas and practices) that signify distinct meanings that result from associations between things, people, events and ideas and are created through communication, discourse and relationships, which may or may not be shared across different WoK. WoK shape the knowledge production of actors and determine what has import and what is given credibility and trust. The (WoK) concept stresses knowing rather than beliefs, interests or images that are the focal points of other closely associated concepts (Sabatier et al., 2005). WoK is a particularly useful concept in understanding how people approach physical phenomena like the environment or water resources where experiential, expert and other avenues of knowledge are so important. Also, WoK are mobile and dynamic rather than fixed and static, as are some definitions of frames and core beliefs, and are malleable as they come to bear on particular policy contexts – in fact WoK are constantly evolving as new policy actors enter into a network and new ideas cross network boundaries. The sharing of understanding and meaning depend largely on diverse processes and tools, many of which are identified in Part I as critical to transitions.

The concept is inclusive because it reflects commitment to multiple WoK that extend beyond the model of self-interest that has dominated so much of political science and policy studies. People have a large repertoire of cognitive and emotional tools such as authority, intuition, moral reasoning, direct experience, logic, belief or faith, mysticism, self-interest, rationality, and so on that contribute to their way of comprehending a problem. Any one of these may become more or less relevant through continuous discourse, engagement and learning. The major assumption that underlies public choice and pluralist theories widely embraced in much policy science scholarship is that self-interest dominates political discourse. As a consequence far too little attention is paid to the construction of policy 'ideas' or 'images' that are emphasized in Chapter 1. While core beliefs (Sabatier, 1993) can capture ideas beyond self-interest, the term 'belief' fails to engage the complex phenomenon of understanding.

Another justification for employing the concept of WoK is that it links with and engages social studies of science scholarship that have important insights for the study of transitions, even though this literature has not received much consideration in policy science. The social studies of science have their own distinct literature that provides useful insights about ways of knowing. Among the foundational ideas of the movement is the contention that knowledge is profoundly shaped by the processes of knowledge creation that are essentially social (see for instance Latour and Woolgar, 1979; Jasanoff, 1990). Social studies of science scholars like Brian Wynne cite cases in which failure to incorporate local knowledge in deference to that of experts reinforces path-dependent policies that are resistant to transitions (Wynne, 1996). Social studies of science suggest several kinds of policy tools that can facilitate collaboration between different ways of knowing. Boundary objects (Star and Griesemer, 1989) provide an artifact such as a model, scenario, template, prototype and the like that draw upon necessary, but different ways of knowing and enhance mutual understanding among different ways of knowing. Other scholars suggest that boundary organizations, like advisory committees or task forces drawing upon science and lay members, may perform similar services across organizational boundaries (Jasanoff, 1990; Guston, 2001).

WoK is related to and builds upon notions of policy image (Baumgartner and Jones, 2002); policy discourse (Hajer, 1995) and policy frames (Schön and Rein, 1994) and worldviews as they are described in Part I of this book. The differences between WoK and these ideas are that WoK is more dynamic, more situated and more dependent upon strategic intervention of policy entrepreneurs that build networks across different ways of knowing through leadership and the strategic use of communications and policy tools. Change in WoK involves a 'novel synthesis' of knowing across networks by integrating existing knowledge and learning from problem-solving efforts (Weber and Khademian, 2008). WoK adds importantly to the framework introduced in Part I of this book by emphasizing the generation, transmission, receipt, interpretation and integration of knowledge leading to collaboration across many kinds of boundaries and the role of policy entrepreneurs in this networking.

4.3 Historical experience of US water resources

4.3.1 Introduction

The historical record of water resources policy in the United States reveals both stability and change. While the big dam era revealed in many other chapters in this book established an early beachhead in the US that has yet to be entirely dislodged, transitions occurred nevertheless. New WoK besides water as a product of development were introduced, and new venues emerged. Importantly the historical transitions were facilitated by policy entrepreneurs, many of whom justified and facilitated changes by introducing and espousing new WoK. The important lesson of this historical overview is that the strategic action of policy entrepreneurs in transitions involved the building of networks across different ways of knowing so as to bring in new values and perspectives.

4.3.2 Post-colonial experience

Federal agency involvement in water resources is as old as the republic. Early on the desire of particular localities to benefit from water development was pursued through the iron triangle of federal agencies, congressional committees and locally based interest. Congress early on expressed a powerful and continuing interest in water issues. Although the principle of reasonable use constrained riparian right holders under state law, water policy was overlain by early assertions of federal authority. The interstate commerce and war power clauses of the US Constitution allowed Congress to fortify coasts, clear and improve harbours and engage in other 'civic improvements'. The Army Corps of Engineers was active during the revolutionary war, and Congress re-established the Corps of Engineers and a military academy at West Point in 1802. Thus water management was rooted in science long before George Perkins Marsh and the Progressive era. 'Until 1866, the superintendent of the academy was always an engineer and during the first half of the 19th century, West Point was the major and for a while, the only engineering school in the country' (US Army Corps of Engineers, n.d.). Thus the Corps of Engineers emerged as the only formally trained body of scientists in the new republic.

As part of the nation's small army, the Corps was available to serve the wishes of Congress and the executive branch. Enthusiasm for federal subsidies was a very early pattern in natural resource law and policy: as the Corps' own official history notes: 'From the beginning, many politicians wanted the Corps to contribute to both military construction and works "of a civic nature" (US Army Corps of Engineers, n.d.). The familiar elements of the Corps' current profile were carved in stone by the 1824 Supreme Court ruling in *Gibbons* v. *Ogden*. The decision clarified that federal authority over interstate commerce included riverine navigation. Arthur Maass has suggested that it is not always clear that the centralized authorities were serving national needs: state and local governments and diverse beneficiaries have long exhibited agility in deflecting federal programmes to local priorities. Nonetheless, the centralization of government authority through concentrated scientific expertise and subsidies began early in water (Maass, 1951). Thus the early and what was to become path-dependent 'water as product' way of knowing water that fitted nicely with growth began as a federal impetus toward nation-building.

4.3.3 Progressive era

According to the Progressives, theirs was the era in which enlightened bureaucrats and politicians transformed ideological conflicts into issues of expertise and efficiency (Walton, 1992). The scientist John Wesley Powell collaborated with political entrepreneurs like Francis Newland to enact the 1902 Reclamation Act that provided federal subsidies for agency-led, expert-based water development intended to turn barren wastelands in the West into thriving communities. A close reading shows that the newly established progressive federal agency, the Bureau of Reclamation, followed the model of the Corps of Engineers in using nationally based expertise to serve local constituencies that took over projects once completed and ran them as they wanted (Smith, 1986). In this way the Progressive era reinforced the 'water as product' WoK water even as it added an additional federal venue. It also associated science-based knowledge and expertise with legitimacy in water resources policy.

4.3.4 The Depression

The Progressive era's centralization of venues at the federal level and faith in scientific expertise actually peaked with the Great Depression (the worst part of the economic downturn occurring during the period 1929–33), wherein the federal government moved front and centre in natural resource planning. This extreme expansion of federal authority was manifest in a new understanding of property: the government had the right, indeed the obligation, to adjust ownership patterns that were not socially and economically optimal. President Franklin Delano Roosevelt, partly under the influence of his uncle, the planner Frederic Delano, used executive orders to create a series of expert-dominated planning boards. These bodies, especially the National Regional Planning Boards (1935–8) advanced expert-based, multipurpose regional planning agencies

(Wescoat, 2000). Federal authority was used to create the Tennessee Valley Authority (TVA) as a federal corporation, and while the form has not been copied, the idea that river basin organizations should be an organizing concept that is layered over federal agency jurisdictions, states and localities persists today. Visionary directors and its semi-autonomous status give the TVA room to experiment with innovative approaches, such as integrated land and water planning, a public power sector and voluntaristic engagement of 'grass-roots' constituencies (Wescoat, 2000, p. 147). All of these characteristics must have a familiar ring to adherents of IWRM, which is being portrayed today by many as a recent transition. Of course the persistent rationale and justification remained that of growth and development.

At the other end of the geographically based planning unit spectrum, the Soil Conservation Service (SCS), another federal agency established by President Roosevelt, launched a lasting programme in smaller upland watershed management that was often in conflict with other federal agencies and river basin planning. The SCS's founding director, Hugh Hammond Bennett, believed strongly in stopping raindrops where they fall and preventing soil erosion. Soil science contributed greatly to the credibility of the small dams built by the SCS, rationalized as preserving the topsoil essential to the flowering of civilization.

4.3.5 The environmental decade

With astonishing speed, the federal government passed one major statute after another in the post-Earth Day fervour of the 1970s, dramatically enhancing the central authority of federal venues over natural resources including, especially, water. While the level of venues did not change, another significant transition took place. A new WoK emerged, founded on the themes of health and ecology. Central to this transition were policy entrepreneurs like Rachel Carson who raised consciousness about the hidden danger of DDT in the 1960s and Ralph Nader, who turned a consumer advocacy movement into national outrage over environmental degradation. Suddenly scientific experts at the newly constituted Environmental Protection Agency were charged with setting acceptable levels of water pollution without regard to cost or industry concern. A political game of one-upmanship between the Nixon administration and presidential hopeful Edmund Muskee resulted in an exceptionally draconian Clean Water Act that empowered the Environmental Protection Agency to take over many of the powers previously afforded to the states (Ingram and Ullery, 1980). While much of the federal approach to water quality had previously been to foster state and local capacities, the Clean Water Act added many new regulations to the federal arsenal to force action. In addition to enlarging federal power, environmental decade legislation also increased public access to decision-making, exemplified by the National Environmental Protection Act, which instituted environmental impact assessment. The Clean Water Coalition, a national environmental network in favour of the expenditure of greater financial and other resources, helped transform the issue of water from an expert-dominated issue to one that attracted broad grass-roots support (Ingram and Ullery, 1980).

Greater public access and local input cut both ways, however. While a new WoK water as an essential aspect of environment gained adherents, the water as product WoK continued to have many subscribers. When then President Jimmy Carter announced that he was cutting funding for 18 ongoing water projects serving various state and local constituencies in the West, the strength of localities wanting their fair share of water subsidies for economic development became clear. Even though 74 congressmen had earlier expressed support for reform of water resources programmes, the congressional reaction to the 'hit list' was overwhelmingly negative. After a great deal of rancour, Carter eventually agreed to restore funding for half of the deleted projects on the original hit list. Nonetheless, water policy was changed in important ways. More arduous economic and environmental reviews were instituted, and a precedent of 'no new starts' of additional water projects was established in 1981, although the Corps did continue to get money for 'preauthorization studies'. The tougher standards of evaluation along with the fact that most good dam sites had already been used compelled many to believe that the era of big dams was over. Water came to be viewed as having an economic value, and water as an economic commodity WoK emerged on the scene.

4.3.6 Emergence of fairness and equity

Because of the importance of water to life and livelihood, equity had always been a matter of some public concern in irrigation in the American West (Maass and Anderson, 1978; Brown and Ingram, 1987). The economic development of the big dam era was never equally distributed among affected people and the inequities accumulated, forming a backlash. Bureau of Reclamation projects had negative consequences for Native American and rural Hispanics whose lands were sometimes flooded by reservoirs and whose water needs were seldom included among the beneficiaries of water project services. Long customary precedent was enshrined in law with the Winters Doctrine that gave Native Americans very senior reserved water rights early in the twentieth century (in contrast to mere appropriative and riparian rights), but the law had little practical application until after mid-century. The spirit of the civil rights movement also infused a 'water and equity' WoK that associated water with human rights and the attachment of people to places.

Legal experts voiced many of these concerns in court venues, and in a landmark Supreme Court case, Native American homelands or reservations were given water rights equivalent to 'practicable irrigable acreage'. What water development construction project did go forward thereafter usually involved some sort of settlement of Native American water claims and also contained provisions serving native communities. For instance when the central Arizona aqueduct began to move water to the City of Tucson, settlement was required with the Tohono O'odham Native American community that had lost groundwater to the city for many years (Brown and Ingram, 1987). Similarly, rural Hispanics in the southwest asserted and sometimes prevailed in legal water claims dating back to Spanish colonization. The John Nichols novel entitled *The Milagro Beanfields War* (later made into a movie) illustrated and dramatically popularized a real case in the upper Rio Grande in the state of New Mexico (Nichols, 1974). A sense of injustice was not limited to minority groups, however, and farming communities designated as areas of origin in water transfers from rural to urban settings believed that with the loss of water came a loss of security and opportunity. The historic water grab of the city of Los Angeles that piped water out of the Owens Valley to fuel urban population growth (in the process turning the valley into a dustbowl) became a kind of rallying cry for many other threatened localities (the Owens Valley was also a theme of the movie *Chinatown*). The association of water with equity and sense of place gained momentum when courts eventually forced Los Angeles beginning in 2006 to give back much of the water it had taken.

4.3.7 Reagan era to the present

Beginning as early as the Clean Water Act of 1972, national study commissions and think tanks criticized command-and-control regulations as inefficient and ineffective, and urged greater use of markets. Congressional committees were hard of hearing when economists like Alan Kneese (see Kneese and Bower, 1968) from Resources for The Future first espoused the use of tradable pollution permits. Yet budget-conscious officials in the President's Office of Management and Budget as well as policy entrepreneurs within Secretary of the Interior James Watt's administration adopted market-based resource policies as a brand of religion and the water as economic commodity WoK gained strength. Today there is widespread use of markets and market-like incentives in water management. At the same time there is continuity in which the federal role as producer of economic development persists, newly cast as preventing Katrina-like flood disasters and promoting expensive environmental restoration and adaptive management projects. The mixture of old and new is well reflected in the movement of water from farms to cities through long-term leases and sales. While today water moves through market-like exchanges rather than through newly built projects, the parties are not individual buyers and sellers, but rather government entities like cities, states and irrigation districts.

The brief history recounted here leaves out a lot of detail and some important events. At the same time it leads to some important conclusions about transitions in water resources administration in the US. First, the state has always been a dominant player, and the federal government has spearheaded many historic transitions. However even in times of great federal dominance, states and localities have played a powerful role. The federal venue was selected in part for its ability to pay and provide expertise for large-scale engineering works. To gain a legislative majority necessary to authorize projects in this venue, many particularistic projects had to be strung together into an omnibus package (Ingram, 1990 [1969]). Second, transitions have depended importantly on entrepreneurship that has come not just from political figures, but also from the ranks of federal agencies and the courts. Third, many contemporary water ideas such as watershed and river basin management and collaborative management have roots in history. Fourth, the dominant use of water as a means of economic development of different regions and localities has become increasingly questioned by people and groups who see the values of water differently, as an essential element of the environment, a human right tied to equity and place and as economic commodity. It is to the different ways of knowing and valuing water that this chapter now turns.

4.4 Multiple ways of knowing water resources

Figure 4.1 illustrates a collective cognitive and emotional map of ways of knowing water in contemporary US water policy. Elements in Figure 4.1 with different shapes denote all those aspects related to water problems that could be part of an interpretation or WoK. The different shades, shapes and sizes indicate that some are physical elements like reservoirs, aqueducts, pipes and the like; others are events like drought or flooding; some are prominent leaders and authorities, organizations and institutions; others are perceptions

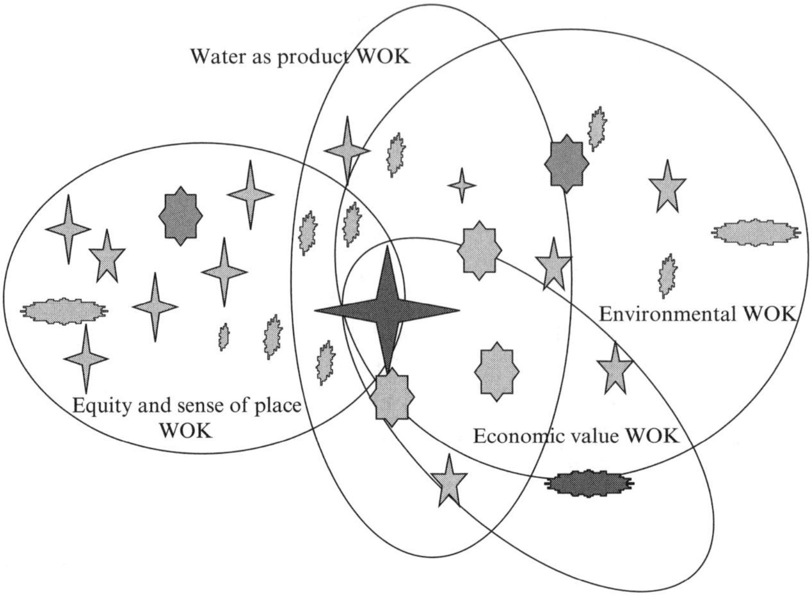

Figure 4.1 Multiple ways of knowing water resources

of risks due to floods, droughts or contamination; yet others may relate to values like efficiency and fairness. Elements within each WoK reflect associated events, people, technologies and other aspects that are important to one or another WoK.

The various elements are plotted on what we depict as a policy space, which is akin to spaces across which interests and political positions vary (as in Thompson et al., 1988). Some elements are shared by several WoK (the star at the centre of the map reflects a common concern with climate change that is viewed differently in different WoK, but has come to be of concern to all).

The most central and inclusive WoK is water as a product for growth and development, and includes the great bulk of water agencies charged with this function including irrigation districts, municipal water supply agencies, regional and federal water agencies. There is a lot of overlap between this WoK and the ecological WoK, but the same elements tend to be viewed differently. The Endangered Species Act is generally regarded as a constraint, and fish and wildlife tend to be seen as yet another set of demands upon a product in limited supply. Global climate change is recognized as quite important and as an emerging challenge that makes managing water as a product for growth and development more difficult. Overlap occurs also with the environmental WoK and the equity WoK because conveyance structures, often man-made, are necessary to deliver water to disadvantaged peoples important to the equity WoK, and to provide water for habitat and recreation important to the environmental WoK. The water as economic value WoK has gained influence as water agencies and cities have come to recognize the usefulness of markets in identifying water with values they care about.

The way of knowing portrayed in the far left oval might represent that of equity and sense of place WoK. This way of knowing is informed by moral and ethical reasoning, logic and direct experience. Many disadvantaged peoples, including Native Americans, the urban and rural poor, lack a sufficient affordable clean water supply (there are over 1 billion people currently without access to drinking water services, and even in the US the differences between the quantity and quality of water at the disposal of different economic, racial and ethnic groups is considerable). Compacts, treaties and other agreements promise to reduce the number of underserved and secure community access to water decision-making. Court decisions uphold treaties and the trust doctrine underlies the US Department of the Interior's role as trustee of tribal water rights. Water lawyers interpret and defend laws and policies related to human rights. Alongside the overlapping institutions and actors, an additional factor is often the infrastructure necessary to deliver the water to people. There may also be fish, such as salmon, and other wildlife that are dependent on water and part of the environmental WoK, and important also to cultural values closely associated with equity and sense of place. The intersection of a range of aspects creates the potential for a web of WoK.

The circle on the left of Figure 4.1 could represent the environmental way of knowing water resources. This WoK comes from moral reasoning, science and experience with nature. It can be held considerably separate from the equity WoK, since it tends to see humans as an invasive species already oversupplied with water. Its narrative would interpret the large star on the figure, which signifies global climate change as clear evidence that humans are out of balance with nature. This WoK is bolstered by many scientific studies that show flooding as a natural process and droughts as an inevitability, making the additional element of the precautionary principle advisable. Other facets include watersheds, aquifers, species, dams, irrigation and electrical power generation. The Endangered Species Act generates many documents and court cases. Government agencies like the Fish and Wildlife Service and their state counterparts grow out of the environmental WoK. Elements of all types are involved in its narrative that casts water as an essential part of nature, never to be objectified and treated as a human right or as property.

The fourth way of knowing is informed by instrumental reasoning and sees water as an economic commodity like all others that can be assigned an economic value and as subject to exchange through markets. Note that there is considerable overlap between this way of knowing and that of the environmental WoK, since many environmentalists have come to see subsidies provided by government to various groups of water users as a major cause of overdevelopment and damage to the environment. Also, many environmentalists have come to think of ecological services, which can translate into commodities, as consistent with their way of knowing. This WoK contains the principles and standards that provide criteria for the evaluation of water resources projects, think tanks like Resources for the Future, which has been a major proponent of this perspective, and the World Bank, a major international proponent.

Overlapping with the water-as-product WoK are elements, such as municipal water supply agencies, that have received increasing proportions of their water supplies through rural urban water sales and leases. Water banks, institutional arrangements that have emerged in California, Arizona and elsewhere, are additional overlapping elements fully accepted in both the product and economic value WoK. A central difference between the product-of-development and economic-commodity WoK is that while the former understands the value of water in terms of use (for example assigning water rights according to senior versus junior priority, contiguity to land, and so on), the latter assumes the value of water will be determined simply as the point of intersection of current supply and demand curves. Water, in the first case, is allocated by right, and in the second by willingness to pay.

The considerable gaps between different WoK underlying US water policy suggest a lack of agreement conducive to policy stasis and lack of action. While this is a time of great ferment, conditions do not exist that would favour wholesale change based on a widely shared, collective way of knowing. However when new events happen, where overlap among different WoK occurs, and in particular contexts, transitions become possible. While each WoK has its own historical basis, new events and problems constantly occur that call out for interpretation. Various ways of knowing are not stable and elements are added, subtracted and rearranged. WoK and the relationships that tie elements together in a problem space are continuously reworked. People and organizations can enrol in or depart from various WoK. WoK are thus plastic, dynamic and evolving.

Learning can be built into a policy through such tools as positive inducements (such as funding grants that require local groups to create and sustain boundary organizations) or through mandates that require such participation as part of funding or regulations. Policy tools that provide flexibility for managers, knowledge brokers and other leaders in networks are necessary if they are to foster ongoing interaction between different ways of knowing. Policy can mandate (or fund) ongoing research that serves the information needs of local participants or that tracks change in the ways of knowing the problem. Statutes or local guidelines can provide for ongoing investigations and techniques for integrating research and programme operation. Capacity-building tools provide for training of agency officials, continuing education, networks and websites. Community participation can be encouraged, or mandated by policy designs.

We have recounted one central finding that threads through US water history – new policy developments emerge from the interaction among different WoK within the policy field (where we understand 'field' as the kind of orthodoxy described in Bourdieu, 1986). Conflict and exchange between competing WoK provide a force tending toward the creation of institutional reforms and new policy instruments.

4.5 Profiles of transitions

4.5.1 Introduction

Elements that overlap more than one perspective are especially important to the possibility of transitions, as the cases in this section illustrate. Overlap occurs when different perspectives consider some of the same elements, but may see them differently. The fluidity of WoK allows for the possibility of change and provides the opportunity for collaboration by holders of different WoK. The most elastic and movable elements lie at the margins of WoK where collaborative relationships can be most easily built. The policy challenge for policy entrepreneurs is to mould elements or to introduce new elements that build relationships where trust increases and shared elements take on characteristics of salience, reliability and legitimacy. In particular contexts various tools discussed in Part I of this volume can come into play as illustrated by the three cases below. New elements of policy design can be introduced that have appeal across several ways of knowing. Policy entrepreneurs can facilitate recognition of elements common to several perspectives. Through reflexive communication, adherents to different ways of knowing can tell that they have been understood (O'Leary and Bingham, 2007).

New venues and boundary organizations may bring adherents to different ways of knowing together so that shared activities and dialogue may increase empathy and understanding across WoK. Cross-WoK networks can be advanced by policy entrepreneurs through such practices as identifying the appropriate players and establishing legitimacy and transparency of policy processes (O'Leary and Bingham, 2007).

Cases have been selected because they reflect transitions at the subnational level in the US where the greatest opportunity for policy transitions now exists. The paradigm-shifting transitions that have occurred at transnational and national levels as documented in other chapters in this book now rarely occur in the US. Transitions are seen at subnational levels, and, in the case of RISA's national administrative agency generated policy, implemented at the river basin level (RISA refers to Regional Integrated Science and Assessments). Cases have been chosen also because they reflect a paradigm shift (albeit at the subnational level) and the instrumental role of policy entrepreneurs in facilitating linkages among different WoK. Overlapping and criss-crossing WoK create new policies and designs for governance. Increasingly, as the CalFed story shows, this involves so-called network forms of governance. Networks cross organizational and sometimes sector boundaries. Linkages involve a host of policy actors, often including international non-governmental organizations (NGOs), federal, state, regional and other agencies, as well as the scientific and academic communities.

4.5.2 The California Bay-Delta estuary system (CalFed)

Since the early 1990s a cascade of decisions has been made in the California Bay-Delta region that have produced what can only be classified as a water transition. The San Francisco Bay/Sacramento–San Joaquin Delta (or Bay-Delta for short) is the largest estuary system on the west coast of the US, draining some 40 per cent of the waters of the state of California, including the watersheds of the Sacramento and the San Joaquin Rivers (see Map 4.1).

What could be seen as 'path-dependence' came to an abrupt halt in 1982, when the voters, worried about costs and environmental impacts, turned down the peripheral canal designed to route waters around rather than through the delta, ending the possibility of responding to mounting problems through construction-based solutions. Clearly, the WoK water-as-product no longer dominated in California. Not only was water development brought to a standstill, but federal agencies including the Environmental Protection Agency and the National Oceanic and Atmospheric Administration (NOAA) challenged the status quo on the grounds of inadequate water quality and threats to endangered species, reflecting an environmental WoK. The whole matter came to a head when the pumps that send water south to 40 million residents of California were stopped in order to protect the delta smelt under the Endangered Species Act.

In the context of WoK, the dominant water-as-product narrative was being pulled apart by the environmental perspective. According to lessons from this perspective, most water projects need to be re-engineered to reflect natural forces more closely. Further, a concerted effort towards environmental restoration is needed before any further development of water resources to serve people. The clash over the delta smelt was only one of

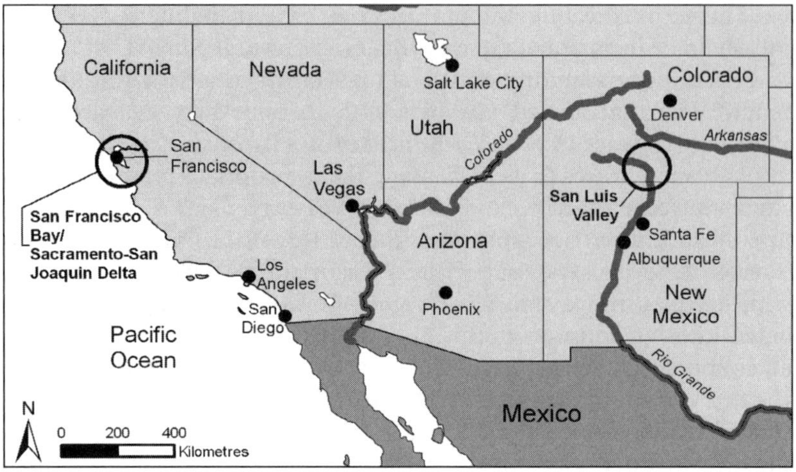

Map 4.1 Western US, the California Bay-Delta and Colorado's San Luis Valley

a large number over the Endangered Species Act which, when invoked, allows no public participation or consultation (Ingram and Fraser, 2006).

The Secretary of the Interior during the Clinton administration was Bruce Babbitt, who clearly took on the role of policy entrepreneur. He was determined to take proactive steps to be certain that the Endangered Species Act would not be repealed by Congress, which was then under the control of the Republicans. The Department of the Interior under Babbitt's administration endorsed a regional approach to water problems that engaged all of the federal and state agencies. Further, it espoused adaptive management, an idea that placed great emphasis on both science and experiential knowledge in guiding public policy. Babbit himself endorsed policy design for continuous public participation. Through many days of negotiation, during which Secretary Babbitt or his representatives were a large presence, some common ground was found between the environmental and product WoK, and a new programme was started, commonly referred to as CalFed (Ingram and Fraser, 2006). CalFed could be termed a boundary organization because it incorporated members from many agencies and levels of government. The management strategy offered by Secretary Babbitt promised flexibility and ways to avoid damaging conflict. The programme has undergone a number of changes since 2002 when the record of decision was signed by a large number of public and private entities. While it is less important now than it was until 2006, CalFed persists as an important regional entity. It is a classic network design, composed of boundary spanning relationships, multiple sectors and, most importantly, multiple WoK (Lejano and Ingram, 2009).

The CalFed Record of Decision contained elements drawn from all of the WoK and for this reason had widespread appeal. It embraced issues from the environmental perspective and promised large investments in environmental restoration. It accepted the need for security in water supplies that was central to the water-as-product WoK embraced by irrigation districts and municipal water agencies. The Record of Decision promised no surprises and stated that whatever costs or water necessary to protect endangered species would not place additional burdens on agricultural and urban contractors. Ethical perspectives were taken into account to the extent that environmental justice representatives were specifically to be included in the new fora established through CalFed. Recognizing water as an economic commodity that could be bought and sold was embodied in the Environmental Water Account that provided fish agencies with water and funding to use in markets to acquire water to release for protection of species.

The establishment of a great number of new venues within which people from different perspectives could come together to discuss problems may be the overarching contribution of CalFed. In addition to management committees that brought together agency representatives from all levels of government, numerous public advisory committees were set up. Finally, science was to play an important role in fostering adaptive management, and an independent chief scientist position was established as well as an independent science board. Numerous other science boards were to monitor and advise specific programmes.

While CalFed's many successes and its failures are too numerous to be described here, it is probably useful to examine how the programme has dealt with climate change. Until recently most water agencies in the US have not taken climate change much into account (Rayner et al., 2005). CalFed has at least raised the issue and forced it onto the governmental agenda at the regional level where federal, state and local governments are all present. Climate change, an element shared by several WoK, has provided an opportunity for a water policy transition. How CalFed has treated sea level rise and below-sea level agriculture in the delta region is a case in point. CalFed offers an opportunity for action based on both science and public participation. The central delta features five county governments in addition to multiple federal and state agencies and non-governmental organizations whose perspectives are being integrated into the management process, which is one of the purposes of the CalFed programme.

A key decision being faced is whether delta interests should invest in building up and repairing levies to protect subsided soils: 400 000 acres subsided due to microbial oxidation of peat soils that have been used for agriculture. A significant number of the islands are below sea level and protected from inundation by dykes that are in relatively poor condition. Continuing sea level rise and regional climate change are expected to have additional major impacts such as flooding and changes in seasonal precipitation patterns. There are concerns that multiple islands would be inundated in a '100-year storm event' – this represents extreme local vulnerability to flooding. Hundreds of millions of dollars of restoration work has been done in the delta and associated watersheds, and more investment is required if native ecosystems are to persist into the next century. There is a need to invest in restoring lands at intertidal and higher elevations so that the wetlands can evolve uphill in response to the rising sea level (Pitzer, 2008).

Climate change-related information is also critical to delta management as it estimates the changing volume and timing of run-off from the Sierra Nevada mountain range. The extent to which snow pack diminishes and snowmelt run-off occurs holds implications for flood control, water supply and conveyance, and seawater intrusion – all of which affect habitat and land use decisions. CalFed has put all of these issues on the governmental agenda.

4.5.3 Water and culture in the San Luis valley[1]

The Nature Conservancy, an international NGO, and San Luis, a rural community, collaborated in 2002 on the purchase of the 97 000-acre Baca Ranch located in the San Luis

Valley in Colorado (see Map 4.1). Widely hailed as an example of the use of markets to protect ecological values (the ranch borders on the Great Sand Dunes National Monument), the purchase was also a victory for the San Luis community and the protection of its culture. This case reflects water transition that brought together the environmental, economic and ethical WoK (see Arnold, 2008).

Situated in the American West, the San Luis Valley was settled rather early. Around 1851, Spanish-speaking settlers from New Mexico established the first outpost. From the beginning, agriculture in the area was touch and go, facing aridity and a short growing season as constant adversaries due to the high elevation of the valley. Critical to survival were the *acequias*, or ditch associations that are as much a communal network as they are water management organizations. Anglo-American irrigation and more large-scale mutual irrigation companies followed, and the area became prosperous. This prosperity, however, was based on uncertain water supplies. In response to the interstate Rio Grande Compact that placed an embargo on additional diversions in Colorado, San Luis residents united across economic and racial lines to establish the Rio Grande Water Conservation District through a popular vote in the valley.

In 1986, the owner of the Baca Ranch filed claims for 200 000 acre feet[2] of valley groundwater. The plan was to drill 100 wells to extract groundwater and ship it to thirsty cities along the front range of Colorado including Denver, Pueblo and Colorado Springs. In response valley residents mobilized into a powerful grass-roots social movement, attracted broad media coverage, and even agreed to tax themselves for resources to fight to retain water in the valley. In publicity, valley residents used the symbol of the Owens Valley in California that lost its water resources, to its great detriment, to Los Angeles in what is famous as an unjust water raid. In response the Baca Ranch owners proposed several initiatives to amend the Colorado Constitution to require water metering and a flow tax for agricultural water users. Valley residents argued that meters and the tax would be prohibitively expensive and destroy the farming community. To raise money to oppose the initiatives that were to be voted on state-wide, valley residents formed Citizens for Colorado's Water that was joined by many conservation organizations, unions and some church groups. The lobbying organization portrayed the ballot measures as greedy subterfuge, trying to undermine justice that was all on the side of San Luis Valley residents. Voters responded accordingly, and defeated the propositions three to one.

Valley residents rejoiced in their victory, but feared it would be only temporary. The manager of the Rio Grande Conservation District opined: 'If you leave Baca Ranch in private hands, the owners will certainly try to export the ranch's water.' The member of Congress for the area introduced legislation to convert the Sand Dunes National Monument to the highest possible status as a National Park, but only if accompanied by federal ownership of the Baca Ranch. Secretary Bruce Babbitt embraced the proposal, and President Clinton signed the legislation. The Nature Conservancy agreed to purchase the land and to convey title to the federal government. While clearly this was a victory for nature, it was also a victory for San Luis Valley rural agricultural values and culture.

Real changes occurred in this case with both substantive and strategic dimensions. Substantively, market mechanisms were used by environmentalists to tame the excesses of the emerging water market in the American West. Strategically, a grass-roots-based social movement was able to attract broad support using the symbol of the Owens Valley injustice. In terms of WoK, cooperation occurred among adherents to equity and environmental WoK, in opposition to the economic commodity WoK that was aligned with the perspectives of those seeing markets and a higher valuation of water as the appropriate responses to scarcity. Again, the combination of WoK resulting in a new network of policy actors was instrumental in causing the transition.

4.5.4 Regional integrated science and assessment teams

This water transition differs from the other two examples in three ways: the initiative came from NOAA, a federal organization peripheral to the water resources community; it is focused on the use of science in water resource decisions; and it involved a WoK outside the water sector – one based on forecasting tools unfamiliar to water professionals, and on perspectives that integrated global phenomena such as ocean temperatures and weather patterns aligned with the more conventional WoK of water-as-product and water as an environmental element. NOAA has invested for decades in a global network of observation equipment and analytical capacity aimed at long-term weather prediction. In principle, knowing that the weather is likely to be hotter or cooler and precipitation is going to increase and decrease should be of great interest to water managers. Yet researchers found that water managers did not use the data provided by NOAA, even though the agency had invested heavily in user-friendly web-based information systems (Rayner et al., 2005). In response NOAA collaborated with the Environmental Protection Agency's Global Change Research Program to set up nine Regional Integrated Science and Assessments (RISAs). These teams create networks among teams of scientists in government and in state universities with state and locally based water managers. The idea is to foster trust and collaboration so that the forecasting products produced by the scientists are truly useful and managers sufficiently understand the uncertainties in the forecasting products so that they can use them with confidence.

Analysts have generally endorsed the RISAs as successful, based on practices that have engaged people with different WoK and from agencies with different missions (McNie et al., 2007; Ingram and Stern, 2008). Such practices include driving research that engages the knowledge and experience of water managers embedded in different physical contexts. RISAs also act as information brokers, networking information from different sources and WoK. In addition RISAs generally use problem-based approaches that focus directly on solving water-related problems related to wildfires, water supply, water quality, flooding, drought, and so on. RISAs see themselves as part of a learning system and promote opportunities for joint learning among stakeholders involved in specific issues. Another example of RISA practice sees basic as well as applied researchers engaged in the RISA process so that contributions are made continually and not just to immediate problem-solving, but also to basic understanding of the climatic links to hydrologic cycles. Even in this seemingly technical field, we see the coming together of more established water sector WoK with a different climate-based WoK. While this case is less illustrative of the action of policy entrepreneurs in the policy-making process, it does highlight the importance of place-based information brokers who can translate across different WoK in policy implementation.

4.6 Conclusion

Water resources history in the United States illustrates that numerous large water transitions have occurred where new WoK have challenged dominant ones. Such transitions were usually abetted and then institutionalized by the emergence of different and often new venues. For most of American history transitions occurred at the federal or national level. Most interestingly, this chapter illustrates that when transitions occur in water resources they overshadow, but do not replace, previous WoK. In consequence multiple ways of knowing exist today in the US. At the federal level at least, contention reigns unless water is associated with another highly visible issue like climate change and networks are built across sectors. At other levels, such as regions and states, transitions are possible when bridges are built among several WoK through the action of policy entrepreneurs to bring together diverse viewpoints using policy instruments and designs that are attractive to different perspectives. The new institutional designs that result from the multiple WoK are increasingly taking the form of boundary-spanning networks, as is the case with watershed management authorities that are becoming more and more common in the US (for example, see Sabatier et al., 2005; Ingram and Stern, 2008).

Water resources history in the US also illustrates that many old ideas such as watershed and river basin management and public engagement in decision-making have been newly packaged and do not represent true innovation. IWRM and adaptive management incorporate elements that are by now long familiar. This finding suggests that transitions do not depend on new ideas as much as messages that appeal across different perspectives and engage ways of thinking about water previously marginalized or absent. Often academic contributions to policy change have been confined to what Kingdon (1984) refers to as the 'policy stream' or have suggested only substantive approaches to problem-solving rather than attending to either the framing of problems or the building of political support. In contrast, policy science as it is used in this volume and chapter suggests that insight into strategies of policy entrepreneurs, communications across different WoK about water issues or problems, and the identification of venues favourable to transitions can make large contributions to policy change.

Policy entrepreneurs who are critical to the generation, transmission, receipt, interpretation and integration of knowledge leading to collaboration across many kinds of boundaries appear to come from many different positions at various levels inside government and, in the case of implementing RISAs, universities. More commonalities can be found in the tools they use: convening parties not previously closely associated; embracing perspectives different from previously operative WoK; interpreting and translating meanings among people from different WoK; introducing boundary objects; and engaging in boundary-spanning actions. While it is helpful to the effectiveness of policy entrepreneurs and change-oriented information brokers to work in contexts where existing 'policy images' and institutions have fallen into disrepute (as they did when the delta smelt shut down the pumps, or when the Baca Ranch threatened to sell groundwater out from under the San Luis Valley), existing WoK need not be considered totally bankrupt. In fact one of the lessons of this chapter is that WoK persist even as new ones are introduced. Rather, it is the requirements of specific contexts and situations that cannot be effectively understood and acted upon through existing WoK that open up the opportunities for new WoK to gain strength. It is the fluidity and malleability of knowledge in practice and its communication and interaction that enable transitions.

This chapter also illustrates the importance of climate change to water policy transitions in the USA. While the USA has been far behind many other countries in taking actions to prevent the build up of greenhouse gases, the likely impact of climate change on water resources has prompted significant transitions in how water is treated in regional and local contexts. In the cases of regional water management in the California Bay-Delta region and in the Regional Integrated Environmental Assessments, climate change is a kind of boundary object that is a common element to the diverse ways of knowing water extant in the USA. It is a topic around which transitions can be built that are inclusive and robust and certainly forward-looking.

Notes

1. The narrative (but not the interpretation) in this section of the chapter relies heavily on Arnold (2008).
2. An acre foot is enough water to cover 1 acre of land 1 foot deep. It is a commonly used measure of water rights in the American West.

References

Arnold, T.C. (2008), 'The San Luis Valley and the moral economy of water', in J. Whiteley, H. Ingram and R. Perry (eds), *Water Place and Equity*, Boston, MA: MIT Press, pp. 37–68.

Baumgartner, F.R. and B.D. Jones (eds) (2002), *Policy Dynamics*, Chicago, IL: University of Chicago Press.

Bourdieu, P. (1986), 'The force of law: toward a sociology of the juridical field', *Hastings Law Journal*, **38**, 814–53.

Brown, F.L. and H. Ingram (1987), *Water and Poverty in the Southwest*, Tucson, AZ: University of Arizona Press.

Dewulf, A., B. Gray, L. Putman, N. Aarts, R. Lewicki, R. Bouwen and C. van Woerkum (2005), 'Disentangling approaches to framing: mapping the terrain', paper for the Eighteenth IACM Conference, Seville, Spain.

Feldman, M.S., A.M. Khademian, H. Ingram and A.L. Schneider (2006), 'Ways of knowing and inclusive management practices', *Public Administration Review*, Supplement, **66** (December), 89–99.

Guston, D.H. (2001), 'Boundary organizations in environmental policy and science', *Science, Technology and Human Values*, **26** (4), 399–408.

Hajer, M.A. (1995), *The Politics of Environmental Discourse: Ecological Modernization and the Policy Process*, Oxford: Oxford University Press.

Ingram, H. (1990), *Water Politics: Continuity and change*, Albuquerque, New Mexico: University of New Mexico Press; updated version of (1969), *The Patterns of Politics on Water Resources: The Case of New Mexico's Role in the Colorado River Basin Bill*, Albuquerque, NM: University of New Mexico Press (Institute of Government Research).

Ingram, H. and L. Fraser (2006), 'Path dependency and adroit innovation: the case of California water', in R. Repetto (ed.), *Punctuated Equilibrium and the Dynamics of US Environmental Policy*, New Haven, CT: Yale University Press, pp. 78–109.

Ingram, H. and P. Stern (eds) (2008), *Research and Networks for Decision Support in the NOAA Sectoral Applications Program*, Washington, DC: National Academies Press.

Ingram, H. and S. Ullery (1980), 'Policy innovation and institutional fragmentation', *Policy Studies Journal*, **8** (5), 664–82.

Jasanoff, S. (1990), *The Fifth Branch: Science Advisors as Policy Makers*, Cambridge, MA: Harvard University Press.

Kingdon, J. (1984), *Agendas, Alternatives and Public Policies*, Boston, MA: Little Brown.

Kneese, A.V. and B. Bower (1968), *Managing Water Quality: Economics, Technology and Institutions*, Baltimore, MD: Johns Hopkins Press.

Latour, B. and S. Woolgar (1979), *Laboratory Life: The Social Construction of Scientific Facts*, Beverly Hills, CA: Sage.

Lejano, R. and H. Ingram (2009), 'Collaborative networks and new ways of knowing', *Environmental Science and Policy*, **12**, 653–62.

Maass, A. (1951), *Muddy Waters: The Army Corps of Engineers and the Nation's Rivers*, Cambridge, MA: Harvard University Press.

Maass, A. and R.L. Anderson (1978), *And the Desert Shall Rejoice: Conflict, Growth and Justice in Arid Environments*, Cambridge, MA: MIT Press.

McNie, E., R. Pielke, Jr and D. Sarewitz (2007), 'Climate science policy: lessons from the RISAs – workshop report – final draft, 15–17 August, 2005', East–West Center, Honolulu, Hawaii, 26 January.

Nichols, J. (1974), *The Milagro Beanfield War*, New York: Holt Rinehart & Winston.

O'Leary, R. and L. Bingham (2007), 'A manager's guide to resolving conflicts in collaborative networks', report, IBM Center for the Business of Government.

Pitzer, G. (2008), 'Finding a vision for the delta', *Western Water*, Water Education Foundation March/April, 1–2.

Rayner, S., D. Lach and H. Ingram (2005), 'Weather forecasts are for wimps: why water resource managers do not use climate forecasts', *Climatic Change*, **69**, 197–227.

Sabatier, P.A. (1993), 'Policy change over a decade or more', in P.A. Sabatier and H. Jenkins-Smith (eds), *Policy Change and Learning: An Advocacy Coalition Approach*, Boulder, CO: Westview Press, pp. 13–40.

Sabatier, P., W. Focht and M. Lubell (eds) (2005), *Swimming Upstream: Collaborative Approaches to Watershed Management*, Cambridge, MA: MIT Press.

Schön, D. and M. Rein (1994), *Frame Reflection*, New York: Basic Books.

Smith, K.L. (1986), *The Magnificent Experiment: Building the Salt River Project 1890–1917*, Tucson, AZ: University of Arizona Press.

Star, S.L. and J. Griesemer (1989), 'Institutional ecology, translations and boundary objects: amateurs and professionals in Berkeley's museum of vertebrate zoology', *Social Studies of Science*, **19** (3), 387–420.

Thompson, M., S. Rayner and S. Ney (1988), 'Risk and governance Part II: policy in a complex and plurally perceived world', *Government and Opposition*, **33** (2), 139–66.

US Army Corps of Engineers (n.d.), 'Brief history', http://www.hq.usace.army.mil/history/brief.htm#1beg, accessed 31 May 2008.

Walton, J. (1992), *Western Times and Water Wars: State, Culture, and Rebellion in California*, Berkeley, CA: University of California Press.

Weber, E.P. and A.M. Khademian (2008), 'Wicked problems, knowledge challenges, and collaborative capacity building in networked settings', *Public Administration Review*, **68** (March/April), 334–49.

Wescoat, J.L. (2000), '"Watersheds" in regional planning', in Robert Fishman (ed.), *The American Planning Tradition: Culture and Policy*, Washington, DC: Woodrow Wilson Center Press, pp. 147–72.

Wynne, B. (1996), 'May the sheep safely graze? A reflexive review of the expert–lay knowledge divide', in, S. Lash, B. Szerszynski and B. Wynne, *Risk, Environment and Modernity: Toward a New Ecology*, Boston, MA: Routledge & Kegan Paul, pp. 44–83.

5 Political and economic apertures and the shifting state–citizen relationship: reforming Mexico's national water policy

Margaret Wilder

5.1 Introduction

For nearly twenty years, national water policy has been a crucial yet little noticed crucible in which to examine the advance of democratization and political opening in Mexico. In December 1992 Mexico adopted sweeping changes to national water policy that presaged the possibility of a new state–citizen relationship. Mexico hosted the Fourth World Water Forum in 2006, and showcased its modernized water sector before thousands in attendance. The three key components of Mexico's strategy – marketization, decentralization, and sustainability – were embedded in a broader context of political opening and economic restructuring. The new policy envisioned both a leaner and meaner role for the state while at the same time promising to institutionalize a new role for citizens as water users. Multiple and often conflicting agendas were built into the new water policy, which sought at once to make the water sector more efficient and more sustainable and more reflective of citizens' needs while transferring the enormous financial burden of managing water systems to water users themselves (Wilder and Romero Lankao, 2006). Mexico has emerged as what one World Bank observer called a 'successful new globalizer' (Easter et al., 1998) while others see the national water reforms as either a mixed bag or an experiment that has failed in its fundamental objectives while further entrenching pre-existing inequalities (Barkin and Klooster, 2006; Moreno, 2006). The 2004 sustainability and decentralization modifications in particular have received very limited attention yet hold perhaps the most potential for transformative change if fully enacted.

This chapter advances four main arguments:

1. That democratic transition in Mexico has been about 'democratizing' water policy, but that the democratic transition is at the same time working itself out through water governance.
2. That the transition to 'a new culture of water' in Mexico focused on efficiency, decentralization and sustainability has only minimally and incompletely fulfilled its goal of establishing a new state–citizen relationship around water policy.
3. That the transformation of Mexico's water policy in its initial phase (1992–2000) was a top-down process due to Party of the Institutionalized Revolution (PRI) party dominance and the ability to link formally (albeit not substantively) marketization and efficiency principles with an emerging international paradigm of sustainability and democratization.
4. That the 2004 legal modifications that would substantively commit Mexico to a more democratized and decentralized water management paradigm have stalled, due

> in part to the political aperture (see below) that has created a more fractured political and governance structure. Mexico's experiment with integrated water resources management (IWRM) and sustainable water policy-making seems to be in pause mode, indicating a lack of consensus about and lack of commitment to full consolidation of the sustainability and democratization reforms.

This analysis draws on the concept of 'aperture' in two of its meanings to illuminate the transition in Mexico's water policy. In its contemporary sense, an aperture is an optical feature used to view an object as through a camera. In this way, the water policy reforms serve as a particular lens through which to observe and interpret the democratic transition in Mexico. The water policy transition is a specific opening onto a set of political processes in transformation. In a rare and broader meaning, an aperture is a breaking open or opening up. In this sense, then, Mexico's political opening to admit new political parties and civil society actors to formalized participation in national electoral and political processes can be understood as political aperture; in the same way, economic restructuring that admitted privatization, domestic and international markets, international trade agreements, corporations and foreign capital can be said to be a process of economic aperture. In Spanish, the concept of political opening is referred to as *la apertura política*. The chapter employs the concept of aperture both as a specific lens via water policy onto processes of democratic transition and as the very breaking open of broad political and economic processes themselves.[1]

Mexico's transition underscores the evolving and dynamic nature of water policy, and the way that new management paradigms must be able to weather political transitions and create deep political resonance to achieve policy stability or risk subversion or discontinuity due to fragmented governance structures and political culture. To the extent that water policy can transition in a formal, institutionalized sense without being fully consolidated or realized, the emerging governance scheme can appear to be window-dressing or a mere discursive distraction. Even as Mexico's water policy has become more 'democratized' in a formal sense – through the admission of subnational actors to the decision-making arena – the process has exposed the ways in which the messy and complex processes of democratic transition can lead to uneven commitment to and implementation of ambitious water policy transition schemes. Paradoxically, the more the political system has opened to admit new actors to the water policy-making arena, the more difficult it has become to achieve the very participation and sustainability goals it originally sought to accomplish. Successful decentralization requires the philosophical, institutional and financial support of a strong state (Wilder and Romero, 2006). The specific nature of water in both its material and symbolic aspects lends it the nature of an 'uncooperative commodity' (Bakker, 2004) that has almost limitless reach in terms of its life-sustaining, productive and cultural materialities (Blatter and Ingram, 2001).

Thus water is not easily contained and addressed within a single policy context; rather, water is an integral feature of multiple social and economic sectors, including urban, agriculture, industry, development, tourism, neighbourhoods, social equity, health and energy, leading to a complex network of overlapping policies that have to be coherently co-managed so as to achieve water policy goals. In Mexico the water policy transition is an aspect of a broader democratic transition, but at the same time democracy is working itself out through emerging water governance institutions and arrangements. Transitions in water governance in Mexico, then, are centrally about the politics of water, but are also about politics (and party politics). Agents working to democratize the water policy arena are reshaping the contours of democratic transition more broadly.

Section 5.2 discusses the transition of Mexico's water policy in 1992 and in 2004 and analyses the implications of Mexico's water policy transformation for the state and citizens in two specific contexts – irrigation management and river basin councils – a decade and a half after its adoption. Section 5.3 discusses four forces behind sweeping policy transformation, including the international water management paradigm, the role of civil society actors, and the processes of economic restructuring and political opening, reflecting both exogenous and endogenous pressures on the state and political system. The section also illuminates key policy entrepreneurs who significantly shaped the 1992 and 2004 transitions. The final section discusses the implications and conclusions that can be drawn from Mexico's experience.

5.2 The emergence of a new water management paradigm

5.2.1 Introduction

The transformation of Mexico's national water policy in 1992 represented the country's political and economic opening, while at the same time reflecting the emergence of a new international paradigm for water management and a raft of exogenous and endogenous pressures on its failing water infrastructure. Given this combination of impulses, it is perhaps not surprising that the resulting national water policy embodied multiple and often conflicting agendas.

In keeping with the spread of political democracy and the movement toward liberalizing the economy, the Salinas administration (1988–94) undertook a wholesale modernization and decentralization of water resource use and management. In February 1989 the National Water Commission (CONAGUA) came into being as an independent agency attached to the Ministry of Agriculture and Water Resources (SARH). CONAGUA was to be the sole authority at the federal level to deal with water issues, and its primary responsibilities included: definition of Mexico's water policies; formulation of the national water programme; development of potable and sewage water systems and treatment; construction, operation and maintenance of dams and irrigation systems; expedition of titles of concession and development of a public registry of water rights; mediation of water conflicts; promotion of the efficient use of water and its conservation; promotion of a culture that treats water as a vital and scarce resource; and promotion and undertaking of research and technological development relating to water (Ley de Aguas Nacionales, Article 8, 1992).

The new National Water Law (Ley de Aguas Nacionales) was published in the *Diario Oficial* (Mexico's Congressional Record) on 1 December 1992 and became effective the following day. The law was the regulatory structure (*reglamentaria*) that had been established in paragraphs 5 and 7 of the Mexican Constitution (Tellez, 1993, p. 104). The regulation implementing the law was adopted on 12 January 1994. It replaced the former national water law, the Federal Law of Waters (Ley Federal de Aguas), that had been in effect since 1972, and which in its turn had supplanted the earlier water legislation, Law of National Property Waters (Ley de Aguas de Propiedad Nacional) dating to 1936.

Mexico's new water governance paradigm included virtually all the features prescribed and financially supported by the World Bank and the Inter-American Development Bank, including: establishment of a public registry of water rights to provide secure water rights; the establishment of formal markets for trading water surpluses; the initiation of full cost recovery water pricing and elimination of subsidies; the transference of the country's 81 major irrigation districts to water user associations; municipal and state management of urban water and sanitation services; the allowance of private sector management of water services via government concession; and an integrated water resources management that institutionalized local participation through the creation of a network of major river basin councils (*consejos de cuenca*). CONAGUA as the federal water management agency assumed a largely policy-making and oversight role, leaving most daily operations and decision-making to municipal and state water managers and irrigation districts, while itself retaining key strategic functions.

The scholarship on Mexico's water transition is growing, with a focus on decentralized and privatized urban water management and its financial aspects (Torregrosa, 1995; Palacios Velez, 1997, 2000; Whiteford and Melville, 2002; Perez Prado, 2003; Barkin and Klooster, 2006; Wilder and Romero, 2006; Pineda Pablos, 1999, 2004, 2006; Pineda Pablos and Camou, 2007, Pineda Pablos et al., 2007) and the implications of water reform for urban areas and for rural–urban trade-offs (Castro, 2006; Moreno, 2006). Most of these assessments are pessimistic about the real outcomes of the water policy transition and remain cynical about the likelihood of transformative democratic change in the water sector. While the focus of this chapter is on the emergence of the water policy transition itself, it is useful to provide a very brief overview of some of the evidence of how the new policy has played out in two specific northern Mexico contexts in the US border state of Sonora, Mexico (see Map 5.1): irrigation districts and river basin councils (see also Wilder and Whiteford, 2006; Wilder and Romero, 2006; Moreno, 2006; Castro, 2006 for detailed discussion of impacts of the water policy transition).

Irrigation uses more than 80 per cent of water supply and is therefore a critically important sector to assess in terms of the water policy transition. At the same time, river basin councils represent the principal institutional innovation of the water policy transition, reflecting the opening of the water sector to more participatory processes of planning and decision-making.

Sonora is a highly vulnerable state due to its rapidly growing urban population, intensive commercial agriculture coupled with assertive industrial and coastal development, and its arid desert climate and landscape (Liverman and Merideth, 2002; Ray et al., 2007). Projected climate change impacts include increased temperatures and drier conditions, leading to reduced water supply and more severe and prolonged droughts (IPCC, 2007; Seagar et al., 2007). As the major irrigating state in the republic of Mexico and a border state sharing a frontier with Arizona in the United States (Mexico's principal trade partner), Sonora is a crucible for a modernizing Mexico with its high concentration of foreign-owned *maquiladora* assembly plants, seven major commercial irrigation districts producing wheat, grapes, citrus, asparagus and melons primarily for export to the USA and Japan, accelerated demographic growth based on domestic migration and massive tourism, and residential resort development along the Gulf of California coast. Yet despite its advantageous resource assets and insertion into the global economy, the achievements of water policy reforms in Sonora have been limited.

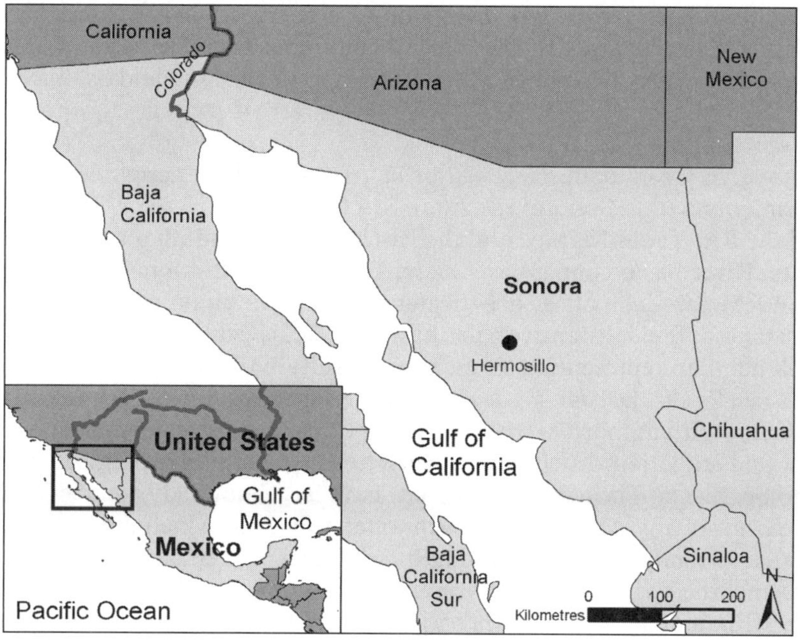

Map 5.1 *Mexico and Sonora*

5.2.2 Irrigation districts

The transference of irrigation districts in Sonora represents the best claim for success in terms of the decentralization components of the water policy package. Sonoran irrigators report satisfaction with a growth in 'political equity' represented by water user management of the district, which allows water modules (subdistrict-level associations) to manage their own operations, including election of their leadership (rotating annually among small, communal farmers and large private producers), establishing their own quotas (beyond a district-wide base fee), and setting strategies for reinvestment of those revenues into specific projects that benefit the module (for example, installation of drip irrigation, cleaning of field-specific canals) (Wilder and Whiteford, 2006; Wilder, 2008a).

Overall, however, irrigators have experienced a net decrease in 'economic equity' due to exposure to the demands of international agricultural markets, lack of credit, trade liberalization resulting in new competitors as well as loss of protective tariffs and quotas, elimination of subsidies (such as electrical subsidies for groundwater irrigators), and rising input costs. Water costs alone increased in the Altar–Pitiquito–Caborca district by 89 per cent for wheat, 118 per cent for table grapes, and by 186 per cent for asparagus in the first six years after the transference (Wilder and Romero, 2006), and water costs in the wheat-producing Rio Yaqui irrigation district increased even faster.

Small, communal producers known as *ejidatarios* have abandoned active production at an accelerated pace and turned more than ever to land and water leasing to private producers, resulting in a kind of de facto privatization of productive assets for the most marginal producers (Luers et al., 2005; Wilder and Romero, 2006; Wilder and Whiteford, 2006). The water reforms have resulted overall in some important gains in political equity for irrigators, but these gains are overwhelmed by losses in economic equity that have resulted in non-competitive conditions even for large producers while marginal producers are struggling to hold onto their water, land and livelihoods.

5.2.3 River basin councils

Of 25 major river basin councils (*consejos de cuenca*) created across Mexico, three are located primarily within Sonora: the Altar-Río Concepción in the northern part of the state, and the Río Yaqui-Mátape and the Río Mayo in the south and southeastern part of the state. River basin councils are one of the major institutional changes associated with the IWRM paradigm of Mexico's water reform programme, yet the Sonoran experience indicates that few substantive gains have been achieved since their creation (Wilder, 2008a). Rather than representing a fundamental shift in the way water policy is made at local and basin levels, the *consejos de cuenca* were superimposed upon an existing institutional structure with authority largely vested in CONAGUA, municipal and state water managers, and irrigation district user associations, leaving little political space for the *consejos* to operate in. The *consejos de cuenca* have large, unwieldy memberships (including the governor of any state within a given watershed) and an inequitable representation structure that virtually guarantees strong state control of agendas, policies and planning processes, since the law stipulates that the director of CONAGUA must chair the *consejo de cuenca*. Although agriculture uses 85 per cent of the water, there is only one voting member from agriculture on the *consejo*, as there is for each of the other much smaller water-consuming sectors (for example, ranching, industrial, energy, urban) (Wilder, 2008a). In the past, workings of the *consejos de cuenca* in Sonora have been largely hidden from public view due to a lack of open meetings, agendas and transparent procedures. A major constraint for the *consejos de cuenca* is a lack of jurisdictional authority to challenge fundamental water allocation and infrastructure arrangements that at root shape the very way water is used and consumed in this arid, drought-prone environment, and for whose benefit. Yet the *consejos* are in their infancy in Sonora and may yet prove to be important instruments of change focused on sustainable and equitable practices in Sonora.[2] The emergence of river basin councils provides a veneer of local participation and sustainable planning practices to national water policy, yet in practice the river basin councils in Mexico, 15 years after their establishment, are embryonic and in search of firm substance, direction and voice. These brief examples from northern Mexico illuminate the disconnect between the legal framework of 1992 and 2004 that instituted decentralized and participatory mechanisms for water governance and the limited, fragmented nature of consolidation of these reforms that has occurred.

The next section turns to the question of how Mexico's water policy transition emerged, focusing on five converging sets of conditions and processes: international influences, exogenous and endogenous influences, national economic opening, the role of civil society actors, and national political opening.

5.3 Accounting for policy dynamics

5.3.1 International water management paradigm

The late 1980s and early 1990s represented a major shift in thinking about global water management (Gleick et al., 2002; Blatter and Ingram, 2001; Conca, 2006; Varady et al., 2008; see also Gupta (Chapter 3 in this volume) on global freshwater discourses).[3] Mexico became an early adopter of a full slate of transformations in water policy promoted and financed by the World Bank and centred around the principles of marketization, decentralization and sustainability, associated with an emerging international water management paradigm that championed these principles of water governance and heralded in Mexico the introduction of a 'new water culture' by then President Carlos Salinas de Gortari. Mexico's adoption of a new water management framework at the end of 1992 dovetailed with the 1992 Earth Summit in Rio de Janeiro and the emergence of the Dublin Principles (also in 1992) for sustainable water management. These called for understanding water as an economic good, rather than a traditional free good, in order to maximize efficiency of its use and promote its conservation. Privatization of water services was held up as a desirable alternative to traditional state control that was seen as overly centralized, wasteful, out of touch with local realities and inflexible. The new policies also institutionalized a participatory role for local government, water user sectors and local citizens in decision-making about water (Gleick et al., 2002; Blatter and Ingram, 2001; Conca, 2006; Varady et al., 2008). Mexico, following in the footsteps of Chile, emerged as a leader among developing countries that were turning to neoliberal, market-oriented and decentralized models for water management, including Brazil, Argentina, Peru and Colombia in Latin America, and other developing countries including India, the Philippines, Nepal, Zimbabwe and Kenya, among others (Easter et al., 1998; Hinchcliffe et al., 1999; Bauer, 2004; Conca, 2006; Lemos, 2008).

5.3.2 Exogenous and endogenous pressures for transition

By the late 1980s Mexico's water infrastructure was in a shambles and the government had no resources to address its expensive and extensive needs. Mexico's irrigation infrastructure, responsible for over 80 per cent of national water use, required 'staggering' levels of investment to rehabilitate irrigation canals and channels, estimated at over US$300 billion annually (Cummings et al., 1989, p. 27). The federal government was subsidizing irrigation water use by 80 per cent, with irrigators contributing only 20 per cent (CONAGUA, 1994, pp. 26–7). The Agriculture and Water Resources Ministry had spent the previous three decades constructing expensive dams to open up new irrigated lands, with limited expenditure on maintenance of existing infrastructure. Between 1965 and 1981 more than a quarter of all irrigated lands required rehabilitation and utilized one-third of the hydraulic infrastructure budget (Cummings et al., 1989, p. 27). By 1980 erosion was silting up irrigation canals, seawater salinization had damaged coastal aquifers, and faulty drainage crippled irrigation systems; half of the irrigation supply never reached cultivated fields due to leaks and run-offs (Yates, 1981; Cummings et al., 1989, p. 26; Buras, 1996, p. 233).

Apart from troubles in the water sector, Mexico was experiencing multiple shocks to its economic and political systems. In 1982 the peso devaluation exploded onto the global economic scene, setting off a chain reaction of defaults on international debts by governments across Latin America. The 1985 earthquake, the 1986 oil shock and the 1987 stock market crash sent reverberations throughout Mexico's government and business classes, and the resulting decline in real per capita income during the 1980s was as large as that of the Great Depression (Gould, 1996, p. 25). The new President elected in 1982, Miguel de la Madrid (1982–88), 'inherited perhaps the worst economic crisis in the country's history' (Gould, 1996, p. 23). Although de la Madrid took steps towards liberalizing the economy, his efforts were insufficient now that Mexico's foreign debt had increased to close to US$100 billion by 1985. Mexico was forced to look for international assistance to restructure its debt burden, resulting in a US$12 billion dollar IMF rescue package (Gates, 1988, p. 299). Mexico's 1982 debt default and subsequent currency devaluation amounted to a 'focusing event' for the country's political and business leaders.

5.3.3 Economic opening

As a result of the focusing event of the faltering Mexican economy, a new advocacy coalition began to emerge among the movers and shakers of the country (Sabatier, 1993). A consensus began to emerge among Mexico's leaders that the free market mechanism was the only way for the domestic economy to become 'viable in the world' (Valdes-Ugalde, 1996, p. 59). In 1986 Mexico reversed its long-standing resistance to participating in international treaties and signed on to the General Agreement on Tariffs and Trade (GATT, later the World Trade Organization, WTO) (Gates, 1988). Late in his term (1987) President de la Madrid led representatives of the labour, business and farming sectors in the first of a series of new economic initiatives that were subsequently taken up by President Carlos Salinas de Gortari after his 1988 election. The two measures, jointly referred to as the Pacto (pact), were designed to combine fiscal monetary restraint with structural reforms and an incomes policy (that is, controls on wages and prices).

From the point of view of water policy, the most significant of the steps decided upon was Mexico's decision to embrace free trade policies via its overall strategy of economic liberalization. However as Meijerink insightfully recognizes, fundamental shifts that initially seem to have been decided upon in a short time frame are often in fact the result of a gradual shift in national mood and the growth of a new coalition around a new set of ideas (Meijerink, 2005; see also Meijerink and Huitema in the theoretical introduction, Chapter 2 in this volume). Mexico's economic emergency had been unfolding for at least seven years before the national water policy transition was undertaken. In the case of developing countries like Mexico, it is also necessary to underscore the financial exigencies under which they often operate due to conditionality requirements for lending from institutions such as the World Bank and International Monetary Fund.

In order to participate in the North American Free Trade Agreement (NAFTA), Mexico was under pressure to modernize its water policy framework and water systems. Given the critical state of water resources by the late 1980s, it became necessary to establish a new water resources policy that would increase industrial and agricultural productivity, ensure conservation of the resource, preserve water quality and address supply and maintenance issues. After years of discussion, preparation and debate, NAFTA became effective 1 January 1994. NAFTA was just the tip of the iceberg of Mexico's move to embrace free trade. To much less fanfare, over the next 12 years Mexico entered into free trade agreements with multiple trade partners, including Israel, Chile, the European Union, Japan and the Central American nations.[4]

NAFTA was structured to phase out tariff protections and import quotas entirely over a 15-year period. NAFTA and the other FTAs intensified water use by stimulating industrial productivity and refocusing Mexican producers on high-value agricultural exports. The agreements forced small-scale Mexican producers to become more efficient and competitive, or to abandon production and transfer land and water resources to private producers and corporate entities (Nadal, 2002; Luers et al., 2005; Wilder and Whiteford, 2006).

To achieve its economic goals, a sweeping revision of Mexico's water policy was called for. Former Subsecretary of Agriculture Luís Tellez argued that the 1992 promulgation of the Article 27 constitutional reforms (that fostered private land titles and promoted land sales for communal properties or *ejidos*) and the adoption of a new Agrarian Law made it unthinkable for Mexico to continue with its outmoded 20-year-old Law of Federal Waters, which was not responsive to the country's modernization programme. A new water law was needed that would facilitate increased user participation, authorize more legal security over the rights to use and exploit water, induce a more efficient use of the 'vital liquid' through conservation of water quality and quantity, and permit better administration of water toward the goal of 'integral, sustainable development' (Tellez, 1993, p. 104). This interpretation is supported by President Salinas's own address to the Mexican Chamber of Deputies in his introduction of the Article 27 reforms: 'Today the countryside demands a new response to give opportunities of well-being to the farming way of life and to strengthen our Nation . . . Modernization responds to a new reality and demands adequate answers . . . The countryside today demands a new attitude and new mentality' (Tellez, 1993, p. 134). The ultimate goal was expressly 'a new culture of water' (Gortari, cited in Tellez, 1993, p. 105; see also Chapter 15 on Spain, this volume).

Luís Tellez, a PRI party stalwart and an MIT-trained PhD in economics, was one of the principal Salinas technocrats responsible for the water modernization strategy. The National Water Commission was placed within the purview of SARH at its 1989 creation. With the national government in the control of the ruling PRI (Partido de la Revolución Institutional – Institutional Revolution Party) party, the top-down imposition of Mexico's national water policy could be efficiently accomplished. Within just eight years, for example, 95 per cent of major irrigation districts had been transferred to water user associations charged with managing them (CONAGUA, 2008).

While the 1992 reforms were carried out by modernizing technocrats like Tellez within the Salinas administration, environmental aspects of water management gained more prominence in the two successive administrations due to a major administrative change and the environmentalists at the helm. Following Salinas, Ernesto Zedillo was the PRI candidate elected to the presidency for the 1994–2000 term after the assassination of the original PRI nominee, Luís Donaldo Colosio. Zedillo created a new Environment Ministry and appointed a respected environmentalist and academic, Julia Carabias Lillo, to head it.

Carabias holds a Masters of science from Mexico's National Autonomous University (UNAM) and serves on the UNAM faculty in Mexico City. Carabias remains a very prominent voice in Mexico for sustainable water and environmental policy. She was recognized with the 2005 United Nations Environmental Programme's Champions of the Earth award and the Cosmos 2004 International Prize. She also served four years as the chair of the science and technology advisory panel of the Global Environment Facility (2002–06). Carabias is the lead author of an important volume on water, environmental sustainability and society (Carabias and Landa, 2005) that made a splash when it was presented and released to a standing-room-only crowd of a few hundred people at the 2006 World Water Forum in Mexico.

During her tenure as Environment Minister, Carabias engineered the transfer of

CONAGUA from the Agriculture and Water Resources Ministry to the new Ministry of the Environment and Natural Resources (SEMARNAP, later changed to SEMARNAT to reflect the movement of fisheries out of the agency). This constituted an important instance of network management through venue manipulation (Baumgartner and Jones, 1991). In the Mexican case, Carabias was a policy entrepreneur interested in wresting water management away from the controlling grasp of the agriculture agency and shifting it into an agency focused on water conservation. According to Carabias, this change portended a broader turn – away from agricultural productivity and towards environmental sustainability – as a principal focus of Mexico's changing water governance paradigm (interview, 18 January 2008):

> Everything was done that could be done under the 1992 LAN (national water law) – increased social participation and decentralization . . . The 1992 LAN represented a major advance (over the previous water law) in terms of the principles expressed, and those principles were ultimately concretized in the LAN of 2004, due to moving water administration to the environment agency SEMARNAT. Between 1992 and 2004, the major change was to incorporate a sustainability vision. Without that, the sustainability ethic might never have occurred. (Interview, 18 January 2008)

Carabias's administration of SEMARNAT achieved some of the principal goals of the national water law, including the transference of irrigation districts to water user associations and the formal establishment of the river basin councils (*consejos de cuenca*) to take on IWRM. As the decentralization programme transferred to the PAN (Partido de Acción National – National Action Party) administration in 2000 at the end of the Zedillo tenure, the vision and gains of the original eight years of the water reform programme came under scrutiny, and additional modifications were pursued, as discussed in a subsequent section.

5.3.4 Civil society actors

Political opening was reflected not only in the *alternancia* of political parties and changes at the ballot box, but also in the emergence of new mobilizations of civil society during the 1990s.The unexpected uprising of the Zapatista movement in the southern state of Chiapas on 1 January 1994, the same day that NAFTA took effect, was undoubtedly the most dramatic sign of public mobilization to demand economic and environmental rights under the shift to neoliberalism (Harvey, 1996). The growth by leaps and bounds of neighbourhood social movements in that decade was part of the broader political aperture reshaping water politics in Mexico. In 1990 the Nahuas indigenous people of the Upper Balsas region of Guerrero mobilized against a large-scale hydroelectric dam (the San Juan Tetelcingo) (Hindley, 1999). In the northern industrial city of Monterrey, housewives in poor *colonias* marched to demand water services, while neighbourhood residents in Mexico City took to the streets to resist privatization of municipal water services (Bennett, 1995; Castro, 2006). Communities on the Mexico–US border participated in environmental improvement processes associated with the Border Environment Cooperation Commission (BECC) that institutionalized public consultation processes in a way not seen before, and environmental activism across the border region expanded (Lemos and Luna, 1999; Liverman and Merideth, 2002). Mexican citizens were demanding transparency, accountability and political space, and emerging as significant actors in the environmental policy arena.

5.3.5 Political opening and the 2004 modification of the National Water Law

The election of Vicente Fox Quesada as President of Mexico, on 2 July 2000, heralded a new era in Mexican politics. The profound story of that night was not only the election of an opposition candidate to the presidency after a 71-year one-party reign by the PRI that caused thousands to throng the streets in celebration across Mexico, but was found in the actions of then President Ernesto Zedillo, who in statesmanlike fashion, calmly announced his successor on national television, thereby proving to Mexico and to the world that it was a democratic nation where opposition party candidates could be elected in a free, open and non-violent process (Otero, 2004; Preston and Dillon, 2004). Fox's election was a culmination of at least 12 years of prior political opening, evidenced by very credible presidential runs by opposition candidates and an increasing number of gubernatorial, senatorial and congressional delegate seats won by opposition parties on both the left and the right (Cornelius et al., 1999).[5]

The elation that accompanied the election of Fox soon dissipated due to his administration's inability to carry out the major promises of change on which he was elected, including sweeping fiscal and taxation reform, anti-corruption policies, and comprehensive immigration reform in the United States.[6] Fox appointed Victor Lichtinger, a high-profile environmental economist and former executive director of the trilateral (Canada–Mexico–USA) North American Commission on Environmental Cooperation as his Environment Minister. Lichtinger was a favourite of environmentalist groups and continued pushing Mexico's water policy towards a more decentralized management structure, emphasizing the need for sustainable development and use of water resources, while at the same time underscoring the importance of the private sector and markets to fix Mexico's water infrastructure. Lichtinger positioned himself as a crusader against corruption within the environmental bureaucracy and worked with Mexico's environmental enforcement agency to rein in special interests that resulted in omissions and acts of corruption, especially at the state level.[7] Due to the controversial nature of his work and the threat he posed to established interests, and despite the outcry of the environmental community, Fox fired Lichtinger midway through his term (September 2003) and replaced him with an engineer with no environmental credentials. One observer noted: 'Fox wanted experts, but he needed politicians' (Vera, 2003).

The 2004 sustainability and decentralization LAN modifications resulted from a course on the environment requested by senators in the period just after Carabias left office (Carabias interview, 18 January 2008). The expert consultations organized before the Senate resulted in a proposal to modify the LAN. In April 2003 the Senate's Water Resources Commission, headed by PRI Senator Ulises Ruíz, presented a bill to reform the National Water Law to the executive branch, based on perceived dissatisfaction among some water using groups with the 1992 law.[8] Preparing himself for a gubernatorial bid in his home state of Oaxaca, Ulises Ruíz used the water reforms as a platform to gain national visibility (Crónica Legislativa, 2004). The 2003 bill specifically:

1. Introduced more substantive decentralization of CONAGUA through 13 'watershed organisms' (organismos de cuenca) that would serve as empowered and authoritative regional CONAGUA headquarters offices.
2. Built upon the 1992 law's creation of *consejos de cuenca* and stipulated specific membership and responsibilities of the councils.

3. Indicated that the environment should be considered as a legitimate water user, although no rights or conditions were specified.

On 1 September 2003 (in the same month Lichtinger was ousted) the Fox administration sent the Senate its decision to veto the proposed law. President Fox objected to the more radical decentralization of CONAGUA, claiming that its proposed areas of authority were not constitutionally valid and overstepped responsibilities designated only to Hacienda (Mexico's powerful Commerce and Fiscal Ministry) (Crónica Legislativa, 2004). Ulises responded to the presidential veto by organizing a series of public consultations over the next several months with water user organizations, non-governmental organizations, agricultural producers, water service providers and government officials to reconsider the reforms. In December 2003 Ulises and the Senatorial Water Resources Commission called for a special congressional session to consider a new set of veto-proof modifications to the national water law, which passed with 469 votes. President Fox signed the newly revised 2004 LAN into law on 19 April 2004. Although the 2004 LAN required the decentralization of CONAGUA into 13 regional headquarters: 'The federal government later decided to maintain [CONAGUA] as a "deconcentrated" body within the SEMARNAT framework, leaving ambiguous many of the LAN's written aspects.'[9] The *Organismos de Cuenca* do not have financial and jurisdictional autonomy from Mexico City. Lending credence to the notion that the decentralization initiative has lost ground in Mexico, the 2004 modifications to the National Water Law should have been followed within 12 months by the adoption of a set of specific regulations to implement it. However, the regulations (*reglamentaria*) have been stalled in the halls of the Mexican Congress, and there appears to be little if any movement or impetus toward adopting them. Thus, today, the status of the 2004 Law is ambiguous and the level of high-level commitment to the advances of 2004 appears uncertain.

The hotly contested 2006 presidential race was between two parties of the traditional opposition, and was won by Felipe Calderón, the conservative PAN candidate.[10] Thus far in his administration Calderón has not had a clearly discernible environmental focus. Calderón's Environment Minister, Juan Rafael Elvira Quesada, is working on reforestation to prevent flooding and on wastewater treatment in Mexico City (Deputato, 2008). The former Subsecretary of Agriculture under Salinas, Luis Tellez, is now Minister for Communications and Transportation and has continued his focus on modernizing the water sector. One of the major initiatives focused on water is a five-year, 270 billion peso national strategy focused on public–private investment in modernizing national infrastructure, with water and environmental projects as two of the four priority areas identified (BBC, 2008). The 2007–12 National Water Plan relies heavily on the sustainability discourse while giving priority to efficiency and productivity measures (CONAGUA, 2008). The new National Water Plan reports a need to double the annual budget to address Mexico's water needs and emphasizes private investment as a targeted source for these resources. Calderón calls sustainable human development a central objective of his government and states that environmental sustainability is one of the Plan's 'guiding axes' (CONAGUA, 2008). The report emphasizes the role of river basin councils in facilitating public participation and promoting 'a new culture of water' (CONAGUA, 2008).

CONAGUA has proposed new changes to the LAN that would roll back some of the environmental gains of the 2004 version of the law. Due to complaints from heavy hitting water users in agriculture and industry, CONAGUA is promoting reductions in the minimum financial penalties that apply to those who use more than their designated water allocation or who discharge untreated sewage waters.[11]

5.4 Implications and conclusions

Despite the potential embodied in the National Water Law adopted in 1992 and modified in 2004, Mexico's transition to a new water culture has been only partial and very incomplete. While the transition to a market logic and efficiency principles has been well consolidated, the decentralization and sustainability initiatives of the 1992 law have become mired due to national political fragmentation and a lack of strong political will to allow the decentralization agenda to develop and take root. The original water law modifications were promoted and imposed in a top-down fashion by the PRI-controlled Salinas administration, while the post-2000 experience has been one of political party *alternancia* (alternating political parties) typified by conflict and lack of consensus over the ultimate direction of Mexico's water policy. Political party dominance is now in a continually shifting state both across and within time and space (Wilder, 2008b). As Meijerink and Huitema point out (in their theoretical introduction to this volume, Chapter 2), water policy transitions have a temporal aspect. In the beginning they require the 'alteration [and] . . . breaking open' of policy communities that have 'crystallized around a policy domain'. As these transitions mature, they must find ways to navigate evolving political structures and coalitions or risk capture and subversion by entrenched elites that may be reluctant to cede authority to emerging decentralized institutions. Policy change would ideally then be followed by policy stability, wherein the 'transition' itself must transform into the substantive new 'stable' policy. Mexico's democratic transition is launched, but by no means fully consolidated (as shown most recently in the fraud allegations after the 2006 presidential election that briefly challenged the legitimacy of Calderón's presidency). At 15, Mexico's water policy appears stuck in a transitional phase.

As a result of the multiple influences of exogenous and endogenous pressures, as well as the political and economic context of the opening of political parties and political spaces and, finally, its coincidence with the emergence of international prescriptions for sustainable resource management, Mexico's national water policy resulted in a kind of enormous soup cauldron into which all the requisite ingredients were mixed. The confluence of economic factors, the rise of the free trade agenda, the desire for efficient modernized institutions and the political aperture allowed for a sustainability and conservation agenda to be grafted onto Mexico's agenda of making water resources management more efficient, productive and cost-effective. Decentralization resulted in a transfer of the huge financial burden of urban water services and irrigation management to local municipalities and water users and an easing of the financial responsibility of the federal government as it retrenched.

Fifteen years after the original reforms were enacted there is a cacophony of influences, but a lack of coherent direction about what ultimate path national water policy will follow. Meaningful implementation of the decentralization agenda would potentially present challenges to the status quo – including possible changes in the distribution of water rights, allocations of available water, and fundamental changes in the federal–municipal relationship. It remains to be seen whether the Mexican government will move beyond the decentralization and sustainability experiment and commit to fulfil in the future what it has begun with some degree of promise. Evidence points to an ongoing battle between the different agendas embedded within the broad water reform package, with efficiency and economic development on one side sparring with equity and sustainability on the other.

The Mexican case holds important lessons that can inform the policy sciences approach, especially with regard to its relevance for developing countries. Perhaps most significant is the extent of profound transformation that Mexico was undergoing at the time it initiated its water policy transition – the transition in Mexico and many developing countries is about much more than water. As dramatic as the transition was, it represented only one broad policy area among many that were being 'modernized' as the result of national economic restructuring. This complexity lends a little intractability or unwieldiness when it comes to applying sleek policy models that seek to reflect how policy change occurred. Mexico's experience also points up the difficulties inherent in a changing political landscape where office holders, institutions and parties are in constant flux, and 'policy entrepreneurs' are at risk of being replaced frequently owing to changes in administration at municipal, state and federal levels. These shifts also represent a great deal of churn and change within the advocacy coalition of PRI and business leaders who led the initial transition. As the galvanizing effects on advocacy coalitions fade over time as the country moves beyond the initial focusing event, it can be difficult to sustain the policy momentum, especially for the most fundamental changes in control over water resources – such as relinquishing centralized control to an array of decentralized, empowered local institutions and civil society, actors, or changing water policy to achieve environmentally sustainable goals.

The Mexican case also points up some potential deficits in the policy sciences approach as set forth in the theoretical introduction of this volume (Chapter 2) as a construct for interpreting policy change in developing countries. First, the policy sciences framework gives prominence to the nation state as the primary locus of national policy change. In the case of Mexico and many other developing countries, structural requirements imposed by powerful international lending institutions and global shifts in thinking about water policy were both important factors in the transition to a new national water policy in Mexico, yet the current framework does not capture this reality very adeptly. Thus the policy sciences framework would benefit from a more nuanced and embroidered understanding of the role of international influences – both structural and ideational – in the construction of problems and solutions in the developing world.

In the same vein the policy sciences approach for the most part assumes that water policy transitions are occurring in well-consolidated democracies; in reality the political landscape is more uneven and chequered, even for the people within particular systems where not only the water transition, but the democratic transition is still being played out and is taking hold unevenly across and within both time and space. The Mexican case exemplifies the nature of water policy transition as a process that occurs over time and space rather than a policy change in a particular historical moment. The policy sciences approach also needs to grapple more forthrightly with water's particular characteristics as an 'uncooperative commodity' (Bakker, 2004). Water is a resource with a keen emotional resonance for individuals and communities, unlike that of any other natural resource (Blatter and Ingram, 2001; Gleick et al., 2002). A challenge awaits: to factor the particular attributes of water and the socio-cultural, socio-economic meanings of water more meaningfully into the policy sciences approach. These enhancements will lend an already agile interpretive framework more flexibility to respond to water transitions in the developing democracies of the world.

Acknowledgements
The author gratefully acknowledges the helpful comments provided by Sander Meijerink, Dave Huitema, Sara Hughes, Jennifer McKay and Helen Ingram on an earlier version of this chapter.

Notes

1. The concept of apertures may bear some prima facie resemblance to Kingdon's policy sciences concept of 'political windows' or 'windows of opportunity', political opportunities made possible by a convergence of streams of thinking about a particular issue. Such windows can be exploited by a savvy entrepreneur who is able to recognize the opportunity for transformative policy change (Meijerink and Huitema in the theoretical introduction to this volume, Chapter 2; Kingdon, 1995). Although both of these concepts speak to the key role of 'openings', the nature of the opening in each concept is particular and distinct.

   The idea of political aperture has the sense of a broad breaking apart of deeply rooted political bindings (such as the 71-year corporatist rule by the PRI party in the Mexican case), and the admitting of new actors to the electoral arena or policy-making table. In contrast, 'windows of opportunity' gives the sense of a defined, transitory and temporary moment of opportunity for a motivated change agent to exploit a change in problem definition and broaden the acceptable range of possibilities to address the problem – by introducing a scheme for water policy transition at just the right political or economic moment. Political apertures refer to a broader opening, more process-like in nature. They signify a more profound transition in thinking about a water (or other) policy issue and are less transitory in nature.
2. For example, one positive sign of a more open and participatory process has been the recent reinvigoration of *consejo de cuenca* processes within the transboundary Upper San Pedro River area communities near Naco and Cananea, Sonora and Sierra Vista/Fort Huachuca, Arizona. In this area where the Altar-Rio Concepción is the major river basin council, some sub-watershed level committees have been actively meeting and working, including a new binational *comisión de cuenca* that was instituted at CONAGUA's invitation.
3. Some background information in this section reviews material on Mexico's water policy reforms published in Wilder and Whiteford (2006) and Wilder and Romero (2006).
4. Data on Mexico's Trade agreements can be viewed on the website Secretaría de la Economía, at http://www.economia.gob.mx.
5. Cuauhtéhmoc Cardenas, son of the populist former President Lázaro Cardenas, made a very serious run for the presidency on the leftist Party of the Democratic Revolution (PRD) ticket in 1988, against Carlos Salinas de Gortari. Many believed he won the popular vote, but lost due to election night fraud engineered by the PRI at the direction of President de la Madrid (Preston and Dillon, 2004, pp. 149–50). He ran again in 1994, but was defeated by PRI candidate Ernesto Zedillo. He won the nation's second most visible elected position, that of Mayor of Mexico City, in the first election ever conducted for that post, holding it from 1997 to 2000.

   The late 1980s and early 1990s were marked by key electoral successes of opposition candidates in traditional PRI strongholds. The conservative PAN (Partido de Accion Nacional – National Action Party) won the municipal presidency of the capital of Chiapas in 1994 (Dresser, 1998, p. 229). Overall, from 1989 to 1995, the PAN gained governorships in three additional states and the mayorships of 189 additional municipalities. The PAN's presence increased by 125 seats in local congresses and by 18 seats at the national level, and the number of PANista senators increased 25-fold to 25 positions during the same period.
6. Faced with a Congress dominated in both chambers by PRI party stalwarts and viewed as an unpopular outsider by his own party, Fox was unable to garner support for key legislative proposals. His campaign promise to work with US President George W. Bush to achieve comprehensive immigration reform and better treatment of migrants came to a screeching halt just one week after 2 September 2001 when the two presidents stood on the steps of the White House shaking hands over the plan. The attacks on the World Trade Center and Pentagon the following Tuesday, 11 September (9/11) not only forced comprehensive immigration reform to Bush's back burner, but led to a tightened, less humane immigration policy that explicitly linked anti-terrorism and anti-immigrant policies (Nevins, 2007).

7. The liberal *Proceso* news magazine reported on the 'irregularities' inherited by Lichtinger of SEMARNAT and his counterpart in the enforcement agency, PROFEPA, which included such things as environmental impact statements awarded on a discretionary basis, distribution of water for political purposes, and preferential treatment given to companies responsible for polluting (Monge and Ortiz, 2001).
8. Ulises Ruíz is an unlikely champion of a more democratized and decentralized water policy. The PRI candidate for Governor of Oaxaca in 2004, he won after a hard-knuckled campaign, which he ran partly on a platform based on the 'new culture of water'. Ulises was Governor during the massive teacher demonstrations of the summer of 2006, when repressive military and law enforcement tactics were used in the central plaza to put down and disrupt public demonstrations.
9. *La Jornada*, 13 August 2007.
10. Felipe Calderón of the PAN ran against the leftist PRD (Party of the Democratic Revolution) candidate, Manuel Lopez Obrador. Although some had predicted a comeback of the PRI party – in the person of its candidate Roberto Madrazo – the PRI was placed third in what became a close contest between Calderón and Lopez Obrador. Lopez Obrador's vision, focused on social development and investment in social programmes, contrasted sharply with the Calderón platform of continued free market orientation and stimulation of foreign investment. Calderon was awarded the election, and allegations of fraud were not borne out by the National Elections Institute nor by international observers. Lopez Obrador, however, claimed victory and in the months just after Calderón was installed as President he tried to establish himself as the 'legitimate President'. But over time his followers and momentum drained away. Lopez Obrador's lasting impact, nevertheless, was to underscore the social concerns of the voting populace and to cause a heightened rhetoric about and focus on poverty and social development in the new Calderón administration.
11. *La Jornada*, 13 August 2007.

References

Bakker, K.J. (2004), *An Uncooperative Commodity: Privatizing Water in England and Wales*, Oxford: Oxford University Press.

Barkin, D. and D. Klooster (2006), 'Estrategias de la Gestión de Agua Urbana', in David Barkin (ed.), *La Gestión de Agua Urbana en México: Retos, Debates y Bienestar*, Guadalajara: Universidad de Guadalajara.

Bauer, C. (2004), *Siren Song: Chilean Water Law as a Model for International Reform*, Washington, DC: Resources for the Future.

Baumgartner, F.R. and B.D. Jones (eds) (1991), 'Agenda dynamics and policy subsystems', *Journal of Politics*, **53** (4), 1044–74.

BBC (2008), BBC Monitoring Latin America, 8 February, 'Mexican President Felipe Calderón launches national infrastructure plan'.

Bennett, V. (1995), *The Politics of Water: Urban Protests, Gender and Power in Monterrey, Mexico*, Pittsburgh, PA: University of Pittsburgh Press.

Blatter, J. and H.M. Ingram (eds) (2001), *Reflections on Water: New Approaches to Transboundary Conflicts and Cooperation*, Cambridge, MA: MIT Press.

Buras, N. (1996), 'The water resources of Mexico: their utilization and management', in Laura Randall (ed.), *The Changing Structure of Mexico*, Armonk, NY: M.E. Sharpe.

Carabias, J. and R. Landa (2005), *Agua, MedioAmbiente, y Sociedad: Hacia la Gestión Integral de los Recursos Hídricos en Méxicoi*, Mexico, DF: Universidad Autónoma de México, El Colegio de México, y Fundación Río Arronte.

Castro, J.E. (2006), *Water, Power and Citizenship: Social Struggle in the Basin of Mexico*, Oxford: Palgrave Macmillan.

CONAGUA (1994), *La Transferencia de los Distritos de Riego en Mexico*, Mexico, DF: Comisión Nacional del Agua.

CONAGUA (2008), *Plan Nacional Hídrico 2007–2012*, Mexico, DF: Comisión Nacional del Agua.

Conca, K. (2006), *Governing Water: Contentious Transnational Politics and Global Institution Building*, Cambridge, MA: MIT Press.

Cornelius, W.A., T.A. Eisenstadt, and J. Hindley (eds) (1999), *Subnational Politics and Democratization in Mexico*, La Jolla, CA: Center for US–Mexican Studies.

Crónica Legislativa (2004), 20 April, Mexico, DF.

Cummings, R.G., V. Brajeer, J. McFarland, J. Trava and M. El-Ashry (1989), *Waterworks: Improving Irrigation Management in Mexican Agriculture*, Washington, DC: World Resources Institute.

Deputato, G. (2008), 'Saving Mexico', Report, 1 April, CNN, http://www.cnn.com/2008/WORLD/americas/04/01/mexico.environment/index.html?iref=newssearch, accessed 24 April 2008.

Dresser, D. (1998), 'Post-NAFTA politics in Mexico: uneasy, uncertain, unpredictable', in Carol Wise (ed.),

The Post-NAFTA Political Economy: Mexico and the Western Hemisphere, University Park, PA: The Pennsylvania State University Press, pp. 221–56.

Easter, W., H. Plusquellec and A. Subramanian (1998), *Irrigation Improvement Strategy Review: A Review of Bankwide Experience Based on Selected 'New Style' Projects*, Washington, DC: World Bank.

Gates, M. (1988), 'Codifying marginality: the evolution of Mexican agricultural policy and its impact on the peasantry', *Journal of Latin American Studies*, **20**, 277–311.

Gleick, P.H., G. Wolff, E.L. Chalecki and R. Reyes (2002), *The New Economy of Water: The Risks and Benefits of Globalization and Privatization of Fresh Water*, Oakland, CA: Pacific Institute.

Gould, D.M. (1996), 'Mexico's crisis: looking back to assess the future', in Laura Randall (ed.), *Changing Structure of Mexico: Political, Economic and Social Prospects*, Armonk, NY: M.E. Sharpe, pp. 15–42.

Harvey, N. (1996), 'The reshaping of agrarian policy in Mexico', in Laura Randall (ed.), *Changing Structure of Mexico: Political, Economic and Social Prospects*, Armonk, NY: M.E. Sharpe, pp. 103–9.

Hinchcliffe, F., J. Thompson, J. Pretty, I. Gujit and P. Shah (1999), *Fertile Ground: The Impacts of Participatory Watershed Management*, London: Intermediate Technology Publications.

Hindley, J. (1999), 'Indigenous mobilization, development, and democratization in Guerrero: the Nahua people vs. the Tetelcingo Dam', in W.A. Cornelius, T.A. Eisenstadt and J. Hindley (eds), *Subnational Politics and Democratization in Mexico*, La Jolla, CA: Center for US–Mexican Studies, pp. 207–38.

IPCC (2007), *Climate Change (2007): The Physical Science Basis. Contribution of Working Group I to the Fourth Assessment Report of the Intergovernmental Panel on Climate Change*, Cambridge: Cambridge University Press.

Kingdon, J.W. (1995), *Agendas, Alternatives and Public Policies*, 2nd edn, New York: HarperCollins.

La Jornada (2007), *La Jornada* newspaper, 13 August.

Lemos, M.C. (2008), 'Whose water is it anyway? Water management, knowledge and equity in Northeast Brazil', in John M. Whiteley, Helen Ingram and Richard W. Perry (eds), *Water, Place, and Equity*, Cambridge, MA: MIT Press, pp. 249–70.

Lemos, M.C. and A. Luna (1999), 'BECC and public participation in the US–Mexico border: lessons from Ambos Nogales', *Journal of Borderlands Studies*, **14** (1), 43–64.

Liverman, D.M. and R. Merideth (2002), 'Climate and society in the US southwest: the context for a regional assessment', *Climate Research*, **21** (3), 199–218.

Luers, A.L., R.L. Naylor and P.A. Matson (2005), 'A case study of land reform and coastal land transformation in Southern Sonora, Mexico', *Land Use Policy*, **23** (4), 436–47.

Meijerink, S. (2005), 'Understanding policy stability and change: the interplay of advocacy coalitions and epistemic communities, windows of opportunity, and Dutch coastal flooding policy, 1945–2003', *Journal of European Public Policy*, **12** (6), 1060–77.

Monge, R. and S. Ortiz (2001), *Proceso*, 8 July.

Moreno, J.L. (2006), *Por Abajo del Agua: Sobreexplotación y agotamiento del Acuífero de la Costa de Hermosillo, 1945–2005*, Hermosillo: El Colegio de Sonora.

Nadal, A. (2002), 'Corn in NAFTA: eight years after', a research report prepared for the North American Commission on Environmental Cooperation, Montreal: North America Commission on Environmental Cooperation, May, available at: http://www.cec.org/files/pdf/ECONOMY/corn_NAFTA_en.pdf.

Nevins, J. (2007), 'Dying for a cup of coffee? Migrant deaths in the US–Mexico border region in a neoliberal age', *Geopolitics*, **12** (2), 228–47.

Otero, G. (ed.) (2004), *Mexico in Transition: Neoliberal Globalism, the State and Civil Society*, Nova Scotia: Fernwood Publishing.

Palacios Velez, E. (1997), 'Benefits and second generation problems: the case of Mexico', paper presented at the International Workshop on Participatory Irrigation Management: Second Generation Problems, Cali, Colombia, 9–15 February.

Palacios Velez, E. (2000), 'Breve Evaluación del Proceso de la Transferencia de los Distritos de Riego en Mexico', in Enrique Palacios Velez and E. Espinosa de Leon (eds), *Procesos de Transferencia y Gestión, Mesa 1*, Mexico, DF: Comisión Nacional del Agua, vol. 1, pp. 121–30.

Perez Prado, L.N. (2003), 'Mexico's irrigation management transfer program: notes for a policy research agenda', *Journal of Environment and Development*, **12** (4), 373–88.

Pineda Pablos, N. (1999), 'Urban water policy in Mexico: municipalization and privatization of water services', Doctoral dissertation, Department of Latin American Studies, University of Texas, Austin.

Pineda Pablos, N. (2004), *Hermosillo y su agua*, Hermosillo: El Colegio de Sonora.

Pineda Pablos, N. (ed.) (2006), *La Búsqueda de la Tarifa Justa: El Cobro de los Servicios de Agua Potable y Alcantarillado en Mexico*, Hermosillo: El Colegio de Sonora.

Pineda Pablos, N. and E.R. Camou (2007), *De las buenas intenciones a las cuentas claras: Planeación, desempeño, y rendición de cuentas en seis municipios de Sonora*, Hermosillo: El Colegio de Sonora.

Pineda Pablos, N., A. Browning-Aiken and M. Wilder (2007), 'Equilibrio de bajo nivel y manejo urbano del agua en Cananea, Sonora', *Frontera Norte*, **19** (37), 143–72.

Preston, J. and S. Dillon (2004), *Opening Mexico: The Making of a Democracy*, New York: Farrar, Straus & Giroux.

Ray, A.J., G.M. Garfin, M. Wilder, M. Vásquez-León, M. Lenart and A.C. Comrie (2007), 'Applications of monsoon research: opportunities to inform decision making and reduce regional vulnerability', *Journal of Climate*, **20** (9), 1608–27.

Sabatier, P.A. (1993), 'Policy change over a decade or more', in P.A. Sabatier and H.C. Jenkins-Smith (eds), *Policy Change and Learning: An Advocacy Coalition Approach*, Boulder, CO: Westview Press, pp. 13–40.

Seagar, R., M. Ting, I. Held, Y. Kushnir, J. Lu, G. Vecchi, H. Huang, N. Harnik, A. Leetmaa, N. Chau, L. Cuihua, J. Velez and N. Naik (2007), 'Model projections of an imminent transition to a more arid climate in southwestern North America', *Science*, **316** (5828), 1181–4.

Tellez, L. (ed.) (1993), *Nueva Legislación de Tierras, Bosques y Aguas*, Mexico, DF: Fondo de Cultura Económica.

Torregrosa, M.L. (1995), 'Aspectos sociales de los procesos de modernización y transferencia de los distritos de riego: un contexto para el planteamienta de un problema', unpublished paper presented at the Jornada sobre el Agua, Hermosillo, Sonora, Mexico.

Valdes-Ugalde, F. (1996), 'The changing relationship between the state and the economy in Mexico', in Laura Randall (ed.), *The Changing Structure of Mexico*, Armonk, NY: M.E. Sharpe, pp. 55–62.

Varady, R.G., K. Meehan, J. Rodda, E. McGovern and M. Iles-Shih (2008), 'Strengthening global water initiatives', *Environment*, March/April, 20–41.

Vera, Rodrigo (2003), 'Lichtinger: Ajuste de Cuentas', *Proceso*, 8 September.

Whiteford, S. and R. Melville (eds) (2002), *Protecting a Sacred Gift: Water and Social Change in Mexico*, La Jolla, CA: Center for US–Mexican Studies, University of California Press.

Wilder, M. (2008a), 'Equity and water in Mexico's changing institutional landscape', in J.M. Whiteley, H. Ingram and R. Warren Perry (eds), *Water, Place, and Equity*, Cambridge, MA: MIT Press, pp. 95–116.

Wilder, M. (2008b), 'Promises under construction: the evolving paradigm for water governance and the case of northern Mexico', paper presented at the Sixth Rosenberg Forum for International Water Policy, Zaragoza, Spain, 23–29 June.

Wilder, M. and P. Romero Lankao (2006), 'Paradoxes of decentralization: neoliberal reforms and water institutions in Mexico', *World Development*, **34** (11), 1977–95.

Wilder, M. and S. Whiteford (2006), 'Flowing uphill towards money: groundwater and free trade in Northern Mexico's Ejido sector', in Laura Randall (ed.), *The Changing Structure of Mexico: Political, Economic and Social Prospects*, Armonk, NY: M.E. Sharpe, pp. 341–58.

Yates, P.L. (1981), *Mexico's Agricultural Dilemma*, Tucson, AZ: University of Arizona Press.

PART III

AUSTRALASIA

6 Implementing integrated river basin management in China

Dorri te Boekhorst, Toine Smits, Yu Xiubo, Li Lifeng, Lei Gang and Zhang Chen

6.1 Introduction

Water-related problems are one of the most pressing issues facing China. They are argued to have the potential ultimately to affect China's social, economic and political stability. In terms of the sustainable development of the country, they could be an important limiting factor (Kreimer and Munasinghe, 1991; World Bank, 1997; Flavin and Gardner, 2006; Turner and Otsuka, 2006). The main challenges concerning water are threefold: water scarcity, water pollution, and flood control (WWF, 2003; Chen, 2005; Lee, 2006; Yin et al., 2006). To meet these challenges, China's river ecosystems have to be properly protected. This will involve the development of an alternative approach to water management and major changes in different areas of society from land use and government to livelihoods.

The new approach to water management is envisioned in the strategy of integrated river basin management (IRBM), which was incorporated into China's 2002 Water Law. In 2004 a major joint report by the World Wide Fund for Nature (WWF) and a task force of the China Council for International Cooperation on Environment and Development (CCICED) stressed to the Chinese government the importance of IRBM (CCICED and WWF, 2004). As a non-governmental organization (NGO), WWF China used its national and international network to stimulate the acceptance and application of IRBM. In this chapter we focus on the role of WWF China as a policy entrepreneur.

First we present a general picture of the institutional set-up and legislation related to water management in China. We then illustrate the strategy and activities of WWF China with two cases: the Central Yangtze: Partnership for a Living River[1] and the WWF–HSBC Yangtze Programme. Finally, applying the concepts described in the first part of this book, we analyse the role of WWF China as a policy entrepreneur.

WWF has been active in China since 1978 and has been able to develop successfully in Chinese society. The cases we present here productively link small-scale bottom-up processes with national top-down measures. The role of the CCICED IRBM Task Force as a high-level governmental advisory body is in part described below using participatory observation by one of the Dutch authors of this chapter. All the Chinese authors are familiar with the cases presented here, either as former project leaders, or through advising on the process of integrating IRBM in China.

6.2 Institutional framework

The State Council of the People's Republic of China directs several ministries, agencies and commissions involved in water management (see also Figure 6.1). The Ministry of Water Resources (MWR) manages all administration concerning water quantity in

Premier Vice-Premiers	**STATE COUNCIL**	State Councillors Secretary General Ministers Auditor-General

MINISTRIES AND COMMISSIONS	ORGANIZATIONS
Ministry of Foreign Affairs Ministry of National Defence	**State Forestry Administration** **Commission of the State-owned Assets Supervision and Administration****
State Reform and Development Commission Ministry of Finance People's Bank of China	
Ministry of Science and Technology Ministry of Education Ministry of Labour and Social Security Ministry of Personnel Ministry of Land and Resources	
Ministry of Railways **Ministry of Communications** **Ministry of Construction** **Ministry of Agriculture** **Ministry of Water Resources** **Ministry of Environmental Protection*** Ministry of Commerce Ministry of Information Industry State Commission of Science, Techn. and Industry for National Defence	

Notes:
Bold font indicates offices involved in water management. Other organizations, administrative offices and institutions under the State Council are not shown.
Top box: state administrative affairs departments; second box; macro control departments; third box: education, science and technology, culture, social security and resource administration departments; bottom box: specialized administration departments.
Box structure is based on the OECD report 'Environmental Performance China' (OECD, 2007).
* Former State Environmental Protection Agency (SEPA).
** Special organization under the State Council.

Figure 6.1 Simplified administrative structure relevant to Chinese water management

China. It is authorized by the Water Law to oversee China's 'water resource management'. Because water and flood control especially have been very important in Chinese history, the ministry is old and powerful. Its main responsibilities include surface and groundwater management, river basin management, flood control, and water and soil conservation. Mediation and arbitration of intersector or interprovince water disputes are also part of the MWR responsibilities. The ministry has water bureaus and departments at both the provincial and county levels (Turner and Otsuka, 2006; MWR official website).

The main laws and regulations providing the MWR mandate include the Water Law (1988, revised 2002), the key legal instrument containing general principles for the management of water resources in China (Chen, 2005; Lee, 2006; OECD, 2007). The Water

Law (revised 2002) reflects current thinking on integrated water resource and demand management. It enshrines the principles that everyone should have access to safe water and that water conservation and environmental protection are governmental priorities. The law focuses on four topics:

1. Water allocation, rights and permits.
2. River basin management.
3. Water use efficiency.
4. Conservation and environmental protection.

It defines river basin management institutions and functions. One of its core goals is to strengthen the administrative rights of river basin management organizations in order to improve the implementation of water conservation and management measures. Furthermore, the law requires the integration of water resource and economic development planning (Turner and Otsuka, 2006; OECD, 2007). In addition to the Water Law, the Law on Prevention and Control of Water Pollution (WPPC), the Water and Soil Conservation Law and other relevant laws and regulations, such as the 1997 Flood Control Law, all support China's institutional framework for water policy.

The MWR is responsible for several other institutes and commissions implementing regulations regarding water resources management, of which the seven River Basin Commissions (RBCs) are the most important (Lee, 2006; Yan et al., 2006; Ministry of Water Resources, MWR, official website). The RBCs were originally created in the 1950s to mitigate flood damage and exploit water resources, for instance to generate electricity and provide navigation facilities. From a historical point of view, the RBCs are extensions of the MWR and accustomed to taking a top-down, sectoral approach to river basin management (Turner and Otsuka, 2005, 2006). This is slowly changing and expected to change further while IRBM is being implemented in China.

In addition to the MWR, the Ministry of Environmental Protection (MEP)[2] is another important ministry involved in monitoring and improving water quality and ecological rehabilitation of the fluvial ecosystem.

6.3 Geohydrological data for the Yangtze River basin

Many river basins in China suffer from water shortage, flooding and pollution. The geohydrological background information given here is limited to the Yangtze River basin as the basin forms the target area for WWF activities.

The Yangtze (Changjiang) River, at 6300 km, is China's longest and the world's third-longest river. The snow-fed waterway starts in the Kunlun mountains of the Tibetan Plateau in Qinghai province, flows southeast and empties into the East Chinese Sea about 20 km north of Shanghai. It has about 700 tributaries and is the largest river system in China with a basin covering an area close to 19 per cent of China's total territory. It provides freshwater resources to around 400 million people (about a third of the Chinese population). The total water availability per year in the basin amounts to 996 billion m^3 on a long-term mean basis, accounting for 36.5 per cent of China's total. Yangtze basin subcatchments larger than 1000 km^2 number 400, eight of these exceeding 80 000 km^2. The area covered by the basin is rich in lakes, totalling a lake area of more than 15 200 km^2, with most of the lakes situated in the middle and lower reaches. The

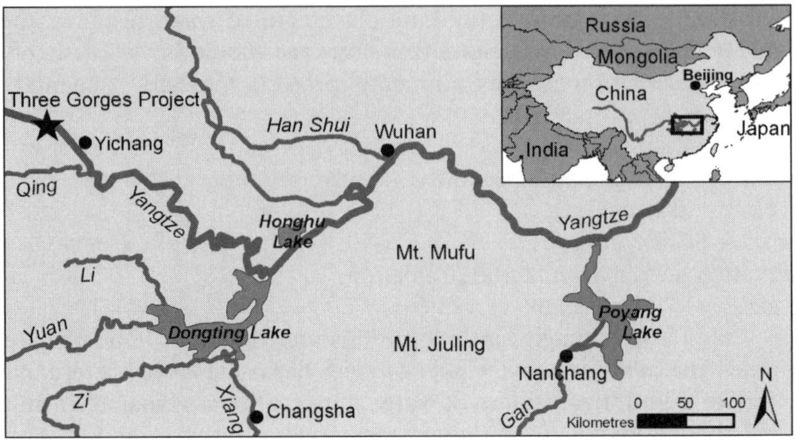

Map 6.1 China and the Yangtze River basin, including Poyang Lake and Dongting Lake

area includes the two largest freshwater lakes in China, Poyang Lake and Dongting Lake (Map 6.1) (Zhang et al., 1998; CCICED and WWF, 2004; Yang et al., 2007).

Located in the centre of China, the Yangtze basin covers 19 provinces, autonomous regions and municipalities. With its many tributaries, the river functions as the most important transportation network through the heart of some of the most densely populated and economically important areas in China, known as the Yangtze River Economic Belt. The Yangtze is also the major inland waterway of navigation in China. Its navigation channel on both mainstream and tributaries adds up to 57 000 km, 52.5 per cent of the nation's total. The 2837 km mainstream navigation channel has a shipping capacity equivalent to four to six railways, each of this same length (Changjiang (Yangtze) River Water Resources Commission, CWRC official website). The Yangtze is a major resource for irrigation, agriculture and hydroelectric power (with a technically exploitable potential of 256 270 MW and an estimated annual power output of over 100 TWh, equivalent to 48 and 49 per cent respectively of the nation's totals). It currently holds 17 hydroelectric dams, including the well-known Three Gorges Dam. The Yangtze agricultural area delivers close to half the country's total crop harvest and contributes about a third of the national grain and gross domestic product (GDP) total. The river basin is recognized internationally as an important ecosystem rich in biodiversity (Yin et al., 2006). Currently six wetlands in the Yangtze River basin are listed as sites under the Ramsar Convention on the Wetlands of International Importance (Ramsar) (Cui and Wang, 2008). Two-thirds of China's plant species occur in the basin, and 370 fish species (several seriously threatened) and a diverse range of birds depend on the river basin system (Zhang et al., 1998; CWRC official website; WWF China official website; WWF International official website; Yangtze Forum official website).

The Yangtze River basin falls under the jurisdiction of the CWRC, one of the seven river basin commissions under the MWR. Together with the prefectural and local branches of the MEP, they are responsible for the water quality and quantity of the river basin. Other ministries under the State Council also have responsibilities related to management of the water system. Regional authorities (provinces, cities), public and private organizations (for example conservation groups and construction firms) can be considered as 'stakeholders' of the river basin given their various claims on land and water use and their views on management.

6.4 Problem definition

Unconstrained deforestation activities in the upstream parts of the river basin (Nakamura, 2003) and gradual, but ongoing, land reclamation activities in mid- and downstream sections (Zhao et al., 2003; Yin et al., 2006) have undermined the hydrological and ecological resilience of the river basin. Originally Poyang Lake and Dongting Lake functioned as giant reservoirs that moderated extremes of high and low water discharges. However during the past decades, especially between the 1930s and 1978, many of the natural lakes and wetlands have disappeared due to land reclamation. Moreover hundreds of natural lakes, essential for spawning fish, feeding and natural retention have been disconnected from the Yangtze River for agricultural purposes and to protect farmers and fishermen against waterborne diseases such as snail disease (schistosomiasis) (Fischer and Heilig, 1997; Li et al., 2000; Utzinger et al., 2005). This has led to a dramatic decrease in lake size and flood retention areas (an estimated drop of 6000 km^2 in surface area of lakes ≥ 1 km^2 in Jianghan Plain alone) (Yin and Li, 2001; Zhao and Fang, 2004; Yin et al., 2006).

The relative abundance of water in the Yangtze River basin and the structural water scarcity problems in the Yellow River basin were the basis for the current South-to-North Water Transfer Project, a far-reaching and enormous engineering project to divert water to the north through three planned routes, east, middle and west (Varis and Vakkilainen, 2001; Turner and Otsuka, 2005).

Apart from water quantity-related developments, heavy pollution of the river system from industrial waste and mainly agricultural non-point sources provides the second serious threat to the river basin system (Wang et al., 2006). In general, water quality in China has deteriorated tremendously, reducing the access of about a third of the Chinese population to safe and clean drinking water and negatively affecting the natural habitat of many species (Varis and Vakkilainen, 2001; Fang et al., 2006).

6.5 The shift to integrated river basin management

It can be argued that many of the water problems in China result from a structural denial of the importance of a healthy ecosystem for a sustainable economy and from a sectoral approach to river basin management. Basically the Five Year Plans (FYPs) were focused on the economic development of the country. With the start of the first one in 1953, economic growth was a top priority and until the sixth FYP there was little attention in these plans to environmental values. In addition, coordination was poor or non-existent among local and regional authorities within the river basin. Most administrators and politicians, and most of the general public, considered the exploitation of the river and its catchment area as an undisputable right. Overarching problems were addressed by the MWR, but mainly from an engineering point of view. The emphasis on engineering solutions, with a 'primary focus on supply-side options that can provide water with a high assurance of supply' (Turton et al., 2007) and with little respect for environmental consequences, is generally referred to as the 'hydraulic mission' (Waterbury, 1979; Turton et al., 2007; 2004; see also Gupta, Chapter 3 in this volume) of a society. This has formed the dominant paradigm in Chinese water management as illustrated by the many vast hydro-engineering projects, including the present South-to-North Water Transfer Project.

However, recent years have seen an increased incidence of floods (major floods occurring every one and a half years between 1988 and 1998, compared to an interval of four to five years between 1852 and 1980), topped by the flood of 1998, and alarming environmental degradation entailing the loss of biodiversity and extinction of some endemic species (CCICED and WWF, 2004; Zhao et al., 2005). These events, on top of existing problems, formed the motivation for the central government to change the situation (Zhao et al., 2006). The launching of the US$1.2 billion '32-character' policy in 1998 to be implemented by local government in the Yangtze River Basin could be interpreted as a first step towards IRBM. The '32-character' policy was launched by the central government after the flood events of 1998 with the aim of stopping further deterioration of the hydrological resilience of the Chinese river basins. The name refers to the number of Chinese characters needed to formulate the various policy measures:

1. Stop logging and promote forestation (the National Logging Ban).
2. Return reclaimed land to the river (make room for the river).
3. Convert cultivated land to wetland and lakes (wetland restoration).
4. Resettle people who live in flood-prone areas.
5. Reinforce the primary flood defence structures.
6. Dredge river sections that have silted up.
7. Give people work instead of relief subsidies for agricultural use of mountain slopes and reintroduce forest and grass (grain for green).

Although the 32-character guiding policy was an important step in recognizing the key functions of nature in the sustainable use of a river basin, implementation has not been easy. Soon after the proclamation of the policy, its shortcomings became clear: it does not properly address sustainable alternatives to 'wrong land use within the river basin' (for example land reclamation for growing rice, extensive logging); and the many complex institutional arrangements between water resource management and land use plans have proved a real hindrance (Wang et al., 2007). Moreover there was little or no awareness among the wider public of how a river (ecosystem) functions and of the urgent need for coordination of up- and downstream land use. The popularity of the concept of IRBM at different policy levels has increased as public and administrative understanding of the problems in China has increased (personal communications – WWF China; GWP China official website).

6.6 WWF China as a policy entrepreneur for integrated river basin management

6.6.1 Introduction

Why is WWF so concerned with the fate of rivers? River basins are the way nature gathers and delivers water. They constitute ecosystems that provide services to humans in the form of hydropower, transport, recreation and tourism, as well as flood and drought regulation, sediment and nutrient retention, and habitat for diverse fauna and flora. Freshwater biodiversity is an important source of food, income and livelihood, particularly for rural communities in developing countries. Studies have estimated the economic value of 'healthy' river basins in the billions of dollars (Schuyt and Brander, 2004; Schuyt, 2005). Therefore one of the primary goals of WWF is to develop strategies to manage the use of rivers and their wetlands without reducing biodiversity and ecological functioning (WWF, 2003).

To use and manage river basins in a sustainable way is complex, but in the experience of the WWF (and others), no strategy will be effective unless it is implemented through cooperation across social, economic and political boundaries (Muldavin, 2000; WWF, 2003; Nakamura, 2006). In this context IRBM can be seen as a tool for communities to balance development and conservation needs. Hence the WWF's actions to support development and implementation of IRBM in China form part of such cooperation.

Against this background WWF China launched two major programmes: The Central Yangtze: Partnership for a Living River, 1998–2008[3] and the WWF–HSBC Yangtze Programme in 2002. Both programmes supported the goal to: 'restore the Yangtze as a living river by securing effective management of a significant area of wetlands in the central Yangtze basin over a period of 25 years' (WWF, 2003). Realization of this goal requires cross-sectoral cooperation throughout the river basin, public awareness and participation. IRBM was seen as an important tool in meeting this need and achieving the programme aims (CCICED and WWF, 2004). Here we describe two cases that illustrate the introduction of IRBM in China.

6.6.2 The Central Yangtze: partnership for a living river

This WWF programme, completed in 2008, had the following objectives:

1. To restore Dongting Lake wetlands and to enhance the management of the Yangtze as a 'living river' through partnership with all stakeholders.
2. To restore biodiversity in the Central and Lower Yangtze River and develop new land use and flood plain management approaches with authorities and other organizations.

In order to realize these goals, pilot sites were selected to demonstrate that working through a bottom-up, co-production model, economic, ecological and safety interests can be brought into harmony. Also, actions were taken to promote IRBM at the highest political level in order to facilitate the creation of a legal framework for the organization of round-table events for the stakeholders in the Central and Lower Yangtze regions (Hunan province official website; personal communications – WWF). Both subprojects and the actions of the CCICED are discussed below.

Pilot sites: Dongting Lake Dongting Lake, China's second-largest freshwater lake, is located in the southern part of the Central Yangtze River basin, in Hunan province. The lake area totals 15 465 km^2 and spans 13 counties. The lake is one of the two remaining retention lakes naturally connected to the Yangtze River (Poyang Lake is the second). It has an important function as a natural reservoir, storing freshwater and reducing the extent and impact of flooding in areas occupied by people. Rich in biodiversity, including important fish and water bird populations and many threatened species (Fang et al., 2006), the lake is of great socio-economic and cultural importance. The Yangtze River and the lakes of its basin, including Poyang Lake and Dongting Lake, are listed by WWF as one of the 200 global conservation priority regions (Olson and Dinerstein, 1998).

Due to land reclamation, the size of the lake has been drastically reduced over the last few decades (Map 6.2). The maximal lake area is set at the 1825 figure of 6270 km^2. By the 1950s, this area had fallen to 4350 km^2, while the following 50 years saw yet another 1650 km^2 reclaimed, shrinking the lake by an additional 40 per cent. Besides reduction, the lake also faces fragmentation. The number of patches of reclaimed land, known as polders, increased by 25 per cent between the 1930s and 1998 (Zhao et al., 2005). To create and sustain the resulting 'polder society' requires technical engineering in the form of dams, dykes, canals, and sluices. Such development weakens the lake's ability to store and safely release floodwaters. The natural siltation of the area adds to this problem, but is likely to reduce, since the Three Gorges Dam will trap a large portion of the sediment (CCICED and WWF, 2004; Schuyt, 2005; Zhao et al., 2005).

Dongting Lake is divided into three sections: east (Dong), west (Xi) and south (Nan). East Dongting was one of the first seven sites to be designated in 1992 as the most important international wetland conservation areas in China under Ramsar (Ramsar Convention on Wetlands (Secretariat) official website). In 2002, as a result of the WWF project, the State Forest Administration (SFA) added south Dongting Wetland and Waterfowl Reserve and West Dongting Lake to the Ramsar list. The WWF project was embarked on its specific aim to restore Lake Dongting and its wetlands to their 1950s maximal extent (4350 km^2) in ten years; this in order to reverse a massive loss of biodiversity and to improve ecological conditions and security for the conservation of the natural heritage, ecosystem and species of the Yangtze. (This proved to be too ambitious and was adjusted in 1992 to the more general goal of 'restoring Dongting Lake wetlands'.) This goal was combined with the aim of identifying alternative livelihoods for the local people that achieve a better balance with the natural environment (for example ecotourism). Improved security was the subject of the second specific objective for the lake, setting out wetland restoration as a means to mitigate catastrophic peak floods that destroy economic, agricultural and infrastructure assets.

WWF China selected the recently reclaimed polder Xipanshanzou, a village in Yuanjiang County, Hunan province, in the Dongting Lake area, to start realizing the above goals. After the 1998 flood the national government decided to give up the Xipanshanzhou polder, which opened the way for WWF China to start a joint transition process with the local people to alternative land use.

Starting in 1999, 147 households were selected to participate in the project. First, WWF China organized meetings for the local people, explaining how the natural lake systems function and how human intervention during the previous decades had led to the deterioration of the ecological functioning and increased flood risks. After these 'awareness sessions', WWF led a joint exploration of alternative livelihoods more in alignment with the natural dynamics of such a huge lake as Dongting.

Eventually a series of workshops produced the main outcome of support by the WWF project for several alternative income-generating activities, including animal husbandry (pig, duck, cow, hen and goat), fish cage farming and organic horticulture (but moving away from rice growing). Biogas as an alternative source of energy was also introduced, and a number of households living in the polder were resettled to higher ground.

The approach created public support for the new way of life, but one result that particularly boosted motivation was an increase of more than 100 per cent of the participants' annual income. In 2000 income attributable to the project was 1465 RMB/household; in 2003, it had reached 2881 RMB/household and in 2004, the figure was 3196 RMB/household. A WWF survey (Schuyt, 2005) demonstrated that households participating in the project had a higher income level than non-participants. The average household income of polder residents totalled 9360 RMB/household in 2003 (up from 2000 RMB/household in 1999). In addition to a rise in income, the project also generated important living standard improvements.

Improved living standards produced a ripple effect: without any intervention by the Dongting Lake project, neighbouring families replicated activities like cooking on biogas stoves and pig farming. The project demonstrated active participation by and benefits for women and the elderly. Another major outcome came in the form of raised awareness of the local community (Schuyt, 2005). Overall the livelihoods approach used by the WWF project at Dongting Lake produced lasting improvements seen beyond the end of funding in 2001.

Various partnerships were built during the project for livelihood development. At the local level, WWF China supported the creation of a community-based organization, in which 110 families participated, and an Organic Farming Association (OFA) managed by farmers themselves. Organic agriculture companies (amongst others, the Hunan Sun and Soil Organic Agriculture Company and the Naniwan Agriculture Company) also became involved during the exploration and building of a solid market for the products of organic farming and horticulture. This public–private cooperation provided support and marketing advice to farmers and the OFA. The MEP's Organic Development Center certified some of the organic products, thereby giving them official endorsement.

Apart from joint efforts with WWF China in the Dongting Lake project to search for and develop alternative livelihoods, actions were also carried out to restore the ecological value of the Xipanshanzhou polder. Success in identifying sufficient alternative livelihoods led to relocation of the residents of the polder and wetland restoration of its entire acreage, a total embanked area of 110 hectares. At present the former polder serves also as a flood retention area during high peak flows on the Yangtze when it can be inundated to reduce the floodwater level (WWF China official website). Preliminary biodiversity monitoring reports from WWF China showed that vegetation species diversity and quantity both increased, bringing an increase also in amphibian, fish and bird species (WWF China official website). As a result of the wetland restoration, major environmental and heavy metal indices for water (including dissolved oxygen, nitric salt, ammonia-salt, coliform, microbe population and fluorine) show lower levels than those set in health standards for water used to irrigate food crops (Schuyt, 2005).

In summary the WWF Dongting Lake project demonstrates that a bottom-up approach that uses the expertise of local farmers makes it possible to identify and adopt alternative livelihoods better suited to the natural environment and conducive to its restoration. Both results derive from important aspects of IRBM.

The role of the IRBM task force of the China Council for International Cooperation on Environment and Development Active involvement of the other major water resource management institutions became more pressing during the Dongting demonstration project. Possibly one of the most difficult and important steps was to bring the CWRC

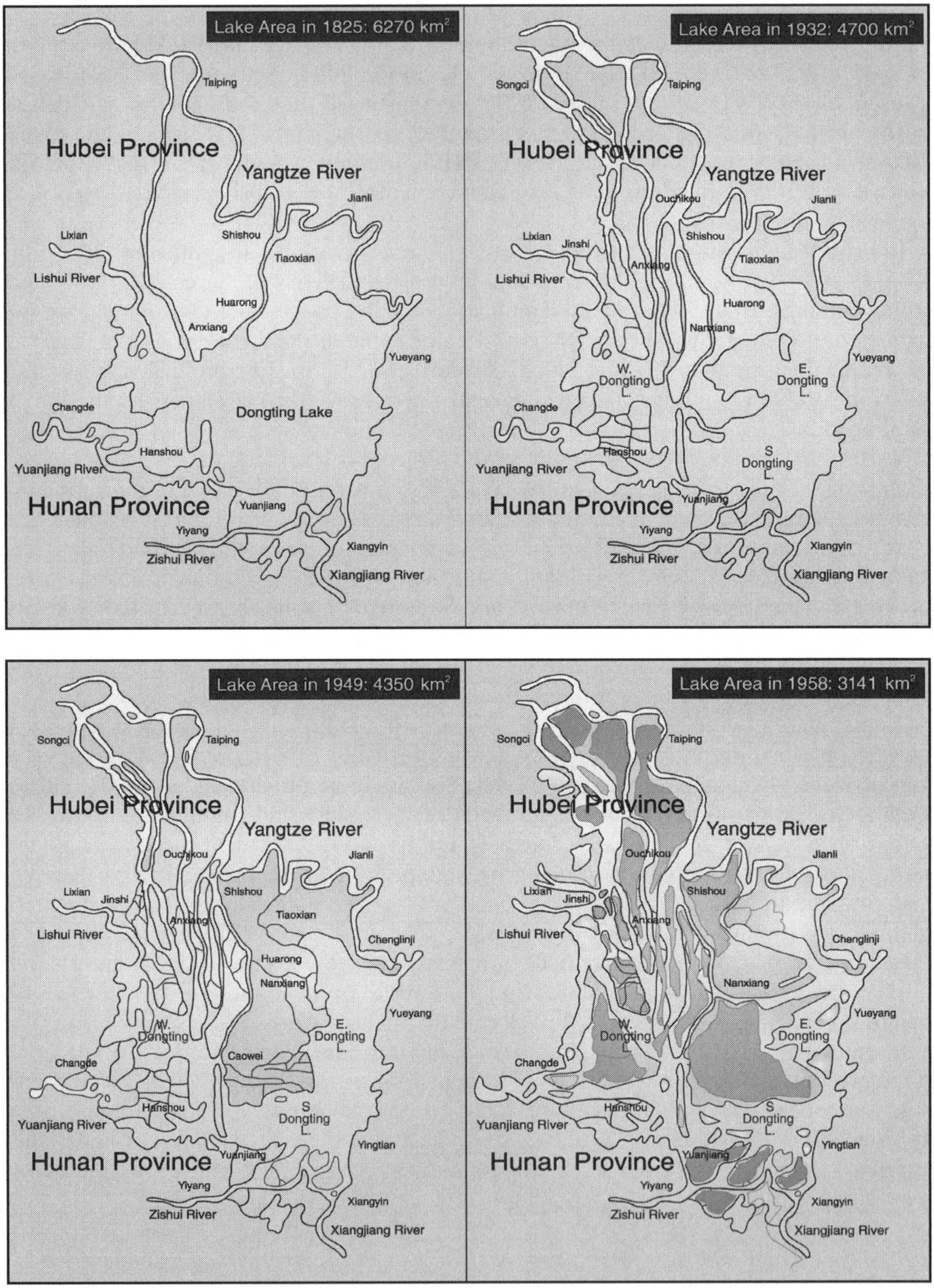

Map 6.2 Shrinkage of the Dongting Lake area due to land reclamation projects

and the MWR into the process. A perfect opportunity occurred with the decision by the CCICED to devote a special task force to IRBM implementation in China. The Chinese government had established the CCICED in 1992 as a high-level non-governmental advisory organization with the aim 'to further strengthen cooperation and exchange between China and the international community in the field of environment and development' (CCICED official website).

For the past several years, the CCICED has been successful in articulating high-level advice and assisting Chinese decision-makers to increase their understanding of the relation between environmental protection and economic development in the river basin context.

The CCICED launched the IRBM Task Force in 2002 to advise and inform the national government on implementation of IRBM in China. At that time the CCICED's WWF representative had succeeded in gaining a chair position. This made it possible to promote two candidates who had a clear view of IRBM and the importance of a healthy river ecosystem to co-chair the IRBM Task Force. The viewpoints of the co-chairs are important because they set the agenda for the task force and thus pave the way for recommendations to the Chinese government. WWF China was also pleased to co-host the task force secretariat as a way to further ground its position.

The co-chairs of the IRBM Task Force delivered their recommendations to the CCICED Annual General Meeting in November 2004. The recommendations went on to the relevant ministries (in particular the MWR) for implementation. In 2006 the CWRC began revisions (expected to take three years) to the Yangtze Basin Comprehensive Utilization and Development Plan to make it more sustainable and integrated. Although this was not a direct result of CCICED recommendations, it shows how a new way of thinking has taken hold to some degree in the Chinese government. Even more important was acceptance of the recommendation to establish a special Yangtze Forum, which would gather all key stakeholders of the river basin (Yangtze Forum official website).

The Yangtze Forum has been established as a precursor to an Integrated Yangtze River Basin Commission and is responsible for developing a vision for the river's management and for coordination of the activities of national and provincial government. Its membership established a multidisciplinary team including representatives from all the major water management institutions, NGOs and scientists.

It is important to note that the work of the IRBM Task Force of the CCICED directly facilitated the local interactions between CWRC and WWF China in the Dongting Lake case. In this way WWF China was able to demonstrate that there are good alternatives to the mainstream water management approaches used previously. It also increased the credibility and trustworthiness of WWF China as a partner. The success of the Dongting Lake Project has received a lot of public attention, involving all kinds of media, site visits, an international conference, publications and so on, creating visibility and impact. Its positive results and reception opened the door for other projects.

6.6.3 The WWF–HSBC Yangtze programme: 'restoring the web of life'

The WWF–HSBC Yangtze Programme is a major freshwater initiative to restore the 'web of life' along the Yangtze River by reconnecting lakes to the river's main flow. The reconnection programme, part of the five-year eco partnership between the WWF and

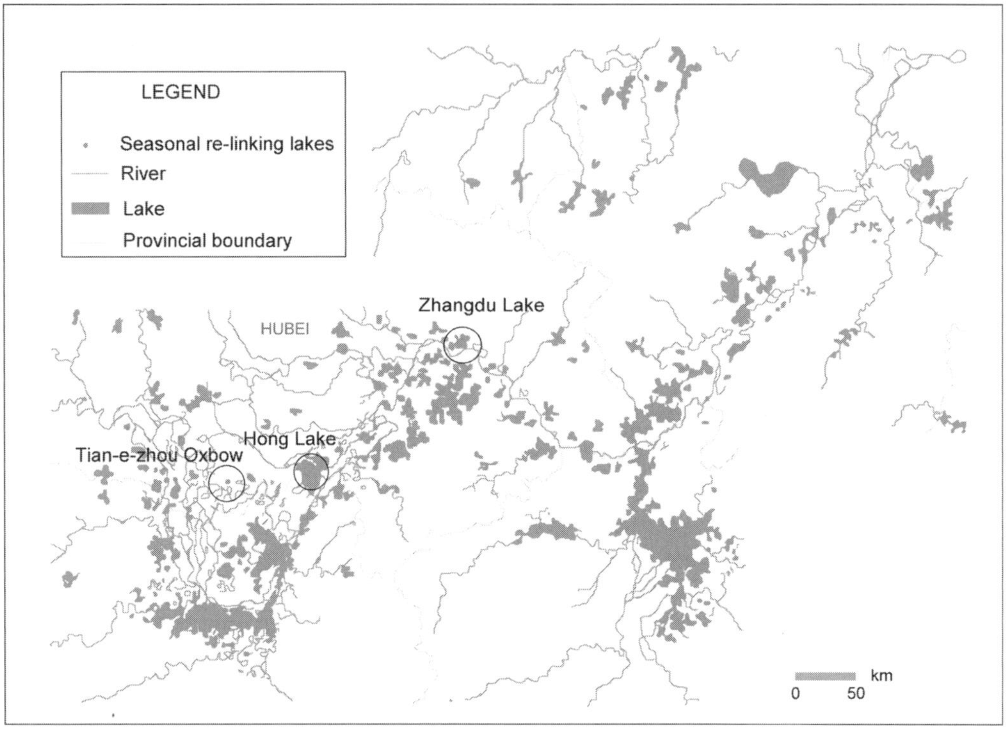

Map 6.3 Yangtze River and Zhangdu Lake, Hong Lake and Tian-e-Zhou Oxbow

HSBC (initials of the Hongkong Shanghai Banking Corporation, one of the original banks merged and rebranded), started in 2002. The goal was set to rehabilitate the food chain and ecological networks in the Central Yangtze River basin through policy work and demonstration projects that undertake reconnection, support local economic development, and create a network of wetland reserves. The Zhangdu Lake, Hong Lake, and the Tian-e-Zhou Oxbow in Hubei were selected as demonstration sites (Map 6.3).

More than 100 lakes once dotted the Yangtze River basin with natural channels linking the river's main tributaries. Even at its present-day reduced level, this network of interwoven streams and lakes creates a unique and complex ecosystem rich in biodiversity, but middle and lower reaches of rivers are often the sites of heavy rainfall and extensive flooding. Farmers who work the rice paddies are at risk of snail fever (schistosomiasis), an infection that leads to gradual, serious damage to the internal organs. Snail fever is caused by a blood fluke (a trematode), which grows in lakes and rivers and swims into the paddies (Wu et al., 2008). To prevent flooding and stop blood flukes from reaching rice paddies, dams and thousands of kilometres of dykes were built in the basin beginning in the 1950s (Utzinger et al., 2005). The dams and dykes often cut off water flow between rivers and lakes. Now only Poyang Lake and Dongting Lake still have naturally occurring links with the Yangtze River.

In recent years environmentalists have begun to study the effect of damming on local biodiversity. They conclude that the Yangtze and its lakes once formed a complex wetland network providing just the right conditions for fish to spawn and feed, but disconnection obstructed the natural flow of migratory fish, dramatically decreasing biodiversity across the whole basin (Fu et al., 2003; Park et al., 2003; Wu et al., 2004; Lopez-Pujol et al., 2006; Yang et al., 2007). Sluice gates were constructed in the embankments, but were used only for irrigation. Few considered the impact on water quality, fish migration routes, flora and fauna. As a consequence, inbreeding among the fish led to inferior-quality specimens of much smaller size. In addition, the number of fish species decreased, dropping, for example, in Zhangdu Lake from more than 90 to 40.

Within a few years, the locals' lucrative fishing business seemed on its last legs. Still, though, resource destruction continued as intensive land reclamation followed the construction of the dykes, with both agricultural and urban settlements springing up on the former flood plains and around the lakes. In the 1990s Zhangdu Lake dwindled to one-quarter of its original 1930s size. Lakes also began to suffer from the inflow of farm run-off and domestic and industrial sewage. Deteriorating water quality affected local ecology and wildlife. Natural fisheries output in the two lake regions declined sharply. Without the natural water exchange with the river, the lakes started to clog quickly with silt. For example, the annual deposit of silt in Zhangdu Lake increased from an average depth of just 1 cm to 1 metre after its disconnection from the Yangtze in the 1950s. All these effects continue today, the worse for their continuation over time.

Within the framework of the WWF–HSBC project, environmental experts propose seasonal opening of the dykes, outside the flood season, to help restore the lakes' natural links to the Yangtze. The seasonal opening coincides with the fish-breeding season to enable the migratory flow, allowing fish to breed upstream in the Yangtze and fish fry to return to the lake where they mature. This measure will also ensure a natural hydrological fluctuation in the lakes, where wetland areas will receive fresh water from the Yangtze, helping to ensure a healthy wetland ecosystem. To prevent the re-entry of blood flukes during the linking period, the local governments now provide funds to build concrete snail retention ponds between the first and second sluice gates linking river and lake.

The programme staff have also been promoting the concept of environment-friendly farms and fisheries in the reconnection scheme, hoping to restore wetland habitats for displaced migratory and endemic birds, as well as reduce water pollution. Local fish farmers have been using fertilizer for a few years to promote the growth of algae, which serves as food for fish, but with the accumulation of poisonous substances in the fertilizer that sinks into the silt, epidemics among fish have become increasingly frequent. Local fishermen were invited to join an eco-fishery programme advocated by the local government and WWF China. Farming began of a more environment-friendly type that demands clean aquatic conditions and uses a special fish species that feed on artificial fish pellets. Fish farmers who have put the eco-fishery programme into practice are currently earning more than two to three times the sums typical of traditional farming practices (WWF China official website; personal communications – WWF China).

The Yangtze programme has so far resulted in the seasonal relinkage of 11 lakes to the Yangtze River. The wetlands covering 450 000 hectares are effectively protected. Water quality has improved, and 34 varieties of birds and 14 species of fish now travel freely back to the river. In all, 15 million bamboo poles and fishing nets have been removed, and the 'falling and rising tide of the Hong Lake' has returned. In addition, members of

296 families have been taught sustainable fishing solutions, which has helped to increase their incomes by more than 20 per cent.

6.7 Analysis

It can be argued that Pierre and Peters's 'strong state', mentioned in Chapter 1 of this book, in many ways still applies to the Chinese political situation. Obviously government takes the leading role, especially in water management. This in itself, and in relation to the institutionalization of water resource management, might not actually be much different than in other, more democratically governed countries. However it can be argued that in Western (democratic) culture a case must become a social movement or at least a social issue before it can become part of a political agenda. There is a vast body of knowledge on the extent of these movements, the capacity to actually change policy and the different ways in which agenda-setting and policy influence can occur (see the theoretical introduction by Meijerink and Huitema, Chapter 2 in this volume). In China, agenda-setting is mainly limited to the central government as the country's leader of innovation. This means that policy-making in China is much less of a 'multilevel game' than in Western nations. However much this may still hold true, clearly Chinese society and the way it is governed are changing, opening avenues for third parties to enter the game, albeit with caution and care. The projects examined show WWF China successfully following new political paths in China in just such a way.

WWF China plays the role of main policy entrepreneur in our case studies. Since the start of China's Open and Reform Policy in 1978, the WWF has been active in the country (launching a panda protection project in 1980). An international NGO, WWF currently has a subdivision in China, but still no status as a Chinese NGO. A lot of time has been needed for it to be able to address environmental issues in China and cooperate successfully with governmental bodies in finding a better balance between nature and the economy. One of the major characteristics of policy entrepreneurship in China seems to be perseverance. Over the years, the WWF has slowly built a coalition with the Chinese Academy of Sciences and the State Forestry Administration. Resulting cooperation led to nature reserve staff training in the 1980s and, as above, to some Ramsar site listings as early as 1992, but it was not until the 1998 flood and the proclamation of the 32-character policy that a problem window was created that gave WWF China the opportunity to bring a new perspective on sustainable river basin management to the table. In this context WWF China was able to launch its Living River programme and organize joint planning events with local farmers, regional authorities and water management experts to identify alternative livelihoods (case one, Dongting Lake), restore wetlands, and implement sustainable fishing programmes (case two, Hong Lake).

The introduction and implementation of new ideas generates a broad spectrum of technical and non-technical questions and often requires seed money to trigger transition processes. In water management and land use, therefore, the need to make optimal use of the available knowledge, experience and finances makes it essential to build coalitions with public, private and governmental stakeholders in the river basin. This is true at both the national and the international level. The successful creation of coalitions led to the discovery of the experts and funding necessary to launch the Living River programme and its demonstration projects.

However, in any society, working from the bottom up to generate policy change is not sufficient. Particularly in a country like China, where agenda-setting for innovation is usually centralized, it is very important to invest time and attention in enlisting administration officials, lawmakers, policy research specialists and journalists. For this reason WWF China worked with the CCICED once it became clear that this organization would formulate recommendations to implement IRBM in China. This 'investment' proved highly profitable. By providing the secretariat of the IRBM task force and lobbying for suitable co-chairs, WWF provided a firm basis for incorporating the approach into the final CCICED recommendations with an effective plan for implementation including, for example, establishment of the Yangtze Forum. WWF China's strategic cooperation also helped to expedite the administrative process, resulting, for instance, in the recommendations being delivered directly to the premier of the State Council without passing through the ministerial levels.

Parallel to this activity, intensive contact with the academic field (for example the Chinese Academy of Sciences) opened the way to media coverage in the form of a special issue in *Chinese National Geography*, articles in the *China Daily* and newsletters for experts dealing with IRBM. This had a huge impact on awareness in- and outside China. Now, following the demonstration projects, the transition to IRBM seems to be gaining momentum and meeting little coordinated effort to block forward movement.

That said, the context for the transition process needs to be considered. The introduction and implementation of IRBM as a new paradigm can be regarded as the emergence of resistance to the Chinese hydrological mission and its unintended consequences. However the ongoing execution of large hydro-engineering projects indicates a competition between two paradigms and an outcome that has still to be determined (Boxer, 2001; McCormack, 2001). As seen, too, in other countries, the difference between policy transition and policy implementation will also be telling.

The adoption of IRBM in China necessarily involves the Chinese political system and consequently its governance system. In this respect we refer to the trialogue model currently refined by Turton et al. (2007), in which governance and especially good governance (as a product) depends on a dynamic interplay between the domains of government, society and science. This model underlines the importance of stakeholder participation, accountability and ultimately democracy in order to achieve good governance. The concept of IRBM (and the related concept of integrated water resources management) is based on the same principles regarding the 'human system' as formulated by the Global Water Partnership (Global Water Partnership, GWP, 2000; Jonch-Clausen and Fugl, 2001; Falkenmark et al., 2004). It requires a major participatory role for all stakeholders involved and appears to root firmly in a democratic policy system. As such, IRBM as developed elsewhere might not be fully applicable in the Chinese situation.

As shown, however, the Chinese government has allowed room for a new paradigm to enter. The rules of stakeholder participation may differ but, as cautiously undertaken in the cases presented above, efforts are being made to involve stakeholders in parts of the policy cycle (Wang et al., 2003; Zhong and Mol, 2007). How the government will attempt to direct the transition to IRBM and the results, both in water management and in governance, will be of great interest for future study.

6.8 Strategies of transition management

In order to characterize the strategies used by WWF China acting as a policy entrepreneur, we use the terminology and categorization presented in Chapter 2.

6.8.1 Development of new ideas

WWF China introduced ideas new to Chinese decision-makers. The importance of wetlands, and the possibility of using them as flood retention areas while simultaneously restoring valuable ecosystems, was a novelty previously only discussed in academic settings in China. The concept of IRBM as an approach where the environment, safety and economic matters can go hand in hand (case one, Living River programme) was likely the most influential 'new' idea. The IRBM conceptual framework provided a basis for rethinking all kinds of engineering solutions and methods that exploit freshwater resources and cause detrimental effects to ecosystem functions.

6.8.2 Build coalitions and sell ideas

Coalition-building is a very important part of WWF strategy and thus also of WWF China. It is a core WWF belief that: 'no solution will be effective in any river basin unless it is implemented through cooperation across social, economic, and political boundaries' (WWF, 2003). All stakeholders in a river basin need to come together around the table, establish respect for each other, number the needed experts among themselves, and bring the financing necessary to identify solutions in which the environmental, safety and economic goals are well balanced. Generally this method of working prevents unfair distribution of effort and costs where only one or two groups of stakeholders would have to 'bear the pain'.

Involvement of the wider public in the transition process takes the formulation of clear messages in plain language and illustrated with practical examples of where the 'new' ideas have proven to be feasible, easily adopted and scaled up by other stakeholders. These messages need to be made available prior to or during public events designed to draw media attention. WWF China accomplished all these steps very effectively in both the Yangtze River basin cases studied.

6.8.3 Recognize and exploit windows of opportunity

That WWF China acted during the right window of opportunity is clear. It did so by first presenting plans already formulated for the Yangtze River basin and Dongting Lake in concordance with and building upon the national 32-character policy. Next, to ensure top-level influence, it capitalized on WWF representation on the CCICED and support for IRBM by other members in order to steer the related task force and hence to influence national policy-makers.

Only once part of the senior hierarchy of the CCICED could the WWF create a 'shadow network', a circle outside direct governmental influence, where there is more room for debate. This network has proven to be very influential in the introduction of IRBM and in bypassing less enthusiastic organizations and players.

6.8.4 Recognize, exploit, create and/or manipulate the multiple venues in modern societies

Many venues exist for policy issue debate. These include, for example, various government processes, from planning to legislative and science fora. As Richardson (2000) argues, these venues are an interest group resource: that is, they can be exploited by policy entrepreneurs.

In the context of introducing IRBM to China, WWF China has also demonstrated a capacity to recognize and make efficient use of international and national venues that are known and visible in society. One of the most important parts of the WWF strategy is its constant advocacy of its ideas, successes, approaches and visions for the future. Clearly use of both international and national media is very important, but informing politicians at all levels of government through field visits, conferences, lectures and so on, might be just as important. The success of the IRBM task force and the amount of positive media coverage of its demonstration projects show that WWF China can be considered a very effective 'venue shopper' (Richardson, 2000).

6.8.5 Orchestrate and manage networks

Upscaling ideas and implementing new policies entails expanding and managing networks at all levels in society. Mustering sufficient public support and the necessary 'entrepreneurship' to envision, draft and implement new policies and/or concepts demands more than one social actor. The WWF sets out, therefore, to finance demonstration projects, but is keen to return ownership to the appropriate level of governance or society, including financial ownership for scaling up the demonstration projects and converting innovation to common practice. This explains why, apart from realizing demonstration projects, the WWF's core business is the orchestration and management of networks worldwide. These networks are managed by individuals highly skilled at addressing stakeholders of all kinds with respect and at communicating solution-oriented strategies.

6.9 Lessons learned and conclusions

In this chapter we have elaborated the role of WWF China as a change agent for the transition process to IRBM in China. We have demonstrated that WWF uses various strategies of different types to support this transition process. Successful deployment of these strategies for change in environmental policy requires special skills, actions and attitudes on the part of the policy entrepreneur (the role taken by WWF China in the cases presented above), including:

1. Broad focus: efforts are needed to understand the interests of partner stakeholders. For WWF China this means a focus not only on biodiversity, but also on the importance of a healthy ecosystem in balance with sustainable livelihoods and economies.
2. Pursuit of win–win situations that respect the interests of all stakeholders: what do they seek to gain or lose from the ultimate goal, and where do they stand regarding the balance of economy, safety and ecology?
3. Focus on the realization of demonstration projects: one successful demonstration project proved more convincing than any number of reports and academic studies.
4. Support for and action on a transition process by players at all societal levels: this requires investment by a change agent in a joint planning process run from the bottom up and from the top down.
5. Use of media as powerful tools for dissemination: messages need to be linked to news events.

6. Attention to the continuation of demonstration projects: help-desks and/or other forms of support need to be organized to assist stakeholders as they take over the lead in the transition process.
7. Frequent, continuous and meaningful communication: reiteration of long-term goals and presentation and analysis of short-term successes proved helpful in the work of WWF China.

The WWF as an IRBM policy entrepreneur contributed considerably to China's current stance on river basin management to the point where hardly any official now defends the traditional development model of high levels of growth and high pollution. According to the newspaper *China Daily*, officials of Shanxi, Henan and some other resource-rich provinces planned to close down local small coal mines and chemical plants managed on nineteenth-century concepts, even though these were erected only a decade ago or so earlier (You, 2007). Specifically in regard to the water sector, *China Daily* mentions that Hunan province recently closed down 230 paper-producing plants, producing an improvement of the water quality from level V to level III (near drinking water standard) within two months (Xu, 2007). Although the newspaper is a state medium and as such gives voice to the official view, nevertheless these kinds of reports are a sign that those who govern China recognize the growing importance of environmental issues. This represents a continuing opportunity for skilled, strategic action by policy entrepreneurs seeking transitions in China to ecologically and economically sustainable policies.

The 17th Communist Party of China (CPC) Congress in October 2007 settled China's political strategy for the next five years and officially confirmed a promising political future for the first time in modern history for sustainable development. As stated by President Hu Jiantao, the CPC aims to:

> [I]mprove energy, resources, ecological, and environmental conservation and enhance China's capacity for sustainable development. Adhering to the basic state policy of conserving resources and the environment is vital to the immediate interests of the people and the survival and development of the Chinese nation. We must give prominence to building a resource-conserving, environment-friendly society in our strategy for industrialization and modernization and get every organization and family to act accordingly. (Hu, 2007)

Such a statement is welcome. However experiences in many countries have taught that the transition towards an environment-friendly society is a complex and difficult process, one that involves an indispensable role for sophisticated policy entrepreneurs like WWF China.

Notes

1. The Central Yangtze: Partnership for a Living River programme was begun by WWF China in 1998. The initiative for this project came from WWF China, and the first financial donations to pay for related activities were generated by WWF Netherlands.
2. As of March 2008 the State Environmental Protection Agency (SEPA) was upgraded to the Ministry of Environmental Protection (MEP).
3. The Yangtze Focal Project builds upon the Central Yangtze: Partnership for a Living River programme.

References

Boxer, B. (2001), 'Contradictions and challenges in China's water policy development', *Water International*, **26** (3), 335–41.

CCICED, official website www.cciced.org/node_7040746.htm, accessed 19 September 2008.

CCICED and WWF (2004), *Promoting Integrated River Basin Management and Restoring China's Living Rivers*, Beijing: CCICED.

Chen, Z. (2005), 'Tackling China's water pollution problem: a legal and institutional perspective from Taihu Lake water pollution control', *Temple Journal of Science, Technology and Environmental Law*, **24**, 325–50.

Cui, L.J. and Y.F. Wang (2008), *Wetlands of International Importance in China*, Beijing: China Forestry Press (in Chinese).

CWRC, official website, www.cjh.com.cn/eng/index.html, accessed 19 September 2008.

Falkenmark, M., L. Gottschalk, J. Lundqvist and P. Wouters (2004), 'Towards integrated catchment management: increasing the dialogue between scientists, policy-makers and stakeholders', *International Journal of Water Resources Development*, **20** (3), 297–309.

Fang, J., Z.H. Wang, S. Zhao, Y. Li, Z. Tang, D. Yu and L. Ni (2006), 'Biodiversity changes in the lakes of the Central Yangtze', *Frontiers in Ecology and the Environment*, **4** (7), 369–77.

Fischer, G. and G.K. Heilig (1997), 'Population momentum and the demand on land and water resources', *Philosophical Transactions of the Royal Society of London Series B – Biological Sciences*, **352** (1356), 869–88.

Flavin, C. and G. Gardner (2006), 'China, India, and the new world order', in L. Starke (eds), *State of the World 2006*, Washington, DC: Worldwatch Institute, pp. 3–23.

Fu, C.Z., J.H. Wu, J.K. Chen, Q.H. Qu and G.C. Lei (2003), 'Freshwater fish biodiversity in the Yangtze River basin of China: patterns, threats and conservation', *Biodiversity and Conservation*, **12** (8), 1649–85.

Global Water Partnership (GWP) (2000), 'Integrated water resources management', TAC Background Papers, Global Water Partnership, Stockholm.

Global Water Partnership China (GWP China) official website, www.gwpchina.org/echinagwp/Publish/News.aspx?NewsID=2bcf1615-f6b9-4525-8d6f-33508483e7dd&DisplayOrder=05, accessed 8 October 2008.

Hu, J. (2007), 'Speech: report to the 17th National Congress of the Communist Party of China (CPC)', Bejing, 15 October.

Hunan province, official website, www.hbj.hunan.gov.cn/dongT1/wzDisplay.aspx?id=130, accessed 8 October 2008.

Jonch-Clausen, T. and J. Fugl (2001), 'Firming up the conceptual basis of integrated water resources management', *International Journal of Water Resources Development*, **17** (4), 501–10.

Kreimer, A. and M. Munasinghe (1991), 'Managing environmental degradation and natural disasters: an overview', in A. Kreimer and M. Munasinghe (eds), *Managing Natural Disasters and the Environment*, Washington, DC: World Bank, pp. 3–6.

Lee, S. (2006), 'China's water policy challenges', China Policy Institute Discussion Paper, Nottingham: China Policy Institute of the University of Nottingham.

Li, Y.S., A.C. Sleigh, A.G.P. Ross, G.M. Williams and M. Tanner (2000), 'Epidemiology of *Schistosoma japonicum* in China: morbidity and strategies for control in the Dongting Lake region', *International Journal for Parasitology*, **30** (3), 273–81.

Lopez-Pujol, J., F.M. Zhang and S. Ge (2006), 'Plant biodiversity in China: richly varied, endangered, and in need of conservation', *Biodiversity and Conservation*, **15** (12), 3983–4026.

McCormack, G. (2001), 'Water margins: competing paradigms in China', *Critical Asian Studies*, **33** (1), 5–30.

Ministry of Water Resources (MWR), official website, www.mwr.gov.cn/english1/, accessed 19 September 2008.

Muldavin, J. (2000), 'The paradoxes of environmental policy and resource management in reform-era China', *Economic Geography*, **76** (3), 244–71.

Nakamura, T. (2003), 'Ecosystem-based river basin management: its approach and policy-level application', *Hydrological Processes* **17**, 2711–25.

Nakamura, T. (2006), 'Development of decision making indicators for ecosystem-based river management', *Hydrological Processes*, **20**, 1293–1308.

OECD (2007), China, Environmental Performance Reviews, No. 5, Paris: OECD Publishing.

Olson, D.M. and E. Dinerstein (1998), 'The Global 200: a representation approach to conserving the earth's most biologically valuable ecoregions', *Conservation Biology*, **12** (3), 502–15.

Park, Y.S., J.B. Chang, S. Lek, W. Cao and S. Brosse (2003), 'Conservation strategies for endemic fish species threatened by the Three Gorges Dam', *Conservation Biology*, **17** (6), 1748–58.

Ramsar Convention on Wetlands (Secretariat), official website, www.ramsar.org, accessed 19 September 2008.

Richardson, J. (2000), 'Government, interest groups and policy change', *Political Studies*, **48** (5), 1006–25.

Schuyt, K. (2005), *Freshwater and Poverty Reduction: Serving People, Saving Nature*, Zeist, The Netherlands: WWF International.

Schuyt, K. and L. Brander (2004), *The Economic Values of the World's Wetlands, Living Waters*, Gland, Switzerland and Amsterdam: WWF International and Institute for Environmental Studies, Vrije Universiteit Amsterdam.

Turner, J.L. and K. Otsuka (eds) (2005), *Promoting Sustainable River Basin Governance: Crafting Japan–US Water Partnerships in China*, IDE JETRO Spot Survey Report No. 28, Chiba: Institute of Developing Economies.

Turner, J.L. and K. Otsuka (2006), *Reaching Across the Water*, Washington, DC: Woodrow Wilson International Center for Scholars.

Turton, A.R., H.J. Hattingh, G.A. Maree, M. Claassen, P.J. Roux and W.F. Strijdom (2007), *Governance as a Trialogue: Government–Society–Science in Transition,* Berlin and Heidelberg: Springer-Verlag.

Turton, A.R., R. Meissner, P.M. Mampane and O. Seremo (2004), *A Hydropolitical History of South Africa's International River Basins*, Pretoria: Water Research Commission.

Utzinger, J., X.N. Zhou, M.G. Chen and R. Bergquist (2005), 'Conquering schistosomiasis in China: the long march', *Acta Tropica*, **96** (2–3), 69–96.

Varis, O. and P. Vakkilainen (2001), 'China's 8 challenges to water resources management in the first quarter of the 21st century', *Geomorphology*, **41**, 93–104.

Wang, Q.G., G. Gu and Y. Higano (2006), 'Toward integrated environmental management for challenges in water environmental protection of Lake Taihu basin in China', *Environmental Management*, **37** (5), 579–88.

Wang, Y., L.F. Li, X.J. Wang, X.B. Yu and Y.H. Wang (2007), *Taking Stock of Integrated River Basin Management in China*, Beijing: WWF China.

Wang, Y., R.K. Morgan and M. Cashmore (2003), 'Environmental impact assessment of projects in the People's Republic of China: new law, old problems', *Environmental Impact Assessment Review*, **23**, 543–79.

Waterbury, J. (1979), *Hydropolitics of the Nile Valley*, New York: Syracuse University Press.

World Bank (1997), *Clear Water and Blue Skies: China's Environment in the New Century*, Washington, DC: World Bank.

Wu, J.G., J.H. Huang, X.G. Han, Z.Q. Xie and X.M. Gao (2004), 'The Three Gorges Dam: an ecological perspective', *Frontiers in Ecology and the Environment*, **2** (5), 241–8.

Wu, X.H., S.Q. Zhang, X.J. Xu, Y.X. Huang, P. Steinmann, J. Utzinger, T.P. Wang, J. Xu, J. Zhang and X.N. Zhou (2008), 'Effect of floods on the transmission of schistosomiasis in the Yangtze River valley, People's Republic of China', *Parasitology International*, **57** (3), 271–6.

WWF (2003), *Managing Rivers Wisely: Lessons from WWF Work for Integrated River Basin Management*, T. Jones, B. Phillips, C. Williams and J. Pittock (eds), Gland, Switzerland: WWF International.

WWF China, official website, www.wwfchina.org, accessed, 19 September 2008.

WWF International, official website, www.panda.org, accessed 19 September 2008.

Xu, X. (2007), 'Hunan Province cleans up its act', *China Daily*, 17 October, p. 24, available at: http://www.chinadaily.com.cn/cndy/2007-10/17/content_6181797.htm, accessed 20 June 2009.

Yan, F., D. He and B. Kinne (2006), 'Water resources administration institution in China', *Water Policy*, **8**, 291–301.

Yang, G.S., L.D. Weng and L.F. Li (2007), *Yangtze Conservation and Development Report*, Wuhan.

Yangtze Forum, official website, www.yangtzeforum.org, accessed 19 September 2008.

Yin, H. and C. Li (2001), 'Human impact on floods and flood disasters on the Yangtze River', *Geomorphology*, **41** (2–3), 105–9.

Yin, H., G. Liu, J. Pi, G. Chen and C. Li (2006), 'On the river–lake relationship of the middle Yangtze reaches', *Geomorphology*, **85**, 197–207.

You, J.M. (2007), 'Transforming the model', *China Daily*, 15 October, p. 5, available at: http://www.chinadaily.com.cn/bw/2007-10/15/content_6173701.htm, accessed 20 June 2009.

Zhang, M., G. Yang, H. Zhang and CWRC (1998), 'Water resource development and utilization in the Yangtze Valley', Los Alamos National Lab, unpublished work.

Zhao, S. and J. Fang (2004), 'Impact of impoldering and lake restoration on land-cover changes in Dongting Lake area, Central Yangtze', *Ambio*, **33** (6), 311–15.

Zhao, S., J. Fang, W. Ji and Z. Tang (2003), 'Lake restoration from impoldering: impact of land conversion on riparian landscape in Honghu Lake area, Central Yangtze', *Agriculture, Ecosystems and Environment*, **95** (1), 111–18.

Zhao, S.Q., J.Y. Fang, S.L. Miao, B. Gu, G. Tao, C.H. Peng and Z.Y. Tang (2005), 'The 7-decade degradation of a large freshwater lake in central Yangtze river, China', *Environmental Science and Technology*, **39** (2), 431–6.

Zhao, S., C. Peng, H. Jiang, D. Tian, X. Lei and X. Zhou (2006), 'Land use change in Asia and the ecological consequences', *Ecological Research*, **21** (6), 890–96.

Zhong, L.-J. and A.P.J. Mol (2007), 'Participatory environmental governance in China: public hearings on urban water tariff setting', *Journal of Environmental Management*, **88** (4), 899–913.

7 Where does policy change come from? And where does it end up? Establishing water user associations in large-scale canal irrigation systems in India

Vishal Narain

7.1 Introduction

In most developing countries the state has made huge investments in large-scale canal irrigation. It has also played a predominant role in managing and operating these systems. However canal irrigation has been plagued by several problems. In India, for instance, the most widely known difficulties have arisen over inequity in water distribution between head and tail reaches, water logging, low recovery from irrigation fees, mismanagement of irrigation infrastructure and a widening gap between irrigation potential created and functions utilized.

Different approaches to addressing these issues have been advocated and experimented with, emphasizing both technological and institutional interventions. The latter have included rights reform and market creation, pricing and decentralization. A need for a greater role for farmers in the management of irrigation systems has also been perceived and advocated as a solution to the above mentioned problems. This trend is referred to variously as irrigation management transfer (IMT), irrigation management turnover or participatory irrigation management (PIM) (Turral, 1995; Brewer et al., 1999; EDI, 1998). IMT is defined as the transfer of responsibility and authority for irrigation system management from government agencies to water user associations or other private entities (IndianPIM, 2001). IMT may include transfer of decision-making authority (or governance); it may also include transfer of ownership of scheme infrastructure (typically called privatization). PIM refers to programmes that seek to increase farmers' direct involvement in system management – either as a complement or as a substitute for the state role (EDI, 1998).

Activities classified as PIM or IMT normally take the form of the establishment of farmers groups called water user associations (WUAs). These WUAs are then made to take over functions pertaining to operations and maintenance of irrigation systems. The extent of user involvement in IMT programmes and irrigation management functions has varied from country to country; it may include turning over to users partial management responsibilities, such as water delivery, canal maintenance and paying for irrigation services (as in Sri Lanka or the Philippines), while final approval of operations and maintenance (O&M) plans and budgets are subject to government approval (as with the first wave of IMT in Colombia) (IndianPIM, 2001).

In the Indian context the use of the term 'IMT' has been confined to the academic community (Brewer et al., 1999; IIM(A)-IWMI, 1999; Pant, 1999, 2000; Kolavalli and Brewer, 1999). The term 'irrigation management turnover' is rarely used, PIM being much more widely preferred.

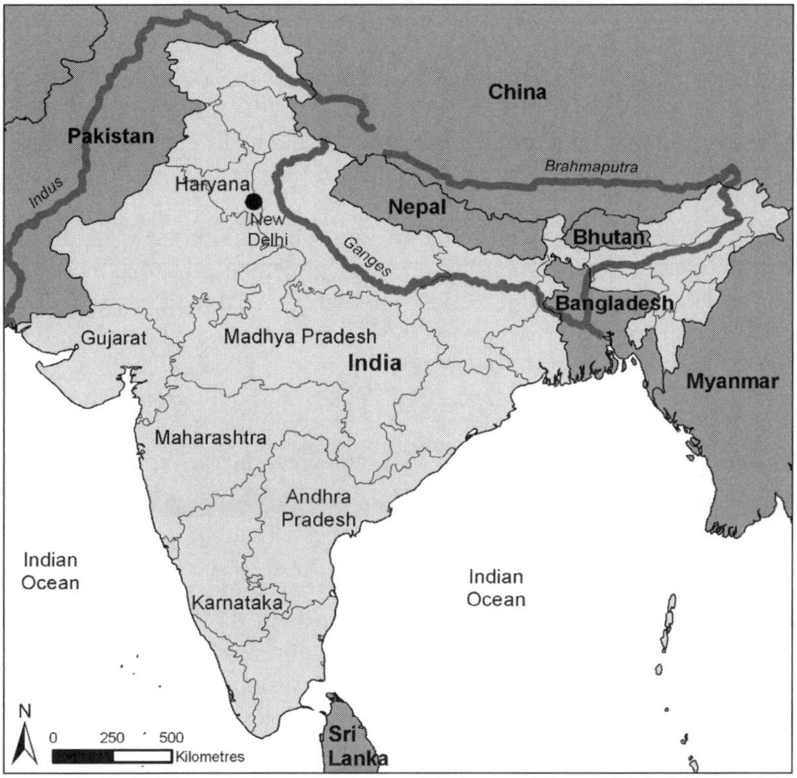

Map 7.1 India: key states that have implemented PIM reforms

Although the concept of PIM has been recognized in the country's National Water Policy, there is considerable variation in the approaches adopted by different states to formulate and implement policies for its achievement. Water is a matter for states in India (see Map 7.1). State governments develop programmes, policies and laws in this domain. At present the states of Andhra Pradesh, Chhattisgarh, Gujarat, Madhya Pradesh, Maharashtra, Orissa and Tamil Nadu have all enacted PIM legislation (Bassi, 2008). Several other states have sought instead to register WUAs under existing state law.

The current interest in PIM in India stems from the 1980s, when some efforts were made by non-governmental organizations (NGOs) in two states of western India, namely, Gujarat and Maharashtra, to organize farmers working with irrigation systems into user groups.[1] This was followed by extensive lobbying of the respective state governments by these NGOs. During the 1990s, especially the latter part of the decade, PIM received a big push and came high on the agenda of several state governments. The state of Andhra Pradesh acquired great visibility in India and abroad; the Andhra Pradesh Farmers Management of Irrigation Systems Act, 1997, sought to create 10 292 WUAs at the stroke of a pen throughout the leading irrigation area of the state. An NGO called IndianPIM, the Indian Network on Participatory Irrigation Management, called for speedy universalization of PIM in India. In this chapter we look at the role played by different individuals or organizations in this process, the strategies they employed and their influence and limitations.

The following section answers the question, 'Where does policy change come from?' It explores the genesis of policies for PIM internationally and in India. These policies are seen to be located in certain policy discourses and development narratives that have been propagated predominantly by multilateral and bilateral institutions, while NGOs and specific individuals have acted as policy entrepreneurs. Policy networks and epistemic communities have also played a role in some states. The next section analyses the implementation of the reforms, focusing on how implementation has been thwarted by vested interests. The policy outcomes have been highly variable, shaped by the exercise of power among the actors involved in implementation processes. I describe the strategies employed by some of the actors to thwart reform efforts. The final section concludes with the main messages emerging from the case of WUAs and irrigation systems in India.

7.2 Where does policy change come from? Development narratives and policy discourses perpetuated by multilateral organizations

7.2.1 Introduction

Transitions relate to the emergence of new ideas that challenge an existing paradigm. Many policy changes are located in specific ideas enshrined in what are called policy discourses (see also Gupta's Chapter 3 on global water discourses in this volume). A policy discourse describes a way of thinking, an outlook, a system of values and priorities that marginalizes other possible ways of thinking (Sutton, 1999). A discourse can be considered to be an ensemble of ideas, concepts and categories through which meaning is given to phenomena; discourses shape certain problems, distinguishing some aspects of a situation and marginalizing others (see also Hajer, 1995). A discourse can also refer to dialogue, language and conversation or to the analysis of language used in policy-making.

Grillo (1997) lists some prominent discourses in the development literature (see Gupta's contribution to this volume, Chapter 3). These are as follows:

1. Discourse of the state-engendered order in development, involving the intervention of experts located in UN agencies and embodied in multilateral and bilateral aid agencies.
2. Market-engendered spontaneous order discourse – institutionally located in the International Monetary Fund (IMF) and the World Bank.
3. Discourse of the public sphere, found in NGOs, research institutes and charities, constructed through the optimistic, reason-informed pursuit of formal and substantive democracy.

These discourses have acquired prominence at different points of time. The discourse of the state-engendered order in development dominated thinking on public policy through most of the period prior to the 1980s. The market-engendered spontaneous order discourse subsequently took precedence. The discourse of the public sphere has acquired greater significance in recent years especially following the 1990s. This sequence corresponds to paradigmatic shifts in thinking on appropriate forms of governance, when emphasis shifted from the state to markets, local institutions and partnerships. Policies for user involvement in irrigation management or natural resource management in general grow out of the discourse of the public sphere.

7.2.2 The 'good governance' paradigm

Policies for the involvement of users in irrigation management have their genesis in a widely felt need to improve governance through democratization and greater involvement of civil society. At the international level such policies have been related to initiatives for 'good governance'.

The 'good governance' paradigm was floated and espoused by the World Bank in the 1990s in response to the unsatisfactory results seen in some of the reform programmes supported by the bank. Governance in this sense came to be identified as a means for bringing about sustainable economic development. It was argued that an infusion of financial resources needed to be matched by appropriate institutional reform and changes within the political system in order to bring about better economic management. The World Bank's vision of good governance includes a bureaucracy imbued with a professional ethos, an accountable executive arm of government, a strong civil society participating in public affairs and citizens behaving according to the rule of law (Shylendra, 2004). More lately there has been a broadening of this concept by official aid agencies incorporating political dimensions so as to include and emphasize participatory development in the form of democratic systems that promote transparent and accountable societies and civil society participation. Good governance has been one of the important notions driving policy change in favour of PIM in many Indian states, particularly where WUAs have been formed as part of programmes funded by bilateral and multilateral organizations.

For instance the most important factor driving WUA formation in the northwestern Indian state of Haryana was the agreement between the government of Haryana and the World Bank for the Water Resources Consolidation Project (WRCP) (Narain, 2003). This project, initiated in 1994, mandated improvement of farmer contributions to O&M in large systems by raising fees and by promoting WUAs below the outlets to take over maintenance of lined watercourses.

The emphasis on WUA formation under the WRCP supports the general thrust toward devolution of functions in the World Bank's programmes. The World Bank's Water Resources Management Policy (1993), stressing the need for decentralization, asserts that:

> [T]he principle is that nothing should be done at a higher level that can be done satisfactorily at a lower level. Thus where local or private capabilities exist and where an appropriate regulatory support can be established, the Bank will support central government efforts to decentralise responsibility to local governments and to transfer service delivery functions to the private sector to financially autonomous public corporations and to community organisations such as water user associations . . . when properly designed, community based programmes can be highly effective in managing natural resources, providing basic infrastructure or ensuring primary social services . . . Successful design requires tapping into local needs, understanding and building on the strengths of existing institutions and defining the changes needed in intermediary implementing agencies to support community action (World Bank, 1996, p. 247).

Many of the irrigation projects initiated also by the government of Maharashtra, such as the Maharashtra Water Utilization Project (supported by the World Bank) and the Maharashtra Minor Irrigation Project aided by the United States Agency for International Development (USAID), incorporated guidelines for the formation of WUAs (Kulkarni and Kulkarni, 1994).[2] 'Thus in a way, the draft agreement . . . as well as certain procedures to enhance farmers' participation in managing minor irrigation schemes are the outcomes of these projects' (Kulkarni and Kulkarni, 1994, p. 60).

The Ford Foundation was instrumental in popularizing the Philippine model of management transfer in the states of Gujarat and Maharashtra.[3] The foundation supported visits of government officials in both these states to the Philippines and gave financial support to the NGOs in this sector for organizing farmers for management turnover. The USAID and the World Bank also promoted PIM initiatives in Gujarat (Mollinga, 2001).

In a similar vein in Karnataka in southern India, policies for WUA establishment received a push under a programme for financial support from the World Bank. The significance of the role of multilateral institutions can be gauged from the fact that the momentum of discussion on irrigation policy in the state later shifted to water tanks because of a World Bank loan for large-scale tank rehabilitation, even though formation of WUAs and WUA federations did proceed on a large scale (Narain, 2003). The formation of WUAs in many Indian states is seen mainly as a response to the ongoing reform programmes initiated by multilateral institutions such as the World Bank. It was commonly held in Haryana by several members of the irrigation bureaucracy that the chief reason for the creation of WUAs was the World Bank's insistence on this measure under its WRCP. The members felt that without this there would be little demand for PIM (Narain, 2003).[4]

7.2.3 The discourse on participatory irrigation management

The good governance discourse triggers the more specific discourse on participation heard echoing in several policy statements circulating in government circles.

The emphasis on PIM in India can be traced back to the realization that the performance of irrigation systems was unsatisfactory, due in good part to the lack of involvement of farmers in their management. Given huge investments in the irrigation sector in most developing countries it was felt that irrigation (mis)management was a problem of social engineering rather than one of technical engineering (Narain, 2008a; see also Chapter 9 by Bhat and Mollinga in this volume). It was assumed that irrigation systems were well designed technically and that their function essentially required an organizational structure involving farmers. The perceived need for management changes also justified reorienting the bureaucracy. These efforts can be seen as a result of a paradigm shift from emphasizing technical improvements in irrigation systems to recognizing the importance of institutional factors in irrigation system management.

In terms of participation in the early 1980s the dominant themes representing this discourse moved on to the organization of farmers in the later 1980s and finally turnover and self-governance in the 1990s (Maloney and Raju, 1994). The wave of economic reforms of the 1990s carried the major ideas of privatization, decentralization and private management.

Two workshops were organized by the Administrative Staff College of India, Hyderabad; the first was in July 1987: People's Participation in Irrigation Management. Conceptual advances were made in the second workshop in January 1992: Farmers'

Management in Indian Irrigation Systems. The word 'participation' was dropped, substituted by the word 'management' (Maloney and Raju, 1994, p. 29). At the turn of the century, PIM came to be considered the most essential irrigation paradigm for the twenty-first century (Hooja and Joshi, 2000). IndianPIM called for speedy universalization of PIM in India.

7.2.4 Development narratives

The argument that irrigation system performance cannot improve in the absence of farmers' participation in irrigation management became a self-perpetuating narrative and conventional wisdom. In the development community, such narratives are essentially stories that individuals, organizations and the community as a whole tell themselves and others in order to find patterns in ambiguity or make sense of the uncertain environment in which development takes place. As a result a development narrative tends to guide policy development by focusing the attention of policy-makers on specific forms of policy intervention (Sutton, 1999).

A narrative in this way can be seen to justify a policy intervention. A common example of a narrative that has certainly been the basis of efforts to build local-level institutions for natural resource management is 'the tragedy of the commons' (Hardin, 1968). This 'story' has influenced efforts aimed at nationalization and privatization of natural resources like land and water. Crafting collective institutions has been seen as a way out of the tragedy and as an alternative to nationalization and privatization of natural resources in order to avert the tragedy.

One critique of the role of development narratives, however, contends that there is a tendency to produce and reproduce oversimplified narratives, especially in situations where difficult and ambiguous decisions have to be made. Development narratives can lead individuals and organizations to make wrong assumptions about the cause of a problem and from this to draw irrelevant lessons from the experience. This happens typically in attempts to transplant development model blueprints from one setting to another where they may not be valid. Such narratives can be misleading when they are used as a 'one size fits all' approach to different kinds of problem situations. Evidence of this can be seen in India where PIM has come to be seen as a 'one size fits all' solution for the ills of the irrigation sector. It has been taken to be some kind of mantra or magic cure for treating the malaises that inflict irrigation systems.

We can see specific evidence of this at two levels in WUA formation in India. The first is in trying to plant models of IMT in Indian settings from abroad.[5] The second is in the way PIM came to be seen as a near blueprint solution to the problems of the irrigation sector, where in practice the potential for reform through WUA formation would vary from one setting to another, affected by, for instance, the design principles of canal irrigation and the irrigation technology in use (see Narain, 2008b).

7.2.5 NGOs as policy entrepreneurs: organizing farmers and lobbying governments

In India it is not hard to recount the names of several activists and leaders who have been instrumental in bringing about social change. Anna Hazare, known for spearheading watershed conservation efforts in Maharashtra, Dr Varghese Kurien, who pioneered India's White Revolution and Sam Pitroda, known for spearheading change in the field of information technology, could all be considered policy entrepreneurs. There are also several NGOs working for the formation and mobilization of self-help groups and in the field of micro credit and natural resource management. They have often lobbied governments for policy change and have been instrumental in organizing people for the same.

Maharashtra has gained a reputation as a state known for the important role played by NGOs in a number of natural resource management activities (Pendse and Bhogle, 1994). Several NGOs are scattered throughout the state. These include Marathwada Sheti Sahaya Mandal in Aurangabad, Vanrai in Pune, Manav Lok in Ambejogai, Beed district and Samaj Prabodhan Sanstha in Pune. They have involved themselves in a wide range of rural development programmes focusing particularly on watershed development. Social workers like Vijay Borade of Adgaon village and Anna Hazare of Ralegaon Siddhi are recognized in India and abroad for organizing resource users for water and soil conservation, social forestry and watershed management.

In the canal irrigation sector, NGOs' strategies for change have had two components: first, organizing users for irrigation management; and second, lobbying governments to bring about appropriate policy changes. They have taken the basic approach of starting with pilot experiments, mobilizing water users and, after initial mobilization efforts, lobbying the state government. The pilot experiments by two NGOs, the Society for Promoting Participative Ecosystem Management (SOPPECOM) and the Samaj Parivartan Kendra (SPK), have made a significant contribution. SOPPECOM, in transferring management to farmers on one of the minors of the Mula Irrigation Project, provided the foundation for a set of guidelines later issued by the Cooperatives Department for transferring management to WUA.

While NGOs in Maharashtra have played an important role on this front, a major weakness of their approach has been their inability to produce reform on a scale larger than the local level. The absence of a consistent state policy and support for NGO-led research and action have further limited the scale of their activity.[6] Another limitation to their role and approach stems from their lack of strategic effort to influence policy reform (Mollinga, 2001; see Chapter 14 by Werners et al. in this volume). NGOs have tended to focus wholly on farmer organization for water management. There has been little attempt to involve the users in advocacy or policy formulation. As a result, users have remained mere recipients of the policy initiatives.

Further, it is important to note that only two NGOs have been involved in the canal irrigation sector, while several NGOs have worked in many other sectors in the state. There are three main reasons for this (Narain, 2003):

1. The perception among the NGOs that irrigation is the domain of the rural elite – a small section of powerful farmers, owners of large farms, who have been the principal beneficiaries of water allocation so far. NGOs are reluctant to support them.
2. Relatively little financial support has been available for research and action in the canal irrigation sector. While several NGOs are involved in other natural resource management activities, they all rely on external sources of funding, including the central government. Such resources have not been forthcoming for canal irrigation.
3. A clear-cut state policy on the role of NGOs in the PIM process has been absent.

With these three factors in play, no more than two NGOs have entered the canal irrigation sector.

The neighbouring state of Gujarat has also seen NGOs play a major role as policy entrepreneurs. As in several other states, the influence of national-level policies such as the National Water Policy of 1987 and 2002 has been important (Narain, 2003). Further, the activities of NGOs in organizing user groups for management of natural resources in several sectors, including lift irrigation, groundwater and forests, have provided evidence of the capability of farmers to manage resources effectively. More specifically in the realm of canal irrigation NGOs have succeeded in lobbying for change and urging the state government to pass a series of resolutions, starting in 1992, to effect management turnover.

An important component of the PIM process in Gujarat was the reliance on process documentation research (PDR) (Shashidharan, 2000; Parthasarathy, 2000). PDR was carried out in six of the 13 pilot projects (Parthasarathy, 2000). Two research and academic institutes – the Institute of Rural Management, Anand (IRMA) and the Gujarat Institute of Development Research, Gandhinagar (GIDR) – undertook field research on the pilot projects. Commissioned by a high-level working group, this process provided feedback that led to mid-course corrections.

In Gujarat, as in Maharashtra, NGOs have demonstrated the feasibility of a farmer-centred management system. However as in Maharashtra, an important weakness of this approach has been its inability, alone, to ensure a scaling up (Shashidharan, 2000). NGOs have proved their ability to experiment and come up with good results, but there has been little evidence to demonstrate their effectiveness on a sufficiently larger scale.

Further, there is some evidence that NGOs in the state have adopted a strategic shift from organizing farmers to lobbying for policy change. The role of NGOs in the PIM process has been questioned: should NGOs pursue steady efforts at forming and replicating local-level resource user organizations in canal irrigation or should they seek to inform policy change (Narain 2003)? There is now a new focus within the NGO sector on policy advocacy and dissemination, and training of farmers and irrigation engineers, where earlier this focus was mainly on creating and replicating local-level resource user organizations.[7]

In Karnataka, too, the process of PIM received some forward thrust in 1997 from the formation of Sahayog, an NGO, to serve as a platform for debating reform proposals and negotiating for further reform. It could be said that Sahayog learned from the weaknesses of the strategies employed by NGOs in other states and sought to involve farmers themselves in influencing policy change and demanding policy reform from the government.

The basic objective of Sahayog was to identify whether farmers were motivated to claim reform from the government and, if so, to facilitate the organization of farmers around a platform to discuss the desired policy changes (Mollinga, 2001). The basic rationale behind the formation of Sahayog stemmed from the exclusion of farmers themselves from the policy process:

> Irrigation reform and larger participation of farmers in irrigation management has been on the Indian policy agenda since the 1970s, but the progress has been slow, and reform design confined to a narrow circle of senior bureaucrats and some water academics and professionals. The lack of a broad-based consultation of all relevant stakeholders has made the reform process vulnerable to personnel changes. Also, the context of the reforms scarcely questions the exclusive government control over allocation and distribution of water and other resources – a

> questioning that is necessary for management processes to improve. The strategic analysis of the actors in this innovation profile was that, unless farmer-irrigators actively demanded and shaped the reform, progress would be slow and haphazard. (World Bank, 2005, p. 96)

Sahayog served as an intermediary between the farmers and the irrigation bureaucracy and as a platform to bring different actors together to facilitate the formulation of an effective strategy and a transition to its implementation. The NGO also sought to sensitize the public through the mass media on the imperatives, objectives and modalities of reform. A follow-up development on this front was the formation of Pragathi, an offshoot of the NGO, by farmers who became disenchanted with a rift among the more professional leadership of Sahayog.

7.2.6 Epistemic communities and policy networks: coalitions that influence policy

In the Indian context the role of epistemic communities and policy networks in the policy formulation process can be clearly seen in two states, namely, Madhya Pradesh and Andhra Pradesh. Bassi (2008) notes the important role played by senior experts from the Water Resources Department (WRD) in the policy formulation process with regard to the passage of the Madhya Pradesh Farmers Management of Irrigation Systems Act. The approach of these experts was to look at the IMT process worldwide and guide the PIM policy drafting for the state. An important factor that facilitated their success was that they enjoyed support of the then Chief Minister who had an interest in participatory irrigation management. However Bassi notes that one weakness of their approach was that several of the other stakeholders were not involved. Nevertheless the strong network of the senior-level bureaucracy and the Chief Minister succeeded in overcoming the resistance at the lower levels of the bureaucracy as well as among the farmers, who were reluctant to take over irrigation management functions on account of the poor state of the irrigation infrastructure.

The network between the irrigation bureaucracy and the political establishment played the most important part in the formulation and implementation of the Madhya Pradesh Farmer Management of Irrigation Systems Act (personal communication, Nitin Bassi, 26 May 2008). However within the irrigation bureaucracy only a few individuals took an active interest in PIM. They were successful in drafting and coming up with the Act because of their higher authority and political support. The staff at lower levels within the Irrigation Department were not much in favour of IMT because of their insecurity over loss of power to control and manage irrigation infrastructure.

Unlike Maharashtra and Gujarat where NGOs acted strongly as policy entrepreneurs, in Madhya Pradesh the policy process did not have any involvement of NGOs or a civil society network in the initial stages. Although four NGOs were involved in the later stages, this occurred only in the implementation of PIM in seven irrigation projects supported by the ICEF (India–Canada Environment Facility). In this case involvement was more a means of fulfilling a project requirement. As in several other states, farmers' involvement in the PIM design process was also minimal. Only a few select farmers in the state were consulted on the idea of PIM. The related meetings or discussions were few and included only select individuals who did not represent the larger majority.[8]

The other state where networks and advocacy coalitions seem to have played an important role was Andhra Pradesh, where a strong partnership between the then Chief Minister and the Secretary of State was instrumental in giving a push to PIM reforms. This culminated in the passage of the Andhra Pradesh Farmer Management of Irrigation Systems Act of 1997.

7.3 Analysing the process of policy implementation: street-level bureaucracies and the games people play

7.3.1 Introduction

While insights into the process of policy formulation are important in terms of devising effective strategies for change, an understanding of policy implementation processes is necessary to examine the existence of what in the policy science literature is called the 'implementation gap'. Thomas and Grindle (1990) note that the real challenges with regard to policy in the context of developing countries lie not at the stage of policy formulation, but instead after the policy decision is made and set for implementation (also see Chapter 9 on Indonesia by Bhat and Mollinga in this volume). The linear model of policy-making has led donors to support substantial efforts to strengthen policy analysis in developing countries in the expectation that good analysis will translate into good decision-making and this into good policy. Operationalizing 'good policy-making' has taken the form of technical assistance contracts to build capacity in planning and policy analysis in many sectors.

In practice however the real issues arise 'after the decision', that is, at the implementation stage. The outcomes of a policy intervention can be extremely variable. The range of outcomes results from the fact that implementation is an interactive and ongoing process of decision-making by policy elites and managers in response to actual or anticipated reactions to reform initiatives. These officials could be potential strategic managers working within complex policy and institutional environments. Characteristics of particular reforms determine the type of conflict and opposition that surround their implementation.

In Thomas and Grindle's (1990) interactive model pressures to put reform issues on the policy agenda are seen to come from many sources, including frequent reform-mongering by policy elites. The agenda represents a stockpile of proposed changes. Some items on the agenda are acted upon, but many are not, often because of the preferences, perceptions and actions of policy elites and their appreciation of the economic and political environment they face.

The policy implementation process is complex, highly interactive and at least as political as it is technical (Brinkerhoff and Crosby, 2002). Besides technical and institutional analysis it calls for consensus-building, participation of key stakeholders, mechanisms for conflict resolution, compromise, contingency planning and adaptation.

Evidence that these steps have occurred with regard to policies for PIM in India is rare indeed. In fact as noted above one critique of the policy process has been that the most important actors – the irrigation bureaucracy and the farmers – have tended to be mere recipients of policies for participatory irrigation management and not involved in the policy formation process itself (Mollinga, 2001; Narain, 2003). They have tended to claim little ownership of the policy initiatives and have sought instead to thwart change, as we shall see below.

7.3.2 Role of street-level bureaucracy

The concept of street-level bureaucracy was developed in Lipsky's (1980) book of the same title and further developed in work by two of his former research students, Weatherly and Prottas (Hill, 2005). Lipsky emphasized the role that street-level bureaucrats play in the implementation of policies, exercising discretion in efforts that resulted in the reformulation of policies in the course of implementation. In his book, Lipsky argues that: 'the decisions of street-level bureaucrats, the routines they establish, and the devices they invent to cope with uncertainties and work pressures, effectively become the public policies they carry out' (Lipsky, 1980, p. xii).

Lipsky argues that this process of street-level policy implementation does not involve, as is typically believed, the advancement of the ideals many bring to personal service work, but rather the development of practices that enable officials to cope with the pressures they face (Hill, 2005). Street-level bureaucrats develop methods of processing in a relatively routine and stereotyped way. They adjust their work habits to reflect lower expectations of themselves and their clients.

The concept of street level bureaucracies is useful in analysing policy implementation in that it draws attention to the role that bureaucrats themselves may play in policy implementation processes. An important factor that shapes policy implementation lies in the motivations, intentions, attitudes and perceptions of those involved in it. This can be an important reason why there is a gap between the intentions and outcomes of a policy, in that policies are not implemented as initially conceived, but rather as street-level bureaucrats modify them in the course of implementation.

The concept of street-level bureaucracy helps to analyse the role that certain individuals in their capacity as field-level implementers play in the irrigation reform process. In several states members of the irrigation department have sought to slow down the implementation process as they saw it hurting their power and interests.

This has been well documented, for instance for the state of Maharashtra, (Narain, 2003; Lele and Patil, 1994) where the IMT model aims at changing relations between the users and the bureaucracy by handing over control over outlets and operating devices at the level of the minor water bodies. This amounts to a significant change in control relations and an altering of the balance of power between the two types of actors. For lower-level staff this means the loss of opportunity to extract illicit payments from farmers for releasing water, which explains the resistance on their part (Narain, 2003).[9]

The bureaucracy is involved in implementing a policy that results in a diminution of its own status, role and importance. This in itself tends to slow down the pace of reform. This resistance is stronger at the lower levels of the bureaucracy that fear a loss of control, power or jobs. This finds expression also at the higher levels of the bureaucracy in the inconsistent behaviour of the Irrigation Department, delays in rehabilitation of projects and intermittent withdrawal of support for effecting management turnover (Narain, 2003). These actions represent the strategies employed by segments of the irrigation bureaucracy to slow down the pace of reform.

In some parts of Maharashtra, for instance in Aurangabad, Jalna, Parbhani and Nanded districts, some WUAs were formed through the efforts of the Irrigation Department officers. Here, too, the rehabilitation of projects took a very long time, about one to two years after registration. This delayed the signing of Memoranda of Understanding (MOUs) between the irrigation department and the users. The main reason cited by the Irrigation Department was the non-availability of funds. Similarly along Distributary 4 of the Mula Right Bank Canal (MRBC) a proposal to form six WUAs found ready acceptance by the farmers (SOPPECOM, 1997). However because of the paucity of funds, rehabilitation could not be completed even after registration and signing of the MOUs. Five out of six WUAs decided to take over the management even without rehabilitation after the drought of 1994–95. One resolved to take over the following year. The MOU was finally executed in 1997.

Lele (2000) reported that turnover in about 400 WUAs covering an area of 150 000 ha was pending due to incomplete rehabilitation of the systems. Although the government of Maharashtra issued instructions to allocate 10 per cent of the O&M grants for rehabilitation, even this was not monitored, and was therefore not fully utilized for this purpose. Besides, Lele argues, this amount was inadequate.

The justification presented by the Irrigation Department and the Directorate of Irrigation Research and Development (DIRD) for the delay in rehabilitation has been that funds are inadequate for the repairs and maintenance that has to be carried out prior to the handover of systems to farmers. They also feel that there is insufficient demand for PIM. Of relevance here is the tug of war that was taking place in Maharashtra between the irrigation bureaucracy and the two NGOs that were working with farmers to effect irrigation management transfer.

7.3.3 Vested interests among farmers

Another impediment is that WUAs have become heavily politicized and farmers who have benefited from the status quo have resisted reforms. The fact that farmers owe allegiance to different political parties deters their collaborative effort. Besides, there is a vested interest among a lobby of farmers that resists the reform effort. This resistance comes from the larger, influential farmers who are able to manipulate water supplies in their favour.

The strategies employed by the farmers to thwart change include threats to policy entrepreneurs through blackmail and letters, withholding payment of fees, and creating and perpetuating political factionalism. Narain (2003) notes that in the Mula Project in Maharashtra the large and powerful farmers who could bribe the Paatkari to get water when they needed it resisted efforts at WUA formation. Since both tended to gain from the status quo, it dampened the WUA formation effort. SOPPECOM, the NGO that was working to implement management transfer, received a threatening letter from a member of the State Legislative Assembly and large sugarcane grower demanding that it 'stop all this PIM business', since he had always got water without paying for it.

There is also reluctance on the part of farmers to meet their financial commitments (Narain, 2003). Farmers are required to deposit share capital, but do not do so. The formation of WUAs creates a relationship of mutual accountability that farmers seek to evade: 'by forming a society, the member farmers will be required to pay irrigation charges regularly, something which they can avoid by not forming a society' (Pendse and Bhogle, 1994, p. 48).

Another aspect of the street-level bureaucracy's involvement in the process, paradoxically, is its complete lack of ownership of the reform process, as seen for instance in Haryana (Narain, 2003). While examining the implementation of policies for participatory irrigation management in Haryana, Narain (ibid.) noted that there were different views circulating within the government on the policy for WUA establishment revealing little real interest or commitment to implementing the policy initiative.

Many field-level officers of the Irrigation Department admitted that they had 'nothing to do' with water user associations. Lining of watercourses – the purpose for which WUAs were primarily set up – had been the domain of the lining wing of the Command Area Development Authority (CADA) and the Haryana State Minor Irrigation and Tubewells Development Corporation (HSMITC).

On the other hand some field-level CADA officers felt that these WUAs were not sustainable because there was limited sustained involvement of other organizations at the village level, in particular the Irrigation Department and the village Panchayat. Proponents of this view held that unless other village-level institutions were involved, the WUAs would not 'deliver results'.[10] Several CADA officials felt that the Irrigation Department, the canal patwaari and the Block Development Officer should take a greater interest in the working of these WUAs. The rationale behind this position was that the CADA withdrew soon after the task of lining the watercourse had been completed and turnover of management of the watercourses had been affected.[11] Thus there is a tendency both to 'pass the buck' to other segments of the bureaucracy, as seen in Haryana, and to look for escape routes, such as the non-availability of funds, as seen in Maharashtra.

Some of the deterrents to the policy process are only indirect and manifest themselves as wavering political commitment and inconsistent proclamations of politicians. High-level political support is now widely acknowledged as a necessary ingredient for the success of irrigation reforms:

> Political patronage for ownership of the programme and spreading the message across will provide necessary impetus to the programme. PIM will have a better mass appeal, when elected representatives promote it. Once ministers and legislatures are committed to PIM and strongly advocate the cause of PIM, it will be far easier for the stakeholders to get convinced to own the programme. The message from this forum will spread faster and have far better results than by the irrigation agency. (Ailawadhi and Bansal 2001, p. 18)

However officers at the CADA and the HSMITC, the two organizations involved in forming WUAs in the state, were quick to point out that the required 'political will' for WUA formation was absent. The policy approach to establishing WUAs was given concrete shape during the reign of the state's former Chief Minister, Bansi Lal. In 2000 a new state government headed by Om Prakash Chautala came into power. Chautala contested the elections to the State Legislative Assembly by promising farmers free water and electricity. In March 2000 he announced his government's commitment to raising agricultural production by repairing watercourses and carrying out free lining of watercourses. Populist announcements such as this contradict the approach of the CADA and MITC. According to several field-level officers they weaken their efforts and dampen farmer enthusiasm. It takes much effort to persuade farmers to contribute. Once achieved but the corresponding share of money from the state government is not forthcoming, the credibility of state agencies is dealt a blow.

7.4 Conclusion

This chapter seeks to elucidate the role that specific individuals have played in formulating and implementing policies for PIM in India. Meijerink and Huitema (Chapter 2 in this volume) note that policy change can be initiated both in- and outside the government. In India actors outside the government, namely bilateral and multilateral organizations and NGOs, have clearly initiated change. The genesis of these policies can be seen to be located in certain development narratives and policy discourses emphasizing the importance of building local-level institutions for irrigation management that mobilize and involve users. There has been an international discourse surrounding participation and user involvement in irrigation management; multilateral and bilateral institutions in different Indian states have propagated this. At the same time NGOs have acted in their capacity as policy entrepreneurs by mobilizing farmers for irrigation management and lobbying the government for policy change. A major limitation of the role of NGOs has been that while they have succeeded, they have not been able to operate at more than a very local level. They have also failed to involve farmers in demanding change. In some states, policy networks involving the bureaucracy and the political establishment have succeeded on their own in bringing about policy change.

In such states as Gujarat and Maharashtra where NGOs have acted as policy entrepreneurs, their dominant strategy has been to work through pilot experiments, organize and mobilize farmers, garner support among them and then to lobby the state governments. While this slow, but steady approach has been successful at the local level, it has constrained their ability to produce reform on a larger scale. NGOs working in states such as Karnataka have redirected their strategy to involve farmers themselves in negotiating with the government for policy change and taking it forward.

Multilateral and bilateral organizations have pushed the PIM agenda down the throat of state governments by mandating user involvement as part of their financial support programmes. They have also provided financial support for visits of state officials to countries whose PIM models they have wanted to see planted in India. NGOs, too, have received support from both types of organizations for work to mobilize farmers for WUA establishment.

Strategies to thwart change become more interesting at the implementation level where street-level bureaucracies play a major role. Strategies used by state irrigation departments include passing responsibility to other sections of the bureaucracy and looking for escape routes. More specifically, irrigation departments have blocked efforts to change by delaying the rehabilitation of irrigation systems and identifying the lack of financial resources and political will as plausible excuses. Segments of the bureaucracy also influence reform outcomes through passivity, as seen in the case of Haryana, where irrigation department officers do not identify with the reform initiatives at all and disown them as 'someone else's baby'. In failing to facilitate reform efforts, non-action retards them.

Farmers often have a vested interest in the status quo. The strategies they have used, as seen in Maharashtra, include threatening policy entrepreneurs, perpetuating political factionalism and refusing to cooperate by not honouring their financial commitments. Politicians influence policy outcomes through pronouncements that contradict the prevailing policy environment, as seen in the case of Haryana, where populist pronouncements go against the approach of state agencies engaged in PIM initiatives.

On the whole, policy implementation has been slow as the most crucial actors – the farmers and the irrigation bureaucracy – have resisted change, employing different strategies or 'non-strategies' as described above. It is not surprising, then, that euphoria about PIM in India is slowly dying out.

Notes

1. This should however not be confused with traditional systems of community-based irrigation management that have existed in India for centuries. See for instance Agarwal and Narain (1997).
2. The USAID and the World Bank have insisted on WUA formation as a reform condition for financial support in the case of their projects on the Jayakwadi Dam (Pant, 2000).
3. For a review of the 'Philippine model' see Korten and Siy (1989).
4. See also Brewer et al. (1999, pp. 280–81), where the authors note that the chief reason for WUA formation being under serious consideration in the state is 'World Bank pressure'.
5. For a discussion of the role of models or prototypes for reform see Mollinga (2001).
6. For a similar analysis of the role of civil society organizations in irrigation reform in the state of Gujarat see Shashidharan (2000) and Ballabh (2002). For a review of issues in NGO accountability see Najam (1996).
7. Parathasarathy (2000) notes other bottlenecks to PIM reforms in the state. There are difficulties in convincing farmers to participate and differences in the understanding of PIM policy at different levels of the bureaucracy. Further, Parthasarathy argues that WUAs need to be more efficient in making allocative and investment decisions for which a clarification of legal rights is imperative.
8. Nitin Bassi (personal communication, 26 May 2008) further notes that most of the WUA executive or management committee members were not sure of their roles and responsibility in irrigation management. There were two reasons for this. First, most of the power functions, like collection of irrigation revenue, distribution of water and administrative control of WUA functioning, remained in the hands of the Water Resources Department thus resulting in less involvement of farmers in the process. In addition, Madhya Pradesh had undergone second WUA elections and the new executive or management committee members seemed confused about or unaware of their roles and responsibilities.
9. A similar observation is made in regard to the implementation process of PIM reforms in Madhya Pradesh (personal communication, Nitin Bassi, 26 May 2008).
10. A related issue is whether the functions that are assigned to WUAs can be handed over to the village Panchayat, the unit of local governance at the village level, intended to be a democratically elected village body with representation from different communities. An interesting observation in the state was that the Jaats, the agriculturists, showed a strong resistance to this. Agriculture, and therefore irrigation, was seen as the domain of those who had land and would be affected, rather than of a village Panchayat (Narain, 2003).
11. In fact the CADA is known to have two structural bottlenecks: (a) its staff are on deputation and there is little institutional continuity; and (b) it is still seen as a technocratic agency staffed entirely by irrigation engineers and technician agronomists, without sociologists, professional trainers, rural economists or rural development experts (Maloney, 2000). Mollinga (1998) argues that in Karnataka the CADA remained by and large ineffective. On account of the weakness of the CADA in its number of staff, it remained extremely dependent on the Irrigation Department. The latter sought to protect its own territory and resisted the formation of a coordinating organization between the CADA and itself. For a further review of the experience with command area development see also Hooja and Kavdia (1994).

References

Agarwal, A. and S. Narain (eds) (1997), *Dying Wisdom: Rise, Fall and Potential of India's Traditional Water Harvesting Systems*, New Delhi: Center for Science and Environment.

Ailawadhi, R.K. and R. Bansal (2001), 'Participatory irrigation management in Haryana', paper written for the Sixth International Seminar on Participatory Irrigation Management, Beijing, 21–26 October.

Ballabh, V. (2002), 'Emerging water crisis and political economy of irrigation reforms in India', *paper prepared for workshop on Asian Irrigation in Transition: Responding to Challenges Ahead*, Asian Institute of Technology, Bangkok, Thailand, 22–23 April.

Bassi, N. (2008), 'Are sector reforms working? Assessing implementation of irrigation management Act of Madhya Pradesh', in *Managing Water in the Face of Growing Scarcity, Inequity and Declining Returns: Exploring Fresh Approaches*, Proceedings of the 7th Annual Partners' Meeting, IWMI – Tata Water Policy Research Programme, Hyderabad: International Water Management Institute, 2–4 April.

Brewer, J., S. Kolavalli, A.H. Kalro, G. Naik, S. Ramnarayan, K.V. Raju and R. Sakthivadivel (1999), *Irrigation Management Transfer in India: Policies, Processes and Performance*, New Delhi, India and Oxford, UK: IBH Publishing Company.

Brinkerhoff, D.W. and B.L. Crosby (2002), *Managing Policy Reform*, Bloomfield, CT: Kumarian Press.

Economic Development Institute (EDI) (1998), *Handbook on Participatory Irrigation Management*, Washington, DC: Economic Development Institute of the World Bank.

Grillo, R. (1997), 'Discourses of development: the view from anthropology', in R. Grillo and R.L. Stirrat (eds), *Discourses of Development: Anthropological Perspectives*, Oxford: Oxford International Publications.

Hajer, M. (1995), *The Politics of Environmental Discourse: Ecological Modernisation and the Policy Process*, Oxford: Clarendon.

Hardin, G. (1968), 'Tragedy of the commons', *Science*, **162** (3859), 1243–8.

Hill, M. (2005), *The Public Policy Process*, 4th edn, Harlow: Pearson Education.

Hooja, R. and L.K. Joshi (2000), 'Introduction', in L.K. Joshi and R. Hooja (eds), *Participatory Irrigation Management: Paradigm for the 21st century*, Jaipur: Rawat Publications, pp. 1–16.

Hooja, R. and P.S. Kavdia (1994), *Planning for Sustainability in Irrigation: Command Area Development and India Gandhi Canal Project*, Jaipur: Rawat Publications.

IIM(A)-IWMI (1999), 'Maharashtra IMT Impact Assessment', a collaborative study by the Indian Institute of Management, Ahmedabad and International Water Management Institute, Sri Lanka, unpublished, January.

IndianPIM (2001), 'Irrigation management transfer: sharing lessons from global experience', Working Paper/Monograph Series No. 3, New Delhi, July.

Kolavalli, S. and J.D. Brewer (1999), 'Facilitating user participation in irrigation management', *Irrigation and Drainage Systems*, **13**, 249–73.

Korten, F.F. and R.Y. Siy (eds) (1989), *Transforming a Bureaucracy: The Experience of the Philippine National Irrigation Administration*, West Hartford, CT: Kumarian Press.

Kulkarni, D.N. and S.Y. Kulkarni (1994), 'Formation of WUAs in Maharashtra: procedural dynamics', in M.V.K. Sivamohan and C. Scott (eds.), *India: Irrigation Management Partnerships*, Hyderabad: Booklinks Corporation, pp. 53–68.

Lele, S.N. (2000), 'Participatory irrigation management in Maharashtra: a case study', in L.K. Joshi and R. Hooja (eds), *Participatory Irrigation Management: Paradigm for the 21st Century*, Jaipur: Rawat Publications, pp. 618–27.

Lele, S.N. and R.K. Patil (1994), *Farmer Participation in Irrigation Management: A Case Study of Maharashtra*, Society for People's Participation in Ecosystem Management, Pune, New Delhi: Horizon India Books.

Lipsky, Michael (1980), *Street-Level Bureaucracy: Dilemmas of the Individual in Public Services*, New York: Russell Sage Foundation.

Maloney, C. (2000), 'PIM projects in Andhra, Kerala and West Bengal: policies to escalate support services', in L.K. Joshi and R. Hooja (eds.), *Participatory Irrigation Management: New Paradigm for the 21st Century*, Jaipur: Rawat Publications, pp. 652–83.

Maloney, C. and K.V. Raju (1994), 'Principles of farmers' organisation for water management: from participation to management', in C. Maloney and K.V. Raju (eds), *Managing Irrigation Together: Practice and Policy in India*, New Delhi: Sage Publications, pp. 15–31.

Mollinga, P.P. (1998), 'On the waterfront: water distribution, technology and agrarian change in a south Indian canal irrigation system', PhD thesis, Wageningen: Wageningen Agricultural University.

Mollinga, P.P. (2001), 'Power in motion: a critical assessment of canal irrigation reform, with a focus on India', IndianPIM Working Paper/Monograph Series No. 1, New Delhi: Indian Network on Participatory Irrigation Management.

Najam, A. (1996), 'NGO accountability: a conceptual framework', *Development Policy Review*, **14** (4), 339–53.

Narain, V. (2003), *Institutions, Technology and Water Control: Water Users' Associations and Irrigation Management Reform in two Large-Scale Systems in India*, Hyderabad: Orient Longman.

Narain, V. (2008a), 'Crafting institutions for collective action in canal irrigation: can we break the deadlocks?' in V. Ballabh (ed.), *Governance of Water: Institutional Alternatives and Political Economy*, New Delhi: Sage Publications, pp. 159–72.

Narain, V. (2008b), 'Reform in Indian canal irrigation: does technology matter?' *Water International*, **33** (1), 33–42.

Pant, N. (1999), 'Impact of irrigation management transfer in Maharashtra: an assessment', *Economic and Political Weekly*, **34** (13), A-17–26.

Pant, N. (2000), 'Impact of irrigation management transfer in Maharashtra', in L.K. Joshi and R. Hooja (eds), *Participatory Irrigation Management: Paradigm for the 21st Century*, Jaipur: Rawat Publications, pp. 579–617.

Parthasarathy, R. (2000), 'Participatory irrigation management programme in Gujarat: institutional and financial issues', *Economic and Political Weekly*, **XXV** (35) and (36), 3147–54.

Pendse, M.D. and S.G. Bhogle (1994), 'Participatory irrigation management in Maharashtra: training and action research', in M.V.K Sivamohan and C. Scott (eds), *India: Irrigation Management Partnerships*, Hyderabad: Booklinks Corporation, pp. 35–52.

Shashidharan, E.M. (2000), 'Civil society organizations and irrigation management in Gujarat, India', in P.P. Mollinga (ed.), *Water for Food and Rural Development: Approaches and Initiatives in South Asia,* New Delhi: Sage Publications, pp. 247–65.

Shylendra, H.S. (2004), 'The emerging governance paradigm and its implications for poverty alleviation and equity', Working Paper 182, Anand, Institute of Rural Management.

SOPPECOM (1997), 'Draft Report on Action Research Programme taken for scaling up Participatory Irrigation Management and conjunctive use of water in Maharashtra', Pune: Society for People's participation in Ecosystem Management.

Sutton, R. (1999), 'The policy process: an overview', Working Paper 118, London: Overseas Development Institute.

Thomas, J.W. and M.S. Grindle (1990), 'After the decision: implementing policy reforms in developing countries', *World Development*, **18** (8), 1163–81.

Turral, H. (1995), 'Devolution of management in public irrigation systems: cost shedding, empowerment and performance. A review', Working Paper 80, London: Overseas Development Institute.

World Bank (1996), *The World Bank Participation Source Book*, Washington, DC: World Bank.

World Bank (2005), *Shaping the Future of Water for Agriculture: A Sourcebook for Investment in Agricultural Water Management*, Washington, DC: World Bank.

8 Averted crises, contested transitions: water management in the Upper Ping River basin, northern Thailand

Louis Lebel, Po Garden, Nutthawat Subsin and Sakkarin Na Nan

8.1 Introduction

Fundamental changes in the content of policy, how policy decisions are made and how management functions are organized have been theorized and explained in many different ways (Meijerink and Huitema, Chapter 2, the theoretical introduction to this volume). This chapter is about four such transitions in water policy in Thailand and how they played out in the Upper Ping River basin in northern Thailand (Map 8.1). In it we explore the roles and strategies of individuals, from a conventional interest in leaders, experts and networks through to an examination of the more diffuse agency of relatively independent actors driven by shared circumstances and objectives.

The national significance of the Upper Ping arises from its strategic position, distinct history and ecological changes. The Ping is the largest tributary of the Chao Phraya (Map 8.1). The Bhumipol Dam, constructed in 1964, marks the lower end of the Upper Ping and is still the largest storage dam and hydropower source within Thailand.

In Thailand national water policy has for decades been driven by elite interests centred in the capital, Bangkok. Dams on the main tributaries of the Chao Phraya River were built and operated to produce electricity for urban–industrial development and regulate monsoonal-varying flows for flood protection and irrigation of the surrounding central plains (Maiklad, 1999; Molle, 2007b). Behind the scenes a centralized technical bureaucracy diligently pursued visions in which it brought standardization and orderliness to diverse locally managed irrigation and rain-fed farming systems.

For centuries the Upper Ping was the central artery of the separate Lanna Kingdom with Chiang Mai as its capital and a key trading town in other periods when under Burmese rule. Royalty and elites have built canals and made rules about water, labour and taxes on harvests that extend back more than 700 years. Throughout this time and up to the present locally built and managed irrigation systems have persisted as another infrastructural and institutional layer (Cohen and Pearson, 1998). The social–ecological legacy is complex landscape and multilevel and polycentric systems of governance for water in which politics of scale are critical (Lebel, 2005b; Lebel et al., 2005; Tan-kim-yong et al., 2005). National governments, powerful as they have been, have not had a monopoly on water governance in northern Thailand. Policies have had to take into account, or risk being completely reshaped by, local institutions, infrastructure and politics.

Over the last five decades (1960–2009) water and land use in the Upper Ping has been transformed by the expansion and intensification of agriculture, urban–industrial

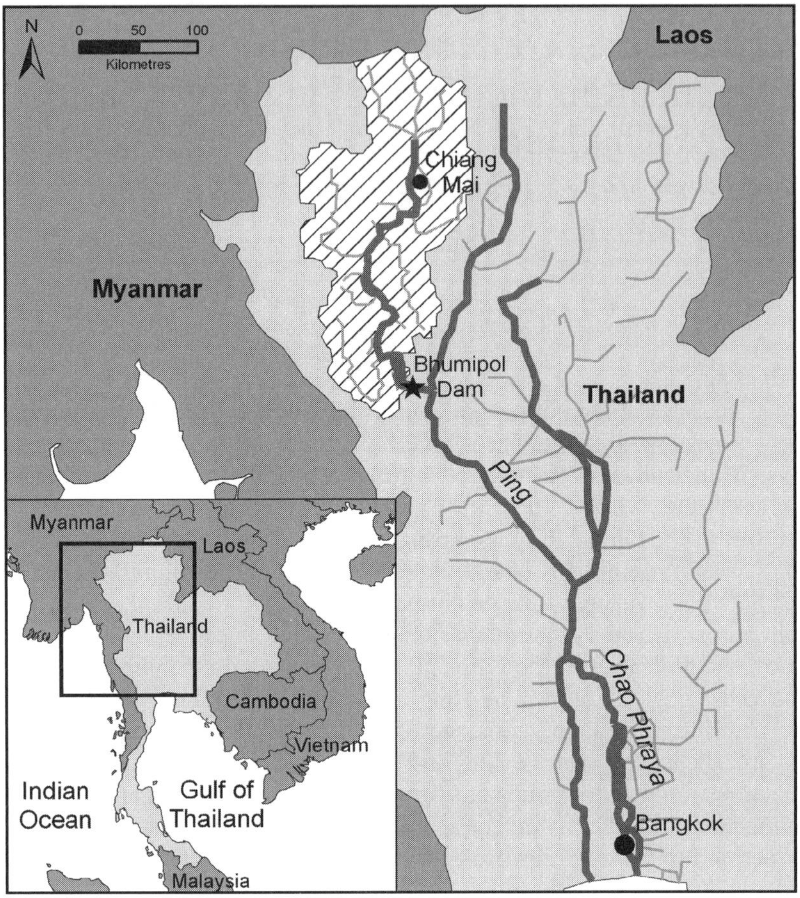

Map 8.1 *Thailand and the Upper Ping basin, including the locations of Chiang Mai, the Bhumipol Dam and Bangkok*

growth and tourism (Rigg and Nattapoolwat, 2001). At the same time interests in forest use and then conservation and upland watershed management have increased. This has produced new conflicts over water allocation, land uses and the ecosystem services from watersheds and floodplains (Walker, 2003). It has also created new incentives, increased technical capacities and political opportunities for pursuing major changes in water management policy. Thus it was not surprising that the Upper Ping was one of the two pilot basins selected from across the country for introducing river basin organizations to promote integrated water resource management (IWRM) (CMU, 2004; Thomas, 2006).

These special features drew us to water management in the Upper Ping as an intriguing case of policy stability and change. On the one hand there were strong social–ecological legacies and patterns of multilevel governance that could both influence and reinterpret national-level policy transitions. On the other hand were diverse social and ecological consequences of rapid economic development that could produce new policy entrepreneurs, coalitions and viable strategies.

Table 8.1 Selected features of three water policy transitions

Transition	Objectives	Infrastructure	Institutions
Wet to dry	All year around irrigation, not just diversion in wet season	Larger storage dams and delivery canals; pumps and wells.	Agricultural export support
Farm to city	Provide secure supplies to urban and industrial users; reduce flood risks	New tap water provision; drainage and flood protection Hydropower	Service agreements and pricing Operating rules for dams and gates Warning systems – compensation
Good to services	Manage multiple services not just allocation of main river flows	Acknowledge ecosystems as service providers	Land use planning Watershed management Multistakeholder engagement

8.2 Transitions

8.2.1 Introduction

Based on our analysis of historical documents and interviews we recognized three policy transitions during the last five decades (Table 8.1). Each transition involved major shifts in the content of policy – such as amounts and types of infrastructure that should be built and priorities for allocation of water – or changes in how policy is made or management functions are organized (Table 8.1). These three transitions were chosen because they were the most significant attempts to influence management of water in the Upper Ping. A case study based on the delta around Bangkok or focused on aggregate national-level policies would not have yielded the same set. In the following sections of this chapter we describe each of the transitions in more detail.

8.2.2 Wet to dry

The wet-to-dry transition (Table 8.1) was driven by the broader objective of intensification, which was met through multiple cropping. Dry season cropping at a large scale first began in the Chao Phraya Delta (Molle, 2004), but was soon followed by demand for the same in the Upper Ping. With the expansion of multiple cropping for export beyond the basin – growing a second rice crop or alternative crops like soya bean, onions, vegetables or garlic in the dry season – pressure on water resources increased greatly in communal irrigation systems (Cohen and Pearson, 1998). The wet-to-dry transition was rationalized and maintained by a development discourse of rural modernization and a dominant policy framework based on state control of water using physical infrastructure. In the initial stages it was strongly supported by international aid programmes.

National development plans The larger water infrastructure projects of the 1960s to 1980s were announced as part of the five-year government development plans. The First National Economic and Social Development Plan (1962–66) identified the export

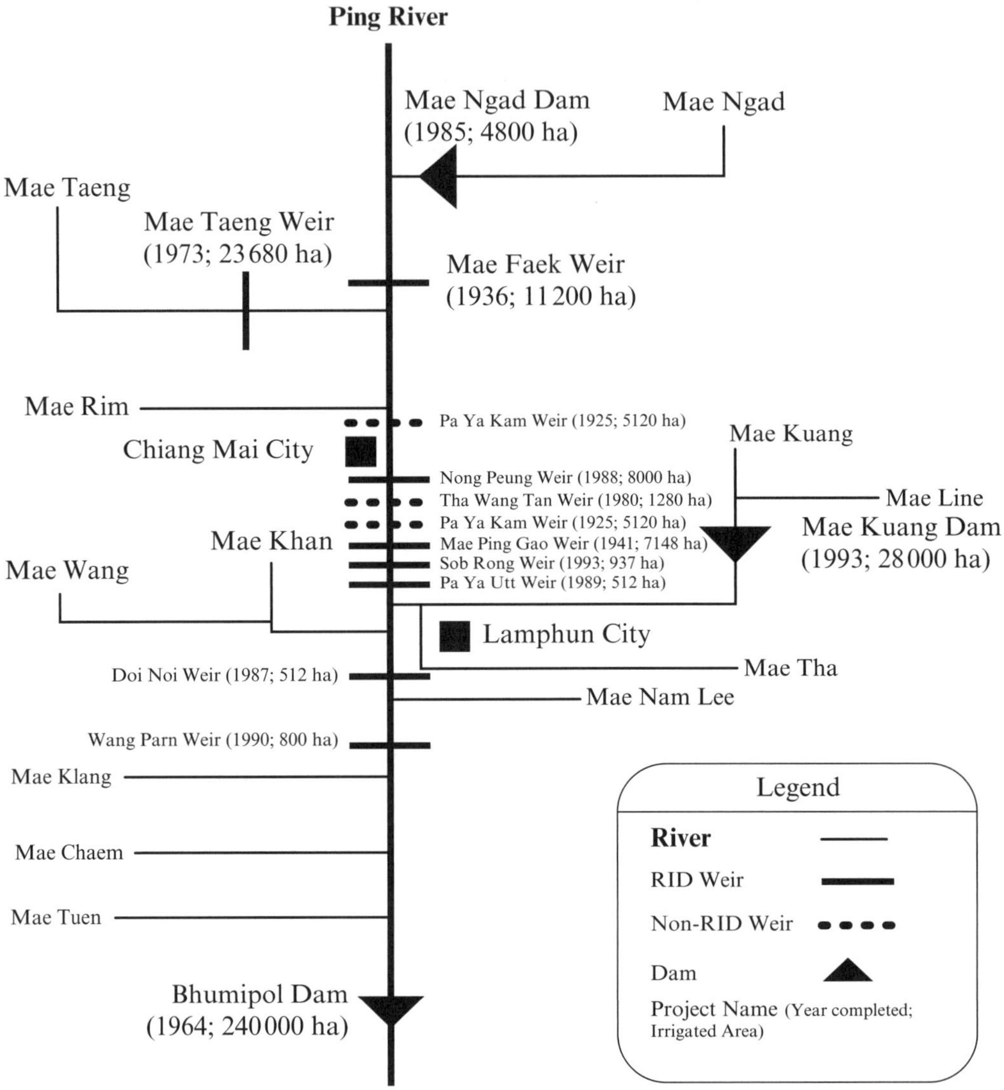

Figure 8.1 Schematic representation of larger water infrastructure in the Upper Ping River basin

potential of the central plains of Chao Phraya Basin. Increasing need for electricity, agriculture and manufacturing led to the construction of a dam at the lower end of the Upper Ping (Figure 8.1) named after His Majesty the King Bhumipol Adulyadej, a critical supporter of investments in water infrastructure projects in Thailand.

In the Upper Ping basin, the Mae Taeng Project was constructed to solve the problem of 'water shortages' for agriculture (Figure 8.1). In the Third National Economic and Social Development Plan (1972–76), major large-scale dams in the Upper Ping basin – the Mae Ngad and Mae Kuang Dam project – began construction (Figure 8.1). In the

Fourth National Economic and Social Development Plan (1977–81) the Royal Irrigation Department (RID) accelerated the replacement with permanent concrete structures on traditional irrigation and established the Northern Industrial Estate in Lamphun Province. There are now four major water infrastructure projects in the upper Ping basin: Mae Faek Weir, Mae Taeng Weir, Mae Ngad Dam and Mae Kuang Dam (Figure 8.1). The cumulative area irrigable in the wet season rose between 1972 and 2005 from about 50 000 to 214 000 hectares.

Community-based irrigation As a result of these projects much of the community-based *muang-fai* infrastructure was replaced in the 1960s–1990s by a system of dams and larger canals run by the state (Surarerks, 1986; Pearson, 1999). The *muang-fai* irrigation system can be traced back to the twelfth century Mon Kingdom of Haripunchai (Sektheera and Thodey, 1974). The weirs (*fai*), which are made from locally available materials – bamboo stakes, logs, leaves and stones – raise the water level for diversion into a canal (*muang*).

Muang-fai systems are managed to cope with dry spells during the wet season, variable timing in onset of the monsoon, and in suitable locations, sharing dry season base flows (Tan-kim-yong, 1983; Tan-kim-yong et al., 2005). Overall this made them amenable, at least in initial phases, to the intensification in cropping which followed with commercial development of first rice and then other crops (Tan-kim-yong, 1983; Ganjanapan, 1984).

Muang-fai irrigation practices have also been adopted and adapted by Lua and Karen communities in small upland valleys (Tan-kim-yong, 1983). Upland irrigation systems are usually technically and institutionally simpler than lowland ones because greater slopes make the gravity-based water system easier to manage (Tan-kim-yong, 1983). In a few low-lying very flood-prone locations the main growing season with *muang-fai* is outside the periods with peak flows.

Today *muang-fai* institutions are diverse: some function like a modern association, some are informal groups undersigned by provincial government, some draw on the power of democratically elected local government. Now about 70 per cent of the irrigated agricultural land is managed and maintained by government agencies while the rest is run by a variety of communal systems (Cohen and Pearson, 1998). Much of the new infrastructure was introduced by the central bureaucracy through the RID with minimal notification or consultation with farmers or local irrigations systems, some of which may be downstream and affected by these interventions (Tan-kim-yong et al., 2005). Positions in the landscape create a diversity of circumstances for water management within the Upper Ping.

The People's Irrigation Act of 1938 recognized the role of local communities in managing irrigation systems while local social practices of construction and maintenance established rights of use. RID went through several iterations of trying to organize farmers, forming water groups and associations. IFAD (International Fund for Agricultural Development) and the World Bank, for example, supported participatory irrigation projects in 1985–90, but these failed to be institutionalized (Tan-kim-yong et al., 2005). Formal groups established and supported by RID to regulate water allocation among farmers in the laterals and sublaterals often collapse from lack of community support because they leave inadequate scope for local management decisions (Molle, 2007a).

The reduction in local autonomy in managing water resources that emerged in the wet-to-dry transition is a point of ongoing contention. The infrastructure built gave greater management control of water to the RID. First, it controls and monitors the flow and distribution of water through a sluice gate along the canal. Second, it imposed rules and regulations that farmers must follow. Third, it limits areas of agriculture as well as the cropping system through water distribution calendars, especially in the dry season.

Farmers are called for a meeting by the RID before the dry season starts in order to inform them about the water budget and areas that could be covered. RID officials, it was explained to us several times, discuss a water budget that is about 80 per cent of the actual amount available because they know plantings will exceed agreed plans. Farmers also know they can bargain for water if their ready-to-harvest crops end up short, as nobody wants to see poor farmers losing their crops to 'drought'. This gaming on allocation is also affected by the farm-to-city transition (see next section) as the dry season allocation for urban consumers leaves less and less water for dry season crops each year.

The policy transition from a management regime focused on supporting wet season rice based on diversion of monsoon-induced floods to a multiseason and mixed-crop calendar required and was supported by large expansion in irrigation infrastructure built and managed by the state. Individual farmers, local *muang-fai* groups and a highly institutionalized RID organization have been the key actors in this transition. Implementation of the new policy lost some of its momentum before it was completed as people left agriculture to get a higher education and then took up work in industry or services: the farm-to-city transition.

8.2.3 Farm to city

The farm-to-city transition (Table 8.1) has two dimensions: first, giving priority to urban and industrial users when water is scarce; second, managing floods to minimize risks to urban and industrial infrastructure.

Allocation Issues of allocation only became a serious policy issue when both land and water started to become scarce as agricultural intensification and expansion proceeded and non-farm uses grew greater. Even in this case national leaders have been reluctant to acknowledge the urban bias explicitly. Instead allocation issues have been left either to sort themselves out as regions urbanize and the mix of users shifts, or to embed reallocation under euphemistic labels like integration (see the section on good to services).

In the Upper Ping agriculture now receives lower priority than urban needs. Officials at Mae Kuang dam in some years 'ask' farmers to stop cultivating rice during the dry season in order to save water for urban consumers (Techawongtham, 2004). The RID also encourages farmers to raise less water-consuming crops. However when asked what is the primary objective of the large-scale water infrastructure development in Chiang Mai, an RID official replied: 'The objective of the irrigation project is to give priority to agriculture.'[1]

Industry, hotels, resorts, golf courses, housing estates, army camps and university campuses all draw on water from rivers and canals or groundwater. The new urban users of water from the Mae Taeng Irrigation Project were not part of the official system, and did not abide by its rules, taking water without contributing fees or maintenance, a problem already existing with some farmers (Wytinck, 1997). Supplying water for the urban and industrial sector has not been seen as the main responsibility of the RID, but as something it reluctantly ends up having to do: 'It is frustrating that those non-agricultural water users, particular government institutes do not have annual plans for water demand or requests to the Mae Taeng project. This disrupts our water management planning.'[2]

Farmers also look for ways to beat the system. Small pumps and groundwater wells are important in key weeks at the end of the dry season for orchard growers. The RID has struggled to control pumping of water by many different groups, including state agencies, from canals under its management, a pattern also observed in the Lower Chao-Phraya basin (Molle, 2004).

At the national level successive development plans placed high priorities on expanding provision of clean tap water, first to cities and then to all towns and most villages, with huge investments in associated infrastructure. In the Upper Ping the growth in provision and consumption of water has been dramatic. Provision of tap water for domestic consumption, for example, grew by about 3.3 per cent per year between 1995 and 2005. Residents of rural areas, however, pay less for water than those in urban areas (Madhoo, 2007).

The farm-to-city transition has ambiguities on the ground where the distinction between rural and urban breaks down spatially and functionally. Urban–rural linkages are physical, as interlocking infrastructure; social, through the mobility of labour; and economic, with two-way remittances and market dependencies (Ganjanapan, 1984; Rigg and Nattapoolwat, 2001; Lebel, 2005a). And industry is not purely linked with urban development. The Lamphun Industrial Estate, a key user of water, was set up in a formerly rural area, with good access to water, and not on the immediate outskirts of Chiang Mai city (Glassman and Sneddon, 2003).

Protection The challenge is not only about growth in water use by urban residences and businesses, but also about the changing burdens and risk from flooding in the wet season (Manuta et al., 2006; Lebel and Sinh, 2007). Drainage is the flip side of irrigation. Whereas farmers in irrigated areas have a strong preference for managing water resources so as to avoid dry season water shortages, urban residents place greater emphasis on eliminating the risks of flooding in the wet season.

Floods create political opportunities: new budget lines are allocated for rehabilitation and recovery that can be captured or directed; crises open windows to transfer control of existing resources. In 2005 Chiang Mai experienced a series of major flood events that acted as 'focusing events' for policy change. Several project proposals for protecting Chiang Mai from floods included in consultant or RID reports years earlier – ranging from notions of making more space for water upstream of the city, and stronger enforcement of river encroachment regulations, through to flood walls and removal of traditional weirs – resurfaced as the flood waters receded.

The flood events were turned into a crisis by public figures, business owners and the media, creating opportunities to get projects implemented. The Prime Minister at the time, Thaksin Shinawtra, was proactive, in part because his power base was in the northern region and Chiang Mai his hometown. By February 2006 Deputy Prime Minister Suwat Liptapanlop who had been put in charge of flood prevention had got projects worth about 13 billion baht approved for longer-term funding. Other projects were led by Minister Newin Chidchob of the Prime Minister's Office. Almost immediately local communities organized opposition to the larger dyke and weir-levelling projects (Phanayanggoor, 2006). The opposition was largely not against more infrastructure of any sort per se, but on the way such decisions were being made: which options, who should manage and how they were to be maintained.

The RID manager of the project to replace communally managed weirs in urban parts of Chiang Mai with a system of water gates, Skol Harnpitakul, argues that the: 'Ping River has been encroached upon and is now too narrow in key places. In addition, weirs prevent water from flowing freely. We have to clear the river. Water gates should replace weirs so that water can flow freely in the rainy season, but still be used for agriculture in the dry season' (Anon, 2006). Removal of traditional weirs and replacement with upstream dams and gates operated by the RID would transfer control of water resources back to a state agency.

In the midst of these heated politics, a key bureaucrat, Thada Sukhapunnaphan, Director of the Hydrology and Water Management Centre for the Upper Northern Region, continued to push for use of data in making better-informed decisions on how to manage floods. Early-warning systems need to be based on useable technologies, and information about how to prepare needs to be simple and targeted to different groups.

The politics of floods changed direction, but did not lose momentum following a military coup. Campaign images for the mid-2007 municipal elections for Chiang Mai pictured most candidates standing waist-deep in floodwaters. Debate continues over whether or not to remove several traditional weirs from near the centre of the city (as these also raise water levels and distribute flood waters), whether high walls should be built to protect the city from peak flows, or whether efforts should focus on restoring river channel width, banks and floodplains.

The conflicts between agricultural and urban–industrial management of water are in part a social–ecological legacy. The wet-to-dry transition created a landscape, infrastructure and supporting culture based on agricultural interests, both within the basin and further downstream. Economic development shifted the goal posts and stranded many of the players. Dependencies created by the past continue to constrain policy change, as evidenced by the resistance to central state agency efforts to remove old infrastructure or replace established institutions.

8.2.4 Good to service

The good-to-service transition is about placing less emphasis on consumable quantities and more on the service-rendering quality of the water flows in rivers and across basin landscapes (see Table 8.1). This includes a shift from management focused on stocks of blue water resources (surface flows in rivers and storage in dams) to management that also pays attention to green water (as moisture in the soil and evaporation from plants back to the atmosphere) (Falkenmark and Rockstrom, 2006). The earliest and strongest expressions of a policy change have been with respect to forest conservation and management of upland watersheds, but in more recent years services from lowland wetlands and floodplains have received more attention, for example in flood protection and as land uses are scrutinized more in IWRM approaches.

Watershed zoning Maintaining upper tributary watershed services to lowland areas where irrigation systems are most developed and urbanization is most advanced has been central to the rationale of the Thai state for decades, especially through watershed classification schemes implemented by the Royal Forest Department (RFD). Claims that widespread deforestation in the uplands of northern Thailand are having a major impact on water resources available to the lowlands have been made repeatedly for decades and used to justify a range of policies (Forsyth, 1998; Walker, 2003). The widely and strongly held belief that more protected areas in the basin will translate into more trees and forest cover and that this in turn would mean higher stream flows in the dry season is not strongly supported by scientific evidence (Walker, 2003). The balance of evidence probably supports the converse: at small and intermediate catchment scales, more trees means less water (Forsyth and Walker, 2008).

Regardless, watershed classification has been used as a strategic tool (Lebel et al., 2008). Thus 'watershed', according to Laungaramsri (2000), is a construct of the state that is used primarily to justify control of upland resources (Vandergeest and Peluso, 1995). The 1985 classification is based on overlays of available maps for soils, topography and forest cover that was used to classify each square kilometre in the watershed 'zone' into six categories (Pratong and Thomas, 1990). In practice rather modest information was available on soils for the 'topographically complex' upper tributary watersheds. Although the proportion of land in classes with severe restrictions appears modest at the national level, this proportion increases rapidly in the upland areas that predominate in northern Thailand, underlining the importance of the spatial level at which conservation objectives are set and monitored (Suraswadi et al., 2005; Thomas, 2006).

David Thomas, while he headed the World Agroforestry Centre (ICRAF) in Chiang Mai, served as a key expert with respect to watershed management and upland land use planning. In his work he has formed partnerships with local government and the RFD (Thomas et al., 2000; Thomas et al., 2004). He told us in an interview about participatory planning that their work: 'showed that the whole basis for land use was that people needed to negotiate among themselves the upstream–downstream issues; we showed that the sub-watershed unit was a very useful management unit at the local level – a lot of local names are based on the streams at the sub-watershed level'.[3] His approach is strongly evidence-based. Researchers working with the ICRAF have helped by: 'putting the maps and data and time series on the table and so lifting the standard of debate'. In his view on the ICRAF's role: 'It was not up to us as a research organization to say what was right and what was wrong, but we were saying that this is the evidence of what is, what has been, and what's going on. Now go compare this to what people say and make your own judgments.'[4]

The controversy over watershed classification is closely related to the issue of community forestry. Efforts to link land use and water use have held up community forestry legislation for more than a decade, thus preventing communities from directly managing their resources (Walker, 2004). In 2002 a draft of the bill did not pass after revisions by the Senate excluded parks, wildlife sanctuaries and headwater areas. Access to and use of land in upper tributary watersheds by farmers continues to be subject to ad hoc negotiations with local agencies and administrations rather than guaranteed or secured through national-level policy and law, because these fail to fit local realities. In November 2007 the bill passed by parliament still had controversial elements that appear to remove the rights of some people to use forests that they had been successfully managing (Janchitfah, 2008).

Basin organization, integration and reforms At the national level a key policy was the 2002 administrative reforms that created new hierarchies of coordination and reformalized basin or watershed management.

The organizational change included establishment of a new Ministry of Environment and Natural Resources with a Department of National Parks now distinct from the RFD with responsibilities for water and land management in upper tributaries and a new Department of Water Resources (DWR). The Department of Water Resources is tasked with IWRM. It prepares master plans and policy advice. Its activities are guided by a National Water Resources Committee. It promotes river basin organizations as the implementation tool for IWRM. Its mandate and position puts it in direct bureaucratic competition with the RID on many issues.

The institutional reforms were in response to conditions placed on loans to the agriculture sector by the Asian Development Bank (ADB) following the 1997–98 financial crisis. In this example changes made to water policy at the national level took advantage of a largely external and unrelated crisis. These included guidelines and requirements from promotion of IWRM concepts through to introduction of river basin committees or organizations (RBOs), and the creation of so-called apex regulatory authorities under new water laws (Abonyi, 2005). Lobbying by farmers with concerns over water pricing has been an important factor in the institutional stalemate on a national Water Law that would give administrative, policy-making and regulatory teeth to RBOs, the DWR and the National Water Resources Committee. As of mid-2009 the Water Law had still not been passed, leaving the position of the DWR and various committees ambiguous. An RBO representative we interviewed did not see this as a serious drawback because: 'We don't need the Water Law. We already have the authority through local government representatives.'[5] The impacts of policy inertia in the Upper Ping were modest.

Since the construction of the Bhumipol Dam the Upper Ping has been managed as a unit, but from the perspective of benefits downstream and not those within the upper part of the basin (Molle, 2007b). The creation of an Upper Ping RBO helped legitimize management objectives at this larger level, well beyond individual irrigation schemes within the main valley and one provincial administration (Molle, 2007a). This downstream motivation for the Thai government to improve the management of the Ping River basin is often forgotten in the enthusiasm for integrated planning and public engagement.

Dr Apichart Anukularmphai is widely acknowledged to have played an important role in introducing IWRM and RBO concepts to the Thai bureaucracy and public. He has done so on behalf of various Thai agencies and the ADB. In his presentations he underlines the importance of viewing change as an evolving participatory process with increasingly greater engagement of stakeholders in RBOs.

Efforts to engage stakeholders have been contained by a process in which only certain types of small infrastructure projects are considered, and a complex planning procedure that allows traditional experts in the bureaucracy to regain control of most of the outcomes. At the river sub-basin level institutionalization has intentionally been much more diverse (Thomas, 2006).

Other networks have proliferated in the uplands. The National Parks Department has its watershed management units and networks. Buddhist monks and their followers have become involved through ordaining trees to 'protect' watersheds (Hirsch, 1997). These networks and alliances have formed, in part, to try and influence the direction of current and future transitions, and sometimes further polarize local conflicts (Tan-kim-yong et al., 2005). The creation of basin committees and other organizations added new layers to an already complex area-based hierarchical administrative structure that is woven through with line agencies, local administrations and elected politicians.

The good-to-service transition is driven by a highly diverse set of individuals both at the national and Upper Ping levels, as well knowledge inputs and other influences from outside Thailand. Understanding and objectives on many key notions of services diverge and are often scale-dependent (Lebel et al., 2008). Seemingly attractive concepts like IWRM that might drive further policy change may turn out to be too ambitious and vague to improve planning and management practices (Molle, 2008). As a consequence it is far from certain that the policy transition will continue to unfold, or whether it will falter.

8.2.5 Overlaps

The three transitions described above overlap in time and could thus jointly influence water management. This notion is summarized schematically in Figure 8.2 as a trajectory describing current practices with respect to two key slow variables at, more or less, decadal intervals. Managing risks of too much or too little water is a key variable in the monsoonal climate of the Upper Ping. The choice between focusing more on water for growing food or for a wider range of services is also fundamental and not easy to reverse once made.

In Figure 8.2, the good-to-service transition is captured in movement to the right, the farm-to-city transition by movement towards the upper right, and the wet-to-dry transition in its direction towards the bottom. These simplified trajectories highlight several other issues important to our consideration of roles of individuals and their strategies in the next section. First, a complex mix of actors may be simultaneously trying to influence pathways of change in actual management within the Upper Ping, captured in the visual analogy of pulling a trajectory towards themselves. Some of these actors, like rice farmers, work primarily at local levels, whereas others, like the RID, are present at both Upper Ping and national levels. Second, some actors have similar preferences (with respect to the two variables shown) and thus might be more likely to form coalitions. Third, some actor's interests have changed over time, drifting across the figure. For instance the municipality's interests in services is increasing and so moving further to the right, while the Irrigation Department has had to become more involved in urban flood protection.

8.3 Strategies

8.3.1 Introduction

In the analysis that follows we emphasize the role of individuals and groups, identifying the strategies they used, the objectives they pursued, and what influence they had. Our observations are organized under four headings: origins, narratives, windows and arenas (see Meijerink and Huitema, Chapter 2, the theoretical introduction to this volume).

8.3.2 Origins

Novel or alternative policy visions had different origins in each of the transitions. The wet-to-dry transition was moulded on images of rural modernization in which food was

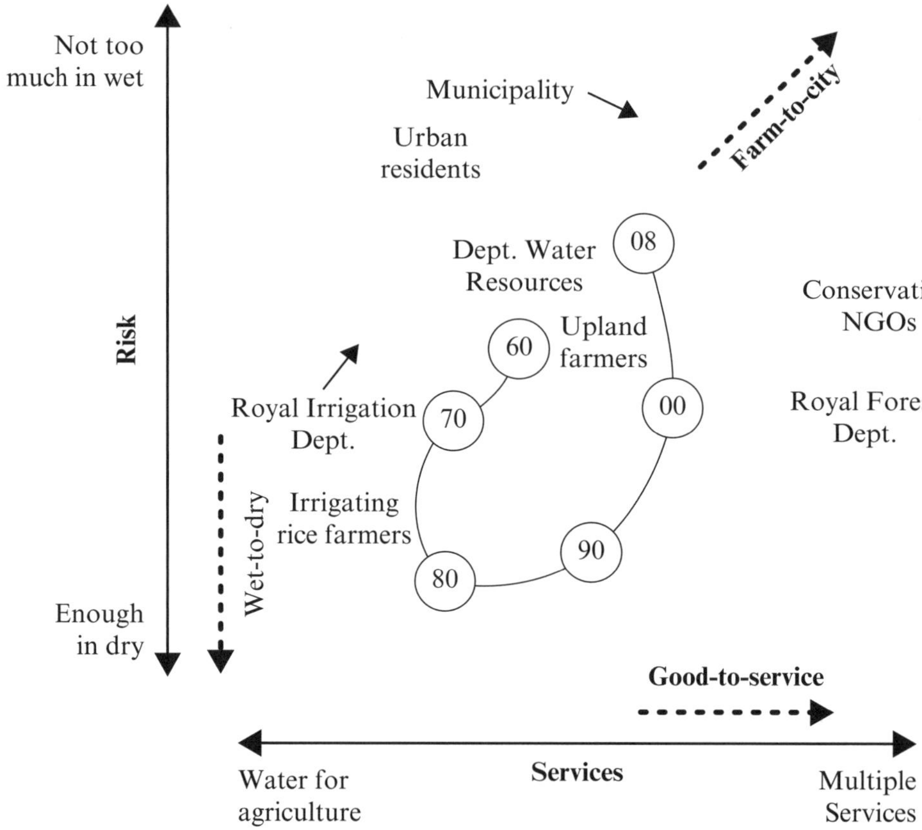

Note: Selected influential actors are plotted approximately where their preferred position lies in relation to these two axes.

Figure 8.2 Schematic summary of pathways of changes in water management, 1960–2008, in the Upper Ping plotted against two key water management variables

grown not just to feed local populations, but also for export, and ultimately became the basis for industrialization. Multilateral banks and then transnational agribusiness firms have played key roles in seeding and cultivating these ideas within successive national governments (Goss and Burch, 2001; Rigg and Nattapoolwat, 2001). Although the Central Plains around Bangkok were the initial focus, a series of national development plans scaled up this vision, making the Upper Ping around Chiang Mai and Lamphun towns the regional centre for northern Thailand (Glassman and Sneddon, 2003). Likewise, multilateral banks had a role in the introduction of the IWRM paradigm in the good-to-services transition which portrayed basin-level management as the solution to the new complexities created by urbanization and industrial growth. As discussed above, domestic experts joined in promoting and helping to refine these ideas in the Thai context.

In more recent years it has not only been experts in bureaucracy or academics that have been proponents of particular ideas important to policy shifts. Consulting firms have grown in importance. Panya Consultants has been particularly prominent in the Upper Ping. It is a professional consulting firm frequently commissioned by the Thai government for water-related projects. It has carried out assessments and feasibility studies for water infrastructure and coordinated participatory planning activities. For example, it conducted the preliminary study of the Mae Ngat-Mae Kuang tunnel that would divert water from the Mae Taeng River to the Mae Ngat Reservoir (Sutiwanich et al., 2006) and was actively involved in the preparatory phases for RBOs and sub-basin organizations (RSBOs). It maintains databases on water resources and its knowledge-organizing capacity has been significant in early phases of projects and policy.

These initial observations fit two common assumptions in the study of policy change: ideas shape interests and policy precedes practice (see the theoretical introduction by Meijerink and Huitema, Chapter 2 in this volume), but in other aspects our exploration of three transitions in the Upper Ping suggests that cause and effect is not always so straightforward.

New ideas may create new interests, but the converse is also true. Ideas about how to manage water in Thailand are strongly institutionalized. The best example is the RID. The RID was central to the wet-to-dry transition, but also a relevant actor in the other transitions. It has been the bastion of expert knowledge on water management for the Thai government. The organizational culture is dominated by engineering professionals and perspectives. A greater differentiation of interests within the agency and how it is represented by its own staff has become more apparent since the 2002 administrative reforms. This can be seen, for example, in the case of flood management roles played by the Hydrology and Water Management Center for the Upper Northern Region. Before this time it did not matter who came to a meeting, as anyone invited would have a good idea what the message would be: we need more water and infrastructure to store and divert it. An RID official we interviewed summarized the organization's logic neatly: 'A dam is a long-term project. It may take decades to overcome obstacles, but it will happen one day.'[6] Strong institutionalized knowledge reduces the role of individuality (cf. Meijerink and Huitema, Chapter 2 in this volume).

The RFD is another key agency in which a parallel process of institutionalized knowledge has taken hold, but with several recent paradigm shifts (Laungaramsri, 2000; Lebel et al., 2004). An initial focus on cutting native trees and then planting trees for forestry was later replaced with concerns for conservation of biodiversity and upper watershed management. The possibilities of local forest management and values of upland rain-fed agriculture, including several different swidden-rotational systems, were dismissed throughout. In part this was a result of larger politics and dominant Thai social norms that view such upland land use as 'obviously' inappropriate. Their identification with minorities of questionable 'Thai-ness' (Vandergeest, 2003) made it simple to translate farming practices into threats to national security, biodiversity conservation and water supply (Forsyth and Walker, 2008). Long-standing organizational interests constrain areas of active inquiry and consequently what knowledge will become available to inform water and land management policy.

Policy change may be more about writing history than making it. In the Upper Ping, upland watershed organizations and lowland irrigation groups existed long before river basin and sub-basin organizations were introduced. Military camps, university campuses and resorts took water from irrigation canals without having a quota or paying.

Table 8.2 Key narratives in the three transitions

Narrative	Transitions
Drought Dry season water shortages are caused by insufficient storage; therefore need to augment supply	Dry → Wet
Forests make water Dry season water shortages are caused by deforestation of upper tributary watersheds; therefore need to protect and reforest	Dry → Wet Good → Services
A natural right Farmers have a natural right to free water. Rice farming is the true, ancestral, roots or foundation of Lanna culture	Farm ← City Dry → Wet
Price is right Water like any other scarce resource should be allocated efficiently, that is, to the highest value uses; sometimes that is no longer agriculture	Farm → City
Living with the river Need to return to living with the river and its flood-ebb cycles as this is the true, ancestral, root of Lanna culture; need to make space for water	Good → Service
Modern life We are a modern society; floods are unacceptable in a city that depends on tourism and related commerce	Farm → City

Note: Arrows (→) (←) indicate support (→) or resistance (←) to the transition.

Industrial estates demanded priority access to water and got it before such policy became explicit. Housing estates raised land levels, redirected irrigation canals and had their own flood protection walls built. Government plans and efforts at coordination came later. Policy transitions at the national level have frequently followed changes in actual practices within the Upper Ping.

8.3.3 *Narratives*

Although new ideas have had some role in policy changes, it has often been through becoming part of more established narratives and has often followed this process. Narratives are influential not because of their accuracy, quality of argument or novelty of ideas, but because they fit beliefs, appeal to interests and deal with relevant social values (Forsyth and Walker, 2008; Molle, 2008).

The drought narrative (see Table 8.2), for example, is very seasonal: politicians and bureaucrats use mass media outlets to 'talk up' the approaching dry season as a 'drought' almost every year (for example Samabuddhi, 2004). This narrative supports a dominant organizational interest within the RID to augment supply rather than moderate demand and the need of politicians to champion projects that appear favourable to the large rural electorate.

Within the Upper Ping, the 'living with the river' narrative (Table 8.2) is associated with part of Chiang Mai civil society, including key individuals like Dr Wasan Jompakdee. In his talks Dr Jompakdee makes use of historical photographs and frequent references to local traditions and Lanna culture. This appeals to nostalgia and the need for identity, but he also introduces understanding about flows, river morphology and sedimentation to argue his points about benefits of riparian restoration (Jompakdee, 2004).

The 'forests make water' narrative – which embodies a core assumption of much forestry and environmental policy in Thailand – has been carefully dissected by Forsyth and Walker (2008). The novel IWRM ideas were quickly justified and embedded in activities targeting upper tributary watersheds in the Upper Ping by the DWR. The notion of watershed protection also fits well with objectives of expanding the protected area system and thus territory under the control of the RFD and the National Parks Department.

Adhering to standard narratives has been a key strategy for individuals wishing to promote and secure national policy changes within the Upper Ping, but they have also had to face resistance in the form of counter-narratives (see the arrows from right to left in Table 8.2).

Most narratives have several proponents. For this reason they can also be thought of as representing 'discourse coalitions'. The drought narrative brings farmers, local government and politicians together with the RID. They are most vocal during dry season conflicts, but also as noted earlier, seed debate with threats of looming drought each year. The 'forests make water' narrative is perhaps the broadest coalition stretching from teachers in schools through to non-governmental organizations (NGOs) focused on conventional approaches to conservation. It also has strong backing from the Thai monarchy who regularly make it a core message in annual speeches.

The 'water is a natural right' coalition is primarily the standard voice for farmers, but with support from rural NGOs and politicians with rural constituencies. The 'natural right' discourse is diametrically opposed by a 'price is right' discourse led by the new water bureaucracy, multilateral banks like ADB and some urban–industrial interests. It should be noted that in Thailand there are, so far, no direct charges for irrigation water; instead the government relies on indirect taxes, for instance on rice exports and land, to finance irrigation projects (Madhoo, 2007). While there is substantial evidence that pricing schemes for domestic and industrial uses can help improve efficiency, the case for enacting such schemes for irrigation projects in Thailand is much more problematic for capacity, institutional and contextual reasons (Molle, 2007c).

The 'living with the river' and 'modern life' narratives also have opposing coalitions, but involving some complex interactions among networks. The issue of removing weirs in the inner city, for example, highlights inconsistencies in what is considered 'traditional' or 'ancestral', and therefore preferable, and what is not.

8.3.4 Windows

One idea in transition management is that it is possible to prepare and wait for windows of opportunity: windows can be exploited by the wise and prescient, but not all windows are recognized beforehand, and sometimes actors only realize that they existed after they have been passed through. Some windows are not seen as opportunities for all the relevant actors, but as threats or looming crises – for some actors a window, for others a brick wall.

Some windows are predictable. In the monsoonal climate of the Upper Ping, rainfall and river discharges are strongly seasonal. Proponents of dams to store or divert water typically begin raising the alarm about 'the drought' in February or March each year and then follow up with solutions involving proposals for new infrastructure. This has been an important strategy in the wet-to-dry transition even as drought has been overtaken by other social–ecological changes in the basin. The converse, talking up floods early in the wet season, also occurs, but involves a very different set of actors, mostly those associated with business and the municipality of Chiang Mai. As observed earlier, turning flood events into crises was an important strategy of local and national politicians following the 2005 events. Crises have opened windows for change that have helped further the farm-to-city transition in the Upper Ping.

Some windows are much less predictable. A convenient alignment of growing public awareness and existing agency interests in upper watersheds has made climate change a useful new issue that has created a window of opportunity. Campaigns to plant trees, protect forests, expand parks, resettle farmers and manage the integrity of the water supply now have another rationale. A similar process can be seen in building up justifications for increasing storage or supply through diversions.

Some windows are obvious only in hindsight and some result from unrelated social transformations and events that fortuitously avert what otherwise would be serious challenges to the launch or continuation of a transition. Several averted crises have been noted in this chapter. Labour shortages for agriculture made possible by irrigation infrastructure investments in the wet-to-dry transition were avoided by an influx of migrant labour from Myanmar and changes from rice to orchards. Serious disputes over dry season water in the farm-to-city transition are being averted by changes in livelihoods and land uses towards non-farm activities, turning peri-urban decision-making bodies more and more towards new urban stakeholder interests. A crisis around loss of upland forests and related services, which the good-to-service transition helped construct, is becoming a non-issue as upland farmers switch to growing trees that maintain many watershed functions or migrate in response to economic opportunities elsewhere (Thomas et al., 2008).

Finally, whether a set of changes looks like a window or not depends on each stakeholder's and observer's own capacities and interests. The 1997–98 financial crisis that struck Thailand and many parts of Asia caused economic hardship and led to immediate cuts in budgets for things like irrigation, but it also helped secure a series of policy reforms triggered by business and civil society concerns after the 1991 military coup by General Suchinda. Various decentralization reforms, including important changes to local government in the mid-1990s, were backed by a participatory and democratically oriented 1997 Constitution. This set the tone for many subsequent political and administrative changes including many related to water. But the crisis also opened windows for the multilateral banks once again to promote their visions of private investment-led development. The Prime Minister at the time, Thaksin Shinawatra, promised numerous mega-projects, like a Thai water grid, as part of his populist platform directed to rural interests in the north and northeast (Molle and Floch, 2008). After the 2006 coup the military regime stepped aside and civil government was restored, and the elected Prime Minister, Samak Sundaravej, followed with more unilateral promises to support a variety of water mega-projects largely as part of a failed attempt at political survival. Windows of opportunity are exploited by many different kinds of individuals acting as entrepreneurs for water policy transitions from which they can benefit.

8.3.5 Arenas

In the Upper Ping, newer institutional forms are more likely to be overlain on pre-existing systems than replace them. This is most obvious when considering management of local irrigation, but also highly relevant for upland watershed management in some locations. This is one mechanism by which established water institutions, both local and state, can be used by individuals to regain power they might otherwise lose through reforms. Such change in governance has occurred despite policy changes aimed at integration and coordination, in the good-to-service transition, or logic of standardization, in the wet-to-dry transition.

The increase in institutional density, complexity and redundancy is not dysfunctional. It provides alternative platforms for deliberation, policy influence at multiple levels in the face of knowledge, and policy and resource uncertainties. Changing circumstances, technologies and environmental conditions as well as shifting values mean that negotiations over water are never over. The flexibility in the current system is higher than in the past and thus potentially valuable to adaptation efforts.

Different actors have responded to complexity in different ways. The strategy of the DWR, as the main agency behind the policy that created the Upper Ping RBOs, has been to push through public consultation processes quickly with little prior information and restriction of debate to small local projects (Lebel and Garden, 2006). Larger schemes, like tunnels and interbasin transfers, were not on the agenda in the various fora in 2005–06. At the river sub-basin level new participants entered deliberations bringing many new issues to the planning tables, but lack of time and resources made it hard to make rapid progress, especially where the mix of stakeholder interests was large (Thomas, 2006).

The number of venues and levels in which an alternative policy idea could be pursued has not grown linearly in the past five decades (1960–2009). Public interest and opportunities to engage with water management ebb and flow. For example, the October 2006 military coup threw out the People's Constitution of 1997. A history of lurching towards democratization and then retreating into centralized, elite rule has been an important overarching context for all three transitions.

Another important contextual factor has been regional water politics (Lebel et al., 2007). It is now difficult for the Thai state to build dams within its territories because of well-organized opposition to such projects by civil society. Instead Thai investors and the government pursue their interests in water and energy projects in neighbouring countries, particularly Myanmar and the Lao People's Democratic Republic, that are under more authoritarian rule. This ongoing politics of scale (Lebel et al., 2005) means transitions at the national level in Thailand influence and are affected by contests over water and projects within smaller areas, like the Upper Ping and the wider Mekong region.

8.4 Conclusion

In the last 50 years three transitions in national-level water policy have been pursued in the Upper Ping (see Table 8.1). These Thai examples show that policy transitions are not tidy or smooth. They may be interrupted, abandoned or left incomplete. It can also be seen that national-level policies often trigger complex interplay among institutions at regional and more local levels. Nevertheless, in Thailand, each national-level transition analysed in this chapter has had impacts upon the use and management of water in the Upper Ping.

The wet-to-dry transition greatly intensified agricultural production by making a second, dry season crop viable for many farmers. This policy shift resulted in construction of substantial new infrastructure for irrigation within the Upper Ping. As a consequence water came under even greater control by the RID.

The farm-to-city transition prioritized growing local urban and industrial needs for water and flood protection against interests of agriculture. This policy has been less explicitly articulated for Chiang Mai then it has historically for Bangkok, but at the operational level urban dry season supply is guaranteed and wet season management aims to minimize risks of flood damage in the city.

The good-to-service transition went beyond viewing water as just a stock that can be stored or a flow that can be allocated or diverted, to acknowledging water as the foundation of many other valuable services to ecosystems and people. This policy initially targeted only land use in mountain watersheds. Now there is much more interest in wetlands, river margins and floodplains. The expanding number of interests has also seen more attention being paid to multistakeholder processes. There are now higher expectations by the public with respect to access to information about projects and opportunities to engage in considering alternative proposals.

Across these three transitions, but in varying combinations and degrees, several common roles of individuals and strategies were observed. For instance, successful strategies include fitting new policy ideas into existing narratives or, conversely, appealing to such narratives to oppose changes (Table 8.2). Some individuals prepare for windows of opportunity, but the viability of this strategy varies with the predictability of openings and capacities. Another strategy was for individuals to engage in several arenas and link issues among them – venue shopping (Meijerink and Huitema, Chapter 2, the theoretical introduction to this volume). Doing so may have helped coordinate or align concurrent transitions at least within the level of the Upper Ping.

Transitions in water policy are an outcome of many social processes at multiple levels. Individuals have variable roles. Sometimes they are the source of new and alternative policy ideals. Sometimes they are institutionalized products of strong organizations. In other cases the influence of individuals comes from a position of authority, the networks they belong to or the coalitions they build. Sometimes the aggregate, relatively independent, practices of individual water users have been driving changes in water policy.

The set of queries raised by Huitema and Meijerink (Chapter 1, this volume) on the role of individuals and their strategies provides a useful framework for understanding multilevel governance and policy processes affecting water management in the Upper Ping. Our analyses, however, suggest that some modest extensions to the current framework are needed.

First, individuals and key agencies can play roles in more than one transition. This creates opportunities for trade-offs and other kinds of interactions if transitions overlap. Thus, in the Upper Ping, several civil society actors and academic experts have engaged in the three transitions. The RID, represented by highly institutionalized experts, has been important as proponent, partner or opponent in all three transitions. Pursuing water for irrigation, while also providing for and protecting municipalities is, on the other hand, likely to involve trade-offs. What strategies are adopted by individuals playing several policy transition games at once?

Second, the boundaries of policy are often ambiguous and may be intentionally so.

What should be considered 'remaking' or just 'implementing' policy when governance is multilevel? A different way of thinking may be warranted if we want to understand more deeply the roles and options for individuals. State, regional and local institutions may be available to actors seeking policy change. How do they use combinations of institutions to empower themselves? When practices precede policy, does that make those farmers and firms taking matter into their own hands the real 'policy entrepreneurs'?

Third and finally, with multiple levels, services and coalitions to take into account, deliberation and negotiation is a never-ending process, but crucial to the pursuit of fair and effective water policy. The processes favoured and content promoted by policy entrepreneurs need more exploration. Was public participation only an instrumental exercise? Or was the engagement meaningful to all stakeholders? In this chapter we did not examine in detail the quality of deliberative engagement in various arenas, but suggest that it is important to do so if the objective is to influence transitions. Those who wish to be 'good', after all, are not the only ones interested or skilled in transition management.

Acknowledgements

Several past staff and students assisted with data collection over the years and were valuable to the preparation of this chapter. In particular we thank Supaporn Khrutmuang, Sophie Carton and Julie Poncet. The early advice, information and inspiration of Francois Molle are gratefully acknowledged. Funding for this work came from: SARCS via the National Central University in Taiwan; the Rockefeller Foundation; the Asia Pacific Network for Global Change Research; and the International Fund for Agricultural Development and Echel Eau through the Challenge Program on Water and Food for Project PN50. Finally, special thanks to Saskia Werners, Zsuzsanna Flachner, Dave Huitema and Sander Meijerink for their constructive comments on earlier drafts of this chapter.

Notes

1. Interview with official from the Royal Irrigation Department, Chiang Mai Province, September 2004.
2. Interview with official from the Royal Irrigation Department, Chiang Mai Province, September 2004.
3. Interview with David Thomas, World Agroforestry Centre, 3 August 2007.
4. Interview with David Thomas, World Agroforestry Centre, 3 August 2007.
5. Interview with river basin organization member, Chiang Mai, 29 August 2005.
6. Interview with official from the Royal Irrigation Department, Chiang Mai Province, December 2004.

References

Abonyi, G. (2005), 'Policy reform in Thailand and the Asian Development Bank's agricultural sector program loan', ERD Working Paper Series, No. 71, Manila: Asian Development Bank.

Anon (2006), 'How and why government tries to get rid of traditional weirs' (in Thai), *Northern Citizen*, **241**, 12–13.

CMU (2004), 'Project to develop a master plan and implementation plan for the conservation and development of environmental and water quality of the Ping River and its tributaries', Final report submitted to the Office of Natural Resources and Environmental Policy and Planning, Ministry of Natural Resources and the Environment [in Thai], Chiang Mai: Chiang Mai University.

Cohen, P.T. and R.E. Pearson (1998), 'Communal irrigation, state, and capital in the Chiang Mai Valley (Northern Thailand): twentieth-century transformations', *Journal of Southeast Asian Studies*, **29**, 86–110.

Falkenmark, M. and J. Rockstrom (2006), 'The new blue and green water paradigm: breaking new ground for water resources planning and management', *Journal of Water Resources Planning and Management*, **132** (3), 129–32.

Forsyth, T. (1998), 'Mountain myths revisited: integrating natural and social environmental science', *Research and Development*, **18** (2), 126–39.

Forsyth, T. and A. Walker (2008), *Forest Guardian, Forest Destroyers: The Politics of Environmental Knowledge in Northern Thailand*, Seattle, WA: University of Washington Press.

Ganjanapan, A. (1984), *The Partial Commercialization of Rice Production in Northern Thailand (1900–1981)*, New York: Cornell University.

Glassman, J. and C. Sneddon (2003), 'Chiang Mai and Khon Kaen as growth poles: regional industrial development in Thailand and its implications for urban sustainability', *Annals of the American Academy of Political and Social Science*, **590** (1), 93–115.

Goss, J. and D. Burch (2001), 'From agricultural modernization to agri-food globlisation: the waning of national development in Thailand', *Third World Quarterly*, **22** (6), 969–86.

Hirsch, P. (ed.) (1997), *Seeing Forests for Trees: Environment and Environmentalism in Thailand*, Chiang Mai: Silkworm Books.

Janchitfah, S. (2008), 'Flaws in the Forestry Bill', *Bangkok Post*, 3 February.

Jompakdee, W. (2004), 'Rivers in jeopardy and the role of civil society in river restoration: Thai experiences', *CMU Journal*, **3** (1), 59–71.

Laungaramsri, P. (2000), 'The ambiguity of "watershed": the politics of people and conservation in northern Thailand', *Soujourn*, **15**, 52–75.

Lebel, L. (2005a), 'Carbon and water management in urbanization', *Global Environmental Change*, **15**, 293–5.

Lebel, L. (2005b), 'Institutional dynamics and interplay: critical processes for forest governance and sustainability in the mountain regions of northern Thailand', in U.M. Huber, H.K.M. Bugmann and M.A. Reasoner (eds), *Global Change and Mountain Regions: An Overview of Current Knowledge*, Berlin: Springer-Verlag, pp. 531–40.

Lebel, L., A. Contreras, S. Pasong and P. Garden (2004), 'Nobody knows best: alternative perspectives on forest management and governance in Southeast Asia: politics, law and economics', *International Environment Agreements*, **4** (2), 111–27.

Lebel, L., R. Daniel, N. Badenoch and P. Garden (2008), 'A multi-level perspective on conserving with communities: experiences from upper tributary watersheds in montane mainland southeast Asia', *International Journal of the Commons*, **2** (1), 127–54.

Lebel, L., J. Dore, R. Daniel and Y.S. Koma (eds.) (2007), *Democratizing Water Governance in the Mekong Region*, Chiang Mai: Mekong Press.

Lebel, L. and P. Garden (2006), 'Managing floods and scarcity in a monsoon climate: politics of knowledge and vulnerability in northern Thailand', USER Working Paper WP-2006-11, Chiang Mai: Unit for Social and Environmental Research at the University of Chiang Mai.

Lebel, L., P. Garden and M. Imamura (2005), 'Politics of scale, position and place in the governance of water resources in the Mekong region', *Ecology and Society*, **10** (2), 18, http://www.ecologyandsociety.org/vol10/iss12/art18/, accessed October 2008.

Lebel, L. and B.T. Sinh (2007), 'Politics of floods and disasters', in L. Lebel, J. Dore, R. Daniel and Y.S. Koma (eds), *Democratizing Water Governance in the Mekong Region*, Chiang Mai: Mekong Press, pp. 37–54.

Madhoo, Y.N. (2007), 'International trends in water utility regimes', *Annals of Public and Cooperative Economics*, **78** (1), 87–135.

Maiklad, P. (1999), 'Development and achievements in flood control and management in Thailand', in UN-ESCAP (ed.), *Regional Cooperation in the 21st Century in Flood Control and Management in Asia and the Pacific*, New York: United Nations, pp. 59–111.

Manuta, J., S. Khrutmuang, D. Huaisai and L. Lebel (2006), 'Institutionalized incapacities and practice in flood disaster management in Thailand', *Science and Culture*, **72**, 10–22.

Molle, F. (2004), 'Technical and institutional responses to basin closure in the Chao Phraya River basin, Thailand', *Water International*, **29** (1), 70–80.

Molle, F. (2007a), 'Irrigation and water policies: trends and challenges', in L. Lebel, J. Dore, R. Daniel and Y.S. Koma (eds), *Democratizing Water Governance in the Mekong Region*, Chiang Mai: Mekong Press, pp. 9–36.

Molle, F. (2007b), 'Scales and power in river basin management: the Chao Phraya River in Thailand', *Geographical Journal*, **173**, 358–73.

Molle, F. (2007c), 'Thailand's "free water": rationale for a water charge and policy shifts', in F. Molle and J. Berkoff (eds), *Irrigation Water Pricing: The Gap between Theory and Practice*, Wallingford: CABI, pp. 126–42.

Molle, F. (2008), 'Nirvana concepts, narratives and policy models: insights from the water sector', *Water Alternatives*, **1** (1), 23–40.

Molle, F. and P. Floch (2008), 'Megaprojects and social and environmental changes: the case of the Thai "water grid"', *Ambio*, **37** (3), 199–204.

Pearson, R. (1999), 'A political economy analysis of the impact of agrarian change and urbanisation on com-

munal irrigation systems in the Chiang Mai valley, northern Thailand', PhD thesis, Macquarie University, Sydney.
Phanayanggoor, P. (2006), 'Local against dyke-levelling', *Bangkok Post*, 6 February.
Pratong, K. and D. Thomas (1990), 'Evolving forest systems in Thailand', in M. Poffenberger (ed.), *Keepers of the Forest: Land Management Alternatives in Southeast Asia*, West Hartford, CT: Kumerian Press.
Rigg, J. and S. Nattapoolwat (2001), 'Embracing the global in Thailand: activism and pragmatism in an era of deagrarianization', *World Development*, **29** (6), 945–60.
Samabuddhi, K. (2004), 'Water crisis looms, says grid study', *Bangkok Post*, 13 June.
Sektheera, R. and A.R. Thodey (1974), *Irrigation Systems in the Chiang Mai Valley Organization and Management*, Chiang Mai: Faculty of Agriculture, Chiang Mai University.
Surarerks, V. (1986), *Historical Development and Management of Irrigation Systems in Northern Thailand*, Chiang Mai: Department of Geography, Chiang Mai University.
Suraswadi, P., D.E. Thomas, K. Pragtong, P. Preechapanya and H. Weyerhaeuser (2005), 'Northern Thailand: changing smallholder land use patterns', in C.A. Palm, S.A. Vosti, P.A. Sanchez and P.J. Ericksen (eds), *Slash-and-Burn Agriculture: The Search for Alternatives*, New York: Columbia University Press, pp. 355–84.
Sutiwanich, C., N. Phienwej, T. Harnpattanapanich, A. Fasching and K. Laubblishler (2006), *Investigation and Preliminary Design of Mae Ngat-Mae Kuang Tunnel Chiang Mai Province*, Thailand: Panya Consultants Co.
Tan-kim-yong, U. (1983), *Resource Mobilization in Traditional Irrigation Systems of Northern Thailand: A Comparison between the Lowland and the Upland Communities*, New York: Cornell University.
Tan-kim-yong, U., P.C. Bruns and B.R. Bruns (2005), 'The emergence of polycentric water governance in northern Thailand', in G.P. Shivakoti, D.L. Vermillion, W.-F. Lam, E. Ostrom, U. Pradhan and R. Yoder (eds), *Asian Irrigation in Transition: Responding to Challenges*, London: Sage Publications, pp. 226–52.
Techawongtham, W. (2004), 'Water, water all about for now', *Bangkok Post*, 14 May.
Thomas, D.E. (2006), 'Participatory watershed management for the Ping River Basin Project', Final Project Report (in Thai), Bangkok: Office of Natural Resources and Environmental Policy and Planning, Ministry of Natural Resources and Environment.
Thomas, D.E., B. Ekhasing, M. Ekhasing, L. Lebel, H.M. Ha, L. Ediger, S. Thongmanivong, X. Jianchu, C. Saengchayosawat and Y. Nyberg (2008), 'Comparative assessment of resource and market access of the poor in upland zones of the Greater Mekong Region', report submitted to the Rockefeller Foundation under Grant No. 2004 SE 024, World Agroforestry Centre, Chiang Mai, Thailand.
Thomas, D.E., P. Preechapanya and P. Saipothong (2004), 'Developing science-based tools for participatory watershed management in montane mainland Southeast Asia', final report to the Rockefeller Foundation for Grant No. 2000 GI 086, World Agroforestry Centre (ICRAF), Chiang Mai, Thailand.
Thomas, D.E., H. Weyerhaeuser, P. Saipothong and T. Onraphai (2000), 'Negotiated land use patterns to meet local and societal needs', in X. Jianchu, A. Wilkes, H. Tilmann, M. Salas, T. Grinter and Y. Shaoting (eds), *Links between Cultures and Biodiversity: Proceedings of the Cultures and Biodiversity Congress 2000*, Yunnan, People's Republic of China: Yunnan Science and Technology Press, 20–23 July, pp. 414–33.
Vandergeest, P. (2003), 'Racialization and citizenship in Thai forest politics', *Society and Natural Resources*, **16** (1), 19–37.
Vandergeest, P. and N.L. Peluso (1995), 'Territorialization and state power in Thailand', *Theory and Society*, **24**, 385–426.
Walker, A. (2003), 'Agricultural transformation and the politics of hydrology in northern Thailand', *Development and Change*, **34** (5), 941–64.
Walker, A. (2004), 'Seeing farmers for the trees: community forestry and the arborealisation of agriculture in northern Thailand', *Asia Pacific Viewpoint*, **45**, 311–24.
Wytinck, S.M. (1997), *Water Management in Northern Thailand: A Case Study of the Mae Taeng Irrigation Project*, Alberta, Canada: University of Alberta.

9 Transitions in Indonesian water policy: policy windows through crisis, response through implementation

Anjali Bhat and Peter P. Mollinga

9.1 Introduction

Indonesia's water sector has been in a state of transition since the mid-1980s. This chapter examines actors and interests that have driven several recent transitions in order to consider the dynamics of these transitions and patterns that comprise seemingly chaotic change. Kingdon's multiple streams framework, Litfin's concept of knowledge brokers, and Grindle and Thomas's notion of incremental implementation are incorporated in the analysis of three particular water policy transitions.

As discussed in the theoretical introduction to this book (Meijerink and Huitema, Chapter 2 in this volume), Kingdon's (1995) multiple streams framework seeks to understand structures and patterns evident in agenda-setting. He saw three process streams as instrumental to developing or framing policy windows. A problem stream involves problem identification and recognition, based upon indicators, focusing events or feedback. Systematically collected indicators can illustrate a problem's presence, while focusing events can more broadly and immediately raise awareness of a problem. Feedback can ensure continued awareness of the problem among policy-makers. A policy stream comprises disparate policy communities that produce alternatives and proposals. Kingdon sees the policy stream as a 'primeval soup' in which policy alternatives are floated among and shortlisted by a community of specialists. The political stream takes into account shifts in public opinion, changes in political administration, and interest group activity to determine actor receptivity. Each of these streams is considered to flow independently, driven by differing forces. Policy entrepreneurs come in at critical junctures to impact upon agenda-setting and to create policy alternatives. A policy window opens in either the political stream or the problem stream, providing an opportunity for action in the form of policy proposals.

Litfin's (1994) focus is on the policy entrepreneurs, or what she calls 'knowledge brokers', operating in international organizations and a low or middle level of government, framing and interpreting knowledge that in turn serves as a source of political power. In problematizing knowledge, she gives greater importance to how much the frame in which knowledge is placed resonates with accepted discourses, than to the actual content of the knowledge. In this way Litfin challenges Haas's (1992) 'epistemic communities' and their influence on policy-making processes as experts having a shared understanding of cause–effect relationships of given problems and particular policy recommendations that come from that understanding. Litfin understands scientific knowledge as being inherently contested, value-laden and inseparable from power.

Grindle and Thomas's (1991) interactive model of the policy process is distinct in that it takes policy change as the central analytical feature in considering the difference between policy adoption and policy implementation. They see policy reform as a process, reversible or alterable at many points of its life cycle by interested parties exerting pressure for change. A crucial part of the policy-making process is the implementation phase, as its outcomes vary in response to characteristics of the reform itself. The quarters from where response to the initiative will come – the public or bureaucratic arena – depend on the reform's distribution of costs and benefits, its technical complexity, its administrative intensity, its short- or long-term impact and the degree to which it encourages participation.

Grindle and Thomas's framework is also useful to this study for its explicit focus on developing-country governance. They see decision-makers in a developing-country context as emerging as central actors in the politics of reform due to such features as information uncertainty, limited resources, large demands on the public sector, extensive state influence within the economy and centralized decision-making – which are due to a legacy of colonial rule, state-building activities, and international and domestic political and economic structural vulnerability. The lack of adequate data gives greater relevance to political factors in policy outcomes and implementation. Decision-makers, moreover, are not as constrained by societal or international interests or conditions and have considerable scope to define the content, timing and sequencing of reform initiatives. They can influence how reforms are shaped so as to make them politically acceptable to a wide range of societal or governmental interests. Lastly, and importantly, Grindle and Thomas consider changes in public policies and institutions as normal and an ongoing aspect of government and its officials.

In Grindle and Thomas's approach, reforms are based on situations of either crisis or non-crisis or 'politics as usual'. Reforms stemming from crisis centre upon issues of political stability and control. Technical or administrative interventions and international pressures, while important, assume less importance to issues of survival, longevity and legitimacy of the regime in power, social stability and costs and benefits to major national interests. In crisis situations, problems with high economic and political stakes are imposed upon a small group of high-level decision-makers by interests external to the government, requiring a major shift in how things are done within a short time frame. Reforms involving 'politics as usual', on the other hand, are chosen by decision-makers rather than being imposed upon them and tend to centre around particular policy issues. Having lower political and economic stakes, middle-level officials tend to be involved. Not being as urgent, they involve incremental changes in existing policies or institutions, with the level of change under reformers' control. At stake are bureaucratic issues such as micro politics, agency power, budgets, careers, incentives for modifying behaviour or procedures, and compliance and responsiveness of implementers.

In this chapter it will be seen that both a problem window and a political window are opened by knowledge brokers from the donor community, working with reform-minded middle-level governmental officials with an interest in the new policy discourse they offer. The reforms are initially formulated as 'crisis' reforms per Grindle and Thomas's characterization, creating opportunities for higher political stakes to be involved. However once the policies go into implementation phase, the reforms are reformulated into those of 'politics as usual' reforms, becoming more incremental and controlled by those responsible for implementation.

Map 9.1 Indonesia

9.2 Background on Indonesia

Indonesia, the largest archipelago in the world, consists of 17 500 islands, of which 6000 are inhabited (Map 9.1). Its population of 207.6 million people (2000 census) is the fourth-largest in the world, including 300 ethnic groups speaking 538 languages and dialects. Population density averages 108 people/km^2, but varies drastically from island to island, with Java being the most densely populated (919/km^2). Climate is characterized by two tropical seasons – rainy and dry – and each is approximately six months long. The growth rate of urban populations averaged 3.88 per cent in 2005, and it is estimated that 57.8 per cent of Indonesia's population will live in urban areas by 2015, as compared with 47.9 per cent in 1990.

Indonesia is characterized as a republic, having governing authority vested in its President, but with parliamentary characteristics. The Constitution provides limited separation of executive, legislative and judicial powers. Since the popular uprising in 1998 that unseated Suharto's authoritarian 'New Order' government, there has been a constitutional reform process, including changes that limit office terms for the President and Vice-President, open election of these offices to a popular vote (previously they were appointed by the People's Consultative Assembly), reconstitution of the People's Consultative Assembly (MPR) and the House of Representatives (DPR), and the institution of various checks and balances. The appointed representation of the armed forces and police in the DPR has ended, and active-duty military and police officers are prohibited from holding political office.

Indonesia is divided into 30 provinces and 315 districts (Kabupaten) and municipalities (Kotamadya), all three of which have their own elected legislative assemblies. Prior to the political reforms, or 'Reformasi', initiated in 1999 the provincial legislative assembly selected the provincial governor who was then appointed by the President. District heads and chiefs of urban centres were selected by the district or legislative assemblies and appointed by the provincial governor. The village chief, who is appointed by the district head, heads the village, the lowest administrative unit.

Law 22/1999 on Regional Government and Law 25/1999, entitled Fiscal Balance between the Centre and the Region, devolved many central government functions and responsibilities to the regional level (provinces, districts, municipalities).[1] Regional governments have the authority to: enact their own regulations; formulate their own plans, programmes and fiscal policies; raise and retain revenues; and exploit natural resources, including water resources. The central government retains responsibility for managing activities that cross provincial boundaries, including roads, irrigation systems, rivers and so on. Provinces, in turn, have responsibility for interdistrict activities. Fiscal decentralization has increased regional revenue share through taxes and fees, locally generated revenues from natural resources, national general allocation funds (DAU) and special allocation funds (DAK). The national DAU allocation has been set at 28 per cent of annual domestic revenue, of which 10 per cent is allocated to provinces and 90 per cent to districts and municipalities. DAK allocation includes external development and sector support services. However due to problems of coordination between the central government and districts, regional disparities in human, economic and natural resource endowments persisted. In response the more recent Laws 32/2004 on Regional Administration and 33/2004 on Fiscal Balance between the central government and regional government were passed to replace the previous Laws 22/1999 and 25/1999. These laws curtail the previous level of decentralization and instate hierarchy between the centre and the regions. For instance, the central government can fire provincial, district or municipal heads on the basis of corruption or threat of security. In 2004 Indonesia held its first relatively fair and free general election for President and members of parliament[2] since 1955. In 2005 elections for provincial governors and municipal and district heads began.

9.3 Background on water resources sector in Indonesia

While water resources in Indonesia are abundant (13 000 m^3/capita/year; DGWR, 2003), seasonal and spatial rainfall variation creates competition between water users. Total annual rainfall varies from 1500 to 4800 mm per year, with 70 per cent of annual rainfall taking place between September and March. Agriculture tends to be completely dependent upon irrigation during the dry season. Groundwater potential in Indonesia is limited, can only partially support urban and rural water supply demand, and provides irrigation water for few areas. Reservoir construction tends to be limited by a lack of good reservoir sites, high population densities that require large relocation resources, and ecological impacts of erosion and pollution from upstream sewage (Ramu, 2004).

Indonesia's 5590 rivers have limited flood-carrying capacity, generally originating from volcanic mountains with steep upper-reach slopes, short moderate middle slopes and flat, meandering lower slopes. Typical high rainfall intensities and upper watershed erosion lead to large sediment loads in river systems and basin mouths, which in turn exacerbate lower-reach flooding. On islands with volcanic activity, watershed erosion is compounded by volcanic debris flows. Other causes of flooding involve human intervention. Urban and rural inhabitants live along riverbanks, especially in areas close to river estuaries, where waterways can be constructed from existing infrastructure. Uncontrolled waste disposal is also a problem (DGWRD, 1993).

Society in Indonesia has long relied upon agriculture. Indonesia's agricultural sector accounts for 22 per cent of its gross domestic products (GDP) and 55 per cent of its workforce's employment (DGWRD, 1993). Rice is Indonesia's staple food crop, except in the drier eastern islands, where maize and cassava are grown. The national importance of irrigation for rice cultivation has made water resources development an important focus of Indonesia's five-year development plans (Repelita), with the first three Repelitas specifically emphasizing irrigation infrastructure development and self-sufficient rice production as main targets. Since 1969, under Suharto's 'New Order' regime, the development of physical infrastructure at the river basin scale has been tied to supporting national rice self-sufficiency and flood control, particularly in affected areas producing staple foods and export commodities, as well as housing large populations. International loans and oil revenues financed the increase in agricultural production initially through rehabilitation of existing physical water resources infrastructure, and then later construction and expansion of new infrastructure and rehabilitation.

For planning purposes, in 1969, Indonesia's rivers were categorized into 90 river territories, or Satuan Wilayah Sungai (SWS), which formed the region for river basin planning, development and management. Of these, 73 were fully located in one province, and termed provincial SWS; 17 were interprovincial or of strategic national importance and thus termed central SWS. In that year nine river basin development agencies were established by the central government in densely populated and economically important river basins to be managed by the Directorate General of Water Resources Development (DGWRD) within the Ministry of Public Works (MPW).

These 'Proyek Induks' used a river basin approach to accelerate development, devising comprehensive integrated multipurpose water resources development plans and carrying out construction projects related to reservoir development, flood control, river improvement, hydropower, new irrigation systems and older system rehabilitation in a phased manner. The Proyek Induks' activities were financed by external loans through development projects. Between 1968 and 1993 about US$10 billion was invested in the irrigation sector alone, of which 70 per cent was financed by external loans. The institutional structures through which international finance investments flowed encouraged a sense of power over profits from rents as embodied in agency budgets, bureaucratic authority, political power and patronage in the provision of services subsidized through these construction projects (Bruns, 2004). Proyek Induks held responsibility for all water resources development activities except for the management of irrigation systems, which remained the responsibility of provinces and their field organizations. In the early 1980s transmigration and regional development in less developed regions were supported through Proyek Induks' construction of new irrigation systems in previously rain-fed areas. After two decades of intensive development efforts within the agricultural sector – mainly expansion and rehabilitation of irrigation systems – Indonesia moved from being the largest rice importer to achieving its target of rice self sufficiency in 1984.[3]

9.4 Crisis reform: irrigation operation and maintenance policy

Oil and gas were the leading sectors of the Indonesian economy during the oil boom of 1973–81, with much of their investment flows going towards irrigation, rural and regional infrastructure development. In the mid-1980s, however, oil prices collapsed and a budget crisis ensued, revealing fundamental fiscal shortcomings for construction-oriented policies. Rising costs for constructing new schemes, low performance of existing schemes in terms of actual irrigated area and farmer benefits, and low budget allocations for operational and maintenance were of particular concern to provincial irrigation officials, the Ministry of Home Affairs (MoHA) and its provincial governments, the National Planning Agency (Bappenas), and the Ministry of Finance (MoF). Evidence of irrigation infrastructure degraded much sooner than its expected lifetime indicated inadequate operation and maintenance as a serious administrative and public expenditure problem.

Meanwhile major international donors grew concerned with investment sustainability in light of recent international attention to operation and maintenance issues in government-constructed irrigation systems. They used their leverage in the time of the fiscal crisis to issue a policy precondition – the 1987 Irrigation Operation and Maintenance Policy (IOMP) – for the first Irrigation Sub-sector Project (ISSP I) in 1988, followed by a second (ISSP II) from 1991 to 1995 (Bruns, 2004). The IOMP conjectured that poor maintenance, rather than poor construction, was the cause of rapid infrastructure deterioration. A separate budget was developed from project funds to be disbursed to the provinces to cover the costs of irrigation system operation and maintenance. The Asian Development Bank (ADB) began similar Integrated Irrigation Sector Projects (IISP I and IISP II). The primary objective of the ISSPs and IISPs, as well as the later Java Irrigation and Water Resources Management Project (JIWMP, 1995–2001) in implementing the IOMP was to improve the quality of operation and maintenance by turning over small schemes to water user associations (WUAs) and collecting and using irrigation service fees to fund larger schemes adequately.

At this time the DGWRD's core interests were already under threat after the achievement of rice self-sufficiency, as the previous growth of irrigation infrastructure development could now be curbed. The DGWRD, in response, outwardly distrusted rice production data and stopped exchanging information, carrying out its development activities independently of agricultural policy (Suhardiman, 2008). The IOMP reforms, developed jointly by a wide group of government agencies and donor organizations,[4] were perceived as a further threat to the DGWRD's sectoral interests. Turning over schemes to farmers would reduce the flow of subsidies coming through its offices to carry out operation and maintenance, which was commonly used for other purposes, like occasional construction assistance (Murray-Rust and Vermillion, 1989; Bruns, 2004). The DGWRD had been blocked by Bappenas from providing project assistance to village irrigation schemes, so DGWRD officials saw that turning over non-government-built infrastructure to villages would deny the DGWRD's involvement and a flow of funds.

However, since the reform's implementation was the domain of the DGWRD, its officials responded to the reform's implications by requiring that all schemes be assessed for improvement needs in advance of being turned over to farmers. Based on the assessment they concluded that all non-government-built schemes required improvements, as did schemes built or recently improved by the government. As such, the programme was swiftly transformed from one promoting operation and maintenance into a construction programme benefiting the DGWRD's particular interests (Bruns and Atmanto, 1992; Bruns, 2004).

As another way of protecting the status quo of the reform in its implementation, the DGWRD was able to restructure its organization and maintain its authority and scope. In the projects as designed, provincial irrigation officials responsible for operation and maintenance were given responsibility to implement turnover of irrigation schemes to WUAs, with the positive effect of combining the knowledge and local-level accountability stemming from their structural position with that of a functional subproject manager. However during the ISSP-II, a new Director General of the DGWRD reorganized the department, dividing up functional project responsibilities of structural positions. As a result provincial irrigation projects were now managed by DGWRD appointees with large-scale construction project management backgrounds and accountable to national-level interests (Bruns, 2004).

Bruns (2004), in his detailed analysis of the policy, concludes that the government was able to continue its role as an operator, rather than transform into an enabler. WUAs were established under ISSP I and II, but these were developed and assisted in a highly bureaucratic manner that did not tap into farmers' particular needs or sense of initiative. Construction funds were highly subsidized, with funds channelled through irrigation bureaucracies that did not necessarily use the allocated budgets for their purposes (Suhardiman, 2008).

The IOMP as a reform sought to shift government roles from construction and rehabilitation to improving operation and maintenance, while developing new institutional roles and responsibilities among provincial and central government as well as for private and common provision of goods (Herman, 2002). Though there was progress in achieving formal targets specified on paper, it was nevertheless evident that little had changed institutionally with regard to how irrigation development was carried out (Bruns, 2004). The government was reducing its role without providing sufficient technical services and regulatory oversight, encouraging further degradation of infrastructure.

The donor agencies involved in the water sector began to urge the Government of Indonesia to develop and implement an integrated water resources management strategy. In 1991 UNDP/FAO created a five-year National Water Resources Policy study, resulting in a draft 25-year National Water Resources Policy Action Plan (1995–2020). It reflected water resources policy objectives as well as cross-sectoral policy coordination in the form of a matrix of requisite government actions. In administrative terms it emphasized a river basin approach and required the development of a National Water Resources Council. The policy action plan was not reviewed by the MPW, however, but Bappenas used it in formulating the sixth five-year national plan (Repelita VI) of 1994–99. The 1992 Cisarua International Water Resources Conference (held in Bogor, West Java, and jointly sponsored by a number of donor agencies on issues related to water resources management) furthered discussion of a shared framework for policy projects developed thereafter. Recommendations from the conference emphasized river basin management in addition to establishing basin management corporations in a few strategic basins (Herman, 2002).

Repelita VI was the first five-year national development plan in which the objectives of the water sector were prepared separately from those of the agricultural sector, recognizing water resources as comprising an independent sector. Moreover, as further evidence of this new emphasis on integrated water resources management, the Bureau of Water Resources and Irrigation was newly established within Bappenas and the DGWRD was reorganized into the Directorate General of Water Resources (DGWR) within the MPW. Indonesia's new second long-term 25-year development plan (PJP II: 1994–2019) had a greater focus on the operation and maintenance of infrastructure and reflected the shift from a single-purpose focus to a multisector river basin approach to promote integrated water resource management. In the PJP II it was decided that authority and responsibility for irrigation management, which had been the primary focus in the previous long-term plan, was to be transferred gradually to the district and provincial levels as part of government policy to increase regional autonomy, while the allocation of water among irrigation and other uses would make up a core function of basin management (Ramu, 1999).

Ideas from the related dialogues and the objectives of the IOMP came together in developing the World Bank's basin planning and water management project, JIWMP, and the ADB's Capacity Building Project (CBP) in 1995. It also supported the development of five pilot river basin management units under provincial authority (Balai PSDA), and basin- and provincial-level institutional arrangements for coordination. The Balai PSDA was expected to collect and house water accounting and allocation data, monitor water quality, and to carry out river infrastructure maintenance for flood control. The DGWR was expected also to prepare guidelines for management in these pilot basins, delineating a water-user rights system, water allocation, water service fees, conjunctive management of surface and groundwater, and river infrastructure maintenance. However the implementation of this project proved unsuccessful. Given its non-construction orientation, the project translated internally within the DGWR into low-grade positions and a lack of financial and human resources at the provincial level. By the project's end in 2000, the pilot Balai PSDAs had been established only within Java.

The ADB's five-year CBP was a more ambitious project seeking to strengthen institutional capacity at the national and regional levels for carrying out water resources management on a multisectoral basis. The aim was to do so through reforming water policies, coordination, human resources management, data collection and management capabilities. This project, too, proved unsatisfactory as it achieved little high-level support or sense of ownership locally. The choice of a project approach, the lack of any effort to seek counterpart funding particularly from the provinces, and the fact that it did not create the means of financing capacity-building from a routine budget were thought to have contributed to this outcome.

The JIWMP and CBP were perceived by the donor community not to have succeeded in achieving the IOMP objectives (Herman, 2002; World Bank, 2005). By 1997 the World Bank decided to suspend the preparation of water sector projects, concluding there was little political will to follow through with them. The ADB was also concerned and sought to review the direction of the irrigation sector's activities.

As a prerequisite for future lending in the water resources sector, the World Bank encouraged Bappenas's Bureau of Water Resources, with Ford Foundation support, to host a national forum for discussion on water policy, comprising five seminars and 13 discussion sessions beginning in September 1997 and concluding in February 1999, with the objective to develop 'an agenda on water resources policy and program reform' and to serve as an input into the policy and strategy development of Repelita VII. Among participants were government ministries and agencies, non-governmental organizations (NGOs), and academics. Leading members were middle-level governmental officials from Bappenas and the DGWR who clearly desired an improved policy and institutional framework. Although the forum's initiation coincided with the East Asian fiscal crisis and the fall of the Suharto regime, it nonetheless helped to draw high-level attention from Bappenas, the MPW, and other senior officials (Bruns, 2004).

9.5 Crisis reform: water sector loan reform

The steady economic growth of the nation over the course of six Repelitas came to a halt in 1997–98 with the East Asian financial crisis. In addition to the acute fiscal difficulties, a severe drought caused by El Niño afflicted Indonesia, leading to major deficits in rice production. Suharto's New Order government collapsed in the ensuing economic and political instability. His top-down, centralized regime was considered to have resulted in weakened legal and regulatory institutions and endemic corruption. The crisis became a focusing event in the political stream, greatly affecting the general public and elites' receptivity to reform initiatives. The MPR pushed for a new political environment of national reform in 1998, calling for greater regional autonomy and transparency, as well as tackling issues of corruption, collusion and nepotism in government (Herman, 2002; World Bank, 2005). Given the crisis and the new environment, the Indonesian government launched a policy and institutional reform strategy, supported by the International Monetary Fund with participation of the World Bank, ADB and a number of bilateral donor institutions. The reforms received popular support, and support from national leaders, elites and NGOs.

Meanwhile, the new, pro-reform President, Abdurrachman Wahid, replaced the MPW with the new Ministry of Settlements and Regional Development (Kimbangwil) in 1999. Kimbangwil's responsibilities now included housing and spatial and regional infrastructure development and organizationally no longer included a DGWR, placing water resources management responsibilities under the Directorate General of Spatial Planning and Regional Development, and irrigation separately under the Directorate General of Rural Development. Kimbangwil's new minister was a reformer with a civil society background who recruited staff through NGO connections. She separated out the former MPW officials into a smaller State Ministry of Public Works (MeNegPU) organization with responsibility for sector policy and institutional development, although they could not manage sectoral development funds. The clear implication of excluding former MPW high-level officials from a decision-making role was the possibility of agreement and of moving forward quickly on sectoral reforms that would have otherwise been in conflict with the construction-oriented interests of the former MPW.

With World Bank staff as facilitators, committed mid-level officials from Bappenas convened an informal task group in late 1998 to formulate ideas for reforming irrigation and water resources policy, drawing on ideas from the earlier discussions and studies. The working group was assisted by university experts to synthesize and clarify key principles for reform, and then held public consultation meetings in Jakarta and several provinces. The resulting policy matrix of indicators that could be monitored was presented to government agencies and donor representatives, with the outcome of a consensus on a major programme of reforms (Bruns, 2004; World Bank, 2005). Bappenas then established a formal task force to finalize a Declaration of Irrigation Management Policy Reform. This task force was headed by the chief of Bappenas's Water Resources Bureau and had 42 members appointed by the Bappenas Deputy Chairman of Infrastructure. All members were government officials except for two NGO representatives (World Bank, 2005).

As the World Bank had decided not to support the water resources sector without larger institutional reforms, it initially issued an Agricultural Sector Adjustment Loan (ASAL), which included a programme of water sector reforms, bringing together deliberations of the Bappenas Forum, the institutional component of the CBP and policy objectives under the JIWMP. However it became apparent that there was insufficient bureaucratic support from the Ministry of Agriculture (MoA) for the agricultural policy aspects of the ASAL, so desiring to speed up the process on the water resources component, the World Bank decided instead to support a Water Sector Adjustment Loan (WATSAL) based on the policy matrix devised by the task force previously convened by Bappenas. This task force became the WATSAL Task Force (World Bank, 2005). In fact WATSAL's governing structure was constructed by Bappenas. The WATSAL Task Force comprised mid- and high-level, pro-reform government officials chosen to participate by Bappenas. A Steering Committee above the Task Force reviewed legal drafts produced by the Task Force and its working teams. Above the Steering Committee, the Coordination Team, comprising ministers from 11 ministries, served as the highest decision-making body. International experts appointed by the World Bank also worked with the Task Force to reframe policy (Suhardiman, 2008).

Kimbangwil's minister agreed to sharing the intersectoral decision-making platform with Bappenas. However the Kimbangwil officials involved in the intersectoral decision-making body, without guidance from the MeNegPU officials, were not very involved in directing the process. Former MPW officials, sequestered in MeNegPU, had no influence on irrigation management transfer (IMT) policy development under the WATSAL. Consequently Presidential Instruction 3/1999 on IMT (Government Regulation 77/2001 on irrigation) was issued without their input. This regulation transferred management responsibility to WUAs and gave them direct access to sectoral development funds, potentially shutting out Kimbangwil's access to these funds completely (Suhardiman, 2008).

The WATSAL Letter of Sector Policy and Policy Matrix initially anticipated the development of small technical amendments to accommodate reforms to be carried out relatively quickly. The Minister of Kimbangwil under the Wahid government, however, decided in 2001 to replace the 1974 Water Resources Development Law with a more modern, cross-sectoral framework law and supporting government regulations (World Bank, 2005). The new law was to consider decentralization laws and focus more on water resources management rather than utilization. Provisions included the development of a National Water Council, stakeholder participation, water use rights and water conservation. About 15 government regulations were expected to be issued to support the law. The government regulations were to be finalized and issued, with further work expected on formulating guidelines for provincial legislation. Formulation of the new law took two years, by which time President Wahid was being deposed, lowering the possibility that the new law would be processed.

After Wahid was impeached and replaced by President Megawati, the WATSAL Task Force's draft of the water law was altered during intergovernmental discussions. Many provisions expected to be dealt with by new government regulations were removed. A change in the framework would inevitably change the drafts of the governmental regulations themselves within the DPR (World Bank, 2005).

Once President Megawati Soekarnoputri came into office in 2001 after Wahid's impeachment, she brought on board a new minister who was a member of her political party and affiliated with the MeNegPU officials. The new minister recombined the two organizations as the Ministry of Settlements and Regional Infrastructure (Kimpraswil) and restored the MeNegPU officials' authority in the new ministry's reinstated DGWR. The change of policy actors at this level resulted in a change in the ministry's interests. The interpretation of the IMT policies that had been signed off on by Kimbangwil was revisited. As Kimpraswil regained power, NGO involvement in the WATSAL Task Force waned.

At this time (2001–02) the WATSAL's IMT programme was being implemented through the JIWMP – Indonesia Water Resources and Irrigation Reform Program (JIWMP–IWIRIP), with a development and turnover component carried out by Kimpraswil's DGWR, which focused on construction and rehabilitation, and an irrigation service fee component belonging to the MoHA, giving the MoHA a stake in the water resources sector and its development funds (Suhardiman, 2008).

Around mid-2003 the Megawati government began to revisit the reform principles and regional autonomy laws, permitting sectoral interests to re-emerge in the process. Kimpraswil was able to recapture its power to further its sectoral interests and sought particularly to reclaim irrigation responsibilities. Although the irrigation objectives of the WATSAL reform were well under way, with programmes in 16 provinces successfully implemented, questioning IMT as an objective, Kimpraswil issued a moratorium on the implementation of Government Regulation 77/2001 on irrigation on the grounds that transferring implied relinquishing government ownership of irrigation infrastructure to farmers, which would financially burden poor irrigation farmers and ensure further infrastructure deterioration. Kimpraswil sought to limit the scope of management transfer to the tertiary level in anticipation of the regional autonomy law revision, which would return deconcentration mechanisms. Once the Water Resources Law was passed in February 2004, irrigation reform principles had been downscaled, requiring revision of Government Regulation 77/2001.

By mid-2003 Kimpraswil's efforts began to pull apart the unified front of the WATSAL Task Force as well as the *Tim Koordinasi* (coordinating team). The MoHA, MoA, and most local governments as well as the National Farmers' Association, which had jointly developed the irrigation policy, disagreed with Kimpraswil's moratorium. Also contributing to the contention was Bappenas's reorganization, which entailed loss of its budget oversight responsibilities to the Ministry of Finance. A new, less powerful Bureau of Water Resources and Irrigation team was appointed to the Task Force, which allowed time for them to come up to speed in coordinating and managing WATSAL.

The WATSAL Task Force sought to move forward in issuing the government regulations and the new Water Resources Law of the third tranche of the loan despite growing pressure from Kimpraswil. Kimpraswil, however, was able to gain political backing from the President during the time of a national election and it succeeded in blocking the Task Force and Tim Koordinasi's leadership. Kimpraswil was able to negotiate terms with the MoHA by ensuring sustained access to sectoral development funds.[5] Bappenas, wanting to regain intersectoral decision-making authority, initially launched a national seminar on IMT to promote an interministerial coalition to counteract Kimpraswil's authority. However in the interest of allowing funds to be disbursed from the World Bank's Water Resources and Irrigation Sector Management Project (WISMP), which was under negotiation, Bappenas decided to support Kimpraswil's position (Suhardiman, 2008).

The World Bank responded to the uncertainty of the government's commitment to the original irrigation reform ideas by not disbursing the third tranche of the loan.

The process of developing the new Water Resources Law continued, but reflected Kimpraswil's continued reluctance to give up authority over sectoral development funds to other levels of government and to farmers. The strategy of the donor knowledge brokers to work with Bappenas and the MoHA in order to counter Kimpraswil's bureaucratic interests did not succeed in reforming the sector, which ultimately remained in Kimpraswil's decision-making domain (Suhardiman, 2008).

The new Water Resources Law was passed in February 2004, and while containing many of the WATSAL reform agenda's objectives, it left the door open for interpretation of laws during implementation. While not including the notion of irrigation management 'transfer', it suggested a more conditional form of delegation to WUAs. Donors became concerned that this interpretation and the new law would require revision of all the new national regulations related to irrigation.

After the new water law was enacted, Kimpraswil halted the IMT programmes being implemented under WATSAL. Provincial and district governments had to fund these activities from their own resources. The MoHA's irrigation service fee programme was relegated, becoming a community organizer recruitment programme. The provinces, however, did not fight with Kimpraswil's stance on WATSAL's IMT programme as long as they were compensated with rerouted funds for their activities.

The 'crisis reform' of WATSAL broadened the arena of decision-making and access to sectoral resources. Given the costs borne by the sectoral bureaucracy to this reform, Kimpraswil sought to reign over the scope of this arena to return it to its sectoral home. As is clear in the above discussion, IMT was the most controversial of the reform components of WATSAL, as it would have greatly curtailed Kimpraswil's access to sectoral funds. The need, however, to revise the drafts of key government regulations that had been prepared in accordance with the Decentralization Laws 22 and 25/1999 and Water Resources Law 7/2004 prior to the issuance of the revised Regional Autonomy Laws 32 and 33/2004 provided an opportunity for the MPW (Kimpraswil was reorganized and renamed MPW in June 2005) to overturn the 'transfer' concept. In 2006 the MPW issued a regulation to replace Government Regulation 77/2001 on irrigation. It classified irrigation areas in a manner that gave the central government authority over those greater than 3000 hectares, while provinces would have responsibility for those of 1000 to 3000 hectares, and districts would have responsibility for those less than 1000 hectares. This new regulation ran counter to efforts promoted by the WATSAL to devolve responsibility for irrigation areas to WUAs, which by now were considered to be 'participating' in irrigation management, the domain of the government.

9.6 Policy-as-usual reform: river basin management

The revisions of the Regional Autonomy Laws 32 and 33/2004 shifted power from the districts to the provinces, reintroducing the deconcentration of central government functions as well as co-management through the transfer of funds to local government. The revisions as well as the new Water Resources Law provided Kimpraswil/MPW with an opportunity to reclaim the authority it had prior to the Reformasi and WATSAL reforms. Overturning the IMT policy with a new government regulation replacing Government Regulation 77/2001 was the final step in a longer process of bureaucratic response to a 'crisis reform'. However in opening up authority for water resources management to provincial- and district-level actors, who were no longer line agents of the ministry, the decentralization laws of 1999 posed other threats to the MPW's scope of authority. For this reason the draft government regulation revision process served as a unique opportunity for the MPW to regain its bureaucratic scope.

Along with the government regulation on irrigation, a number of ministerial regulations were issued in September 2006. Among these was a ministerial regulation issued to redraw basin-level jurisdictions. Previously Indonesia's 5590 rivers were grouped into 90 river territories by ministry regulation in 1969, for purposes of river basin planning, management and development. Of these 90 territories, 17 were considered to be under national jurisdiction as cross-provincial (15) or 'strategic' (two).

The new ministerial regulation, passed in 2006, sought to re-designate Indonesia's rivers into 133 territories, of which 69 were to be under the jurisdiction of the central government as cross-national (5), cross-provincial (27) or strategic (37). Meanwhile 51 of these are now under provincial jurisdiction, and 13 belong to the district.[6] The implication is that provinces that had previously developed capacity to manage basin water resources within their jurisdiction have had their authority suddenly transferred to the central government. A number of provinces are not content with the new categorization scheme, considering it a 'recentralization' of authority thanks to the interpretation of 'strategic'.

To manage these basins under central government jurisdiction, the MPW issued another ministerial regulation to create a new MPW line agency. This institution, called the Balai Wilayah Sungai (BWS) or the Balai Besar Wilayah Sungai (BBWS) – depending on the size and strategic importance of its offices – manages water resources within an administrative jurisdiction comprising a main river basin and smaller surrounding ones. These are legislated to take on a comprehensive range of responsibilities for water resources management. The development of this institution was a response to external legislation that directly impacted upon the MPW's operations. In 2003 the Finance Ministry passed legislation requiring projects to be folded into a permanent body within institutions for budget management purposes. Thus ministries with projects had to develop new bodies to fold them into. Another 2003 law on state-owned enterprises cut off the technical ministry supervision role of the MPW from the operations of the state-owned company for basin management Perum Jasa Tirta I (PJT I).[7] Officials within MPW saw the need both to bring their Proyek Induk operations into a permanent body and to exercise greater control over water resources management than is possible through the PJT I institution given the new policy developments. Since 2006, these MPW officials have sought to develop a new financial instrument, the Badan Layanan Umum (BLU), which would introduce a central government organization to charge water users to cover operational costs.

In a basin like the Brantas in East Java with a standing management organization, this new institutional development has created confusion. The Directorate General of Water Resources at the time of the BBWS's development was not in favour of the Perum Jasa Tirta model of management, and sought to have the BBWS replace it on the grounds that PJT I had low operation and maintenance cost recovery (generally 30 to 60 per cent), meaning that it was not a sound business. The BLU model differs vastly from that of the PJT I, which, as a state-owned company, has to show a budget surplus to be a healthy corporation. Prior to 2005 part of this surplus went to central government development funds (55 per cent), the corporation's general fund (20 per cent), its benefit fund (20 per cent) and its reserves (5 per cent) (Ramu, 2004). However, more recently, state-owned enterprises collectively requested to be able to access more funds for operational and maintenance costs. Revised legislation as of 2005 requires only 10 per cent of after-tax profit to go to central government development funds, while 90 per cent could go to PJT I reserves, 80 per cent of which would go to investments. PJT I's annual budget is approved by the Ministry of Finance, ensuring government oversight of its finances.

The BBWS Brantas and its possible BLU financing mechanism was being considered as a competing model to the PJT I in the Brantas basin and caused great confusion for PJT I, which had to actively promote its institutional features and accomplishments and pursue strategies that ensured its competitiveness as a model. However following the retirement of the DGWR and the installation of his replacement, the attitude toward the PJT I model has become less negative. PJT I is expected to continue to manage water resources in the Brantas basin, although continuing overlaps between the PJT I, the BBWS and the province with respect to roles and responsibilities have yet to be sorted out.

Although many provincial- and district-level governments are not necessarily displeased with the funds promised from the central government via the BBWS to carry out operation and maintenance work in their regions, their work has been severely curtailed by the new definition of responsibility that has emerged in the implementing regulation on irrigation. This will have an impact on local staffing as staff are loaned to or recruited by the BBWS. There are struggles on the part of District and Provincial Water Resources Departments that have developed capacity and programming through decentralization reforms; they take the actions of the central government as a threat to their new authority and are actively communicating their discontent and seeking to regain authority in the implementation phase through negotiation with regional BBWS offices and communication with government officials in Jakarta in the MPW, as well as the MoHA, through which they invoke their authority.

The PJT I model was developed using lessons drawn from a number of organizational models. The model was promoted within Indonesia for basins where payment from industrial, hydropower and municipal water users had the potential to cover operational costs. Prior to the development of the BBWS river basin public corporations were being planned for four other economically developed and strategic basins, with an expectation that with good economic performance they could put their profits toward capital improvements and developing public–private partnerships. For basins that were not considered to be strategic to the state's interests, the provinces' Balai PSDAs were expected to collect and house water accounting and allocation data, monitor water quality and carry out river infrastructure maintenance for flood control.

The new, more comprehensive BBWS model functions directly under the MPW and has responsibility primarily for infrastructure development, having assimilated the project objectives. Policies favouring the BBWS model reflect the concern that too much authority for management was going to the Ministry of State-Owned Enterprises rather than to the MPW. The new vision being developed was that all river basin organizations would be governmental agencies and that these could utilize the BLU means of financing. If it proves economically successful, a BLU institution can be upgraded to the status of a state-owned corporation. Thus two central government-affiliated models are competing in the policy arena of river basin management: one taking a conservative approach to management, bringing it closer to the responsible ministry, distrusting the incentives of any organization not within its direct control; the other employing a more liberal economic perspective on management that assumes that public corporations have greater incentives to manage effectively.

To address serious shortcomings in staffing in the BBWS offices,[8] instruments like task assistance and deconcentration are used.[9] These instruments decentralize responsibility downward in a contractual manner (justified by the new Decentralization Law that has replaced the original one passed in 1999) and consequently leave programme planning and design in the hands of the central government. This risks making the line of accountability unclear for those carrying out the tasks as well as for water users.

9.7 Lessons learned and conclusions

The reforms discussed in this chapter illustrate the fluidity of the policy process and the points at which those bearing the costs of reform can intercede and alter the reform. There is nonetheless a distinct shape to the policy process and the means by which agenda-setting is carried out to prepare for a focusing event that will bring various streams together.

The development of the IOMP reform illustrates Kingdon's (1995) 'problem streams' and the development of what Grindle and Thomas (1991) consider to be a crisis-related reform. The IOMP was devised at a time when governmental officials and donor interests recognized a clear problem with the operation and maintenance of irrigation infrastructure, manifesting in sudden financial deficits, entrenched construction-oriented practices and poor infrastructure performance. The policies devised, while involving a number of sectors, were ultimately to be implemented by the DGWR through various projects. This gave the DGWR an opportunity to resist the IOMP reform by making changes in implementation regulations that, while acceptable on paper for project fund disbursement, no longer threatened its bureaucratic interests. The donor community responded to the lack of fundamental change by initiating a 'policy stream', involving interministerial dialogues and aiming to build a coalition of like-minded governmental officials seeking reform. The Cisarua and Bappenas dialogues provided an opportunity for a range of relevant government ministries to come to consensus on the nature of the problem, develop policy alternatives and strategies and, importantly, to create a cross-sectoral platform to address the issues. Bappenas officials were able to gain a stake in the water resources sector, confirmed with the establishment of the Bureau of Water Resources and Irrigation after the Cisarua dialogue.

The financial crisis and rice production shortfalls due to El Niño were separate problem streams that together brought a distinct focusing event for the donor agencies and pro-reform middle-level government officials, mainly in Bappenas, to put forward the plans developed in the preceding policy stream. Further activities to encourage dialogue and consensus around a more participatory, intersectoral and integrated means of managing water were promoted in the 'political window' opened by the crisis. This led to the development of a more sweeping sectoral reform, tied to sizeable loan disbursement at a critical time when these funds were needed. It is evident that through the dialogues and participatory fora, the donor community sought to develop a local epistemic community that could influence relevant governmental officials through cause–effect knowledge that is part of the international discourse on participatory and integrated water resources management. Although progress was made in devising the reforms by excluding the main policy actors of the MPW from the process, once these actors were able to recapture power, they used their sectoral decision-making authority and political connections initially to block and then eventually to overturn the reforms through strategic means in a non-crisis or 'politics as usual' reform. They used a discourse of direct sectoral oversight, arguing that this was necessary as the best way to supervise irrigation and basin management and that devolving these responsibilities to other entities would be a burden to those entities and a risk to infrastructure health. Strategically, they were able to gain the agreement of other ministries by offering those ministries access to sectoral development funds if the status quo were upheld. Through such negotiations, they were able to work around the intersectoral decision-making platform and regain the authority and access to sectoral funds they almost lost to the reforms.

It is interesting to note that the approach of building an epistemic community, while unsuccessful for the purposes of implementing the WATSAL reform, may have an impact on the future of Indonesia's water resources sector, as pro-reform middle-management government officials from various ministries, including the MPW's DGWR, have had an opportunity to come to consensus on specific issues and develop a common approach. These officials will likely gain promotions and replace those with sectoral interests who represent construction-based policies. Recruitment of staff outside of civil engineering to represent a wider set of skills in water resources management, and restructuring incentives within the MPW to value these skills, build on the efforts made in these policy streams.

The implementing regulations issued by the MPW in 2006 to support the new Water Resources Law illustrate the MPW's strategy to regain authority lost in the process of Reformasi and the WATSAL. This pendulum-swinging 'policy as usual' response to decentralization illustrates the ideas of Grindle (1980) and Grindle and Thomas (1991) that policy formulation processes tend to be centralized and closed in a developing-country context, with limited scope for input except at the implementation phase. The formulation of the implementing regulations was a closed affair, involving bureaucratic interests. However the policies are currently being implemented and it is at this stage that the uniform measures are being tested and reformed on a case-by-case basis to suit political, social, environmental and management realities on the ground. As the costs of these policies to parties outside of the MPW come to light, this dynamic policy process continues.

Notes

1. Functions excepted from devolution are defence, foreign affairs, judiciary and national finance.
2. However members of parliament are still elected from national party lists.
3. Rice production has declined since the mid-1990s due in part to drought, relative price increases in non-paddy crops, reduced subsidy and availability of inputs, conversion of paddy lands to urban use, poor maintenance of irrigation systems due to resource constraints, and competition for water.
4. Carried out under the New Order regime, the preparation of the projects and the IOMP, although intending to involve the participation of farmers, was conducted in a confidential process, with little publicity or public consultation (Bruns, 2004).
5. The Ministry of Home Affairs's Sub-directorate of Water Resources, however, continued to support the idea of the NGO and university members consortium to halt the process of the new Water Resources Law.
6. It can be noted that the classifications put forward in the new regulation had been included in the elucidation of the 2004 Water Law, but without a detailed definition of 'responsibility', leaving the implications for financing, ownership of assets and operation and maintenance to be worked out.

7. PJT I has sought to have a clarification made to Law 19/2003 to allow for sectoral oversight, as this law was intended for *Persero* organizations to make them more competitive, but was applied generally to *Perum* organizations, even though they have the fundamentally different role of public service provision.
8. A significant number of staff from projects had been transferred to provincial and district offices as required by the decentralization reforms of 1999, designed to develop capacity at more localized levels.
9. According to Law 32 of 2004, Task Assistance (or *Tugas Pembantuan*) provides funds from the central government to the provincial-, district- or village-level government for them to carry out earmarked tasks related to physical aspects, such as irrigation maintenance – and not planning, capacity-building, and so on. Deconcentration (or *Dekonsentrasi*) refers to funds provided from the central government to the Governor, serving as a regional representative, to carry out activities. These funds can be used for design, planning and capacity-building.

References

Bruns, B. (2004), 'From voice to empowerment: rerouting irrigation reform in Indonesia', in P. Mollinga and A. Bolding (eds.), *The Politics of Irrigation Reform: Contested Policy Formulation and Implementation in Asia, Africa, and Latin America*, Burlington, VT: Ashgate, pp. 145–65.

Bruns, B. and S.D. Atmanto (1992), 'How to turn over irrigation systems to farmers? Questions and decisions in Indonesia', Irrigation Management Network Paper, No. 10, London: ODI.

DGWRD (Directorate General of Water Resources Development, Ministry of Public Works, Republic of Indonesia) (1993), *25 Years Water Resources Development in Indonesia, 1969 to 1993*, Jakarta: DGWRD.

Grindle, M.S. (1980), 'Policy content and context in implementation', in M.S. Grindle (ed.), *Politics and Policy Implementation in the Third World*, Princeton, NJ: Princeton University Press, pp. 1–12.

Grindle, M.S. and J. Thomas (1991), *Public Choices and Policy Change: The Political Economy of Reform in Developing Countries*, Baltimore, MD: Johns Hopkins University Press.

Haas, P.M. (1992), 'Introduction: epistemic communities and international policy coordination', *International Organization*, **46** (1), 1–35.

Herman, T. (2002), 'Development of effective water management institutions – Indonesia', Regional Case Study, August, Colombo, Sri Lanka: International Water Management Institute (IWMI).

Kingdon, J.W. (1995), *Agendas, Alternatives, and Public Policies*, 2nd edn, New York: HarperCollins.

Litfin, K.T. (1994), *Ozone Discourses: Science and Politics in Global Environmental Corporation*, New York: Columbia University Press.

Murray-Rust, H. and D. Vermillion (1989), *Efficient Irrigation Management and System Turnover*, Vols 1–3, Colombo, Sri Lanka: International Irrigation Management Institute.

Ramu, K. (2004), 'Brantas basin case study – background papers', paper prepared as part of World Bank River Basin Management Study, Washington, DC: World Bank.

Suhardiman, D. (2008), 'Bureaucratic designs: the paradox of irrigation management transfer in Indonesia', dissertation, Wageningen: Wageningen UR.

World Bank (2005), 'Implementation completion report: Indonesia: Water Resources Sector Adjustment Loan', Report No. 32665, Rural Development and Natural Resources Sector Unit, Indonesia Country Office, East Asia and Pacific Region.

10 The contribution of actors to achieving sustainability in Australia through water policy transitions

Sara Hughes and Jennifer McKay

10.1 Introduction

Water has always been the critical resource for Australia's development. Indeed: 'no commodity has played a more significant role in the nation's development, and no belief appears to be more widely shared than the need for the country to grasp every opportunity to conserve more water' (Pigram, 2006, p. 41). Achieving a high level of conservation is the goal of Australia's current transition in water management, which has proven to be contentious and highly political.

Following in the wake of the infrastructure expansion of the 1970s, social and environmental concerns about the consequences of dam-building and overextraction gained support and attention at all levels, prompting a broad-scale re-evaluation of how, and by whom, water should be managed (Archer, 2001; Pigram, 2006). In 1994 the Council of Australian Governments (COAG), a relatively new high-level intergovernmental forum in Australia, responded to concern about the state of many of Australia's rivers by developing a national policy for 'the efficient and sustainable reform of Australia's rural and urban water industries' (COAG, 1994). Four areas of water management were particularly emphasized in the 1994 COAG Water Reform Framework: efficient pricing of water; separating water and land entitlements to create water markets; allocating water for the environment; and environmental regulation (COAG, 1994). Since this initial step broad-scale policy, legal and organizational changes have taken place within both federal and state governments to implement these. Initiatives have included regionally based water allocation plans in the states and increasing local and private sector involvement in water management. Building from these reforms, in 2004 COAG developed the National Water Initiative (NWI), which focused even more explicitly on reversing the trend of overallocation to growers through such measures as providing water specifically for the environment and developing institutions capable of planning for and administering change coherently, adaptably and with effective, sustainable results.

While water management has typically been left in the hands of the state governments in Australia (McKay, 2007), these federal-level reforms have introduced new objectives for water planning and are encouraging new institutional arrangements and policy for their implementation. Their effects have been felt at every level of water planning in Australia and are proving to be a significant divergence from the past. The focus of Australian water management from the time of European settlement in the early 1800s until the early 1980s was on expanding infrastructure and water works projects, most of which were funded by the government directly or through loans (Smith, 1998). Water management goals largely focused on continued expansion for supplying water for both urban and rural development (McKay, 2002). The transition in Australia toward long-term sustainability in water management is both challenging and timely, as predictions of increasing demand and climate change pose serious threats to business-as-usual policy models.

The framework of transition management developed by Meijerink and Huitema (see the theoretical introduction to this volume, Chapter 2) suggests that individuals play a powerful role as policy entrepreneurs in shaping the direction and success of policy transitions and employ a range of strategies at different stages of a transition. The goal of this chapter is to examine the extent to which this framework helps us to understand the role individuals as policy entrepreneurs have played in achieving policy change and implementation in Australia at the state level in response to federal transitions. Our interest lies in the challenge decision-makers and managers face in negotiating the introduction of new users when allocating water resources, while simultaneously grappling with new and often interacting institutional contexts. We focus less on the federal pricing and trading aspects of the Australian transition and instead concentrate on state-level changes in water allocation in one state, South Australia.

To test the framework we chose two cases of policy change that took place in South Australia in response to the transition in national-level water management: the state's efforts to provide environmental water allocations for the Murray River and to introduce the newly developed water allocation licensing system for forestry in the southeast (Map 10.1). The state of South Australia provides particularly valuable opportunities for analysis for three reasons: it is considered a leader in reform within Australia, taking the lead on many policy issues; it has a challenging climate, receiving the least amount of rain of any Australian state; and it is not as well studied as other areas of Australia, adding value to our analysis here. South Australia may also have the most coherent approach of the Australian states to water management. The State Department of Water, Land and Biodiversity Conservation (DWLBC) is charged with assisting regional Natural Resources Management (NRM) Boards to draft water allocation plans and issue the necessary licenses. The Minister of Environment and Conservation oversees the operation of seven of the eight NRM Boards, the Minister for the River Murray oversees the eighth, and both are provided with advice and recommendations from the DWLBC. The water allocation plans are developed by the Natural Resources Management Board to address environmentally sustainable development principles of the Natural Resources Management Act of 2004. These plans are subject to significant contributions from the community in the region; the plans often go through several iterations based on community feedback. However the relevant minister is responsible for approving the final version of the water allocation plans and the community can then implement water sharing as agreed on in the plan. Other states have developed less comprehensive and formalized policy responses.

The data used in our case studies are based on intensive interviews with key actors and in-depth document and policy analyses. In each case we focus on the strategies we observe individual agents using to prepare, instigate, prevent or implement the specific policy change. We use these case studies in the final section of the chapter to make conclusions about the role of policy entrepreneurs in South Australia and how this information may be of use to others in Australia and elsewhere involved in water policy transitions.

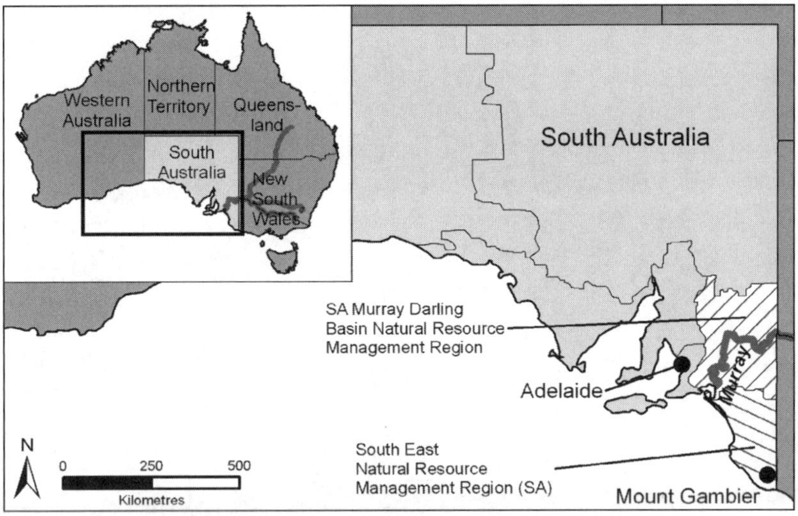

Map 10.1 Australia and case study areas

10.2 Case one: environmental water allocations for the Murray River

10.2.1 Introduction

National water management reform in Australia has often been triggered by conditions in the Murray–Darling Basin, the water source the country relies upon most for drinking water, irrigation and ecosystem support. In turn, federal reforms have significant ramifications within the basin, as it includes four states and the Australian Capital Territory (ACT) (Map 10.2). The Murray River alone provides 70 per cent of the country's irrigation water and between 60 and 90 per cent of Adelaide's (the capital of South Australia) drinking water. It also supports five United Nations Educational, Scientific and Cultural Organization (UNESCO) Ramsar-designated wetlands of critical importance, including the Coorong Delta at the river's mouth in South Australia. The basin is currently overdeveloped and overallocated; it no longer functions as it used to, with less frequent flooding and drying events and significantly reduced base flows (Maheshwari et al., 1995).

In this case we see individuals playing a key strategic role in the establishment of an environmental water allocation, acquiring the water specifically for the environment, and in subsequent efforts to change the amount of the environmental allocation. The strategies individuals use vary for each of these stages of policy development and implementation. However the ambiguity of the science surrounding a target volume for environmental flows means that the most successful strategy throughout the process was to build coalitions that supported returning a given volume of water to the river.

10.2.2 Phase one: setting the target

The 1983 Commonwealth report titled, 'Water 2000' identified instream uses as a major issue of high priority in water resources management to be addressed by the Commonwealth and the states (Department of Resources and Energy, 1983). This was

Source: Department of the Environment, Water, Heritage and the Arts (2008).

Map 10.2 The Murray–Darling Basin catchment area in Australia

significant in marking this as a time when 'water resources management [was] undergoing a fundamental change in direction from a basically developmental thrust to a more balanced thrust that gives proper consideration to environmental, economic, and social goals as well' (Day, 1986, p. 148). However it was not until Australia's 1994 Water Reform Framework, along with continued degradation in the Murray–Darling Basin, that the Murray–Darling Basin Commission (MDBC), the intergovernmental management authority in the basin, initiated the Living Murray programme in 2002. The programme's aim is to recover water for the Murray River in order to address water overallocation and achieve environmental improvements in six 'icon sites' along the river (MDBC, 2006).

The Living Murray programme provided a window of opportunity for scientists and managers who had been eager to see base flows return to important wetlands and river channels. However because the river was already fully allocated, setting the target for water recovery was a largely political decision and is the first step where strategic intervention and negotiation can be seen at work. In 2002 the Ministerial Council directed the MDBC to undertake a comprehensive analysis of the environmental, social and economic costs and benefits associated with returning water to the River Murray system as part of the 'First Step Decision' (MDBC, 2004). They used three environmental flow reference points for analysis: an additional average 350, 750 and 1500 GL per year in the River Murray system. These numbers represent 3 per cent, 6.3 per cent and 12.5 per cent of the volume of water diverted from the river for irrigation, respectively; these values were predicted to have low, low-to-moderate and moderate likelihoods of fostering a healthy river system (Jones et al., 2001).

Eventually these volumes came also to represent the interests of various stakeholder groups. Scientists, such as leading national water expert Peter Cullen, supported the 1500 GL option while calling it a 'good start' (quoted in *The Age*, 28 July 2003). The then opposition leader, Simon Crean, also endorsed the 1500 GL option as a key plank of his and the Labor Party's environmental policy. Environmentalists were in favour of 1500 GL because it had the highest probability of achieving ecological benefits for the river system, although they too pushed for an even higher allocation. In fact the Green Party supported a 3000 GL environmental flow (personal communication). David Paton, a leading ecologist at the University of Adelaide, believes that 1500 GL to the environment is enough perhaps to slow the decline of environmental quality in the river, but probably not to lead to recovery. Finally South Australia favoured the 1500 GL option largely because of the benefit it stood to gain in terms of restoring the Coorong Delta and improved water quality and reliability for Adelaide; Don Blackmore, then Chief Executive of the MDBC, was quoted as saying that Adelaide 'turns out to be the biggest economy in the basin'.[1]

These groups all worked to develop and sell the idea that the health of the River Murray was rapidly declining and that as much water as possible was needed to restore ecological integrity. However the support of scientists, environmentalists and two of the country's three leading political parties was not enough to overcome the persuasive power of economic interest and the political tradition of providing water for irrigation. The Liberal Party, which held power in the federal government from 1995 to 2007, always supported the option that provided the least amount of water for the environment. As one prominent figure in South Australia put it:[2]

> I think the science community has been effective in getting across the idea that there is such a thing as environmental flow, and they're getting better at working out how to sell that message politically, but when it comes down to it, people screaming in newspapers and talk-back radio saying 'my livelihood is coming to an end because you won't give me water and we're going to take to the streets with guns if you don't give us some water to keep us going', the politicians followed every single time. There's never been a case where, when things got tight, the environment got water instead of people.

Interests at the local, state and federal levels, such as irrigation, mining and industrial development, were able to contest the legitimacy of reallocating larger volumes of water from existing users to the environment, such that the final volume agreed to by the Ministerial Council was only 500 GL over five years.

The strategies we see being used to support or contest different volumes of environmental water flows during this window of opportunity are primarily issue-framing and the promotion of ideas. Those who support returning larger volumes of water to the environment frame the numbers in ways that make them sound small and manageable. For example, a leading ecologist at the University of Adelaide says:[3] 'South Australia reduced [its water] consumption by 15 per cent without restrictions in cities. Everybody else should too and then we would have 1500 GL.' Another strategy is to talk about the percentage of total natural flow in the river the volumes represent, which presents even 3000 GL, 25 per cent of median flow, as a reasonable target. In contrast, those with opposing views reference volumes in terms of existing use. For example, on 28 July 2003, *The Age* reported:

> While it is only a little more than a tenth of what is now diverted, 1500 GL is still a lot of water. It is almost all of Victoria's total annual allocation of 1600 GL – which supplies dozens of small towns and regional hubs and thousands of hectares of irrigated pasture and horticulture. '1500 GL is a huge amount', says the state's head of water resources, Campbell Fitzpatrick, 'Impacts on the community will actually be quite large.'

However there is enough momentum and consensus behind the idea of environmental flows that they have become included in policy. This seems to be leading to a new strategy on the part of opponents. When asked who opposes environmental flows, an upper-level DWLBC employee said:[4]

> I think organizations that support that kind of business [irrigation] like the Farmers Federations and the National Party politically. I wouldn't even say that they're opposed now. Ten years ago I would've said that they were opposed, but now it seems not so much that they're opposed, but that they want to manage that process so that the people they represent get a good deal out of it.

This may manifest as managing networks to achieve the greatest benefit possible or utilizing new windows of opportunity during the implementation process.

10.2.3 Phase two: getting the water

One step taken by South Australia to implement the Living Murray First Step Decision to recover 500 GL of water for the environment was to create the River Murray Environmental Manager (RMEM) in 2006. This was largely in response to demands from individuals from within the DWLBC (one of whom became the RMEM) and the public for an independent authority to be responsible for environmental water recovery. It was important to the community and the government that this position be perceived as independent of the state bureaucracy and to have direct access to the minister. According to an interview with the RMEM: 'The community said, "We want somebody to manage these flows for us independently." The Minister is switched on and so created the position within the [Natural Resources Management] board.'[5] The RMEM's position was designed to be largely independent of the agency and to liaise directly with the public, farmers, business and the Minister for the River Murray through the regional NRM Board in the Coorong Delta area. Working through the Board, the River Murray Environmental Manager can have access to both the community and the minister. Support staff and resources were also transferred from the DWLBC to the NRM Board. Funding for the position was through the federal National Water Initiative (with 50 per cent matching funds from the state), with AU$2.4 million (US$2.1 million) dedicated to the position over three years (2006–09). The state support comes from the River Murray levy collected by SA Water, the state's water service agency for urban residents, and amounts to AU$18 million (US$15.5 million) per year. These funds can be used for water recovery projects as well.

The River Murray Environmental Manager is now considered the key decision-maker for the delivery, allocation and management of environmental flows in the River Murray for South Australia. This also includes unregulated flows – those not captured in storage facilities and released. The River Murray Environmental Manager is also able to set priorities for where and when environmental flows will occur. The manager's programmes are still in the early stage of development, but include management plans for the Coorong Delta, a 'strategic watering plan' to get past the drought, wetland protection and an overarching annual watering plan for the River Murray in South Australia. According to a response from the Minister for Environment and Conservation:[6]

> The South Australian Murray–Darling Basin NRM Board has the function of the South Australian River Murray Environmental Manager to provide clear accountability for delivering environmental flow outcomes for the River Murray in South Australia. The Environmental Manager oversees environmental flow management decisions and determines priorities for State-based environmental watering projects. The purpose, objectives and functions of the Environmental Manager are consistent with the scope outlined in the strategy 'Environmental Flows for the River Murray'.

The creation of the River Murray Environmental Manager position by the DWLBC can be seen as both the development and the promotion of a new idea – political independence as legitimacy for environmental water recovery – as well as an aspect of managing networks, as the position was designed to meet the needs of the agency, ministers and the public. Developing a position of authority and political independence was a novel idea that proved central in gaining public support and private engagement. The idea likely developed as a response to the increasingly politicized development of targets for environmental flows and the political nature of deciding where and when these flows take place.

The River Murray Environmental Manager herself is engaged in a number of strategies to secure environmental water allocations. While she obviously has proscribed responsibilities, she has also been passionately proactive in seeking new opportunities for securing environmental flows. Describing her job, other DWLBC staff state that: '[she] coordinates and advocates for environmental water'.[7] Many of these efforts are specifically designed by the River Murray Environmental Manager to depoliticize the environmental flow decision-making process. This has resulted in the manager creating new venues for the public to bring projects forward and for public–private partnerships to develop. For example, she is actively seeking donations of both money and water licenses from local corporations through a programme called 'Water Find'. This is designed to 'attract donations for projects from the [South Australia] Strategic Plan, and people can choose which project to direct their donations to'.[8] She is interested in developing the ability for private organizations to hold environmental water licenses that can be managed and prioritized by the River Murray Environmental Manager. She has also initiated several public meetings and workshops to help build strategies and thinks that the River Murray Environmental Manager idea 'has captured people's attention'.

This move to depoliticize environmental flows by involving private organizations and the public appears to be a successful example of orchestrating and managing networks. The River Murray Environmental Manager has been able to capture a broad base of support for further institutionalizing her post as the key actor for environmental flows for the River Murray. The manager spent considerable time and energy managing and creating a network of supporters, potential supporters and other agency departments. For example, the group at the DWLBC involved in The Living Murray would often come into conflict with the River Murray Environmental Manager's goals because both were looking to restore environmental flows, but sometimes at different times and places. One way of dealing with this overlap was to sign a Memorandum of Understanding (MOU) delegating authority and decision-making over various aspects of these decisions. The manager works with private organizations, the public and government to find ways to secure water licenses and donations to supply environmental water. The diversity of groups and strategies she uses is partly responsible for her success.

The success she has had in forming networks and satisfying the public's demand for political independence highlights the strategic benefits of venue-shopping when seeking participation and implementation. However it may still be too early to gauge her success in actually delivering environmental water, since her position is a vulnerable one: the current drought in Australia means that most efforts have been significantly stalled; it had taken the MOU to calm the jurisdictional tensions within the DWLBC; and further, it is unclear how a successive River Murray Environmental Manager would operate in this role. The social and political benefits of the position and the strategies it allows must be weighed in light of these concerns.

10.2.4 Phase three: changing the target?

A final opportunity to observe the strategic actions of individuals in the context of South Australia's provision of environmental flows in the Murray River comes with the interim sharing agreement developed among the states in 2007. The interim arrangement includes a share for the environment equal to that received by each of the states – essentially adding the environment as a fourth state. The 'Murray–Darling Basin Dry Inflow Contingency Planning Overview Report to First Ministers', created in September 2007, included a recommendation for interim sharing rules of unregulated flows in the River Murray (Department of the Environment and Water Resources, 2007). The extremely low inflows to the Murray–Darling Basin required interim sharing arrangements among the states. This provided scientists at the DWLBC and ministers with a rare opportunity to renegotiate allocations for the environment. As the least-populated state in the basin, South Australia is typically not in a very powerful negotiating position (Figure 10.1), and the window created by the low-flow contingency plan was critical.

The manager of the River Murray programme at DWLBC, who was highly involved in providing support for the development of the report, was optimistic that it could be a first step toward more equal recognition of the environment in 'normal' times. He also credits scientists at the agency as the primary reason why these negotiations were able to favour the environment, given their readiness with accurate and useful information for ministers.

VIC.'S GENEROSITY.

Source: Printed in the *Critic* in October 1904; reprinted with permission from the State Library of South Australia.

Figure 10.1 'Vic's generosity'

These modified sharing arrangements were developed during one of the most severe droughts Australia has ever faced. The success of this negotiation was due in large part to the willingness of hydrologists in the DWLBC and the relevant ministers to take advantage of the opportunity presented and to make a strong case for the requirements.

Ministers and DWLBC staff had been working behind the scenes for many years to make the case for the necessity of environmental flows to South Australia, but the necessity of consensus among the Council for a change in sharing rules had essentially blocked all of these efforts. If this had not already been part of the political agenda in South Australia, the negotiation of temporary allocation measures would likely not have included such provisions.

10.3 Case two: licensing interception losses and direct use of groundwater by commercial forestry in the southeast of South Australia

10.3.1 Introduction

The southeast of South Australia is home to 4.3 per cent of the state's population: a total of 63 000 people live in the region, 26 000 of them in the city of Mt Gambier (SENRM, 2006). The region relies on two groundwater aquifers for domestic water supply as well as for irrigation of wine grapes, lucerne, potatoes, dairy cattle, sheep and beef. Natural resources-based tourism is also a significant economic activity, with unique caves for diving and Ramsar-listed wetlands in the area. Many of the management zones are overallocated. Farming, food processing and agricultural marketing are the traditional employment opportunities in the region, but its booming forestry industry, and the paper products manufacturing and pulp mills it supports, are becoming central to the economy.

The water prescription process in the southeast was put into place in the 1980s and requires a water plan that specifies the amount each grower is allocated. The most recent plan (2005) was drafted by the South East Natural Resources Management Board. While irrigators have always been subject to the water planning process, forestry has only recently been recognized as a 'water-affecting activity', per the 1997 South Australia Water Resources Act, due to a new understanding of the relationship between forests and the water cycle. It is now known that forests in the southeast affect the region's water balance both in terms of direct use of groundwater by tree roots as well as through interception losses from the tree canopy that prevents water from recharging underground aquifers. Key actors involved in the forestry industry, the region's NRM Board, the state government, lobby groups and others have been highly influential in shaping this process. The contributions of their strategies in promoting understanding of the relationship between forests and the water cycle, to the process of developing the science behind forestry water use and to the subsequent implementation of appropriate policy, are the subject of our second case study.

This case study was conducted through many interviews with past and present members of the NRM Board and individuals from the forestry industry and the diverse horticultural groups. All interviewees mentioned that a key factor in the regulation of forestry was leadership of the NRM Board and the credibility lent to it by its chair, an ex-forester. Science also was cited as playing an important role in influencing the decision-making process. We found evidence of 'bureaucratic entrepreneurs' within both the NRM Board and the DWLBC. Such players are described in Teske and Schneider (1994) as: '[A]ctors who help propel dynamic policy change in their community.' In this case the policy concepts were sustainable water management promoted from within the government, and fairness as supported by the diverse horticultural groups.

10.3.2 Phase one: understanding the relationship between forests and water

There are two kinds of forest plantations in the southeast of South Australia: softwood (pine) and hardwood (blue gums). Pine plantations have been in the southeast of South Australia for years and have grown fairly slowly. The state government originally planted them in the 1920s but many have now been sold to private investors. The blue gums, on the other hand, expanded very late in 2000 and have increased in area by 35 per cent. This was largely due to Managed Investment Schemes (MIS) developed by the Australian government in the 1970s as tax vehicles to allow investors in urban areas to provide funds for agricultural development. Over a period of ten years (1997–2007) some of these investments were applied to blue gum forestry development in the south-east of South Australia. This investment resulted in an explosion of commercial forestry development in the region (together with the southwest of Victoria known as the 'Green Triangle'), and by 2008 the forest estate had grown to occupy 14 per cent of the lower southeast landscape that is considered accessible to forestry.[9]

The competition for land and water that resulted from the investment and subsequent expansion quickly raised the community's awareness of the industry's impact on the region's valuable groundwater resources. The resulting concern was not evenly distributed toward the forestry sector as a whole, but was primarily targeted at the blue gum subsector. Local farmers cited rapidly falling groundwater tables as the main source of evidence for these impacts and were hostile because they felt the MIS growers and owners were not following the stewardship principles practiced by local farmers. In 2006 the region began to experience severe drought conditions, further increasing awareness of the groundwater decline as it coincided with the expansion of forestry plantations (SAFF, personal communication). The dairy, lucerne, potato and some wine grower groups got together (some of them members of the South Australian Farmers Federation) and lobbied to make forestry account for its extraction of water from shallow aquifers. This coalition worked to increase awareness of forestry's water use and demanded that plantations should be licensed by the NRM Board. Urban water users also became involved. The citizens of the town of Mt Gambier in the southeast are supplied by the groundwater-fed Blue Lake and significant water level declines were beginning to be noticed by the community and local media. This helped create even greater pressure within the agency and the above agricultural water users to hold forestry accountable for its potential water impacts through interception and direct use of water in shallow aquifers.

In response to these changing perceptions of forestry's relationship to groundwater, on 17 February 2004, South Australia's Minister for Environment and Conservation (Hill, 2004) announced that the state government would adopt a management system prescribing commercial forestry as a water-affecting activity in the lower southeast of the state. On 7 February 2008 the Chairman of the National Water Commission, Ken Matthews, endorsed South Australia's approach: 'South Australia is to be congratulated on its significant effort to address interception by large-scale forestry activity in the lower south-east.'[10]

The Minister for the Environment and Conservation was then empowered by the 2004 Natural Resources Management Act to direct local district councils to refuse foresters a development application for commercial use of forest land if the minister believed that the development would have detrimental impacts on the groundwater resource and its ongoing sustainability. The regulation applied to recharge impact only and not to direct groundwater use. On 31 July 2007 a new Minister for the Environment and Conservation, as part of a state government in the hands of the newly elected Labor Party, extended this regulation to extraction by forests over shallow aquifers (Harvey, 2008). Exceeding the allowed planting level in an area where there are shallow aquifers will require an offsetting water allocation. This was to be regulated through individual arrangements with growers, which are described as: 'cumbersome and inefficient administrative arrangements which take the form of individual agreements between forest managers and land owners, water owners, and the Minister' (Harvey, 2008).

A variety of venues was used by the grower coalition, local citizens and the government to convey the message that forestry should be a regulated water use. These included media coverage, public meetings, Natural Resources Management Council presentations, written reports and networking within government agencies. Including forests as water users was an extremely novel concept developed jointly by non-forestry growers, the NRM Board and concerned scientists at the DWLBC and the Commonwealth Scientific and Industrial Research Organization (CSIRO). The idea arose in part due to water scarcity, making water use by forests more obvious, as well as to improvements in data collection and monitoring, and to the water allocation planning process being specific to the region and allowing for a detailed accounting of water users. These three strategies – playing multiple venues, orchestrating and managing networks, and developing new ideas – seem to be the most prominent in terms of achieving policy change that recognizes forestry as a water-affecting activity and therefore in need of licensing.

10.3.3 Phase two: developing the science

Determining how to license forestry's water use required new science and negotiations. Fortunately agency staff members at the DWLBC were capable of developing figures that all sides were comfortable with. With encouragement from the forestry industry groups, a study was conducted by the federal research agency thanks to a long-standing scientist- and farmer-based committee called the South East Natural Resources Committee, that was supported by the Government of South Australia, the CSIRO, the DWLBC and the South East NRM Board (Benyon and Doody, 2004). The study produced unequivocal evidence that there was some recharge of groundwater under forest canopy, but also that forestry extracted groundwater from shallow aquifers. The acknowledged recharge rate was 0.20 for softwood and 0.30 for hardwood.[11] The forestry lobby group expressed the finding as a recharge benefit. The agency staff at the DWLBC expressed it as a reduction in recharge and considered it a 78 per cent reduction in recharge volumes on those achieved if the land had been under grass for short-rotation hardwood, and an 83 per cent reduction for long-rotation softwood. The recharge impacts are 160 GL per year out of total licensed extractions of 400 GL, nearly half of all licensed extractions. In relation to extraction rates there are differences between blue gums and pine, but the total impact is 80 GL per year. The rigorous and compelling science behind these numbers lends a very high level of legitimacy to the study, whose numbers were accepted by all sides of the debate.

10.3.4 Phase three: implementing the policy

Where planting was taking place, the southeast region's water was fully allocated prior to including forestry as a water-affecting activity. Therefore, despite the fact that there are water markets operating in the region, any change to the water balance would likely require existing water allocations to be altered as well. The original water budgets were developed by the DWLBC under the assumption that there was no groundwater recharge under plantation softwood forest, meaning that forested landscapes intercepted all rainfall and that extraction need not be accounted for. Accepted wisdom, and the assumption also in other regions of Australia, this was recognized as wrong, and in 2004 the National Water Initiative (Clauses 55–57) required water use by forestry to be accounted for.

Coalitions were a central force in determining the path of forestry regulation in the southeast of South Australia. The forestry lobby groups in the two states joined forces to deflect attention, stall the process, provide alternative science, influence politicians and attempt to dispel the idea that forestry uses water. However it was reported that policy change advocates, the 'other side', also benefited greatly from coalition-building. Media, conservation groups and the NRM Board growers formed different lobby groupings, and the South Australian Farmers Federation (SAFF) often worked with them to maintain public support, apply pressure on decision-makers and promote the idea of making forestry account for its water use and hence for it to be licensed and regulated. A key person at the time who still works in the forestry industry said: 'The horticultural sector put pressure on government to regulate forestry by taxing them.'[12]

Forestry lobby groups, policy entrepreneurs from the DWLBC and their interaction via the water planning process of the South East NRM Board have been crucial for the outcomes of this policy process. Once it became clear that the DWLBC and the Catchment Board were seriously considering regulatory options for forestry, a planning committee was formed within the forestry group called the Regional Plantation Committee. The foresters' goal in forming this group was to create cohesion within the industry (joining the hard- and softwood industries) and to present a position statement to policy-makers on the issue of groundwater and forestry regulation. The deliberations among the group resulted in a shared understanding that forestry needed to be included in the water management and planning process, but they also felt that licensing was not the appropriate policy mechanism for forestry.

By presenting this view as a united front, the forestry industry (a relatively strong lobbying group overall) was able to deflect the first round of policy attention away from regulating the direct interception of groundwater (Figure 10.2). A type of compromise position was then reached. The CSIRO study on the recharge benefits of forestry plantations showed that a further expansion of 59 000 hectares in the area could be supported. The resulting regulation – developed by the state and considering the position of forestry – states that for any expansion of plantation forest beyond the 59 000 hectares, or in water management areas where the area set aside for forest development has been reached, the forest developer must offset the impact on the water budget by securing an appropriate water allocation. The exclusive dedication of a share of the regional water resource to the forest industry (for the 59 000 ha expansion) provides for an increase in the estate by approximately 45 per cent (based on data as at 2002) before there will be any need to offset further development against water allocations. This grandfathering mechanism provides industry with significant certainty regarding its opportunities to expand, allows the state the potential to regulate growth beyond this, and was the only way that the groups would accept the proposed changes.

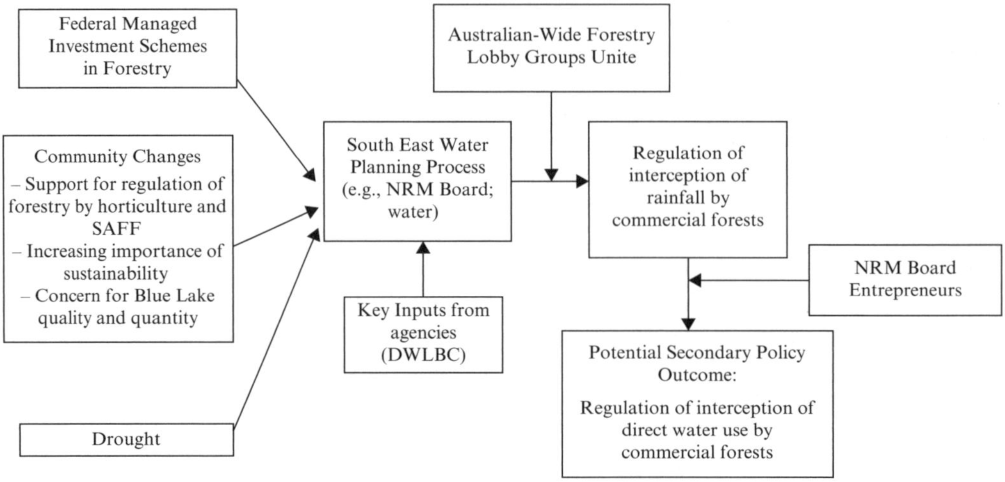

Figure 10.2 Framework for the emergence of southeast South Australian regulation of rain interception and direct extraction

The process of including direct extraction has created divisions in the community between farmers, foresters and NRM Board members. Network management in the forestry case was extremely difficult. The leader of the forestry lobby dealt with treacherous politics in a small community; one reason she was so successful was because she had existing networks and experience to lean on. The South East NRM Board has sought support from the state's relevant Minister for Natural Resources Management to push through the policy and to change the Natural Resources Management Act 2004 to require forestry to obtain a water allocation. As of July 2008 this proposal is still before the current state minister. Further, key organized interest groups – the SAFF and conservation groups – also support the South East NRM Board in seeking to have interception and recharge accounted for. Whether these efforts will result in the kind of radical policy change they are seeking is still unclear, as the forestry groups, and others who stand to profit from forestry expansion, continue to seek ways to prevent the changes. Other plans in other regions are also looking at copying the southeast in relation to water use by forestry. It is notable that the adjacent state, Victoria, was described to us as watching the outcomes in South Australia rather than moving to adopt the same regulation in its own adjacent forestry plantations. The process of consulting on the proposed changes is now under way. The Victorian officials use different procedures to allocate water and a South Australian official reported that: 'the Victorian officials are against volumetric conversion as they feel it increases our [South Australia] allocations'.[13]

A key aspect of the achievements so far has been the fact that the SAFF, the many grower groups and the conservation groups have worked together and that the chair of the NRM board was an ex-forester. In addition the DWLBC staff have been providing groundwater information to all members of the community and working actively to ensure that water users are well informed of the state of the aquifers. The leader of the forestry lobby believed that if this kind of coalition-building had not taken place, forestry regulation would not exist as it does today. When the foresters united, they were able to deflect the focus of regulation to rainfall interception. However the strong community involvement and the science advancement and its promotion have been able to counter the powerful influence of the federal-level forestry groups.

Finally, it is worth noting that a window of opportunity may be created by the water allocation plan development process. Every five years when these plans are developed there is another opportunity for change to be initiated through participation in the process. It will be seen in the future whether these are valuable opportunities for further negotiation and refinement of the policy details and scientific understanding.

10.4 Conclusions: how actors help shape water policy transitions in Australia

Our case studies of water transitions in Australia illustrate the diversity of ways in which actors are playing a role in shaping policy outcomes. They often reframe water policy reforms by initiating new strategies, collaborations and ideas. They act as policy entrepreneurs in their use of windows of opportunity and orchestration of networks to achieve their ends. Actors play a key role at multiple levels of governance and sectors where it is important never to underestimate the power of a key person to drive through reforms.

The type of legislation that exists in South Australia requires the water planning and allocation body (primarily NRM Boards) to have all major players around the table when making changes to existing policies, and that scientific evidence from third parties is available to consider scenarios such as business as usual or radical changes, such as the new forestry allocation. While accounting methods for environmental water are still being developed in Australia, their strength and scientific validity will be a determining factor in the programme's success. It is likely that without the addition of forestry and the environment as new and legitimate water users in Australia, these developments would not have taken place or, at least, would have occurred at a much slower rate. More policy change is under way at the federal level, with a new management plan for the Murray River (and the basin as a whole) to which South Australia will need to adhere. For South Australia this will only affect the River Murray NRM Board and not the southeast; this means that the NRM Boards will play the major role in the southeast in water management and water allocation through the development of water plans.

The case studies also demonstrate that 'bureaucratic entrepreneurs' are extremely influential players in Australia. Like other policy entrepreneurs, they engage in the act of 'creative discovery' by creating or exploiting new opportunities to push forward their ideas (Teske and Schneider, 1994). Within the DWLBC alone we repeatedly observed scientists taking actions to steer science and research toward policy goals (ambitious flow targets and forestry regulation) quite independently and with great success. The role of public administrators in the DWLBC cannot be underestimated in initiating and forwarding water transitions as well as in dismantling strong political opposition. These individuals seem to push the boundaries of bureaucratic authority with innovative science.

Our cases also highlight the diversity of strategies used by actors to influence the transition process. The ability of the forestry groups to reach consensus on their policy preferences allowed them to shift the policy vector, but only temporarily, due to strong community and NRM Board support for the inclusion of forestry as a water user. Entrepreneurs were involved on both sides in policy change. They were able to manage the existing network and exploit venues for change in ways that guided the policy toward an outcome more favourable to their own interests. Often these aligned with long-term requirements to establish sustainable water allocations, a significant feat in a system with such a long history of overallocation. Established groups like the forestry lobby and the NRM Board are not the only ones positioned to act strategically. In the case of environmental allocations new actors have also played a strategic role. In South Australia, a Riverland-based minister for water resources who was committed to river health along with strong community support helped to create a new strategic entity to act on the river's behalf. The new River Murray Environmental Minister is unique in Australia, has the resources and community support to influence significantly environmental water management in the river and is dedicated to environmental health. Their insertion of new ideas directly into the existing institutional framework could change policy directions (Haas, 1989).

Our case studies of policy transition show how substantive change in the water sector, such as new water-accounting methods, unlikely to occur without a transition, can lead to the development of governance strategies and collaboration that are also unlikely to result independently of transition. Collaboration has expanded in both cases. In the forestry case, forestry lobby groups presented a consensus position for the first time and the private sector is beginning to be involved in the provision of environmental water; pressure for policy change helped the South East NRM Board develop a closer working relationship with the Minister for Water Resources and to realize that this relationship was strategically critical to overcome the pressure from forestry lobby groups. In the case of securing environmental flows, new relationships between the public, the DWLBC, private industry and the ministers are changing river management and the political legitimacy of environmental flows.

The cases also highlight the innovation that individuals were able to bring to the policy process. The method of forestry groundwater allocations and accounting has resulted in a completely new accounting method for groundwater in the region, and has been actually used elsewhere in Australia due to increasing understanding of and new data analysis methods for recharge and extraction rates of plantation forestry. Creating a politically independent authority for environmental flows is also part of the federal reforms for the Murray–Darling Basin. These may be innovations of the individuals themselves or, more likely, they are examples of how individuals work to establish innovative ideas within an institutional context.

The processes discussed here arise in case studies with many local and specific variables relevant to understanding the role of the entrepreneur in achieving a sustainable water management transition. These local events are indicative of the way national-scale policy transitions also are negotiated and directed by strategic actors for implementation. Often they may have value as a precedent, in that other regions or states may look for guidance to the success or failure of strategic action in a particular location. For example, the foresters in South Australia may prove instrumental in forcing a similar transition in neighbouring Victoria. Horticultural growers in forestry regions may be inspired to work together, forming a coalition and cooperating with state agencies to insist on a similar transition.

The framework developed by Meijerink and Huitema (Chapter 2, this volume) has proven very useful in understanding the strategies of actors in South Australia and how they have influenced the direction and pace of policy change. We would add two areas of emphasis based on our experience in Australia. First, we would expand the strategy of developing new ideas to include a defined role for science. We have seen that science can be a compelling force for either developing or supporting a policy idea, particularly in the area of water resources. In both of our cases, scientific advancements played a key role in helping to forward policy change. The mechanisms of science's influence in these cases include lending credibility to an idea, allowing actors with an understanding of science to participate in the policy process, and allowing for novel policy options based on new understanding. Second, we would emphasize the importance of the interaction between actors and their institutional context. While strong leaders and passionate advocates are often successful in creating policy change, the range of options available to them, the pace at which they may be able to effect change and the longevity of the changes they do achieve will likely be affected by the rules and boundaries within which they operate. Understanding this relationship may be an interesting area for further research.

Notes

1. *The Age*, 28 July 2003.
2. Interview with an independent consultant on water policy in South Australia, 25 October 2007.
3. Interview with University of Adelaide ecologist, 29 June 2007.
4. Interview with upper-level Department of Water, Land and Biodiversity Conservation staff member, 9 October 2007.
5. Interview with Judy Goode, River Murray Environmental Manager, 18 July 2007.
6. Email response from staff speaking for Minister Gail Gago, Minister for Environment and Conservation, received 17 January 2008.
7. Interview with upper-level Department of Water, Land and Biodiversity Conservation staff member, 9 October 2007.
8. Interview with Judy Goode, River Murray Environmental Manager, 18 July 2007.
9. Land considered accessible under the Development Act 1993 planning provisions is that land currently zoned as general farming or commercial forest. It excludes areas of lakes, wetlands, native vegetation, parks, urban development and rural living. Accessible land is not necessarily suitable land, but it allows for a way of expressing land use distribution and scale.
10. National Water Commission (2008).
11. Notional recharge benefits have been adopted for each forest type and they are 17 per cent for softwood and 23 per cent for hardwood. These relate to the impact on groundwater recharged, averaged over the full rotation. In the case of softwood plantations, it is assumed that the forest uses 83 per cent of the recharge that would occur had that land been used for traditional agricultural purposes. An alternative expression is that the plantation allows 17 per cent of the mean annual recharge that would have occurred under an agricultural landscape. Mathematically this results in a plantation previously assumed to intercept all water to provide a recharge benefit in the ratio of 17/83, which accounts for an additional area of 0.20. In the case of hardwood, the mathematics is 23/77 = 0.30. In October 2001, groundwater and forest scientists agreed on a general acceptance of notional scientific impact values for forestry on groundwater recharge for hardwood and softwood forest cycles in the lower southeast. For management purposes these notional values will remain until refined or confirmed.
12. Interview with Diana Lloyd, Australian Forest Growers, former Executive Officer of the Green Triangle Regional Plantation Committee, 10 July 2007.
13. Interview with upper-level Department of Water, Land and Biodiversity Conservation staff member, 12 July 2007.

References

The Age (2003), 28 July.

Archer, J. (2001), *Australia's Drinking Water: The Coming Crisis*, Pearl Beach, SW: Pure Water Press.

Benyon, R. and T.M. Doody (2004), 'Water use by tree plantations in South East SA', Technical Report 148, Australia: CSIRO.

Council of Australian Governments (COAG) (1994), *Water Reform Framework*, Hobart, Australia: Council of Australian Governments, 25 February, http://www.environment.gov.au/water/publications/action/pubs/policyframework.pdf, accessed October 2008.

Day, D.G. (1986), 'Instream flow provisions: politics and management ', in National Committee on Hydrology and Water Resources (ed.), *Hydrology and Water Resources Symposium 1986: River Basin Management*, Griffith University, Australia.

Department of the Environment and Water Resources (2007), *Murray–Darling Basin Dry Inflow Contingency Planning: Overview of Report to First Ministers*, Department of the Environment and Water Resources, Australia.

Department of the Environment, Water, Heritage and the Arts (2008), 'Murray–Darling Basin as described in the Water Act 2007', created April, http://www.environment.gov.au/water/publications/mdb/pubs/mdb-map.pdf, accessed October 2008.

Department of Resources and Energy (1983), *Water 2000: A Perspective on Australia's Water Resources to the Year 2000*, Department of Resources and Energy, Australia.

Haas, P.M. (1989), 'Do regimes matter? Epistemic communities and Mediterranean pollution control', *International Organization*, **33** (3), 377–403.

Harvey, D. (2008), 'Managing the impact of plantation forests on groundwater', Presentation to Natural Resources Management Council, Government of South Australia, 5 May.

Hill, J. (2004), Ministerial statement to the Hansard House of Assembly, Tuesday 17 February, The Hon. J.D. Hill, Minister for Environment and Conservation.

Jones, G., T. Hillman, R. Kingsford, T. McMahon, K. Walker, A. Arthington, J. Whittington and S. Cartwright (2001), Independent Report of the Expert Reference Panel on Environmental Flows and Water Quality Requirements for the River Murray System, Canberra, Australia: Cooperative Research Centre for Freshwater Ecology.

Maheshwari, B.L., K.F. Walker and T.A. McMahon (1995), 'Effects of regulation on the flow regime of the River Murray, Australia', *Regulated Rivers: Research and Management*, **10**, 15–38.

McKay, J.M. (2002), 'Encountering the South Australian landscape: early European misconceptions and our present water problems', Hawke Institute Working Paper No. 21.

McKay, J. (2007), 'The quest for environmentally sustainable water allocations: issues for federal, state and local governments', *Journal of the Australian Law Reform Commission*, **89**, 21–25.

Murray–Darling Basin Commission (MDBC) (2004), *Murray–Darling Basin Commission 2003–04 Annual Report*, Canberra, ACT: Murray–Darling Basin Commission.

Murray–Darling Basin Commission (MDBC) (2006), *The Living Murray Environmental Watering Plan*, Canberra, ACT: Murray–Darling Basin Commission.

National Water Commission (2008), press release, 7 February 2008, http://www.nwc.gov.au/www/html/875-major-water-interception-report-released.asp?intSiteID=1, accessed 12 November 2008.

Pigram, John J. (2006), *Australia's Water Resources: From Use to Management*, Victoria, Australia: CSIRO Publishing.

Smith, D.I. (1998), *Water in Australia*, Melbourne: Oxford University Press.

South East NRM Board (SENRM) (2006), 'Regional natural resources management plan concept statement', http://www.senrm.sa.gov.au/OurPlans/RegionalNRMPlan/ConceptStatement.aspx, accessed October 2008.

Teske, P. and M. Schneider (1994), 'The bureaucratic entrepreneur: the case of city managers', *Public Administration Review*, **54** (4), 331–40.

PART IV

AFRICA

11 South African water and mining policy: a study of strategies for transition management

Anthony Richard Turton

11.1 Introduction

South Africa is both famous and notorious for many aspects, including the dismal history of apartheid, which saw institutionalized human rights abuses. A less-known aspect is related to water quality and human health, specifically as it pertains to a water-constrained country with a mining-based economy. This chapter presents five policy paradigms (or policy monopolies) as they pertain to the evolution of water policy in the South African mining sector, each of which emerged as a result of a specific shock event. The evolutionary process is about three distinct overarching policy perspectives and the orbit of policy brokers, policy entrepreneurs and policy opponents, like electrons around a nucleus of an atom, connected yet separate.

Overarching policy perspective I gears policy in support of extraction; overarching policy perspective II generates a policy environment that protects the goose (extractive industries) that lays the golden eggs; overarching policy perspective III builds the constitutional and other legal elements that may provide the legal foundation for a new policy paradigm. This third overarching perspective has yet to resolve. Its two opposing dynamics remain in play and will determine the success or failure of current attempts to bring about a governance transition that seperates government as regulator and industry as the object of regulation.

Overarching perspective I, supporting policies centred on extraction, has proceeded through five phases, each embodying a separate policy paradigm. The last of these phases concerns a policy paradigm of redistribution of wealth and privilege. It is this paradigm that provides the context for the present struggle to push through a transition that would anchor South African water governance in a third overarching perspective resolved in the direction of strong regulation of extractive industry.

Of necessity this chapter focuses on this fifth policy paradigm (paradigm V), giving only a brief overview of the earlier phases of policy evolution, helping to contextualize the analysis within the oldest and persistent overarching policy perspective, one geared to resource extraction. Policy paradigm V yields the type of data of interest within the context of this book – the role of policy entrepreneurs and strategy in transition management. Paradigm V offers useful information about the limitations and potential for success of strategies in a high-stakes game revolving around naked human greed and the pursuit of profits set against an emerging human rights culture.

11.2 The relevance of water policy in South Africa

South Africa (Map 11.1) is a water-constrained country (Conley, 1996; Basson et al., 1997). The earliest writers on the subject described the aridity of the country, which has always been one of the key limitations on economic growth and development and hence

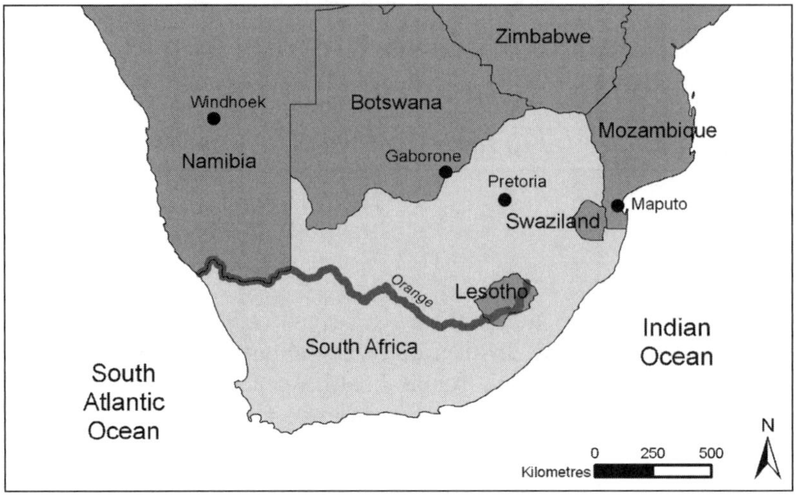

Map 11.1 South Africa and the Orange River

also on political stability (Turton, 2007). The National Water Resource Strategy, the official government blueprint on water resource availability, indicated that in 2004 around 98 per cent of the total national water resource had been allocated at a high assurance of supply level (NWRS, 2004), with many water management areas being overallocated and thus 'closed' (Turton and Ashton, 2008). This means that future economic growth is unlikely to be sustained without a significant shift in thinking about the role of water as a foundation of the economic development. South Africa has thus reached the hydrological limit for economic development using current technologies, which was predicted during the Commission of Enquiry into Water Matters in 1966 (RSA, 1970; Turton et al., 2004). Future economic growth will be determined by how water policy is changed and how technical capacity is built during the current transition to reflect this reality.

South Africa has a mining-based economy (Figure 11.1). Given the geology of the country, gold deposits are often associated with other heavy metals including uranium (Werdmüller, 1986). To date there are 8000 derelict and ownerless mines on record, all of which are unrehabilitated and were costing the taxpayer ZAR 100 billion in 2008 (1 euro = 10 ZAR) (Brown, 2007). There is no clear policy on mine closure in South Africa that meets both the human rights standards set in the Bill of Rights of the National Constitution (1996) and the 'polluter pays' principle enshrined in the National Water Act (1998). There is an unresolved question as to whether mine closure is about mining, water resource management, human health, economic development, legal liability or simply ecological rehabilitation.

South Africa has a history of human rights abuse, of which there have been a number of specific phases. This includes the Great Cattle Killing of 1856–57 (Meer, 1990; Welsh, 2000; Peires, 2003); the ethnic cleansing of the *Mfecane* from 1816 to 1828 (Van Jaarsveld, 1975; Thompson and Lamar, 1981; Edgecombe, 1986; Welsh, 2000; Turton et al., 2004); the Second Anglo-Boer War from 1898 to 1902 that saw the first use of concentration camps (Fawcett, 1901; Hobhouse, 1901; Krebs, 1992; Pakenham, 1992; Farwell, 1999; Hasian, 2003; Mills and Williams, 2006; Turton et al., 2006); and the

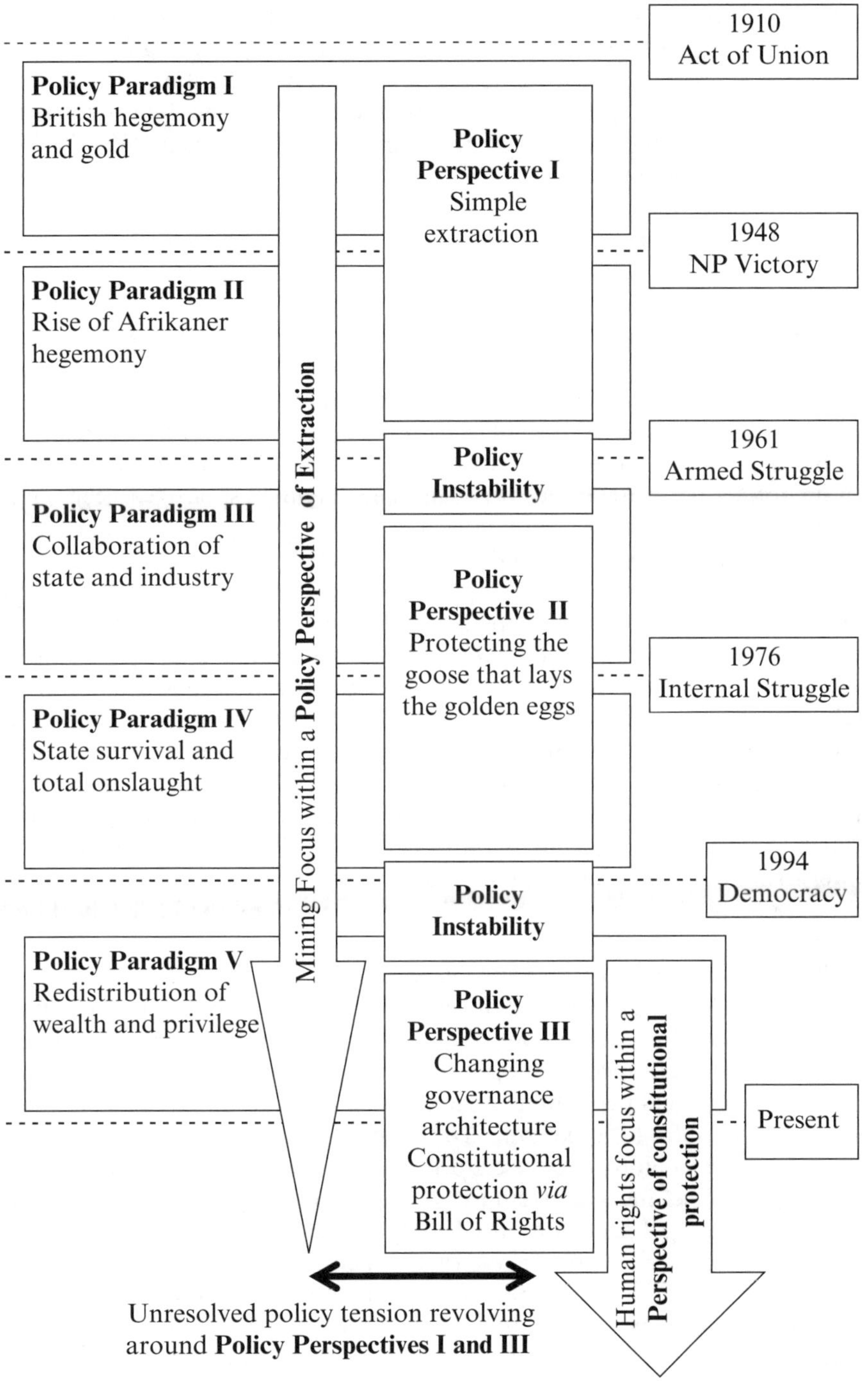

Figure 11.1 Schematic representation of the evolution of mining and water policy showing key transition periods

dismal policy of apartheid (Meer, 1990; Du Toit, 1994; Lodge, 1994; Odendaal, 1994; Ross, 1999). This has left a legacy of winners and losers, which is highly relevant to any contemporary policy debate, as the policy paradigm of redistribution takes root. The 2008 xenophobic violence is simply a new manifestation of the issues that need to be addressed in the unresolved policy debate (Johnston and Bernstein, 2007; Dixon, 2008; Sibanda, 2008; Walker, 2008).

11.3 Five phases of the overarching policy perspective of extraction

11.3.1 Introduction

The essential problem is that the existing mining policy reflects a historic legacy in which powerful financial interests coincided with the interests of a racially defined political elite, which saw government becoming a collaborator rather than a regulator of the mining industry. This manifests itself as the policy paradigm of extraction with five distinct phases (named for analytical purposes as policy paradigms I–V). The relevance of this becomes apparent in noting that water policy per se has been highly responsive to changing political demands, embracing human rights concepts that are considered to be progressive (de Coning and Sherwill, 2004). This is totally at odds with the mining policy, which still reflects the historic power structures that underpinned it for so long, currently manifesting as a poorly defined hybrid of an earlier expression of the policy paradigm of extraction. This needs to be understood in the context of growing disillusionment at the way some political elites have allowed their gatekeeping positions to delegitimize the democratic political process by introducing corruption at the highest echelons of government (Feinstein, 2007; Johnston and Bernstein, 2007; Muller, 2007), a context that is underscored by a growing public awareness of future ecological disasters (Eybers, 2007; Avni, 2007a, 2007b, 2007c; Gosling, 2007; Groenewald, 2007; NNR, 2007; Secretary to Parliament, 2007; Tempelhoff, 2004, 2007b, 2007c) and human health problems (CSIR, 2008; Turton, 2008).

In order to analyse these complexities in a systematic way, we need to understand the process by which the five phases of the policy paradigm of extraction evolved.

11.3.2 Policy paradigm I: British hegemony and gold (1910–48)

South Africa was born as a country out of the ashes of the Second Anglo-Boer War, fought over gold (Turton et al., 2006). Water policy at this time was driven by the need to establish a high assurance of supply for the mines and therefore initially saw the British military engineers performing related tasks until the Rand Water Board (RWB) was established in 1903 (Tempelhoff, 2003). The Act of Union in 1910 established a self-governing British colony composed of four smaller units, two of which were former British colonies (Cape and Natal) and two of which were vanquished Boer republics (Transvaal and Orange Free State) in which gold was found (Geldenhuys, 1984). The rationale for the union was resource capture, which thus became a driver in the policy paradigm of extraction still relevant today. A key element of this phase was a period known as the 'Midas Touch' or 'Long Economic Boom', lasting from 1934 to the mid-1970s (Geldenhuys, 1990; Ross, 1999) (see Figure 11.2). Policy paradigm I was driven by the British with the sole objective of extracting mineral wealth from South Africa, giving rise to the policy paradigm of extraction, which took place in the total absence of any human rights culture.

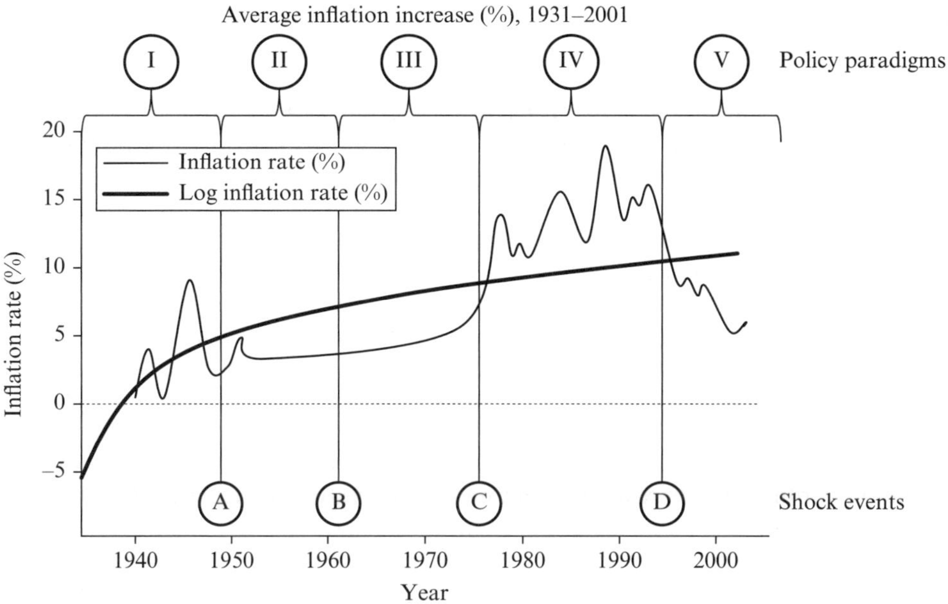

Notes: During periods when inflation was above average there were periods of acute political conflict (1938–45 and 1975–94). Periods of below average inflation coincided with relative stability. The 'Midas Touch' or 'Long Economic Boom' period is visible as a below average inflation period from the early 1950s to 1975.

The vertical lines and alphabetical circles coincide with the shock events, and the numbered circles above the brackets coincide with the policy paradigms illustrated in Figure 11.1 and listed in the text.

Source: Data from Statistics South Africa (2002), as presented in Turton et al. (2004), p. 74.

Figure 11.2 Inflation trends in South Africa track political instability

11.3.3 Policy paradigm II: Rise of Afrikaner hegemony (1948–61)

The transition to this phase occurred when a group of Boer elites, many of whom had been interned during the Second World War for their pro-Nazi sentiments, seized the moment and won the 1948 general election (Liebenberg, 1987; Turton et al., 2004; Mills and Williams, 2006; Turton, 2006) ('A' in Figure 11.2). A characteristic of policy paradigm II was a split between economic power (in British hands) and political power (in Afrikaner hands), but with an invisible undercurrent of growing black resistance (Liebenberg, 1994). The end of this phase occurred during a period of policy instability, begun by a speech made by Harold Macmillan, then British Prime Minister, in February 1960, in which he referred to the 'Wind of Change' that would sweep across Africa (Turton et al., 2004; Turton, 2006). A series of protest marches took place and one became a major crisis known as the Sharpeville Massacre (Spitz and Chaskalson, 2000), triggering a loss in investor confidence ('B' in Figure 11.2). Government response to this was brutal, giving rise to the 'armed struggle' as passive resistance to apartheid gave way to a new, armed militancy, driven by the founding of the Pan Africanist Congress (PAC) in 1959 (Liebenberg, 1994; Lodge, 1994) and the formation of Mkonto we Sizwe (Spear of the Nation) (MK), the armed wing of the African National Congress (ANC), a short while later (Williams, 1994). South Africa started to become a pariah state as a result of these events, making state survival a significant driver of policy.

11.3.4 Policy paradigm III: collaboration of state and industry (1961–76)

The policy instability that gave rise to this transition resulted in the banning of the various liberation movements, which went into exile. On the economic side, it resulted in policy of rapid growth, driven by gold, but also aimed at diversifying the economy and restoring investor confidence. The Jordaan Commission of Enquiry recognized that the only way to mine the deep ore bodies of the Far West Rand was to dewater the dolomites above the reef (Jordaan et al., 1960). A Commission of Enquiry into Water Matters was launched in 1966 (during a period coinciding with the end of the era of the 'Midas Touch' or 'Long Economic Boom' – see Figure 11.2), with the objective of completing a strategic-level study into the national water need related to rapid economic development (Turton et al., 2004).

In 1963 agreement was reached between the government and the Chamber of Mines on dewatering of the dolomites of the Far West Rand (van Eeden, 2007). Two organizations were formed to deal with issues pertinent to the dewatering. The first was the State Coordinating Technical Committee on Sinkholes and Subsidence (SCTC), tasked to deal with the loss of life and property from sinkhole formation and land instability arising from the dewatering process (Kleyweght and Pike, 1982; Adler et al., 2007b). The second was the Far West Rand Dolomitic Water Association (FWRDWA), charged with dealing with compensation matters arising from dewatering (van Eeden, 1992; Adler et al., 2007a), consistent with some of the recommendations of the Jordaan Commission.

During the early portion of policy paradigm III, the armed struggle was starting to take effect, with attacks by the African Resistance Movement (ARM) (Du Toit, 1994), by MK using explosives set at key points (Williams, 1994) and by the *Poqo* wing of the PAC using machetes and axes to kill and maim white people who had been labelled 'settlers' (Lodge, 1994), similar to the Mau-Mau uprising in Kenya. Despite these sporadic attacks, the relative political stability arising from state repression kept the armed struggle outside the borders of the country.

Three events heralded in a period of instability toward the end of the dominance of policy paradigm III. The military coup in Portugal had the effect of changing the regional balance of power by converting the Wars of Liberation in Angola and Mozambique into power vacuums (Turner, 1998; Turton, 2006). The Yom Kippur War in Israel (October 1973) triggered a global oil crisis (Turton et al., 2004). The Soweto Riots took place on 16 June 1976, leaving around 575 people dead in an act of violence reminiscent of the Sharpeville Massacre a decade and a half earlier (Spitz and Chaskalson, 2000; Welsh, 2000). These three events magnified each other's effects, ending the 'Long Economic Boom' (Geldenhuys, 1990; Ross, 1999) (see Figure 11.2).

Policy paradigm III was characterized by the extraction of mineral wealth, using water as a strategic resource, in order to grow the economy rapidly for purposes of political survival. Thus the resource capture of minerals coincided with resource capture of water, a coincidence that became a key element of the South African hydraulic mission with a merging of the interests of the mining houses with the interests of the apartheid state. Oversight was weak, with the two parties (industry and state) becoming collaborators, with environmental degradation being ignored and the costs of mining externalized onto

Note: KwaZulu Natal made a disproportionately large contribution to the national death toll. After the Battle of Mmabatho in 1994 (see Turton, 2006 for more details) the violence tapered off dramatically as South Africa recoiled from the abyss of full-scale civil war.

Source: Data from South African Institute of Race Relations as shown in Turton (2006), p. 232.

Figure 11.3 Deaths by political violence in South Africa and KwaZulu Natal from 1985 to 1996 during policy paradigm IV

an uninformed public (Adler et al., 2007a, 2007b). Under these circumstances there was no space for individual policy entrepreneurs.

11.3.5 Policy paradigm IV: state survival and total onslaught (1976–94)

The policy instability phase triggering the emergence of policy paradigm IV was the eruption of the armed struggle inside South Africa, beginning on 16 June 1976 with the Soweto Riots ('C' in Figure 11.2). This started a rise in inflation, pushing it into a realm where it became a driver of instability in its own right (see Figure 11.2) (Turton et al., 2004). Increasing levels of violence engulfed the entire country as the insurgency grew (Geldenhuys, 1984; Barber and Barratt, 1990; Nortje, 2003) (see Figure 11.3). Water and mining policy must be understood against this background.

In March 1980 Prime Minister P.W. Botha made a speech announcing what he called a 'total national strategy', needed to protect South Africa from the 'total onslaught' being waged by local surrogates of the Soviet Union (Geldenhuys, 1984). The total national strategy saw the mobilization of South Africa's entire financial, natural and human resources on a war footing in an endeavour to thwart this perceived total onslaught (Frankel, 1984). Policy paradigm IV was about state survival during the total onslaught period and came about through the ascendance of overarching policy perspective II, which ushered in policies to protect the goose that laid the golden eggs. This translated into minimal oversight of the mining industry. A recent study of minutes from meetings of the FWRDWA and the SCTC concludes that this lack of oversight allowed the mining companies to maximize their profits by: building in complex mechanisms that avoided liabilities; countering any attempts by the state to regulate the industry; and externalizing costs wherever possible (Adler et al., 2007b). A lasting consequence of this has been the systematic externalization of costs onto an unwilling society (Adler et al., 2007a) that is relevant to the current situation and remains unresolved at the time of writing in early 2009.

Policy paradigm IV was therefore about state survival in which the gold-mining industry was protected because revenue streams were needed to finance the war against the armed struggle at a time of soaring inflation driven by a variety of factors. Two main parties formed a coalition, connected by their common interest – an increasingly embattled and illegitimate state concerned only with its own survival, and an industry concerned only with profit maximization. As with policy paradigm III, there was no space for individual policy entrepreneurs to act.

11.3.6 Policy paradigm V: redistribution of wealth and privilege (1994–present)

The key shock events that opened the policy window related to paradigm V were both internal and external ('D' in Figure 11.2). Internally the recognition that a military solution was simply not possible was being driven by a small, but elite group of political insiders under the leadership of Dr Neil Barnard of the National Intelligence Service (NIS) (Turton, 2006). Externally the Battle of Cuito Cuanavale was the final hot confrontation of the Cold War in Africa, when Russian armoured forces redeployed from Afghanistan into Angola, joined with Cuban reinforcements to the Angolan Armed Forces (FAPLA), resulting in an epic battle that only the veterans seem to remember today (Turner, 1998; Nortje, 2003; Mills and Williams, 2006; Turton, 2006). The tactical defeat of the FAPLA allowed the emergence of negotiations, with the Convention for a Democratic South Africa (CODESA) convening for the first time on 20 December 1991 to start the work of drafting an interim Constitution. This was adopted after much bloodshed in the country (Spitz and Chaskalson, 2000), with South Africa teetering on the very abyss of full-scale civil war when on 11 March 1994 heavy fighting broke out in Mmabatho as armed forces under the broad banner of the Concerned South Africa Group (COSAG) attracted intervention by the South African Defence Force (SADF) (Turton, 2006). This was the last military engagement of significance as South Africa started to transition from war to peace (see Figure 11.3).

These shock events were extreme, but were mitigated by far-reaching constitutional reform. Central to this was the scrapping of all apartheid-related legislation. However, as an unintended consequence, all the administrative and legal precedent needed for subsequent service delivery was lost, becoming a driver of the current unrest (Johnston and Bernstein, 2007; Dixon, 2008; Sibanda, 2008; Walker, 2008). A major thrust of the reform was the nationalization of strategic resources that had been the subject of resource capture over the past century.

A few key pieces of legislation emerged. The Constitution (1996) enshrines the right of an individual to an environment that is not harmful to their health or well-being (Paragraph 24a of Chapter 2: Bill of Rights). Sustainable development is legislated in Paragraph 24b of Chapter 2 of the Bill of Rights, by stating that the environment must be protected for the benefit of present and future generations through reasonable legislative and other measures that: (1) prevent pollution and ecological degradation; (2) promote conservation; and (3) secure ecologically sustainable development and use of natural resources while promoting justifiable economic and social development. The public is granted the right of access to information in Paragraph 32 of Chapter 2 of the Bill of Rights.

Of great importance is Chapter 3 of the Constitution mandating cooperation between all government departments. This has major ramifications for the management of mine water and other environmental and health implications rising from the mining sector, given the privileged status of this industry under overarching policy perspectives I (extraction as central) and II (protected extractive industry). It is the constitutional element that brings policy perspective III into play, since the policy transition at issue is all about a changed architecture of governance, which could clearly establish government as a regulator of industry, being separated from that industry and thus no longer a collaborator.

Reforming the Department of Water Affairs and Forestry (DWAF), the National Water Act (1998) nationalized water resources under private control (DWAF, 1996, 1997; de Coning and Sherwill, 2004). This places heavy emphasis on historic redress by using water as a vehicle for other reforms in society. Implicit in this is the land ownership issue arising from the inequity of the 1913 Native Land Act. Given that water gives land its value, it is believed that through control of water rights signals can be sent out to precipitate a managed land reform process as well, avoiding the Zimbabwe-styled economic meltdown that occurred after land was nationalized and banks lost their collateral. Embedded in this legislation is the 'polluter pays' principle.

Reform on the mining side has been less dramatic. The Mineral and Petroleum Resources Development Act (Act 28) (2002) defines a residue stockpile, but as with the earlier apartheid-era legislation, does not define mining waste, thereby exempting this from waste management practices deemed to be 'normal' for other industries. Mineral residue (that is, 'waste') is considered to be a potential future source of minerals and therefore not waste (Godfrey et al., 2007). This gives the industry enormous protection, so the legacy of past inequity has not been addressed, despite calls to rectify this from the Department of Environmental Affairs and Tourism (DEAT) (Oelofse, 2008). The Mine Health and Safety Act (Act 29) (1996) regulates the human environment underground, but exempts mineral residue from being defined as waste. This also reflects the apartheid past, because the Hazardous Substances Act (Act 15) (1998) defines Group IV hazardous substances as being radioactive material outside a nuclear installation as defined by the Nuclear Energy Act (Act 46) (1999), leaving an undefined grey area around radioactivity produced from mining activities (Godfrey et al., 2007). The National Environmental Management Act (Act 107) (1998) requires remediation of environmental damage, while supporting the reuse of mineral waste, nullifying its classification as waste. The National Environmental Management: Air Quality Act (Act 39) (2004) determines the management regime for dust and air pollution arising from mining. The National Nuclear Regulator (NNR) and the Nuclear Energy Act (Act 46) (1999) both recognize that mineral waste from gold and uranium mines might be radioactive (Godfrey et al., 2007).

Regarding economic reform, the main thrust is for economic growth to address historic inequity by means of redistribution. This process has evolved along three distinct lines.

The first in the immediate post-apartheid era is the Reconstruction and Development Programme (RDP), which focuses on the initial needs for reconstruction in a society torn apart by more than a century of endemic violence (RDP, 1994). The second is the Growth, Employment and Redistribution (GEAR) programme, which focuses on the need for rapid economic growth to make up the backlog of economic stagnation caused by the violence of the internal struggle that initiated policy paradigm IV and the economic sanctions flowing from that time. The third is the Accelerated and Shared Growth Initiative for South Africa (ASGISA), based on first-order principles (human dignity, social equity, justice, fairness, democratic governance), substantive principles (sustainable use of resources, protection of ecosystem integrity, balance between development and protection), and process principles (integration, innovation, consultation, participation, phased implementation) (ASGISA, 2006).

11.4 Actors and strategies of transition management

The net result of the legal and political reforms is that the strategic landscape in which mining is embedded has changed, but is still complex. For the first time in South African history, there is space for action by individual policy entrepreneurs. Using the five strategies of transition management postulated in this book as a conceptual lens, the South African case study offers some useful insights. The opening of the policy 'space' in the immediate post-apartheid period released five identifiable sets of policy entrepreneurs, each with different interests and thus strategies for transition management.

The liberation leaders created the legal, moral and logical foundation for the policy paradigm of redistribution to act as a counterbalance to the policy paradigm of extraction (see Figure 11.1). The most notable here are Nelson Mandela, who created a climate in which reconciliation could be a viable alternative to endemic violence, and Kader Asmal, who became the first democratically elected Minister of Water Affairs. It was under Kader Asmal's inspired leadership that legal reform took place in the water sector (de Coning and Sherwill, 2004). It is significant that no policy entrepreneur of similar status and credibility was appointed to lead the mining sector, probably accounting for the disparity in policy reform between these two sectors. The strategy used here was the development of new ideas and their embedding in legislation.

The mining industry immediately sought to protect its vested interests when democracy came in 1994. Falling back on the experience gained from 30 years of self-regulation during policy paradigms III and IV, it had already prepared by learning how to outmanoeuver government attempts to regulate it (Adler et al., 2007b).

Mining industry players can be regarded as policy opponents, and their strategy was based on destroying any evidence of liability and/or culpability. This has resulted, for example, in attempts to disband the FWRDWA and SCTC because both these bodies have detailed minutes of past meetings and masses of data that can be used to apportion specific liability. The apparent logic is that if those data disappear, then so too does any evidence that might form the legal foundation of future litigation. One tactic for a specific mine known to be a major polluter has been to appoint Mariette Liefferink (see below) as public participation manager for one of their operations, ostensibly to drive a new corporate social responsibility process, but then to control her by giving her an inadequate budget. Another tactic has been to attack the credibility of scientists challenging the status quo (Coetzee et al., 2006), most notably by dubbing one particular report as the 'Infamous 1214'.

Civil society is emerging as a key policy entrepreneur (van Eeden, 2007), deriving its energy from public opinion, inflamed by decades of abuse by both government and the mining industry, bent on retribution and insistent on holding 'perpetrators' accountable for human rights abuses (Avni, 2007a, 2007b, 2007c; Brown, 2007; Gosling, 2007; Groenewald, 2007). Public sentiment is captured in a newspaper article written in response to the government's proposal for compulsory clean-up by mines of their own pollution, stating that: 'To ask the polluter to fix the mess [they had created] was like appointing a convicted paedophile to manage the orphanage' (Tempelhoff, 2007b). Under the doggedly persistent leadership of Mariette Liefferink (Eybers, 2007), an environmental activist and self-proclaimed whistle-blower, a range of non-governmental organizations (NGOs) are now taking on the issue of acid mine drainage (AMD) (Marshall, 2007b). Probably the most prominent of the NGOs, at least in a leadership sense, is the Federation for a Sustained Environment, in which Mariette Liefferink plays a coordinating role. The strategy used here is coalition-building and the selling of new ideas. Other NGOs include the Wonderfontein Action Group (WAG), the Potch Petitioners, the Randfontein Environmental Action Group (REAG) and the Public Environmental Arbiters (PEA). Due to the radioactivity aspect of gold-driven AMD, there is also a strong anti-nuclear component to this coalition.

The main impetus of this coalition-building is AMD, initially from the gold-mining areas of the Far West Rand and thus associated with the 30 years of activity of the FWRDWA and SCTC. A principal tactic is to oppose attempts to disband these two organizations and to preserve the data as public documents commensurate with the constitutional imperative of public access to information. Coalition members rely heavily on scientific outputs from the epistemic community, and are avid consumers of this information. The coalition activities are now spreading to embrace communities affected by coal-driven AMD, which has a slightly different chemistry, but affects a larger geographic extent of the country (Hobbs et al., 2008).

The major concern of civil society is liability, with a case being made for the prosecution of directors of mine companies and/or senior civil servants for failing to meet the constitutional mandate to provide an environment that is safe (Marshall, 2007a; Tempelhoff, 2007a). The lawyer involved is a heavy hitter, having represented Nelson Mandela on many occasions. His credibility is similar to that of the liberation leaders noted above. Policy entrepreneurs from civil society have also successfully linked water to politics (Mmemezi, 2008; Nyathi, 2008; Secretary to Parliament, 2007), tapping into the broader political energy of failing service delivery (Dixon, 2008; Johnston and Bernstein, 2007; Sibanda, 2008). Significantly, civil society policy entrepreneurship shows no signs of recognizing the government's economic development policies, most notably the ASGISA (2006). This means that while government is driving a policy of economic growth, certain portions of the civil society cluster are opposed to growth, saying it is unsustainable in terms of existing water availability, and calling instead for environmental rehabilitation.

The epistemic community is emerging as a major actor, with two specific national science councils and one university directly involved. The first is the Council for Geosciences (CGS), which has played a leading role in generating the first high-confidence peer-reviewed public domain reports indicating that AMD is associated with radionuclide and heavy metal contamination (Coetzee, 1995; Coetzee et al., 2002a, 2002b, 2005;

Kempster et al., 1996; Wade et al., 2002). The most important of these has undoubtedly been the so-called 'Report 1214' (Coetzee et al., 2006), which can be regarded as being a pivotal document, given the fundamental way that it has changed the policy debate. This pioneering work has been eagerly taken up by civil society actors who have actively popularized it (Avni, 2007a, 2007b, 2007c; Brown, 2007; Carte Blanche, 2007; Gosling, 2007; Groenewald, 2007), even using international media (IRIN, 2008).

The second is the Council for Scientific and Industrial Research (CSIR), which was slow to react initially, but once the CGS research had been published, the CSIR became deeply involved. The CSIR work is focused on two specific areas of perceived scientific importance. The first is the need to build scientific capacity to conduct the first national-level epidemiological study of people potentially affected by radionuclide and heavy metal contamination using human deciduous teeth as specimens (the so-called Tooth Fairy Project) (CSIR, 2008). The second is to access the funding needed to complete this work. As of 2009 efforts have failed to bring the Medical Research Council (MRC) on board, ostensibly because of its own lack of capacity. Central to this is the need to build national capacity around water quality management, specifically arising from AMD, but also related to human health in a population that has the highest burden of HIV/AIDS in the world. This is being developed into a proposed National Water Quality Science, Technology and Policy Support Program (Turton, 2008). The North West University has also played a leading role (Winde and van der Walt, 2004; Winde, 2005).

The strategy used by the epistemic community has been the development of a more robust scientific platform on which the new ideas inherent to the Policy Paradigm of Redistribution can be operationalized, thereby hoping to build on the efforts of the liberation leaders noted above by putting their constitutional ideals into enduring practice. Significantly, the CSIR recognizes not only the governmental economic policy (ASGISA, 2006), but also the constitutional imperatives of human rights and cooperative government (Turton, 2008). Attempts at coalition-building by the epistemic community are hampered by the civil society insistence on holding the mines liable, the failure of civil society actors to recognize the need for economic development, and their position that there have been too many reports generated and not enough 'action'. In terms of the CSIR and local government relations there is some convergence between their needs, specifically with respect to the Tooth Fairy Project (Mmemezi, 2008; Smillie, 2008) brokered by the FES (Federation for Environmental Sustainability) (the civil society actor noted above).

Research funders are important policy entrepreneurs. The Water Research Commission (WRC) is a statutory body, created by the Commission of Enquiry into Water Matters in 1966 (RSA, 1970; Turton et al., 2004) and mandated to fund strategic water research. The WRC opened up public funding to research matters previously classified and thus out of the public domain. The first known work relevant to this analysis was on human health implications related to groundwater (Toens et al., 1999). Other significant projects include 'Report 1095' (Wade et al., 2002) that developed a methodology for the assessment of radioactivity in sediment of rivers downstream of gold-mining activities. 'Report 1214' is undoubtedly the most important (Coetzee et al., 2006). By virtue of the impact it is likely to have on policy reform, it can now be considered a pivotal document of the same calibre as the Jordaan Commission of Enquiry (Jordaan et al., 1960) and the Commission of Enquiry into Water Matters in 1966 (RSA, 1970).

The WRC is currently considering funding the Tooth Fairy Project (CSIR, 2008; Smillie, 2008), which would be a national first. The strategy used by the WRC is merely to support initiatives developed by the epistemic community, so it is not an agenda-setter in the policy field. Its role is thus important, but not proactive. When approached, the WRC has been willing to support ground-breaking (and hence potentially controversial) research projects, such as 'Report 1214'. The Council has been hampered by the non-existence of a coherent national response to AMD, a situation that has led the CSIR to help with the development of the proposed National Water Quality Science, Technology and Policy Support Program (Turton, 2008).

11.5 An assessment of the strategies of water policy entrepreneurs

An assessment of the strategies for transition management used by the various policy entrepreneurs reveals an absence of action beyond the development and selling of new ideas and building coalitions and, as shown by the key overarching legislation emerging in paradigm V, recognizing and exploiting windows of opportunity. It could be argued that more water-specific policy change on paper is needed before the transition can be considered complete, but an overall shift in policy paradigm has undoubtedly occurred. At least four significant areas of either convergence or divergence can be identified.

Leadership of the calibre of the liberation leaders noted above has been severely lacking. The contribution of the liberation leaders was to establish a foundation for the policy paradigm of redistribution and to create overarching policy perspective III. The absence of inspired leadership on a continual basis has limited the impact of policy reform aimed at industry regulation. The ruling party has unfortunately become mired in corruption scandals (Feinstein, 2007), most notably over the acquisition of a large arms shipment, the repercussions of which are still reverberating around the country, to the point that the second democratically elected President, Thabo Mbeki, was removed from office on 20 September 2008. This has diverted resources and attention from policy reform, creating a crisis of service delivery, including the national energy crisis that brought the economy to a halt in 2008.

Government restructuring has happened so many times in succession since the transition to democracy in 1994 that some government departments are no longer capable of meeting their mandate for service delivery (see Johnston and Bernstein, 2007; Nyathi, 2008). The restructuring has had a range of unintended consequences, one of which has been the loss of scientific capacity in the CSIR because of an inappropriate funding model (Walwyn and Scholes, 2006). This has eroded the capacity of the CSIR to generate the technical solutions needed to solve the AMD problems facing the country today. There are concerted efforts to change this situation, however, with plans being tabled to create enough synergy to rebuild the lost capacity (Turton, 2008). Significantly the mooted National Water Quality Science, Technology and Policy Support Program has a key policy support dimension, and if approved by government new energy would flow into the policy reform process.

The epistemic community has demonstrated its willingness to take a leadership role in policy reform by generating high-quality science (capacity constraints notwithstanding) that is mostly uncontested and thus useful as a foundation for transition. In this regard the contribution of the CGS in the form of 'Report 1095' and 'Report 1214' has been important. The CSIR, arriving belatedly in the policy arena, has tried to act in a coherent and responsible manner (Godfrey et al., 2007; Adler et al., 2007a, 2007b; Hattingh et al., 2007; Hobbs and Cobbing, 2007; CSIR, 2008; Hobbs et al., 2008; Mmemezi, 2008; Oelofse, 2008; Smillie, 2008; Turton, 2008). Other policy entrepreneurs, however, are sending out mixed signals. This is true in particular of civil society actors who are calling for 'action' and a cessation of what they perceive to be an endless stream of reports and 'worthless' research. In this way they fail to distinguish that many of the previous reports were generated by consultants paid to protect the status quo.

More high-impact, credible science is needed that shifts the factual basis on which policy reform is based. While it is true, for instance, that there has been a plethora of reports about the Wonderfontein Spruit, for example, few of them have been structured in a way that moves the policy debate forward. It has to be conceded that 'Report 1095' and 'Report 1214' have been exceptions. This may also be true of the 'externalization of costs model' that shows how mines have maximized profits by shifting their costs for environmental rehabilitation onto society (Adler et al., 2007a). The mixed signals coming from civil society players are hampering the attempts by the CSIR to get the Tooth Fairy Project funded (CSIR, 2008) because some potential funders have cited the call from some in civil society to stop the flow of research.

The right of access to information poses a specific dilemma. The epistemic community has many reports, papers, dissertations and other scientific literature in its archives that are restricted (classified) by virtue of previous funding agreements. Some of these indicate that both government and the mining industry were aware of the existence of radioactivity and heavy metals in AMD long before it was public knowledge (Avni, 2007a). This hampers policy reform because the knowledge in the archived literature might be used in litigation against perceived 'perpetrators of human rights abuses'. This creates a climate of uncertainty for scientists hesitant to become embroiled in litigation, and for industry leaders averse to investing money in research for solutions for fear of bringing the past into the present and future. It is unclear at this stage what would happen if a subpoena were to be issued to any member of the epistemic community to provide access to literature known to be in their possession.

11.6 Resultant policy dynamics

The dynamics of policy paradigm V are being driven by two incompatible forces. The policy paradigm of extraction, which has a long history of human rights abuse and corruption, has manifested as the Black Economic Empowerment (BEE) programme, with a small group of super-rich elites benefiting at growing cost to the recently enfranchised working class, many of whom are worse off because of a failure in service delivery. This bears an uncanny resemblance to the earlier phases where elites colluded in order to protect their own self-interest. A possible important indication of this is the slow but deliberate movement offshore of the head offices of major mining houses, possibly as a strategic move to distance themselves from the environmental and human health liabilities that are likely to arise from mine closure. This conclusion is speculative, but not implausible.

The second major force is focused on human rights and occurs in the form of the still embryonic policy paradigm V focused on the redistribution of wealth. The moral high ground inherent to this paradigm is being used by the NGO community to attack government on its perceived policy failures. The conflict of liability avoidance by the mining companies and support for human rights as enshrined in the Bill of Rights of the Constitution produces the unsettled state of perspective III. Will the main focus of this emerging overarching policy perspective become economic growth, self-enrichment and corruption of the gatekeeping elite? Or could it be a genuine redistribution of wealth and privilege in a society so long exploited? The answer to this is far from clear. The growing unrest in Khutsong, along with the 2008 wave of xenophobic rioting that occurred in response to failed immigration policies and poor service delivery, is acting as a crucible in which these burning questions generate considerable friction and agitation (Johnston and Bernstein, 2007; Dixon, 2008; Nyathi, 2008; Sibanda, 2008; Turton, 2008; Walker, 2008).

Reforms arising from the tension between these two policy poles rotate around four main axes. First, the issue of liability arising from human health and environmental damage caused by a century of mining is significant. One response to this appears to be the movement of the centre of gravity of major mining operations offshore. Another response is a set of cosy deals under the banner of BEE that seek to entrench the status quo as far as possible, but under a new collusion. Yet another is a lukewarm response to the Tooth Fairy Project, which might well find radioactivity and heavy metal contamination in off-mine communities, triggering liability claims.

Second, the classic transboundary problem about whose rules apply when water crosses a jurisdictional boundary is also relevant. This is manifest in the issue of the definition of mine waste in which the Department of Minerals and Energy (DME) is likely to push for a maintenance of the status quo and keep mine waste as a 'separate' issue under its sole jurisdiction. This would probably see the DEAT, as the line department responsible for 'normal' pollution and specifically wetlands, sidelined in all key areas where mine interests are concerned. The NNR has shown its willingness to act as a broker, bringing the factions together in a way that ensures their compliance with the cooperative governance mandate in the national Constitution. These efforts are being undermined by the political instability arising from the corruption scandal that led to the sacking of President Mbeki, making many fearful officials reluctant to take the lead.

Third, the sustainable development debate centres on the question of what it actually means to develop in a sustainable manner. Central to this is balance between resource use and resource protection, but always with the constitutional imperative of human rights in the background. A key element of this is oversight in a country with no entrenched culture of governance (Hattingh et al., 2007). The CSIR has taken the lead by developing what it calls 'sustainability science' (Burns et al., 2006; Burns and Weaver, 2008), which now needs to be operationalized as policy reform (Turton, 2008).

Fourth, the role of science and technology with the CSIR acting in partnership with stakeholders to develop technical solutions to the problems of AMD is now critical (Hobbs and Cobbing, 2007; Hobbs et al., 2008; Turton, 2008). While the West Rand AMD decant is significant, the next major event is predicted for the East Rand. What is learned during the former can be applied proactively to the latter. The predicted total decant of AMD in the Gauteng area is around 400 megalitres a day, so the volume is large and the impact significant if left unmanaged. As a result of the existing West Rand AMD decant, a company has been formed with the sole objective of developing a new water utilities corporation (Theunissen, 2008). This move is designed to centralize the bad-quality water, stripping it of its radioactivity and heavy metals, and then selling it on to industries that need non-potable process water. The heavy metals harvested are projected to be of significant value, producing revenue to pay for the treatment after the mines close down. The public are sceptical of this, however, because they are tired of being abused by the mining industry. This makes the roll-out of the process a major challenge, needing a fusion of technology, finance, public relations and governance to succeed. CSIR technology, arguably more trusted than most, is probably an essential element of success.

11.7 Conclusion

The South African gold mining and water sectors offer a rich history of policy continuity and change where resource capture has been a dominant force. It has a history of major human rights abuses going back centuries, with three of the dominant social groupings having been decimated in the past by the British. These three groups – the amaXhosa, amaZulu and Afrikaners – have all tried to recover, but their communal memory of violence and human rights violations has not been adequately addressed despite the democratic Constitution of 1996. The failure to deal adequately with the legacy of history has prevented the deployment of strategies for transition management much beyond coalition-building. Where the current process of policy reform will take South Africa, nobody really knows. Will the overarching policy perspective focused on extraction merely corrupt the next generation of political elites at the expense of the long-oppressed masses, as the Khutsong case suggests (Johnston and Bernstein, 2007)? Or will the emerging fifth policy paradigm of redistribution build on the moral high ground and create a just society for the first time in the 350 turbulent years of modern South African history? What about national assets that are in state hands, but generating private wealth? Will funding constraints hamper the generation of science capable of supporting water policy reform and taking it in the direction of sustainability (Walwyn and Scholes, 2006; Turton, 2008)? These are complex issues where only time will tell, and where the jury is still out and the dynamics too fluid to predict with any confidence.

References

Adler, R.A., M. Claassen, L. Godfrey and A.R. Turton (2007a), 'Water, mining and waste: an historical and economic perspective on conflict management in South Africa', *Economics of Peace and Security Journal*, **2** (2), 32–41.

Adler, R., N. Funke, K. Findlater and A.R. Turton (2007b), *The Changing Relationship between the Government and the Mining Industry in South Africa: A Critical Assessment of the Far West Rand Dolomitic Water Association and the State Coordinating Technical Committee*, Pretoria: Council for Scientific and Industrial Research (CSIR).

ASGISA (2006), *People – Planet – Prosperity: A Strategic Framework for Sustainable Development in South Africa*, Pretoria: Department of Environmental Affairs and Tourism (DEAT).

Avni, J. (2007a), 'Lives at risk: state knew about danger for 40 years', *Sowetan*, 24 July.

Avni, J. (2007b), 'Report exposes high water contamination', *Sowetan*, 3 August.

Avni, J. (2007c), 'Poisoned by mines', *Sowetan*, 20 August.

Barber, J. and J. Barratt (1990), *South Africa's Foreign Policy: The Search for Status and Security, 1945–1988*, Cambridge: Cambridge University Press.

Basson, M.S., P.H. van Niekerk and J.A. van Rooyen (1997), *Overview of Water Resources Availability and Utilization in South Africa*, Pretoria: Department of Water Affairs and Forestry.

Brown, J. (2007), 'Derelict mines to cost state R100bn', *Business Report*, 23 May.

Burns, M., M. Audouin and A. Weaver (2006), 'Advancing sustainability science in South Africa', *South African Journal of Science*, **102**, 379–84.

Burns, M.J. and A.v.B. Weaver (eds) (2008), *Advancing Sustainability Science in South Africa*, Stellenbosch: Stellenbosch University Press.

Carte Blanche (2007), *Environment and Conservation*, Presenter Devi Sankaree, Producer Carol Albertyn, M Net TV Programme, 12 August.
Coetzee, H. (1995), 'Radioactivity and the leakage of radioactive waste associated with Witwatersrand gold and uranium mining', in B.J. Merkel, S. Hurst, E.P. Löhnert and W. Struckmeier (eds), *Uranium Mining and Hydrology: Proceedings of the International Conference and Workshop in Freiberg*, Germany, October, Geo Congress I, Cologne: Verlag Sven von Loga, pp. 34–9.
Coetzee, H., J. Venter and G. Ntsume (2005), 'Contamination of wetlands by Witwatersrand gold mines – processes and the economic potential of gold in wetlands', Council for Geosciences Report No. 2005-0106, Pretoria: Council for Geosciences.
Coetzee, H., P. Wade, G. Ntsume and W. Jordaan (2002a), 'Radioactivity Study on Sediments in a Dam in the Wonderfonteinspruit Catchment', *DWAF Report*, Pretoria: Department of Water Affairs and Forestry.
Coetzee, H., P. Wade and F. Winde (2002b), 'Reliance on existing wetlands for pollution control around the Witwatersrand gold/uranium mines in South Africa – are they sufficient?' in B.J. Merkel, B. Planer-Friederich and C. Wolkersdorfer (eds), *Uranium in the Aquatic Environment*, Berlin: Springer, pp. 59–65.
Coetzee, H., F. Winde and P.W. Wade (2006), 'An assessment of sources, pathways, mechanisms and risks of current and potential future pollution of water and sediments in gold-mining areas of the Wonderfonteinspruit catchment', WRC Report No. 1214/1/06, Pretoria: Water Research Commission.
Conley, A.H. (1996), 'A synoptic view of water resources in Southern Africa', in H. Solomon (ed.), *Sink or Swim? Water, Resource Security and State Co-operation*, IDP Monograph Series No. 6, Durban: The Institute for Security Studies, pp. 17–69.
Constitution (1996), *The Constitution of the Republic of South Africa*, Act 108 of 1996, Pretoria: Government Printer.
CSIR (2008), 'High confidence study of children potentially affected by radionuclide and heavy metal contamination arising from the legacy of mine water management practices on the Far West Rand of South Africa', project concept note, Pretoria: Council for Scientific and Industrial Research (CSIR), 26 February.
de Coning, C. and T. Sherwill (2004), 'An assessment of the water policy process in South Africa (1994 to 2003)', Water Research Commission Report TT232/04, Pretoria: Water Research Commission.
Dixon, R. (2008), 'Migrants targeted for fiery deaths in South Africa', *Los Angeles Times*, 20 May.
Du Toit, A. (1994), 'Fragile defiance: the African resistance movement', in I. Liebenberg, F. Lortan, B. Nel and G. van der Westhuizen (eds), *The Long March: The Story of the Struggle for Liberation in South Africa*, Pretoria: HAUM, pp. 96–103.
DWAF (1996), 'Discussion document on water law principles', Pretoria: Department of Water Affairs and Forestry.
DWAF (1997), *White Paper on Water Policy. South Africa*, Pretoria: Department of Water Affairs and Forestry.
Edgecombe, R. (1986), 'The Mfecane or Difaqane', in T. Cameron and S.B. Spies (eds), *New History of South Africa in Word and Image* (in Afrikaans), Cape Town: Human & Rousseau.
Eybers, T. (2007), 'SA "Erin" widens toxic water fight: water too dangerous for farming', *Citizen*, 14 August.
Farwell, B. (1999), *The Great Boer War*, London: Wordsworth Editions.
Fawcett, M.H. (1901), *The Concentration Camps in South Africa*, London: Westminster Gazette.
Feinstein, A. (2007), *After the Party: A Personal and Political Journey Inside the ANC*, Johannesburg: Jonathan Ball Publishers.
Frankel, P.H. (1984), *Pretoria's Praetorians: Civil–Military Relations in South Africa*, London: Cambridge University Press.
Geldenhuys, D. (1984), *The Diplomacy of Isolation: South African Foreign Policy Making*, Johannesburg: Macmillan South Africa.
Geldenhuys, D. (1990), *Isolated States: A Comparative Analysis*, Johannesburg: Jonathan Ball Publishers.
Godfrey, L., S. Oelofse, A. Phiri, A. Nahman and J. Hall (2007), 'Mineral waste: the required governance environment to enable reuse', Report No. CSIR/NRE/PW/IR/2007/0080/C, Pretoria: Council for Scientific and Industrial Research (CSIR).
Gosling, M. (2007), 'The environmental impact of uranium waste', *Cape Times*, 25 June.
Groenewald, Y. (2007), 'SA's eco-prophesy', *Mail and Guardian*, 26 July.
Hasian, Marouf (2003), 'The "hysterical" Emily Hobhouse and Boer War concentration camp controversy', *Western Journal of Communication*, March, http://www.accessmylibrary.com/coms2/summary_0286-23546431_ITM, accessed November 2008.
Hattingh, J., G.A. Maree, P.J. Ashton, J. Leaner, J. Rascher and A.R. Turton (2007), 'A trialogue model for ecosystem governance', *Water Policy*, **9** (2), 11–18.
Hazardous Substances Act (1998), *The Hazardous Substances Act*, Act 15 of 1998, Pretoria: Government Gazette.
Hobbs, P.J. and J.E. Cobbing (2007), 'A hydrogeological assessment of acid mine drainage impacts in the

West Rand basin', Gauteng Province, Rep. no. CSIR/NRE/WR/ER/2007/0097/C, CSIR/THRIP, Pretoria, Republic of South Africa: CSIR/THRIP.

Hobbs, P., S.H.H. Oelofse and J. Rascher (2008), 'Management of environmental impacts from coal mining in the Upper Olifants river catchment as a function of age and scale', in M.J. Patrick, J. Rascher and A.R. Turton (eds), *Reflections on Water in South Africa*, Special Edition of *International Journal of Water Resource Development*, **24** (3), 417–32.

Hobhouse, E. (1901), *Report of a Visit to the Camps of Women and Children in the Cape and Orange River Colonies*, London: Friars Printing Association.

IRIN (2008), 'South Africa: paying the price for mining', UN Office for the Coordination of Humanitarian Affairs, *IRIN Humanitarian News and Analysis*, http://www.irinnews.org/Report.aspx?ReportId=76780, accessed November 2008.

Johnston, S. and A. Bernstein (2007), *Voices of Anger: Protest and Conflict in Two Municipalities. Report to the Conflict and Governance Facility (CAGE)*, Johannesburg: Centre for Development and Enterprise.

Jordaan, J.M., J.F. Enslin, J.P. Kriel, A.R. Havemann, L.E. Kent and W.H. Cable (1960), 'Finale Verslag van die Tussendepartmentele Komitee insake Dolomitiese Mynwater: Verre Wes-Rand, Gerig aan sy Edele die Minister van Waterwese deur die Direkteur van Waterwese' (Final Report of the Interdepartmental Committee on Dolomitic Mine-water: Far West-Rand, directed at His Excellency the Minister of Water Affairs by the Director of Water Affairs), Pretoria: Department of Water Affairs.

Kempster, P.L., H.R. Van Vliet, U. Looser, I. Parker, M.J. Silberbauer and P. Du Toit (1996), 'Overview of radioactivity in water sources: uranium, radium and thorium. Final report', IWQS-No:N/0000/00/RPQ/0196, Pretoria: Institute for Water Quality Studies.

Kleyweght, R.J. and D.R. Pike (1982), 'Surface subsidence and sink-holes caused by lowering the dolomitic table on the far West Rand gold field of South Africa', *Annual Geological Survey of South Africa*, **16**, 77–105.

Krebs, P.M. (1992), 'The last of the gentlemen's wars: women in the Boer War concentration camp controversy', *History Workshop Journal*, **33**, 38–56.

Liebenberg, B.J. (1987), 'Botha and Smuts' Rule, 1910–1924", in C.F.J. Muller (ed.), *Five Hundred Years: South African History*, Pretoria: Academica, pp. 22–35.

Liebenberg, I. (1994), 'Resistance by the SANNC and the ANC, 1912–1960', in I. Liebenberg, F. Lortan, B. Nel and G. van der Westhuizen (eds), *The Long March: The Story of the Struggle for Liberation in South Africa*, Pretoria: HAUM, pp. 8–21.

Lodge, T. (1994), 'The Pan-Africanist Congress, 1959–1990', in I. Liebenberg, F. Lortan, B. Nel and G. van der Westhuizen (eds), *The Long March: The Story of The Struggle for Liberation in South Africa*, Pretoria: HAUM, pp. 104–24.

Marshall, L. (2007a), 'Top lawyer joins forces with Greens', *Sunday Tribune*, 9 September.

Marshall, L. (2007b), 'New alliance to fight environmental degradation', *Sunday Independent*, 9 September.

Meer, F. (1990), *Higher than Hope*, London: Harper Collins.

Mills, G. and D. Williams (2006), *Seven Battles that Shaped South Africa*, Cape Town: Tafelberg.

Mine Health and Safety Act (1996), *The Mine Health and Safety Act*, Act 29 of 1996, Pretoria: Government Gazette.

Mineral and Petroleum Resources Development Act (2002), *The Mineral and Petroleum Resources Development Act*, Act 28 of 2002, Pretoria: Government Gazette.

Mmemezi, H.M.Z. (2008), Official letter of support from the West Rand District Municipality under reference 12/2/1/3 dated 12 May, 'Epidemiological study: Tooth Fairy Project', signed by Councillor Mmemezi in his capacity as Portfolio Chairperson, Infrastructure.

Muller, M. (2007), 'DG reports: a view from the other side', *Weekly Mail and Guardian*, 20 November, http://www.mg.co.za/articlePage.aspx?articleid=325301&area=/insight/insight_comment_and_analysis/, accessed November 2008.

National Environmental Management Act (1998), *The National Environmental Management Act*, Act 107 of 1998, Pretoria: Government Gazette.

National Environmental Management: Air Quality Act (2004), *The National Environmental Management: Air Quality Act*, Act 39 of 2004, Pretoria: Government Gazette.

National Water Act (1998), *The National Water Act*, Act 36 of 1998, Pretoria: Government Gazette.

NNR (2007), 'Radiological impacts of the mining activities to the public in the Wonderfonteinspruit catchment area', Report No. TR-RRD-07-0006, Pretoria: National Nuclear Regulator.

Nortje, P. (2003), *32 Battalion*, Cape Town: Struik Publishers.

Nuclear Energy Act (1999), *The Nuclear Energy Act*, Act 46 of 1999, Pretoria: Government Gazette.

NWRS (2004), 'National water resource strategy', Pretoria, Department of Water Affairs and Forestry (DWAF), http://www.dwaf.gov.za/Documents/Policies/NWRS/Default.htm, accessed November 2008.

Nyathi, S. (2008), 'No water – no vote', *News 24*, http://www.news24.com/News24/South_Africa/News/0,2-7-1442_2326211,00.html, accessed November 2008.

Odendaal, A. (1994), 'The roots of the ANC', in I. Liebenberg, F. Lortan, B. Nel and G. van der Westhuizen (eds), *The Long March: The Story of the Struggle for Liberation in South Africa*, Pretoria: HAUM, pp. 1–7.

Oelofse, S.H.H. (2008), '*Mine water pollution – acid mine decant, effluent and treatment: consideration of key emerging issues that may impact the state of the environment*', Pretoria, Department of Environmental Affairs and Tourism (DEAT), http://soer.deat.gov.za/docport.aspx?m=97&d=28, accessed November 2008.

Pakenham, T. (1992), *The Boer War*, London: Harper Perennial.

Peires, J. (2003), *The Dead Will Arise. Nongqawuse and the Great Xhosa Cattle Killing of 1856–7. History of Political Thought*, Johannesburg: Jonathan Ball Publishers.

RDP (1994), *The Reconstruction and Development Programme: A Policy Framework*, African National Congress, Johannesburg: Umanyano Publications.

Ross, R. (1999), *A Concise History of South Africa*, Cambridge: Cambridge University Press.

RSA (1970), 'Report of the Commission of Enquiry into Water Matters', Document No. R.P. 34/1970, Pretoria: Government Printer.

Secretary to Parliament (2007), 'Internal Question Paper No. 28 of 2007. Question posed by Mr G.R. Morgan (DA) to the Minister of Environmental Affairs and Tourism', Cape Town: Hansard.

Sibanda, B. (2008), 'Falling economic standards cause xenophobia in South Africa', *Afrik.com*, 14 May.

Smillie, S. (2008), 'Tooth Fairy Project may reveal effect of uranium', *Star*, 18 March, p. 10.

Spitz, R. and M. Chaskalson (2000), *The Politics of Transition: A Hidden History of South Africa's Negotiated Settlement*, Johannesburg: Witwatersrand University Press.

Tempelhoff, E. (2004), 'Inwoners moet Weet van Giftige Gebied' (Residents must know of toxic area), *Beeld Newspaper*, 10 August.

Tempelhoff, E. (2007a), 'Myne Misken Menseregte, meen Bizos' (Mines flout human rights, says Bizos), *Beeld Newspaper*, 11 September.

Tempelhoff, E. (2007b), 'Mense en Omgewing Stik oor Goudwelvaart: Mynmonopolie se Gru-Moerass' (People and environment choke over gold wealth: mine monopoly's gruesome mess), *Beeld Newspaper*, 2 February.

Tempelhoff, E. (2007c), 'Gif in die Water: Besoedeling Naby Goudmyne Bedreig Duisende, Besmet Voedsel' (Poison in the water: pollution near goldmines threatens thousands, contaminates food), *Beeld Newspaper*, 24 April.

Tempelhoff, J.W.N. (2003), *The Substance of Ubiquity: Rand Water 1903–2003*, Vanderbijlpark: Kleio Publishers.

Theunissen, N. (2008), 'The big picture solution: acid mine drainage', *Water Sewage and Effluent*, **28** (5), 11–17.

Thompson, L. and H. Lamar (1981), 'The North American and Southern African frontiers', in H. Lamar and L. Thompson (eds), *The Frontier in History: North America and Southern Africa Compared*, New Haven, CT, USA and London, UK: Yale University Press, pp. 12–28.

Toens, P.D., W. Stadler and N.J. Wullschleger (1999), 'The association of groundwater chemistry and geology with atypical lymphocytes (as a biological indicator) in the Pofadder area, North Western Cape, South Africa', Water Research Commission, Report Commission, Report Number: 839/1/98, Pretoria: Water Research Commission.

Turner, J.W. (1998), *Continent Ablaze: The Insurgency Wars in Africa 1960 to the Present*, Johannesburg: Jonathan Ball Publishers.

Turton, A.R. (2006), *Shaking Hands with Billy: The Private Memoirs of Anthony Richard Turton*, limited edition, Krugersdorp: JFA Printers.

Turton, A.R. (2007), 'Can we solve tomorrow's problems with yesterday's experiences and today's science?' Des Midgley Memorial Lecture presented at the 13th SANCIAHS Symposium, Cape Town, 6 September.

Turton, A.R. (2008), 'Three strategic water quality challenges that decision-makers need to know about and how the CSIR should respond', CSIR Report No. CSIR/NRE/WR/EXP/2008/0160/A, Pretoria: Council for Scientific and Industrial Research (CSIR).

Turton, A.R. and P.J. Ashton (2008), 'Basin closure and issues of scale: the Southern African hydropolitical complex', *International Journal of Water Resources Development*, **24** (2), 305–18.

Turton, A.R., R. Meissner, P.M. Mampane and O. Seremo (2004), 'A hydropolitical history of South Africa's international river basins', Report No. 1220/1/04 to the Water Research Commission, Pretoria: Water Research Commission.

Turton, A.R., C. Schultz, H. Buckle, M. Kgomongoe, T. Malungani and M. Drackner (2006), 'Gold, scorched earth and water: the Hydropolitics of Johannesburg', *International Journal of Water Resources Development*, **22** (2), 313–35.

van Eeden, E.S. (1992), *Ekonomiese Ontwikkeling en die Invloed Daarvan op Carletonville,1948–1988: 'n Historiese Study* (*Economic Development and the Influence Thereof on Carletonville, 1948–1988: An Historic Study*), PhD thesis, Potchefstroom University for Higher Christian Education.

van Eeden, E.S. (2007), 'An historical assessment of NGO efficiency in progressing towards a sustainable environmental heritage focus, with as case study the Wonderfontein Spruit catchment, Gauteng', *New Contree*, **53**, 55–78.

Van Jaarsveld, F.A. (1975), *From Van Riebeeck to Vorster 1652–1974: An Introduction to the History of the Republic of South Africa*, Johannesburg: Perskor Publishers.

Wade, P.W., S. Woodbourne, W.M. Morris, P. Vos and N.W. Jarvis (2002), 'Tier 1 risk assessment of selected radionuclides in sediments of the Mooi River catchment', WRC Project No. K5/1095, Pretoria: Water Research Commission.

Walker, P. (2008), 'South African mobs hunt down immigrants', *Guardian*, 19 May.

Walwyn, D. and R.J. Scholes (2006), 'The impact of a mixed income model on the South African CSIR: a recipe for success or disaster?' *South African Journal of Science*, **102**, 239–43.

Welsh, Frank (2000), *A History of South Africa*, London: HarperCollins.

Werdmüller, V.W. (1986), 'The Central Rand', in E.S.A. Antrobus (ed.), *Witwatersrand Gold: 100 Years*, Johannesburg: Geological Society of South Africa, pp. 7–47.

Williams, R. (1994), 'The other armies: writing the history of MK', in I. Liebenberg, F. Lortan, B. Nel and G. van der Westhuizen (eds), *The Long March: The Story of the Struggle for Liberation in South Africa*, Pretoria: HAUM, pp. 22–34.

Winde, F. (2005), 'Impacts of gold-mining activities on water availability and quality in the Wonderfonteinspruit catchment', in H. Coetzee (ed.), *An Assessment of Current and Future Water-pollution Risk with Application to the Mooirivierloop (Wonderfonteinspruit)*, WRC Report No. K5/1214, Pretoria: Water Research Commission, pp. 14–38.

Winde, F. and I.J. van der Walt (2004), 'The significance of groundwater–stream interactions and fluctuating stream chemistry on waterborne uranium contamination of streams – a case study from a gold mining site in South Africa', *Journal of Hydrology*, **287** (1–4), 178–96.

12 Past, present and future landscapes of water policy in Tanzania

Jaqui Goldin and Deusdedit Kibassa

12.1 Introduction

12.1.1 Background

Competition for water has become an increasingly important issue in Tanzania, where the demand for water for irrigation, livestock and industry has doubled in the last ten years due to rapid population growth. The concerns that have emerged around water date back to pre-colonial times, although tensions between demand and supply are starker in the current water landscape. As Mutayoba (2002) cautioned, demand exceeding water availability was leading to local, national and even international conflict. In the past, increases on water demand and ongoing disputes and confusion around issues of water encouraged the promotion of ideals of cooperation that would pre-empt potential conflict over water.

At the time of writing this chapter water stress in Tanzania is acute because rapid population growth and increased economic activity exert pressure on the scarce water resources. Recurring drought due to climate change exacerbates the problem and, because 80 per cent of Tanzania's economy is agriculture related, water availability and the management of the resource are critical. Existing policies attempt to address these pressing issues, and due to ongoing external stressors Tanzania moved towards a new phase in water management in 2002 with a new National Water Policy (Mutayoba, 2002). The 2005 National Water Sector Development Strategy (NWSDS) should ideally help facilitate the proper implementation of a revised 2002 Water Policy, but problems in implementation imply the work of a 'cunning' state that pursues a policy framework that is a best fit with global rather than local socio-political landscapes and where policy networks remain closed. Work by Randeria (2007) rejects notions of a passive state that is subjugated to legal regimes and policy principles in favour of the idea of a 'cunning' state that chooses policy prescriptions and programmes advocated and financed by international donors such as the World Bank. The resulting global agenda produces a fuzzy political landscape with multiple legal systems or what Randeria (2007) refers to as sovereignties.

12.1.2 Facts about Tanzania

Tanzania is one of the East African countries that gained independence in 1961 and became the United Republic of Tanzania in 1964 following the political union of the mainland (Tanganyika) and the isles (Zanzibar). It has a population of approximately 36 million growing at a rate of 2.8 per cent per annum (URT, 2002) and a total area of 945 000 km^2. The mainland occupies 939 702 km^2, while the islands of Zanzibar, Pemba and others in the Indian Ocean make up an additional area of 2643 km^2. Tanzania has a coastline that stretches 800 km (Map 12.1).

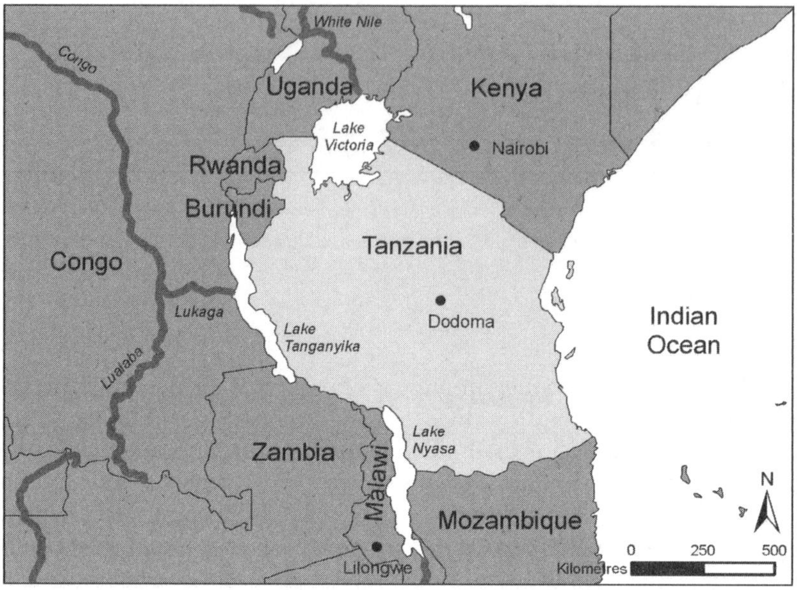

Map 12.1 Tanzania and its major lakes

Dry land in Tanzania forms 61 per cent of its total area. The country has a diversity of climatic conditions with mean daily temperatures ranging from 24°C to 34°C. Mean annual rainfall varies geographically between 500 mm to over 2500 mm with most rainfall generally seen in the higher altitude areas. Levels of access to safe water are generally low, with an estimated 68 per cent of the urban population having access to piped water, but less than half of them receiving a 24-hour service. In rural areas, only 45 per cent have access to safe water and in these non-urban areas, poor infrastructure affects assurance of supply, with reportedly a third of the facilities underperforming. Sanitation services are available to about 80 per cent of the rural population and 85 per cent of the urban population (Kibona et al., 2006).

12.1.3 Water policies in the country

Introduction Water management institutions have evolved from pre-colonial and colonial times through to the contemporary post-independence era. The changes that have taken place during these different periods have been marked by specific demands pertaining to each era. Today most water management imperatives are focused on technical aspects and institutional strategies that incorporate good governance are short in supply, causing institutional fragmentation and a lack of coordination (Sokile et al., 2002). Evidence exists also of shifting allegiances where national elites are embedded more tightly with international organizations than their own citizenry.

Pre-colonial Tanzania was governed by informal rules that determined resource use and protection. These rules were subject to changes as a result both of peaceful interaction among various ethnic groups and of conflicts, warfare and conquest arising between contesting groups. Other causes for changes in the way communities related to natural resources included the impact of long-distance trade and population movements (Boesen et al., 1999). Odgaard and Maganga (1994) explain how experience, changes in leadership regimes, increased pressure on resources and internal competition for scarce resources led to new rules and regulations within the customary systems of water users living in pre-colonial societies. They also note marked demographic changes and as Yudelman (1964) describes at the time of writing *Africans on the Land*, less than one-third of the senior posts in administration were held by Africans; the rest were filled by non-African officials with little or no experience of the adaptive capacity that formed an essential part of the traditional norms and values of Africans.

Pre-colonial landscape In pre-colonial times the different parts of Tanzania were governed as chiefdoms. Famous chiefdoms included those of the Hehe, who occupied the southern highlands of Tanzania under the leadership of Chief Mkwawa; the Nyamwezi, occupying central Tanzania under chief Mirambo; the Chaga in the northern region, led by Mangi; and the Sukuma, occupying the northwest.[1] Within the boundaries of the chiefdoms, water resources were collectively owned: the streams, rivers and wells within a given territory belonged to everyone in that chiefdom. Where rivers crossed chiefdoms, creating transboundary water issues, resolution was achieved through peaceful negotiations among the chiefdoms concerned. As the legitimate authority, the traditional leaders controlled water resources in the chiefdoms and made decisions regarding transboundary water allocation.

Today the challenge lies in bringing customary traditions that have guided and informed water use patterns in Tanzania over time into contemporary, formal institutional settings. Customary norms and values that have determined management of the resource remain embedded in the historical memory of water users – particularly those operating closest to the resource. Current legislative activity, while recognizing the institution of traditional leadership and the plurality of legal systems, has over time restricted and inhibited customary law. Randeria (2003) observes that an ecological and sustainable agriculture or pastoralism has no place in state plans and policies that are driven by notions of capitalism and accumulation of surplus. This author notes in addition that, given the allegiance of the 'cunning' state to the agendas of foreign donors, it is unsurprising that principles underpinning these pre-colonial landscapes are today only vaguely integrated in current state policy.

Nonetheless the 'traditional' systems retain power bases and their own sovereignty, creating a clear case of legal pluralism. Rarely based on written laws and lacking a governing board with powers to elect or replace managers, customary systems maintain a basis for power that relies on an elite individual leader – whether the leader is benevolent or not. Debates continue around the legitimacy of traditional leaders and concerns are voiced about patron–client relationships fostered by colonialists and continuing in different forms in the current political landscape. Nevertheless for many ordinary users working closely with water, the voice of traditional authority is heard over and above the written ordinances proclaimed by the state. Randeria (2007) refers to the coexisting systems as 'overlapping sovereignties'. In Tanzania they sit more or less comfortably alongside one another. As Randeria (2007) suggests, however, the state has dominion insofar as there has been a rapid increase in the alienation of the citizenry from common property resources caused by state appropriation of these resources as 'national assets'.

At this stage in the history of Tanzania it also appears that a cunning state has shifted responsibilities to citizens who must pay for their right to water. Despite 'overlapping sovereignties' it is the sovereignty of the nation – along with its allegiances with international rather than local sovereignties – that dominates. The voices of traditional leaders are muffled in favour of regulating mechanisms that form part of current state policy in alignment with the voices amplified by a global agenda. Social formations, such as village water committees (VWCs), have in the past successfully protected both the social system and the ecosystem and remain valuable institutional settings for water management. These localized bodies, however, are alienated from water policy networks embedded at the scale of the nation state. Mainstreaming spokespersons for customary laws and informal institutions, like VWCs, can be risky because, as argued in discussion below, spokespersons who negotiate within dominant policy networks are unable to challenge the status quo because they are likely to be absorbed into mainstream discourse.

In the past, legal systems deployed by traditional leaders could be relied on to resolve tensions about water use and allocation before conflict arose, thus averting threats that might arise from shared water sources. Such systems were localized, and their rules and values applied to a local socio-political landscape where they demonstrated their ability to adjust and adapt and thus to maintain socio-ecological system equilibrium. Decisions taken by chiefs were appropriate responses to ecosystem requirements. The way that water was protected, used and conserved formed remarkable dynamic, change-sensitive and community-based resource management featuring a high level of what has become commonly referred to as 'social adaptive capacity'. This capacity is the ability of a society to adapt the way it relates to its ecosystem in order to maintain or achieve a sustainable measure of social stability (Malzbender et al., 2005; also see Chapter 17 by Olsson and Galaz in this volume). The measures involved entail well-developed social resources for effectively managing risk and disasters or other threats to the social environment. The guiding principle behind such adaptive capacity lies in local responses to water as a public good. The notion of water as a public good, where collective social bonds determine social action, has its roots in pre-colonial times as an established legal principle in various societies and legal systems ranging from Roman law to African customary law. It has now become an internationally accepted concept (Malzbender et al., 2005; see also Gupta, Chapter 3 in this volume).

Political change and colonialism The coming of the colonialists (German and then British) did not immediately affect informal water management arrangements, but in the early 1900s the demand for water was increasing as economic activities expanded under the colonialists. The government curbed water use, and for the first time in the history of Tanganyika the state intervened in water management – an intervention between the people and a resource. As a result the control of local communities, and the jurisdiction and power of the chiefs in particular, was diminished. The new elites with rules proclaimed by their sovereign empire challenged the informal, customary rules about decentralized water use. In 1910 the German colonial government investigated the feasibility of irrigation agriculture (URT, 2002). In 1923 the Water Ordinance marked a new era with the promulgation of the Statutory Water Law in Tanzania (then Tanganyika). Water bylaws drawn up by the British colonial ruler Sir Richard Turnbull to enforce water management were first introduced in 1926 and unsurprisingly favoured the colonialists (Sokile et al., 2002).

12.2 Water laws and policies transition

12.2.1 From colonial to post-independent

Introduction The demise of colonial rule did not lead to immediate changes in state policies, but 1967 saw the Arusha Declaration, heralding a more socialistic economy that discouraged private ownership of natural resources and promoted collective ownership of national resources. Principles of water use and ownership began to realign with customary rules and led into a diffusion of power and investment of responsibility resulting in Randeria's (2007) observation of intertwined structures of rules and overlapping sovereignties.

Following global trends in watershed management Tanzania adopted a River Basin Management Approach in 1980. In accord with the policy prescribed in Act No. 10 of 1981 (an amendment of the Principle Act No. 42 of 1974), the country was then divided into nine water management areas or basins with priority assigned to the Pangani and Rufiji basins. As a large area defined by macro-hydrological boundaries, a basin can provide the geophysical framework for national water planning. The Minister of Water announced official recognition of the nine river basins in 1989. The ideas behind hydrological rather than political boundaries is that water consumption – and allocation by and to multiple uses and users – in a basin requires a holistic approach to resource management in order to maintain principles of social justice for the users and principles of eco-equilibrium for the environment. Ilcan and Phillips (2008) identify the related processes where knowledge is passed on, advances and mutates as a 'knowledge apparatus'. The emphasis in river basin discourse – now mainstreamed as integrated water resources management (IWRM) – is the need for cooperative governance, where all stakeholders should be involved in the decision-making processes around resource allocation. What has emerged, though, is rationalization by actor networks of the way in which river basins are ordered in line with a sovereign discourse that today maps and shapes the way in which water is used and abused.

Current frameworks pay little or no heed to customary values and norms, although as noted above, such norms and values live on in the historical memory of water users. Malzbender et al. (2005) note that current legislative activity, while recognizing the institution of traditional leadership and the plurality of legal systems, has over time been known to restrict and inhibit customary law.

Notwithstanding the debates about the legitimacy of traditional leaders, for many of the ordinary users working directly with water, the voice of the traditional authority as the recognized sovereignty resonates louder than the written ordinances proclaimed by the state. Oomen (2005) established that the people in rural areas feel a strong need for some form of traditional governance to maintain order and ensure peaceful coexistence. He found this expressed in the saying: 'No herd of cattle can look after itself' (Oomen, in Malzbender et al., 2005). Much evidence exists to show that the pre-existing knowledge apparatus for adjusting and adapting to water stresses has not been well incorporated within current policy networks. One restriction on entry into current policy networks is the inherent difference between the worldviews concerning water and the seemingly incompatible notions of social ownership and private ownership.

In addition to institutional incompatibility Tanzanian water management also involves institutional fragmentation. Sam Shemsanga of the Ministry of Water in Tanzania[2] remarks that regardless of existing attempts to coordinate efforts, a more coherent strategy is needed to make sure that all bodies involved in water resources management are synchronized with one another in order to safeguard this natural resource. Shemsanga is concerned about the contradictions apparent in the duties and responsibilities of different departments that act in isolation with little to no coordination between them. This means that water, a multisectoral concern, is not treated as such, restricting progress in solving water problems in the country. In Randeria's view (2007) the hybrid landscape created by overlapping, coexisting sovereignties is to be expected.

There are consequences of competing sovereignty and an unwillingness to level the playing field or change network settings in the way that Huitema and Meijerink discuss in Part I of this book. In early 2006, for instance, the then newly elected government under the current President, His Excellence Jakaya Kikwete, declared that pastoralists were to move from the Ihefu catchment, source of the long and renowned Ruaha River that provides supplies of water critical for irrigation and hydropower. The decision was driven by policy agents concerned about low flow rates that had a negative impact on downstream irrigation and hydropower production. According to these experts the reason for diminished water flows was pastoral practice. Others protested that the pastoralists had been in the region for many years and had a right of access to the water of the Ruaha.

Beside discordant voices on the matter, both institutional fragmentation and blurred lines of control and sovereignty aggravated an already tense situation. Institutional fracture and closed policy networks resulted in a shift in governmental institutions as the Ministry of Water and Livestock Development was split into two separate ministries – a Ministry of Water and another of Livestock Development. Policy was formulated at the national level. Local-level institutional actors, who had a better understanding of the activities and water use patterns on the ground, were unable to shape policy that affected them through this act of venue manipulation. The interventionist impulses of the state resulted in conflict and distrust. Rainey (1991) claims that public organizations are constantly changing and although institutionally capable of social adaptiveness are resistant to change – in this case raising barriers to entry into policy networks by local experts. Such resistance constricts social adaptive capacity and sees control exerted within the national hegemony and restricted social agency at the local level.

Institutional opportunities Overall, the setting up of institutions to manage the new geographical water units has been slow. The regulation of water extraction remains a problem and the customary institutions inadequately define the rules that deal with extraction. It has been suggested that the matter could be legally regulated in a number of different ways, including by the creation of village and district council bylaws; by the formation of companies under the Companies Ordinance, Cap.112; through cooperative societies, water user associations (WUAs), under the Water Utilization Act (Control and Regulation); or through courts of law or ward tribunals (Sokile et al., 2002). The 1974 Water Companies Act governs district councils and WUAs that used to exist as multistakeholder bodies concerned with common water issues and which are now more or less all defunct. Private companies have been instrumental in bridging the divide between some users as they have the right to extract volumes from sources where individual users do not. Private companies have the status of 'middleman', and their role as policy entrepreneurs is discussed below. They serve as distributors of water and ideally as fair distributors of the resource because – in theory – they are not aligned politically or ethnically with any particular group of users. In fact private companies have aligned themselves with powerful supra-national bodies and are driven by the principles of water as an economic good and cost recovery. For these reasons their social legitimacy and their 'fairness' are questionable.

In summary, demographic changes and the consumption needs of a growing population led to obvious pressures on water resource use and created a pressing need for an integrated approach to water resources management. The water supply systems of the larger urban centres such as Dar es Salaam, Arusha, Mwanza and Moshi felt the strain. In this context, after a lengthy consultation process, the National Water Policy of 1991 was revised and the cabinet approved the new National Water Policy (NAWAPO) in July 2002 adopting policy prescriptions and programmes propelled into water policy networks by international donors. The NAWAPO appeared to address previous policy shortfalls and introduced decentralization of water supply management in line with Agenda 21 of the United Nations Environment Meeting, held in Rio de Janeiro in 1992. The 2002 NAWAPO was followed by the 2005 National Water Sector Development Strategy that sets out how the NAWAPO will be implemented and describes the institutional and legislative changes necessary to ensure its success. The timeline below reflects the various transmutations from 1900s to 2005 that affected the way in which water was governed:

- 1910: German colonial government investigates the feasibility of irrigation agriculture.
- 1923: Water Ordinance by German colonial government.
- 1930: Systematic state interventions in the water sector.
- 1945: The Department of Water Development was founded.
- 1967: Abolition of water user fees by the then independent Tanzania under President Mwalimu Julius Nyerere.
- 1974: Introduction of Water Utilization Act (control and regulation).
- 1981: Amendments to Water Utilization Act (control and regulation).
- 1981: Designation of nine water basins in Tanzania.
- 1989: Minister announces Tanzania's nine river basins.
- 1991: Institution of National Water Policy.
- 1994: Rapid resources assessment.
- 1999: Draft new water policy.
- 2002: Launching of new National Water Policy (NAWAPO).
- 2005: Water sector development strategy.

Influences from the region With the fall of the Berlin Wall and the global influences of trade liberalization, privatization and deep reforms that affected not only the water sector, the decade of the 1990s was a decade of change. The Dublin Principles shaped water resource management reforms and, as was the case elsewhere, most African countries reviewed the way in which water management institutions could be constructed. In the region, South Africa emerged as a leading example of such reforms. It established a national panel in 1994 for the lengthy water law review process with the aim of drafting a new Water Act. South Africa's National Water Act (Act 36 of 1998) and Water Services Act, 1997 (Act 108 of 1997) together provide for the establishment of institutions with responsibility for management and distribution of these services. The objective of these acts, as claimed by the Department of Water Affairs and Forestry (DWAF) is to attain equity, efficiency and equality in water matters through an 'iterative process' (DWAF, 2002). The state laws determine the way in which citizens order their relationship with water. Policy in South Africa has been widely acclaimed, but implementation has been slow.

In Tanzania the National Water Policy has been influenced by the same principles and supports the continuation of current formal WUAs and the rules that govern these bodies. The Tanzanian government (URT, 2002) regards the associations as the lowest level of water resource management and these are imbued with a wide range of responsibilities for managing use, such as allocating water rights, preparation of water utilization plans, enforcement of the law, pollution control and protection of water sources, resolving local water conflicts and facilitating data collection. They are statutory bodies unlike village water committees, which do not have legal status. However while the WUAs are currently dysfunctional in Tanzania, village water committees are operative with social legitimacy among ordinary water users at the village level.

12.3 Dynamics of policy formulation

12.3.1 Introduction

The 1990s as a decade of social reform confirmed the arguments of theorists such as Scheff (1990) that advanced technology, global connections and the demise of autocratic and top-heavy states was taking humanity into the realm of individuality, where collective social bonds have been threatened. It also resulted in states of the developing world who had insufficient funds and were forced to take on loan adopting an economic liberalization paradigm. Traditional and dense bonds of close neighbourhoods and norms and values based on customary law enjoyed social legitimacy because of their guiding principles of social ownership. Such bonds are being replaced by new forms of social interaction. The new social configurations are dominated by supra-national legal regimes, international donors and a cunning state (Randeria, 2007). The new forms of social interaction and new policy networks overlap more or less smoothly with one another, but not particularly comfortably with previous governance systems that ordered the way citizens lived in relation to natural resources such as water. The following section scrutinizes more closely these different influences and the way in which overlapping interests help to shape policy.

12.3.2 Policy influences: state and non-state policy entrepreneurs

The most persistent drivers for change in policy in Tanzania have been donor organizations such as the World Bank and the International Monetary Fund (IMF). They have ensured that a policy of privatization is firmly embedded in Tanzanian water policy.

There were advantages for the state to be cunning in its adoption of these reforms. The utility DAWASA (Dar es Salaam Water and Sewerage Authority) had failed to keep up with population growth in the city where only 26 per cent of water was being billed, 60 per cent was lost through leaks and a further 13 per cent through unauthorized use, illegal taps and non-payers. ActionAid reports on the proposal to introduce 'a form of quasi-privatization' as a concession whereby DAWASA's assets and responsibility for all its functions would be transferred to the private sector (ActionAid, 2004, p. 5). Under the current situation the City of Dar es Salaam is managed by a public–private body as a result of pressure from the World Bank and the IMF on the state to privatize. Money was promised on condition that the state drew up legal frameworks to support the divestiture of parastatals, including public utilities such as DAWASA. In exchange the state coffers would enjoy US$300 million per year. If Tanzania were to slow down on its policy of privatization, it would only receive US$100 million per year.

As the report by ActionAid concludes: 'This provided cash-strapped Tanzania with a 200 million dollar incentive to privatize as quickly as possible' (ActionAid, 2004). IMF officials joined forces with the World Bank to ensure that policy was in place whereby the DAWASA would assign its assets to a private management company as a condition of receiving maximum benefits from debt relief. According to the same report, although Tanzania was not able to meet these conditions and the DAWASA was not privatized, the World Bank and IMF did not let up their pressure on the Tanzanian policy-makers. The Dar es Salaam Water Supply and Sanitation Project (DWSSP) was implemented whereby the Tanzanian government, 'faced with a failing water system had little choice, but to accept money under the Bank's terms' (ActionAid, 2004, p. 6). According to the report, the UK government has played a strong supporting role in pressurizing the Tanzanian government to cement pro-privatization laws and the British company BiWater is fully backed by the UK government.

Importantly, according to this report, government officials themselves supported notions of privatization and welcomed their relationship with the World Bank, IMF and the UK government. The Ministry of Livestock Development described the private sector involvement as 'seriously welcome' while staff at the Presidential Parastatal Sector Reform Commission thought that private sector involvement was essential for the success of water reform (ActionAid, 2004, p. 8). This process reflects what Meijerink and Huitema (Chapter 2, this volume) identify as strategies to maintain policy influence – in this case a clear hegemonic policy discourse where major changes resulted from venue shopping and the adoption of ideas that suited dominant policy networks.

Supported by enough government officials to make sure that 'the government was not dragged into water privatization kicking its feet' (ActionAid, 2004, p. 8), the World Bank – and the IMF – were key mediators between the state and non-state actors and played a dominant role in shaping water law reform. The extract below reflects the strategies of venue-shopping and concept-borrowing that are used to push forward an agenda that restricts new policy networks or policy agendas that contradict the hegemonic policy discourse:

> Donors come and talk about water in Zambia, water in Sweden. Well, I have never been to Sweden and I have never looked at water in Zambia, but they still used this to make suggestions about our policy. The World Bank will often bring the aide-mémoire already written and

> present it to me with no time to digest . . . some donors come with big money and this can skew priorities.[3]

Policy entrepreneurs have a role to play in informing policy-makers about the choices they have that would result in a 'best fit' policy – one that would reconcile sovereign interests of the nation with those of all citizens. In the case of current Tanzanian policy formulation, the 'giants' such as the World Bank and IMF have restricted space for non-state policy entrepreneurs whose information is not in line with dominant discourse proposed by these powerful elites, whose discourse frames water only as an economic good.

There are other spaces where over time scientific knowledge is generated and it is likely that this will filter through into the corridors of power where knowledge transfers can take effect. The media, for instance, expose contemporary debates and act as non-state policy entrepreneurs.[4] Scientific research is also widely disseminated in conferences and academic journals and over time forms a body of collective knowledge that can influence policy. This involves the shadow networks referred to by Meijerink and Huitema (Chapter 2, this volume) where boundary objects are working papers, conference presentations or peer-review journals.

12.3.3 External policy entrepreneurs struggle to break network hegemony

The discussion that follows examines more closely the interesting role that policy entrepreneurs who act as intermediaries play in between state and citizen as 'middlemen' or 'network managers' (Kickert and Koppenjan, 1999, p. 58). It is the role of these agents to distinguish among diverse target groups (ibid.) and, tasked with implementing programmes or communicating information from one set of actors to another, they contribute, at least according to Kickert et al. (1999), to the establishment of patterns of social relations between interdependent actors.

Policy entrepreneurs, particularly those who produce and disseminate 'scientific' knowledge, can be seen as network leaders or what Hardin (1999) calls 'men of honor' or 'trustworthy' others who play an advisory intermediary role and whose actions are not questioned because they are assumed to be grounded in the interest of the person who is trusting. Policy entrepreneurs occupy legitimate and accepted positions of power as, at least in theory, they would be able to delegate, organize and retain full support from state officials as well as local stakeholders (Kickert and Koppenjan, 1999).

Ideally, trustworthy others are able to navigate through the corridors of power because they are able to use 'technical' and 'objective' vocabulary that resonates with politicians, but there are barriers of entry into policy networks because some trustworthy others, for instance academics, are more likely to speak the science 'of the people', which at this stage in the history of policy-making in Tanzania is not mainstream, but rather what Thompson refers to as the 'objective reality' or 'science of people' (2002, p. 233). Such science does not fit well with the dominant paradigm of science driven by the global discourse on privatization and individualism (see also Chapter 3 by Gupta, this volume). Faced with the 'hard knowledge' rooted in the 'real' world dominated by the 'big players' – holding what Thompson (2002) describes as a masculine normative vision of immutability and impenetrability – the science of the people has been slow to penetrate the corridors of power.

Nonetheless the independence of 'trustworthy others' is perpetuated in so-called 'centres of excellence'. It becomes more powerful when there is a collective rather than individual voice of science, for instance from 'men of honor' with allegiances to networks such as the University of Dar es Salaam Staffs Association (UDASA) or Suasa-Sokoine University of Agriculture Staff Association, because it is within such social networks that the pursuit of business as usual (academic) is superseded by the political will of network leaders to influence policy.[5]

It has been the pressure of network leaders – academics – on politicians that has changed the way in which urban water supply is controlled and planned, while urban sites have grown and the demand for water has placed pressure on already over burdened systems of supply. The Act of 1997 was a direct result of pressure on policy-makers by network groups demanding the adoption of an alternative set of principles and rules that would result in a more efficient urban water supply system. Pressure from such groups resulted in the formation of the semi-autonomous Urban Water and Sewerage Authority, the DAWASA (Dar es Salaam Water and Sewerage Authority). In 1999 under the Miscellaneous Amendments Act of 1999, the DAWASA retained full responsibility for urban water supplies, but as is evident from the discussion above, the dominant voice of the World Bank and IMF drowned out the muffled voices of local scientists, and the DAWASA became an instrument through which a government policy of cost recovery would be assured, entailing the concept of water as an economic good and so completely overriding ideals of social ownership.[6]

Research findings can potentially challenge policy and create conditions for change in a resource sector. Examples are provided, for instance, by Mayo and Masahauri (1991) on the benefits of rainwater harvesting; May (2000) on the impact of water vending and metering on revenue collection; and Lankford (2003) on irrigation-based livelihood trends in river basins. Such work provides information on critical issues and challenges in the sector. Other noteworthy research includes investigations into irrigation-based livelihood trends, water fees, free water policy and on gender, poverty and participation.

The authors are examples of experts who are able to influence policy and who, as trustworthy others navigating between state and non-state actors, level the playing field as they aim to bring their set of principles to practical effect. Whether or not this kind of influence will bring about change remains uncertain, but expert voices over time together form a body of collective knowledge about people and are likely to penetrate deeply into policy networks. It is these voices that ideally claim and reclaim principles of human rights, cultural diversity and spaces for indigenous as well as new knowledge and social learning.

In the current political landscape in Tanzania, the most notable barrier to entry into policy networks is non-compliance with the dominant global discourse. Tension exists between those who are influenced by the global discourse and those whose tendency is to think local and to question domination by the cunning state and its international partners. The hegemonic discourse frustrates entry into policy networks where a better understanding of local responses to increased demands on water could be voiced. Grindle and Thomas (1990) are helpful in their analysis of why policy formulation is not just the action of individuals who shape policy, and requires a set of conditions for successful implementation. At any stage in the life cycle of a policy, opposition can reverse or alter its course (ibid.). Grindle and Thomas ignore the influences of sovereign powers who can corrupt or divert opposition from subaltern groups who might threaten a project of reform as laid down by national – and international – sovereignties. In so doing, sovereign powers can also thwart social adaptive capacities.

There has been a global shift where the state is no longer expected to play the role of supplier, but instead becomes a facilitator for delivery of services. This is in line with the move from strong central states to decentralized states. Alongside these changes, the state of Tanzania has retained its stronghold on decision-taking and, despite rhetoric about participation and local decision-making, has left little room for substantive input from below. The state voluntarily limits its sovereignty by devolving responsibility for delivery to realms outside its central orbit and by conceding to a regime of privatization with its own set of rules and regulatory mechanisms. In this way the state opts for strategies proposed by outsiders, and that are often too lofty to be practical. Another risk of poor policy fit is that the proposed institutions to manage water may be shaped to fit a global ideal that is ill-suited to the social, geographical and political landscapes within their jurisdiction.

Over the past decades developing countries have focused on the bottom-up dynamic of development, proposed as a replacement for failed top-down efforts that did not benefit the poor. These concepts are not unique to IWRM, but are fundamental to democracy, a process that ideally closely involves local citizens and communities in finding ways to meet their needs and improve their quality of life. In reality, 'modern' legal systems are superimposed on those systems that were best able to govern the commons, and can thus have negative effects on the way in which the commons are governed. Control and decision-making powers in Tanzania, although supposedly decentralized and scaled down to the lowest levels, remain only partially in the hands of the sovereign state with its constitutional mandate to protect the most vulnerable of its citizens, and partially in the hands of private companies and international partners (Randeria, 2007). The scaling-down and devolution results in a hybrid of mechanisms that increase complexity and can obstruct rather than facilitate good governance of natural resources.

In this complex landscape of shifting allegiances where the state in particular has made clear its position of placating donors rather than vulnerable citizens, non-governmental organizations (NGOs) and transnational network organizations, such as WaterNet, WaterAid and the Global Water Partnership (GWP) or facilitating agents such as the Deutscher Entwicklungsdienst (DED) or German Development Services, the Deutsche Gesellschaft für Technische Zusammernarbeit (GTZ) and the UK Department for International Development (DfID), are expected to support their respective government while at the same time fulfilling their mandate to reduce poverty and better the lives of citizens. This group of intermediaries act as implementers with a commitment to strengthening the water sector, but they are invited to play a role in supporting policies that they were not part of formulating and with which they might not agree.

Finally, to conclude our discussion on policy entrepreneurs, it is useful to record that these actors are expected to equalize the basis of power and control by drawing stakeholders into the fora and by guiding and informing state officials of the needs and preferences or problems that have been identified by the non-state actors and vice versa. A concern with these actors, for instance the DfID, DED or GtZ, which serve in Tanzania as bilateral executing agents, is that when there are cost recovery or other structural failures due to the dominant paradigm of privatization, foreign policy entrepreneurs will retreat and Tanzanians will be left to solve the problems on their own (ActionAid, 2004). The support of these international agencies is critical to the success of water strategies in Tanzania as in other developing-country contexts. However, this support is maximized when coherent systems are put in place, in particular governance structures that are able to cope with an exit strategy.

12.3.4 Obstacles to and opportunities for substantial input from below

An active civil society, involving lobby groups, NGOs and community-based organizations (CBOs) can drive change over time. Notions of multistakeholder involvement and engaging with those who are closest to the resource, in both governance and direct management of water, are central to IWRM. The engagement of civil society in water policy formulation is likely to produce trust (Goldin, 2005, 2008),[7] but this is a risky process because the building of trust has everything to do with the equalizing of power and knowledge. Those following a discourse promoted by the dominant knowledge regime are able to influence policy in ways that those without that knowledge are not.

Although Tanzania has embarked on a process of revising water policies to address the uses and users of water, and conflict and cooperation around water, and to establish appropriate institutional and legal mechanisms with multistakeholder participation, there has been very little attempt to level the playing field and 'virtually no meaningful public participation, nor public discussion of policy alternatives' (ActionAid, 2004, p. 9). The Tanzanian Networking Gender Project (TNGP) said that action on the promise to include all water resource stakeholders 'never happened' (ibid.) and according to WaterAid more than half of the community who were consulted were extremely poor and did not know that privatization was going on. The Africa Youth for Development Group said that they, too, were not asked to give input into policy and confirmed that people did not know what was going on.

This has everything to do with unequal relations of power and knowledge (Goldin, 2003, 2005, 2008) where there is no substantial input into policy by those who are most affected by the policy. NGOs complain that decisions were taken before they were consulted, and local NGOs found it difficult to assess the reforms because of the secrecy surrounding them. They maintain that this exclusion from information was a deliberate strategy to consolidate positions of authority and to take control over policy reform. Ironically, key politicians, such as members of parliament (MPs), were marginalized and according to an MP from the Labour Party in Tanzania had there been more consultation: 'Alternatives to privatization would have been suggested, and the government might have been put to task for failing to spend enough money on repairing the water system' (ActionAid, 2004, p. 10).

Other policy entrepreneurs more or less 'benign' in their allegiances act as mediators between state and non-state actors; for instance, in the Moshi Municipality in the region of Kilimanjaro, the Energy and Water Utilities Regulatory Authority (EWURA), formed in 2001, is instrumental in regulating tariffs and charge adjustments in the water and sewerage sector:

> EWURA acts as a referee, to make sure that there is a fair game between water services providers and the consumers. EWURA was formed after the public concerns through media, seminars

> and conferences, on the high price of water and sewerage services . . . there are various regulatory authorities whose role is to control the pricing of various services in the country.[8]

These mediating bodies negotiate a space and try to equalize relations of power by providing a platform for public expression and by soliciting input from below. Public spaces, however, are restricted, making it difficult for the majority of Tanzanians, who are opposed to privatization because the proposed reforms are unlikely to meet their stated goals, to make their voices heard. In any case there has been very little public discussion or consultation and as a result the opinions of the poor, particularly those of poor women, are ignored during the process of the policy reform (ActionAid, 2004).

Another impediment to substantive input from the poor is the usurping of power by local elites. The imposition of a global discourse in the realm of water management – including the adoption of IWRM for water resource management and notions of privatization for water supply – have together resulted in a shifted and dominant paradigm. As clearly shown above, the state no longer takes on the role of providing for its vulnerable citizens. Instead it plays the role of facilitator of service delivery. In line with notions of decentralization the onus is now on local communities to operate and maintain their water supply systems and citizens are expected, ideally, to provide substantial input into policy from below. This has not, however, resulted in improved capacity for decision-making for all citizens, but rather only for a few. The role of local elites has taken on new, puzzling patterns where these elites serve the interest of the state at a cost to their own constituents. Local elites who were invited into policy networks supported notions of cost recovery and privatization of water because these policies could bring them personal economic gains.

Mfangavo (2005) draws attention to Toner's study of village water supply in the Kilimanjaro region that shows that the poorest were not represented in WUAs where they could have influenced policy. Local elites voted for the installation of private taps, but as Fjelstad (2003) notes, communities that have not been involved in policy formulation are less likely to comply with the rules where water is charged as an economic and not public good. Community participation is at the heart of current water policy, yet communities are unable to operate and maintain water schemes or supervise all activities on their own (Mfangavo, 2005). The absence of equalized power relations means that those at the lowest levels are unable to shape policy and that as a result policy is unlikely to be a 'best fit' for local conditions. 'Evidence from detailed ethnographic research shows that in the GTZ-funded project in Uchira, only a few people in the community dominate decision-making on water use, despite a heavy project emphasis on broad community ownership and that a majority of village residents consider that the water has been privatized by a village elite' (Toner in Mfangavo (2005, p. 22).

The gender dimension of water is commonly acknowledged, but few projects have a high-level involvement of women (Mfangavo, 2005). There are bylaws that are pro-poor and exempt the poor from cost recovery for water, but there are unfair targeting subsidies because once again the poor have been defined by policy-makers and not by the poor themselves (ibid).

Including poor people and their experiences with local conditions has a multiplier effect on social capital as people learn what works and what does not, and are then better able to monitor the accountability of their own members. It is not just a question of learning what works and what does not work, but also of synergy between state and non-state actors as these lessons and experiences shape policy (Goldin, 2008). What is critical here is that there are power influences at different levels that exclude the voices of many, and we see a hierarchy rather than a hybrid of sovereignties. At the village level, communities are subjected to elites who are able to influence policy for their personal gain as they obtain favourable contracts and other income-generating opportunities by supporting notions of privatization.

Drawing communities into decision-making processes is not proposed as a replacement for state involvement, and the sobering paper of Habib (2008), with the evocative title 'Speaking "truth" to all forms of power', draws attention to power relations at all levels – both within the state and within subaltern groups. Synergistic relationships between state and local actors are championed by Evans (1996) as the answer, but synergy between state and non-state actors requires a two-way engagement, one of which is participation from the bottom up, and the other of which is engagement from the top down. It requires a levelling of the playing field and more equal relations of power and opportunities to influence policy, but this is not yet happening in Tanzania.

Intermediaries between the state and its citizenry – in the form of either NGOs or individuals, and including local elites – have the potential to influence formal institutions and to guide the formulation policy. Mainstreaming informal institutions or/and individuals can be risky, and caution is needed to ensure that this does not result in what Selznick (1949) called 'co-optation', a process that absorbs actors into the system in order to avert threats to the stability of the system. Actors can be brought on board so that they do not protest against decisions made in the interest of the state and its international partners. Participation itself has become a risky process because the voice of the dominant actors more often than not prevails. Involvement of communities closest to the resource is more likely to achieve sustainable solutions for natural resource management. Such involvement has become a logical progression evolved from lessons learned, showing that the ideological phasing out of top-down strategies can encourage better water resource management.

According to Kickert and Koppenjan (1999) bureaucrats need to know about the other actors involved and their idiosyncrasies, but for others (Randeria, 2007; Goldin, 2008; Habib, 2008) this know-how has everything to do with power and strategic allegiances that are formed among those who broker power and encourage – or discourage – entry into policy networks. The debate around privatization, for instance, is embedded in discourse that is driven by the state and lodged in global allegiances where powerful elites, such as those of the World Bank and the IMF, intervene in the process of policy formulation and co-opt local elites or outsiders who speak the same language.

Social formations, such as village water committees that actually operate at the level of the village, are closest to the resource and more likely to embrace traditional norms and values that are valuable for water resource management. These committees governed water matters prior to the formation of current water policy. Although their existence has been acknowledged in the National Water Policy of 2002, care must be taken to ensure that their role is not limited and that their voice becomes loud.

Experience in South Africa has shown that despite excellent policy frameworks, stakeholder participation is a difficult process because, as Goldin (2003, 2008) describes, strategic allegiances formed over many years have resulted in closed networks. These strategic allegiances are also apparent in Tanzania where local elites and state actors collude with powerful 'others' to drive forward policy that does not favour vulnerable citizens. Tanzania's National Water Policy of 2002 recognizes the importance of decentralized decision-making and of devolution to the lowest practicable level in the planning, design, management, decision-making and implementation of water issues. It is clear that lower levels shall be given a greater degree of responsibility in the management of water resources subject to appropriate regulatory frameworks. The policy names village water committees as valid institutional settings. As these bodies operate within the framework of the Water Utilization Act of 1974 and the National Water Policy of 2002, they are legitimate management institutions, but without power to influence policy networks they are unable to fulfil their function of improving village-level access to and use of water. The norms and values that operate at the village level do not neatly align with formal rules and the enforcement and the performance capacity of the latter are problematic.

In most cases, local institutions successfully reflect the needs of local societies and for this reason the norms and values that govern local-level organizations should be reflected in water policy networks. This would also mitigate for non-compliance (Fjelstad, 2003). The integration of customary norms and values helps resolve conflicts and misunderstandings and helps legitimate newer institutional frameworks.

12.4 Conclusion

Attributes of social adaptive capacity at the local level, which have fostered community responses to change, have been interrupted because the formulation of water policy in Tanzania has been driven by outside mediators and dominated by a global rather than local agenda. Change is being constructed top-down. Those institutions that are operational at the local level have not formed strategic allegiances with the state and are unable to make a contribution to policy. There is a hybrid institutional framework with overlapping dominions, but the sovereignty of the state in the domain of natural resource management has led to a preoccupation with international partners and a top-down policy implementation. This is risky as it jeopardizes the effectiveness of water users to respond to change.

The arguments presented above propose that traditional societies were governed by dynamic, change-sensitive and community-based resource institutions. This is seen to affect the ability of a society to adapt the way it relates to its ecosystem in ways that maintain or achieve socio-ecological stability. The chapter examines the role of policy entrepreneurs, such as network leaders, who act as intermediaries between state and non-state actors. Network leaders include the traditional leader who was the legitimate authority to hold the responsibility to control water resources in a chiefdom and to make decisions about transboundary water allocation and the manner in which water users related to the resource. The way that water was protected, used, conserved and managed overall reflected a dynamic social adaptive capacity as decisions taken by sovereign powers embedded locally, namely traditional chiefs, were appropriate responses to ecosystem requirements as they were based on norms and values formed at the local level.

Today these network leaders do not make substantive input into policy formulation and their sovereignty is being undermined. The state has aligned itself with principles of decentralization where those closest to the resource are supposedly involved in decision-making that affects their everyday living. This is not happening, however, because state elites have negotiated within their corridors of power in ways that have cemented their sovereign interests in allegiances with international partners. This has occurred as a calculated, voluntary measure by a cunning state and has directly influenced policy in favour of global rather than local discourse, resulting in risky notions of cost recovery and privatization that do not benefit the poor. International donors were intermediaries between state and non-state actors in Tanzania, where the state and international agent allegiances drove policy and strongly influenced the way in which the legal framework was shaped.

Ideas of IWRM are not easily applied, and one of the impediments to implementation is the hybrid of legal and institutional bodies that overlap, but carry different sets of knowledge, ideals and experience that do not sit comfortably together. This makes it difficult to bridge the divide between ideal and practical applications of water resource management. A successful application of IWRM in Tanzania would depend on the ability of the reform process to better integrate knowledge regimes and voices from policy entrepreneurs who operate outside the state. This would mean more purposeful embedding of the state-burdened bureaucracy within civil society. In this way the correctness of the formalistic rules would not carry more weight than the generation of substantive policy input from below. Current Tanzanian water policy remains top-down and local village committees are not yet gaining significant influence within the existing water policy decision-making network.

Unequal relations of power restrict opportunities to challenge the way in which policies are formulated, and this asymmetry blocks entry into policy networks. Although institutions for water management have been put in place as drivers of change and transformation, not all institutional forms result in tangible gains for their members or lead to practical action or change. Social networks and neighbourhood allegiances struggle to create the right conditions for entry into water policy networks. The undersupply in social capital together with the guile of a cunning state mute the voices of those who do not operate within the dominant state–donor-driven knowledge regimes.

Notions of decentralization, participation and inclusion of stakeholders at the local level do not readily result in more equal power, and it is significant that Williams (2004) speaks of participation as a new 'tyranny' whereby the previously marginalized are mainstreamed and local elites in fact become bound more tightly into the structures of power that they are then unable to question. Local elites as network leaders have been drawn into water policy networks, but becoming absorbed into the hegemony of policy networks they support global rather than local knowledge regimes. This is because these mainstream regimes offer economic and social status. In this way the marginalized – or their representatives – who become 'mainstreamed' lose, rather than gain, power. Potential policy agents – who used to represent the interests of the poor – are drawn into forms of discourse where they are no longer able or willing to influence the way that water is managed, allocated or used in favour of the poor. Their position of authority has been replaced by the state, whose priorities are to placate influential brokers, in this case the donor community, whose pressure has been crucial in pushing government to privatize the water system in Dar es Salaam.

The policy environment is itself dynamic and change-sensitive, and although the World Bank and IMF have been largely instrumental in driving forward privatization in

Tanzania, the formal rules of the game change over time as what Schaap and van Twist (1999) name the 'filtering effect of the frame' shifts. At the time of writing this chapter, there have already been radical shifts and adjustments including an unprecedented court ruling against the World Bank and its role in privatizing water, which has been considered a compromise for water supply in Tanzania. As the playing field changes due to this kind of move, new mediators between state and non-state actors are becoming more or less visible.[9] This kind of result now being observed in Tanzania is bound to influence the country's future water policy and may result in the emergence of more policy entrepreneurs promoting greater integration of customary norms and values.

Notes

1. For the Sukuma, residing in the northwestern part of Tanzania, as well as other chiefdoms in Tanzania, all water resources were free for domestic use by anyone. Water needed for cattle was allocated through the tribal leaders. This water is publicly owned and cannot be appropriated by individuals. Information about water management is not available for all chiefdoms, but it is likely that the same set of rules applied.
2. Interview with Sam Shemsanga, 7 December 2007, Dar es Salaam, Tanzania.
3. Interview with Ministry of Livestock Development official in Action Aid, 2004, p. 8.
4. Television programmes such as *Okoa Mazingira Yetu* (Save our environment) that raise public awareness.
5. These organizations have been influential particularly in issues relating to health, education and infrastructure development, but have not yet taken the lead in water policy. This is unsurprising, as water is not yet seen as a high political priority. Global discourses on climate change and other environmental concerns are reversing this rapidly and it is these discourses that are slowly but surely bringing water to the political forefront.
6. Interview with Miss Madaka Tumbo, Institute of Resource Assessment (IRA), University of Dar es Salaam, 14 December 2007.
7. See additional comment on the importance of trust in the water sector; Goldin (2003, 2005, 2008).
8. Interview with Miss Nasrah Yassir Hussein, Environmental Supervisor for MosUrban Water and Sewerage Authority (MUWSA), Dar es Salaam, June 2008.
9. In South Africa, the court ruling that the installation of pre-paid meters (an instrument of privatization and symbol of achievement of notions of water as an economic good) is illegal came about as a result of pressure groups, such as the Phiri Group in Soweto led by the Anti-Privatization Campaign, that strengthened social networks and successfully shifted the balance of power.

References

ActionAid (2004), 'Turning off the taps: donor conditionality and water privatization in Dar es Salaam, Tanzania', report, ActionAid International.

Boesen, J., F. Maganga and R. Odgaard (1999), 'Rules, norms and actual practices: land and water management in the Ruaha River basin, Tanzania', in T. Granfelt (ed.), *Managing the Globalized Environment, Local Strategies to Secure Livelihoods*, London: IT Publications, pp. 114–32.

Department of Water Affairs and Forestry (DWAF) (2002), 'Draft white paper on water services: water is life, sanitation is dignity', October, available at http:// www.dwaf.gov.za/docs/Oth.pdf.

Evans, P. (1996), 'Government action, social capital and development: reviewing the evidence on synergy', *World Development*, **24** (6), 1119–32.

Fjeldstad, O.H. (2003), 'What has trust got to do with it? Non-payment of service charges in local authorities in South Africa', Working Paper WP 2003, no. 12, Chr. Michelsen Institute, Bergen, Norway.

Goldin, J. (2003), 'Washing away the sins of the past', *International Journal of Public Administration*, **26** (6), 711–30.

Goldin, J. (2005), 'Prepacked trust in the water sector', in S. Askvik and N. Bak (eds), *Trust and Public Administration in South Africa*, Aldershot, UK and Burlington, VT, USA: Ashgate, pp. 137–52.

Goldin, J. (2008), 'It take two to tango: steps towards change in the water sector', in D. Hemson, K. Kulindwa, H. Lein, A. Mascarenhas (eds), *Poverty and Water: Explorations of the Reciprocal Relationship*, CROP International Studies in Poverty Research, London: Zed Books, pp. 47–64.

Grindle, M.S. and J.W. Thomas (1990), 'After the decision: implementing policy reforms in developing countries', *World Development*, **18** (8), 1163–81.

Habib, A. (2008), 'Speaking "Truth" to all forms of power: reflections on the role of the public sociologist in South Africa', *Current Sociology*, **56** (3), 389–99.

Hardin, R. (1999), 'Do we want to trust in government?', in M. Warren (ed.), *Democracy and Trust*, Cambridge: Cambridge University Press, pp. 22–41.

Ilcan, S. and L. Phillips (2008), 'Governing through global networks: knowledge mobilities and participatory development', *Current Sociology*, **56** (5), 711–34.

Kibona, E., A.V. Orindi and M.D. Moindi (2006), *Addressing Climate Change in the Development Process in Tanzania*, Nairobi, Kenya: African Centre for Technology Studies.

Kickert, W., E.H. Klijn and J. Koppenjan (1999), 'Introduction: a management perspective on policy networks', in W. Kickert, E.H. Klijn and J. Koppenjan (eds), *Managing Complex Networks: Strategies for the Public Sector*, London: Sage Publications, pp. 1–13.

Kickert, W. and J. Koppenjan (1999), 'Public management and network management: an overview', in W. Kickert, E.H. Klijn and J. Koppenjan (eds), *Managing Complex Networks: Strategies for the Public Sector*, London: Sage Publications, pp. 35–60.

Lankford, B. (2003), 'Irrigation-based livelihood trends in river basins: theory and policy implications for irrigation development', *Physics and Chemistry of the Earth*, **28**, 817–25.

Malzbender, D., J. Goldin, A. Turton and A. Earle (2005) 'Traditional water governance and South Africa's "National Water Act": tension or co-operation?' paper presented at the International Workshop on African Water Laws: Plural Legislative Frameworks for Rural Water Management in Africa, 26–28 January, Gauteng, South Africa.

May, J. (ed.) (2000), *Poverty and Inequality in South Africa: Meeting the Challenge*, Cape Town: David Philip Publishers.

Mayo, A. and D. Masahauri (1991), 'Rain water harvesting for domestic use in Tanzania: a case study', *Water International*, **16** (1), 2–8.

Mfangavo, C. (2005) 'Policy overview: a review of policy and practice in relation to water and HIV/AIDS in Tanzania', CDD Working Paper No. 2, University of Bradford.

Mutayoba, W.N. (2002), 'Management of water resources in Tanzania through basin management', paper presented at the 3rd Waernet/Warfsa Symposium, Water Demand Management for Sustainable Development, Dar es Salaam, 30–31 October.

Odgaard, R. and F. Maganga (1994), 'Local informal land and water management systems in the Ruaha River Basin', draft report, Copenhagen: Centre for Development Research.

Oomen, B. (2005), *Chiefs in South Africa: Law, Power and Culture in the Post-Apartheid Era*, New York: Palgrave Division of St Martin's Press.

Rainey, H.G. (1991), *Understanding and Managing Public Organization*, San Francisco, CA: Jossey-Bass Publishers.

Randeria, S. (2003), 'Glocalization of law: environmental justice, World Bank, NGOs and the cunning state in India', *Current Sociology*, **51** (3–4), 305–28.

Randeria, S. (2007), 'The state of globalization: legal plurality, overlapping sovereignties and ambiguous alliances between civil society and the cunning state in India', *Theory, Culture and Society*, **24** (1), 1–33.

Schaap, L. and M. van Twist (1999), 'The dynamics of closedness in networks', in W. Kickert, E.H. Klijn and J. Koppenjan (eds), *Managing Complex Networks: Strategies for the Public Sector*, London: Sage Publications, pp. 62–76.

Scheff, T.J. (1990), *Microsociology: Discourse, Emotion, and Social Structure*, Chicago, IL: University of Chicago Press.

Selznick, P. (1949), *TVA and the Grassroots: A Study of Politics and Organization*, Berkeley, CA: University of California Press.

Sokile, C., J. Kashaigili and M.J. Kadigi (2002), 'Towards an integrated water resources management in Tanzania: the role of appropriate institutional framework in Rufiji basin, RIPARWIN', in DANIDA/World Bank (eds), *Water Resource Management in the Great Ruaha Basin: A Demand-Driven Management of Land and Water Resources with Local Participation*, Dar es Salaam, Tanzania: Rufiji Basin Water Office, Ministry of Water, Energy and Minerals.

Thompson, L. (2002), 'Discourses everywhere and not a drop to drink: water as a lens on environmental security', in T. Benjaminsen and L. Thompson (eds), *Contested Resources: Challenges to the Governance of Natural Resources in Southern Africa*, University of Western Cape, Cape Town, South Africa: Programme for Land and Agrarian Studies (PLAAS), pp. 231–54.

URT (2002), *United Republic of Tanzania, Census*, Dar es Salaam, Tanzania.

Williams, G. (2004), 'Evaluating participatory development: tyranny, power and (re)politicization', *Third World Quarterly*, **25** (3), 557–78.

Yudelman, M. (1964), *Africans on the Land*, Cambridge, MA: Harvard University Press.

PART V

EUROPE

13 European Union water policy: to transition or not to transition? Coalitions as key

Lena Partzsch

13.1 Introduction

The European Union (EU) is a significant player in water politics. EU member and accession states are subject to its water policy formulation, in particular, the EU Water Framework Directive (WFD). In addition, as an international policy leader and a major global donor, the EU also influences water programmes outside its geographic boundaries. Announced at the Johannesburg Summit in 2002, the EU Water Initiative (EUWI) has become the key strategy guiding EU policy on tackling global water problems (Map 13.1).

In both EU domestic and external water policy, there have been important shifts in direction and approach that are referred to here as transitions. According to Sabatier (1993), such changes are only possible if new advocacy coalitions are able to pressure for change. I test this assumption by comparing success and failure in incorporating transitions in the WFD and EUWI. The WFD is particularly known for transitions to river basin management, public participation and water pricing and full cost recovery. These transitions are also sought by some participants of the EUWI. While adopted through the WFD, however, transitions in particular to water pricing and full cost recovery have so far not been accomplished and even face major opposition in external water policies. I aim to explain this difference by analysing and comparing the main policy entrepreneurs and strategies involved in the processes of the WFD and the EUWI. Thus this chapter contributes to the understanding of policy change and policy stasis.

My research shows that significant differences exist in domestic and external water politics in terms of policy entrepreneurs and institutional context. Policy entrepreneurs in the cases analysed emerged from a range of actors that I divide into three types: state actors, the private sector (agriculture, chemical industry, water companies and so on) and civil society (social movements and non-governmental organizations – NGOs) (Shaw et al., 2000; Hummel, 2002). All three actor categories accommodate experts possessing relevant knowledge and influencing policies through epistemic community networks (see Meijerink and Huitema, Chapter 2 in this volume).

In the water sector the relevance of non-state actors in particular has grown since the early 1990s. Private water companies, such as Veolia, now operate in more than 100 countries with annual turnovers in the double-digit billions of euros. These corporations have become powerful new players in water politics (Barlow and Clarke, 2002; Finger and Allouche, 2002). The non-profit civil society sector is often considered as a counterpart to the profit-driven private sector (Walk and Brunnengräber, 2000; Brand et al., 2001). However this is not always the case and, indeed, cooperation between the private sector and civil society does occur and leads to transitions as discussed below.

This chapter examines water pricing as a policy option in the WFD and EUWI. Water

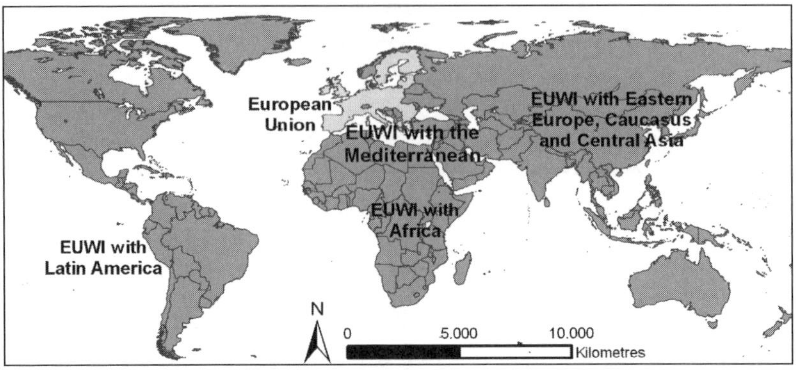

Map 13.1 European Union and the European Union Water Initiative (EUWI)

pricing involves the recovery from users of the costs of water services and the phasing out of subsidies. Water pricing has become a major aim of the EUWI (Dimas, 2005), but has yet to be accomplished as a policy transition. Under the WFD water pricing has become an officially adopted transition even though it formed a main point of controversy in the policy process leading up to the Directive (Kaika, 2003).

My findings stem from analysis of participatory observation and interviews conducted for previous research (Partzsch, 2007). To create a balanced sampling, proponents as well as opponents of water pricing approaches were interviewed. As all interviewees preferred to remain anonymous, only the actor category and date of interview are given as reference.

13.2 EU domestic water strategies

The WFD forms the heart of EU internal water policy (Kaika, 2003; Hödl, 2005). Since its adoption in October 2000 the Directive has caused a general revision of European legislation on water at the local, national and regional level. The WFD has unified the patchwork of different, partly contradictory Directives and separate regulations and has created a legal framework for the administration and protection of water resources in the EU.

The Directive takes a primarily ecological orientation (Mostert, 2003). The Preamble states: 'Water is not a commercial product like any other, but, rather, a heritage which must be protected, defended and treated as such' (Preamble, para. 1). The WFD directs that environmental pollution shall be prevented and an ecologically sustainable water policy established. The aim is to keep European waters in 'good status', improving water quality by 2015 where it falls below standard (Preamble, paras 26, 28, 33; art. 1, 2, 4, 6). For this purpose an innovative classification system has been introduced, factoring in biological, hydro-morphological and physical–chemical quality aspects (Annex V). EU member states are required to submit reports to the European Commission, which then monitors them for consistency.

As mentioned, the WFD is particularly known for transitions to river basin management, public participation and water pricing and full cost recovery (Hödl, 2005; Huitema and Bressers, 2006). River basin management means that measures are no longer applied to an area defined by administrative and political boundaries, but to those naturally defined by a river basin. The river basin unit is the basis for assessment of the status quo and the setting of specific objectives as well as the application of any instrument of governance (SRU, 2004). This entails the water management, which is achieved by river basin commissions or 'appropriate competent authorit[ies]' identified specifically for the purpose (art. 3.2, 3.4). As many river basins reach beyond national borders, governments often also establish transboundary authorities in order to cooperate with neighbouring states (European Commission, 2003; Dimas, 2005; Dworak and Kranz, 2005).

The transition to public participation that is also seen in the WFD (Mostert, 2003; van Ast and Boot, 2003) applied to the process of developing the Directive as well as to the legislation finally adopted. According to article 14: 'member states shall encourage the active involvement of all interested parties in the implementation of this Directive, in particular in the production, review and updating of the river basin management plans', In order to guarantee public information and consultation, the Directive even prescribes contact points and procedures for the public to obtain background documentation and information (Annex VII, A.11).

The discussion below analyses the stakeholder involvement in the WFD negotiation process regarding the transition towards water pricing and full cost recovery. The trend towards water pricing and full cost recovery constitutes the third major transition in EU water policies. The recovery from users of the costs of water services has officially come into effect along with the phasing out of indirect subsidies, which still exist in many EU member states.

More than ever before, the world faces serious challenges in terms of quantity and quality of fresh water. The hydrological cycle yields a limited quantity of water per time period, a quantity that is currently overexploited. In the EU, water scarcity affects 33 river basins covering a total area of 460 000 km^2 (about 10 per cent of the total EU area) and hosting a total population of 82 million (about 16.5 per cent of the total EU population) (European Commission, 2007, p. 8).

The causes of water overuse and pollution often lie beyond the reach of water politics due to the cross-cutting nature of water resources and their relationship to multiple sectors. In the EU, agriculture is the major user of water (64 per cent), followed by energy (20 per cent), public water supply (12 per cent) and industry (4 per cent) (European Commission, 2007, p. 11). Ecological costs of water abstraction and pollution are mostly externalized, that is, not included in prices of traded products and services. Following the 'polluter pays' principle, water pricing is a strategy to internalize environmental costs and provide incentives for reduced water use and effluence. Policy has generally neglected these costs (Finger and Allouche, 2002; UNEP, 2006). The water pricing and cost recovery approach recognizes the cross-sectoral nature of water. Investments in more efficient water technologies and less water-intense alternatives in any sector become more attractive with (higher) water pricing. In this way pollution prevention pays, and business can profit by protecting the environment.

The WFD includes the 'recovery of costs for water services' in article 9 (and 11, Annex VI part B), which states that: 'member states shall take account of the principle of recovery of the costs of water services . . . in accordance in particular with the polluter pays principle' (art. 9.1.). Water pricing policies are considered to provide adequate incentives for users to use water resources efficiently and thereby contribute to the environmental objectives of the Directive (art. 9). Since adoption of the WFD, the EU member states have been obliged to introduce full cost recovery through water pricing for all sectors, including agriculture. Again, assertion and reality have diverged and implementation is still proceeding.

13.3 EU external water strategies

The WFD applies only to the territory of EU member (and accession) states. However some believe it could serve as a worldwide model for integrated water resource management (IWRM), that is, sustainable water use based on the needs of all relevant sectors (agriculture, industry, shipping traffic, water power and so on) (Dworak and Kranz, 2005; UNESCO, 2006). For the management of transboundary river basins in particular, the EU needs to cooperate with countries outside the EU in order fully to undertake a river basin management approach (Berbalk, 2004; Dimas, 2005).

Outside the EU the WFD is promoted as a successful policy example that informs programmes such as the Joint Framework Directive Process (under the EUWI) (European Commission, 2003; Dimas, 2005). The EU claims that: '[The] European experience was essential in addressing an emerging global water crisis that . . . was threatening lives, sustainable development, peace and security' (Europe Aid, 2007). Assistance portfolios include efforts that target infrastructure and technology improvements and that initiate political reform. Collectively the EU and its member states spend close to €1.5 billion per annum on foreign water projects (Europe Aid, 2007).

More than 1 billion people worldwide lack access to potable water and twice as many do not have basic sanitation (UNDP, 2006). The UN Millennium Development Goals (MDGs) and the Johannesburg Plan of Implementation aim to halve those numbers by 2015. Furthermore the Johannesburg Plan requires the elaboration of national plans for improved water efficiency and IWRM. At the 2002 Johannesburg Summit the EU announced the EUWI as a common initiative to contribute to the MDGs and IWRM. The initiative is expected to continue until 2015, the deadline for both the MDGs and the Johannesburg water targets. Four geographical regions are targeted: Africa; the Mediterranean; Latin America; and Eastern Europe, Caucasus and Central Asia (see Map 13.1) (European Commission, 2002).

As a political initiative the development of the EUWI has not followed a clearly defined institutional procedure. Unlike the legislative procedure that led to the adoption of the WFD, the EUWI did not originate with documents of a constitutive order. The WFD was adopted as legislation, which required the formal EU process, including the agreement of the European Parliament. In contrast, the EUWI is a non-legislative strategy with less formal requirements. Basically it provides a platform for single projects of partners at the regional, national and local levels. The EUWI does not even have a set spending budget, an element that usually forms the basis of any development cooperation programme. The EUWI thus represents an open process that brings together state and non-state actors in- and outside the EU from the sectors of development, environment, external relations and research. The annual plenary meeting, the so-called Multi-Stakeholder Forum, is facilitated by the European Commission and settled in the context of the Stockholm World Water Week.[1]

The transitions to river basin administration, public participation and water pricing and full cost recovery are on the EUWI agenda, but with different motivations than under the WFD. Examples of EUWI river basin management planning developed for some of the partner countries and transboundary river basins include the EUWI country dialogue with Kazakhstan that incorporated diverse stakeholders in the development of strategies for sustainable water management (Staudenmann, 2005; UNDP Kazakhstan, 2006), and the EUWI with Africa, which focused on IWRM in transboundary river basins. Five basins were selected, and management plans negotiated for each: the Orange–Senqu basin in Southern Africa, the Lake Chad basin in Central Africa, the Lake Victoria–Kagera River basin in East Africa, and the Niger and Volta basins in West Africa (EUWI, 2004).

Stakeholders from the private sector and civil society have been incorporated at all stages in the EUWI process including, for instance, the development of IWRM plans. The EUWI's public participation approach goes further than the efforts of the WFD: it not only entails 'information' and 'consultation', but also formal, equal partnership with non-state actors, that is, 'active involvement' (Huitema and Bressers, 2006). Civil society and private sector stakeholders have been among the initial EUWI partners (European Commission, 2002). In the EUWI, partners have aimed mainly to improve coordination and cooperation, for example through technological innovation and diffusion, in order to create synergy especially in the field of infrastructure for water supply and sanitation with IWRM as a lesser (and, in practice, often neglected) priority. The efficiency of existing EU aid flows should be improved by mobilizing all relevant actors (European Commission, 2002; Dimas, 2005).

While the WFD was adopted as EU environmental legislation, the EUWI hails to a greater extent from a development policy context. These different institutional contexts affect the constellation of policy entrepreneurs and their strategies regarding water pricing and full cost recovery. While water pricing is an issue of EU external as well as internal water policy, the underlying motives of the EUWI partners and particularly of the European Commission have been different. Water pricing in external policies is not primarily driven by environmental concerns. The EUWI seeks to generate new financial sources for constructing water infrastructure in order to further its aim of increasing the efficiency of existing EU aid flows (Dimas, 2005).

In order to achieve the MDGs, 275 000 new water connections need to be built and financed every day (Rechkemmer and Schmidt, 2006). Pricing water supply and sanitation offers an opportunity to generate new capital and recover investments. In addition the EUWI has tried to mobilize private investments instead of increasing the amount of official development aid (European Commission, 2002; Dimas, 2005). Following the failure to find private funding, the European Commission set up the EU Water Facility as a new funding instrument in 2004. The Facility funds €500 million in total for projects in African, Caribbean and Pacific (ACP) countries (European Commission, 2004). Projects that apply for funding must bring in additional funding from other, possibly private, sources. The expectation is that private investment would be recovered by pricing water services. Establishing water pricing has hence continued to be a major aim of the EUWI despite opposition to it by key stakeholders. Such key stakeholders from the field of water therefore refused to participate in the EUWI.

The following section outlines the perspectives of institutions and actors underlying the different motivations for water pricing in the political processes of the WFD and EUWI. My analysis shows that the main agents pushing the transition to water pricing

Table 13.1 Comparing transitions in EU domestic and external water policies

	EU Water Framework Directive	EU Water Initiative
Legal status	Binding domestic legislation	External political initiative
Water pricing	Adoption of full cost recovery for any sector, incl. households	Aiming for water pricing for households, in particular
Main change agents	Environmental NGOs and water industry	Donors and water industry
Main blockade agents	Farmers, consumer organizations	Majority of civil society

are different in EU domestic and external strategies (Table 13.1). Only policy entrepreneurs in domestic policy have been successful so far.

13.4 Water Framework Directive: the adoption of a transition towards water pricing

Significant differences exist between the institutional contexts of the WFD and the EUWI (see Table 13.1). The WFD is clearly settled within the hierarchical and legislative structure of the EU. Stakeholder participation might contribute additional resources (for example information for the river basin management plans), and thereby influence political output. However even in light of the WFD's call for 'active involvement', state actors have the final say; they are accountable to the public for decisions made under the auspices of the WFD. The most influential non-state stakeholder groups in the WFD negotiation process were environmental NGOs, the private water industry, agriculture, consumer organizations and the chemical industry (Kaika, 2003).

Negotiations for the WFD officially began with a European Commission first draft communication in 1996, and ended when the Directive was finally adopted in 2000 by the European Parliament and the European Council according to formally prescribed procedures. The environmental departments within each institution were responsible for preparation and negotiations. Immediately after the first draft communication, the European Commission launched an open call for participation and invited specific groups and organizations to become actively involved (Kaika, 2003). Different working groups were formed in which typically one or two representatives of each stakeholder group came together.[2] Stakeholders voiced their opinions and consulted the Commission. The Commission was fully open and carefully considered input and suggestions from not only officially invited actors, but also from stakeholders who participated on their own initiative. Expertise from all actor categories was welcome (Kaika, 2003).

Civil society representatives did not feel particularly disadvantaged compared to private sector representatives in this regard. A civil society representative explained: 'One could indeed influence the process if one had good arguments. This is what was tried by different sides. One or the other point in the final document can be traced back to what the groups brought in.'[3] The statement underpins the relevance of expert knowledge ('good ideas') (see Huitema and Meijerink, Chapter 1, this volume), which in the case of the WFD was provided by all three types of actor categories, including civil society. However the Commission was not obliged to take on board suggestions and views, and non-state actors could not share in the decision-making. Stakeholders were primarily incorporated in the WFD negotiation process in order to prevent potential opposition at a later stage (Kaika, 2003).

The European Commission tried to balance the interests of different lobbying groups. The documents it produced, however, were not binding since it falls to the European Parliament and the Council to negotiate and co-decide on EU environmental legislation. The European Parliament is generally more detached from, and therefore more resilient to, national networks of influence, whereas the Council represents EU member states and national lobbying groups. Hence the European Parliament supported far stricter requirements for environmental protection and implementation timelines (Kaika, 2003). Members of the European Parliament (MEPs) filed amendments written by the diverse lobbying groups, including environmental NGOs.[4] MEP amendments became the second access point (or 'venue' per Meijerink and Huitema; see Chapter 2, the theoretical introduction to this volume), after the Commission's working groups, for lobbying groups to the policy formulation process.

Because the Council acted on behalf of the different EU member states and national lobbying groups, member states' positions were not necessarily formed only at the nation state level, but also in Brussels. Thus the Council provided a third potential venue for those lobbying groups with good contacts to national government representatives. Atypical alliances resulted: Ireland, which had introduced a new charging system with zero domestic charges a short time before the WFD negotiations started, joined with the European South against full cost recovery. Portugal, however, allied with the European North on the issue of strict river basin management, partly because the imminent Spanish national hydrological plan will affect Portuguese water flows (Kaika, 2003; for a discussion of the Spanish water plan see also Font and Subirats, Chapter 15 in this volume).

Which lobbying groups could best use their influence potential? Was the transition to water pricing made possible because new advocacy coalitions were able to pressure for change? I pose these questions in my analysis.

Environmental NGOs were among the most active Brussels lobbying groups. These organizations had a common position in support of full cost recovery as a means to provide incentives for more efficient water use. The European Environmental Bureau (EEB), World Wide Fund for Nature (WWF) and Greenpeace were the NGOs most actively involved. Their activities and involvement reflected their variation in financial and staff resources. The WWF, for instance, sent one of its own expert representatives to almost every working group set up by the European Commission, while the EEB, an umbrella organization of 143 European environmental NGOs, had trouble recruiting qualified experts from the national member organizations, such as the British Royal Society for the Protection of Birds or the German Grüne Liga (EEB and Breyer, 2001; Kaika, 2003).[5] The EEB nevertheless managed to establish a strong lobbying presence that joined with others to significant effect.

Taking a pro-pricing position, the environmental NGOs challenged the notion that the civil society sector acts only as a counterpart to the private sector (Walk and Brunnengräber, 2000; Brand et al., 2001). Predictably the water industry also supported water pricing and full cost recovery since it allows increases in the price of water in the name of 'environmental protection'; the prospect of better raw water quality through the implementation of the Directive would also reduce treatment costs, producing a double benefit for water supply companies (Kaika, 2003). In consequence, water industry representatives built a coalition with environmental NGOs.[6] The water industry, like the WWF and EEB, had strong lobbying representation in Brussels, this in the form of EUREAU (European Federation of National Associations of Water and Waste Water Services).[7] These policy entrepreneurs, all in support of full cost recovery, thus formed a particularly strong coalition in the working groups and campaigns that accompanied the negotiation process.

The coalition, however, also faced opposition and blockade. Water pricing has not proved beneficial to every sector represented by stakeholder groups. The transition provoked opposition in particular from the agricultural sector and consumer organizations (Kaika, 2003). The chemical industry took a stand in between, not opposing water pricing in general, but arguing that environmental costs, including the cost for cessation of discharges of priority hazardous substances, should be met by EU citizens and not by the chemical industry (Kaika, 2003).

The WFD negotiations hence became a tough process, often resulting in only vague formulations about possible compromise. This means, in fact, that some decisions were passed on to the implementation phase at the national level of member states. At the EU level, however, the coalition of policy entrepreneurs in support of water pricing succeeded: the text finally adopted includes the 'recovery of costs for water services' (art. 9). Costs of water services are supposed to be recovered by 'an adequate contribution of the different water uses, disaggregated into at least industry, households and agriculture' (art. 9.1). The final formulation thus leaves no doubt that households or EU citizens are not the only polluters that pay, representing a successful result in domestic water policy for environmental NGOs and the water industry against blockade agents from the agricultural sector, consumer organizations and the chemical industry. Below I describe how policy entrepreneurs in support of water pricing have not prevailed, though, when it comes to water policy external to the EU.

13.5 EU Water Initiative: failure to adopt a transition towards water pricing

Unlike the WFD, the EUWI is a political, non-legislative initiative that has included non-state actors in decision-making from its very beginning in 2002 (even the first draft communication was edited by state and non-state actors). The participatory approach of the EUWI is much more extensive than the WFD policy process. The most active stakeholders involved in the EUWI have been representatives of the water industry, political consultancies and NGOs primarily dealing with water issues, including WaterAid, Green Cross and Tearfund (European Commission, 2002). In the absence of a fixed legislative procedure, the EUWI began with a design phase in which all state and non-state participants had to agree on an overall structure.[8] Institutions and procedures were adopted at the EUWI annual meeting, the so-called Multi-Stakeholder Forum (MSF), in Stockholm in 2004 (in fact, two years after the EUWI was announced in Johannesburg).[9] The European Commission developed a key role in organizing and maintaining the EUWI process. A multidisciplinary EUWI team of different Directorate-Generals (DGs) was established, including DGs for the Environment, Development, External Relations, Research and Europe Aid (EUWI, 2008).[10] In this way, a broad range of policy departments manages the EUWI, while development of the WFD occurred only under the auspices of the environmental departments. As DG Development had previously been the European Commission department in charge of water management in developing countries, it has retained a key role. The difference in institutional context, then, is marked as the EUWI emerged from a development policy perspective, and the WFD proceeded as an environmental legislation initiative.

The different institutional contexts in turn lead to different configurations of actors in the EUWI and the WFD. In addition to the water industry, the main policy entrepreneurs seeking a transition to water pricing in the EUWI are state actors, namely the European Commission, EU member states and governments from the partner regions (EUWI, 2005). While the WFD actors are focused on environmental issues, the EUWI actors concentrate on water development matters. NGOs tend to support the expansion of infrastructure for the poor rather than for mitigation of environmental problems such as saving water and preventing pollution.[11] The WWF is the only NGO actively involved in both processes (although to a lesser degree in the EUWI) (EUWI, 2005).

On behalf of the private sector, representatives only of the water supply industry engage in the EUWI. This means that some powerful actors (although less influential on water pricing in the WFD process) are not participating in the EUWI, the agricultural sector in particular, which made the strongest stand against full cost recovery in the WFD process. It would be easy to assume, therefore, that the coalition of NGOs and water industry representatives supporting a water pricing transition would be even stronger in the EUWI than in the domestic context, but the reality is different.

The non-legislative (and broader) institutional context and contrasting thematic approach of the EUWI mobilized and antagonized a wider range of civil society actors. In addition to environmental NGOs, like the WWF, civil society actors from other policy fields, in particular development aid, have been involved in the EUWI. Such actors include WaterAid and Tearfund. Some movements and groups, including the Anti-Privatisation Forum (APF), mobilized against the initiative because they consider themselves a counterpart to the profit-oriented private water sector. This kind of opposition came out of the anti-globalization movement and targeted global water politics because water privatization has failed in many countries. Cases of failure underpinned the arguments of groups like the APF accompanying the EUWI negotiations.[12]

Even some of the environmental NGOs that strongly supported full cost recovery in the WFD take a different standpoint when it comes to developing countries and 'pricing poor households and farmers'. Friends of the Earth (FOE), for instance, a member of the European Environmental Bureau (EEB), has taken a strong stand against water commodification and privatization (Friends of the Earth, 2001).[13] In the WFD process, FOE pushed for full cost recovery as a means to internalize environmental costs in prices of traded products and to provide incentives for precautionary resource management, that is, for more efficient water use and less pollution. In the domestic context, FOE had no concerns about coalition-building with representatives of the water industry.[14] In the international context, to the contrary, FOE finds fault with the environmental and social records of multinational water companies, and therefore turns against water commodification: '[M]ultinational water companies . . . which have a duty to maximize profits . . . do not necessarily have long term sustainability, water conservation, environmental protection and fair access to water in mind . . . Water is not a commodity to be traded . . . but a resource essential to all life' (Friends of the Earth, 2001, p. 10).

Many civil society groups particularly criticize the practice of public–private co-financing, accusing the EU Water Facility of channelling development aid into foreign investment subsidies for European water corporations (Deckwirth, 2004).[15] Those actors who object in this way refuse any form of cooperation with the private sector in the context of EU external affairs, even rejecting the EUWI as a whole for the reason of private sector participation.[16] Moreover the European Parliament questions whether the European Commission's water policy efforts actually benefit the poor and has made inquiries as to whether only French and German water corporations receive the benefits of approved projects (European Parliament, 2003). Indeed only European water companies are involved in the EUWI (European Commission, 2002). In partner countries, the dominance of European actors discredits the initiative and spurs general opposition against water commodification (EUWI, 2005).

The few NGOs actively participating in the EUWI have been primarily interested in contributions to the MDGs, specifically in the form of expanding water infrastructure for poor households (WaterAid and Tearfund, 2005). They accuse other social movements of using water as a 'Trojan horse'[17] for the voice of fundamental anti-privatization interests instead of lobbying for expanded infrastructure. When the WFD was negotiated, the water industry situation was actually the same. Private water companies were among the most influential lobbying groups, and water pricing and full cost recovery was of most benefit to them. Cost recovery justified price increases and water supply companies benefited from better raw water quality required due to stricter regulation.

The water industry's attempts to build coalitions with civil society groups are more difficult to realize in the context of external water policies because the majority of civil society has taken a firm stand against water pricing and privatization. Therefore, in the EUWI, the industry's coalition-building was based on contributions to the MDGs and an increase in the number of households connected to the water system. As one representative of a private water company explained: '[W]e have a more natural contact with NGOs that are active on the ground – and also with some global NGOs like WaterAid in the UK that are specializing in water.' Unlike the WFD, it was not possible for industry to cooperate with the broad range of NGOs. Thus cooperation and EUWI participation has been limited to specific NGOs 'working on the ground' who do not question the aims of treating water as a commodity and privatizing water services. Outside the EUWI, however, civil society opposition to water pricing and privatization has continued. The EUWI has failed to actively involve these opposition groups and interests, which indeed represent the majority of civil society organizations working on international water issues.

Objectives of IWRM and environmental protection have been progressively hindered by EUWI partners, for instance, by the dissolution of additional EUWI working groups on IWRM (instead IWRM needs to be mainstreamed by other working groups) (EUWI, 2005). The stakeholder selection represented by the EUWI partners (water industry and NGOs focused on expanding infrastructure on the ground) helped shift the emphasis to infrastructure for household water supplies, which account for only 10 per cent of worldwide water use (whereas industry represents 20 per cent and agriculture 70 per cent) (UNESCO, 2006). Pricing household water supply therefore has only a marginal impact on the overall demand for, and pollution of, water resources. While in the WFD process, many civil society groups were mobilized in support of water pricing for environmental reasons although they do not support the EUWI's water pricing strategy. Thus, while full cost recovery could be implemented as binding domestic legislation by the WFD, pricing strategies in EU external water policies face major opposition. The powerful coalition of water industry and NGO participants that succeeded in the WFD process does not exist in the context of external water policies. Here the majority of civil society actors have turned against the water industry and a transition to water pricing and full cost recovery.

13.6 Conclusion: explaining the adoption of or failure to adopt transitions

Differences exist between EU domestic and external water policies that can help to explain how a transition to water pricing comes to be adopted or not. The contrast in institutional contexts plays a major role: the WFD is environmental legislation formulated from a primarily ecological perspective. In the WFD process, state actors' responsibilities were tied to the environmental departments of EU member states, and stakeholders were invited to participate in a way designed to focus on environmental needs. Levels of private stakeholder participation were very high and included not only water suppliers, but also polluting sectors. In contrast, the EUWI includes diverse EU policy departments in a way that prioritizes development issues, particularly expanding water infrastructure for the poor. Consequently the water supply industry has become the sole representative of the private sector in the EUWI. The different processes of origin and management of the two initiatives have led the EUWI strategy to focus on (re) financing infrastructural investments ('charging users'), while environmental concerns gave rise to the strategy of water pricing in the WFD process ('charging polluters').

Water pricing for environmental reasons found support from the environmental NGOs involved in the WFD negotiations. As cost recovery was viewed as a weapon against polluters, the NGOs adopted the strategy of coalition-building with water industry representatives in order to fight pollution and protect ecosystems. I found that their relative strength of expert resources was influential in the working groups set up by the European Commission. The very strong coalition that resulted succeeded and became a policy entrepreneur for the transition to water pricing.

The question I raised was whether it is correct to assume that transitions are possible only if a new advocacy coalition is able to pressure for the change. The case of successful adoption of a transition in the WFD shows the formation of a new coalition in favour of the transition. The coalition was new and atypical as the non-profit civil society sector did not act as a counterpart to the profit-driven private sector. Thus the case of the WFD proves Sabatier's approach. The role of coalitions in the EUWI provides a counter-test.

Yet a transition to full cost recovery has failed to occur despite the intention in EU external policy for water pricing to refinance the expansion of infrastructure for the poor – expansion that would contribute to achieving the MDGs. The strategy has provoked resistance from major players in civil society. Although some NGOs are cooperating with the water industry in the EUWI, the NGO community has not found common ground on full cost recovery in developing countries. This, combined with doubts about the beneficiaries of mechanisms of co-finance applied by the EU Water Facility, has ruled out a coalition in the EUWI process of water industry representatives and most NGOs. The lack of a powerful new advocacy coalition has led to the failure of the initiative to begin a transition to water pricing and full cost recovery as seen in domestic policy.

Transitions are only possible if new advocacy coalitions are able to pressure for change. Based on the cases studied, I can amplify this finding. I can conclude, first, that the institutional context is a significant factor in whether or not a policy transition will occur. Second, non-state actors play a key role in policy change and policy stasis. Third, a policy transition is more likely to be adopted if the relevant civil society groups find common ground and form an alliance with a strong actor from the private sector.

Notes

1. Interviews with state actors on 29 November 2004 and 21 April 2005.
2. Interview with civil society representative on 8 November 2004.
3. Translated interview from 8 November 2004.
4. Interview with civil society representative on 8 November 2004.
5. Interview with civil society representative on 8 November 2004.
6. Interview with civil society representative on 8 November 2004.
7. Interview with private sector representative on 25 November 2004.
8. Interview with state actor on 29 November 2004.
9. Interview with state actor on 29 November 2004.
10. Interview with state actor on 2 December 2004.
11. Interview with civil society actor 13 April 2004.
12. Interviews with civil society representatives on 13 November 2004 and 19 January 2005.
13. Interviews with civil society representatives on 8 and 13 November 2004 and 19 January 2005.
14. Interview with civil society representative on 8 November 2004.
15. Interview with civil society representative on 19 January 2005.
16. Interviews with civil society representatives on 13 November 2004 and 19 January 2005.
17. Interview with civil society representative on 13 April 2005.

References

Barlow, M. and T. Clarke (2002), *Blue Gold: The Fight to Stop the Corporate Theft of the World's Water*, New York: New Press.

Berbalk, D. (2004), 'Integrated water resources management: how to use the benefits of the EU experience in implementing the WFD for the EUWI? What does "integrated" mean?' EUWI Session 1, World Water Week, Stockholm, 16 August.

Brand, U., A. Demirovic, C. Görg and J. Hirsch (eds) (2001), *Nichtregierungsorganisationen in der Transformation des Staates*, Münster: Westfälisches Dampfboot.

Deckwirth, C. (2004), *Sprudelnde Gewinne? Transnationale Konzerne im Wassersektor und die Rolle des GATS*, Bonn: Weltwirtschaft, Ökologie und Entwicklung.

Dimas, S. (2005),'EU Water for Life Initiative, meeting the MDGs and the JPOI goals on water and sanitation', The European Commission Side Event, European Union Delegation of the European Commission to the United Nations, New York, 21 April.

Dworak, T. and N. Kranz (2005), 'Die EU-Wasserrahmenrichtlinie als Ansatz für ein integriertes Flussgebietsmanagement', in S. Neubert, W. Scheumann, A. van Edig and W. Huppert (eds), *Integriertes Wasserressourcen-Management (IWRM). Ein Konzept in die Praxis überführen*, Baden-Baden: Nomos, pp. 45–60.

EEB (European Environmental Bureau) and H. Breyer (eds) (2001), *Handbuch zur EU-Wasserpolitik im Zeichen der Wasser-Rahmenrichtlinie*, Brussels: European Environmental Bureau.

Europe Aid (2007), 'Europe aid Annual Report 2007 on the European Community's development policy and the implementation of external assistance in 2006', http://ec.europa.eu/europeaid/multimedia/publications/documents/annual-reports/europeaid_annual_report_2007_en.pdf, accessed 15 May 2008.

European Commission (2002), 'EU Water Initiative: water for life health, livelihoods, economic development, peace and security, working document', Draft 21 August 2002, www.johannesburgsummit.org/html/sustainable_dev/p2_managing_resources/2508_eu_water_initiative.pdf, accessed 15 May 2008.

European Commission (2003), 'Water for life. EU Water Initiative. International cooperation – from knowledge to action', http://ec.europa.eu/research/water-initiative/pdf/water-for-life_en.pdf, accessed 15 May 2008.

European Commission (2004), 'Communication on the future development of the EU Water Initiative and the modalities for the establishment of a Water Facility for ACP countries', COM (2004) 43 final, http://eur-lex.europa.eu/LexUriServ/LexUriServ.do?uri=COM:2004:0043:FIN:EN:PDF, accessed 15 May 2008.

European Commission (2007), 'Addressing the challenge of water scarcity and droughts in the European Union Impact Assessment', Commission Staff Working Document, COM (2007) 414 final, http://eur-lex.europa.eu/LexUriServ/LexUriServ.do?uri=COM:2007:0414:FIN:EN:PDF, accessed 15 May 2008.

European Parliament (2003), 'Report on the Commission communication on water management in developing countries and priorities for EU development cooperation', A5-0273/200322, July 2003, www.europarl.europa.eu/registre/seance_pleniere/textes_deposes/rapports/2003/0273/P5_A(2003)0273_EN.doc, accessed 15 May 2008.

EUWI (European Union Water Initiative) (2004), 'Terms of Reference for the Working Group on the Integrated Water Resources Management Component of the EU Water Initiative', approved by the working group on 17 December, Brussels.

EUWI (European Union Water Initiative) (2005), 'Participant observation at multi-stakeholder forum', 22–25 August', Stockholm.

EUWI (European Union Water Initiative) (2008), 'Secretariat of the EU Water Initiative', www.euwi.net/index.php?main=1&sub=1&id=134, accessed 26 May 2008.

Finger, M. and J. Allouche (2002), *Water Privatisation. Transnational Corporations and the Re-Regulation of the Water Industry*, London, UK and New York, USA: Spon Press.

Friends of the Earth (2001), 'Dirty water: the environment and social records of four multi-national water companies', www.foe.co.uk/resource/briefings/dirty_water.pdf, accessed on 15 May 2008.

Hödl, E. (2005), *Wasserrahmenrichtlinie und Wasserrecht*, Juristische Monografien, Band 32, Vienna: Neuer Wissenschaftlicher Verlag.

Huitema, D. and J.T.A. Bressers (2006), 'Scaling water governance: the case of the implementation of the European Water Framework Directive in the Netherlands', paper presented to the Synthesis Conference of the Institutional Dimensions of Global Environmental Change program, Bali, Indonesia, December, Enschede, CSTM-reeks nr 304, pp. 1–22.

Hummel, H. (2002), 'Global pluralism? Merging IR and comparative politics traditions in developing a theoretical framework for analyzing actors in global governance', paper prepared for 43rd ISA Annual Convention, 21–27 March, New Orleans.

Kaika, M. (2003), 'The Water Framework Directive: a new directive for a changing social, political and economic European framework', *European Planning Studies*, **11** (3), 299–316.

Mostert, E. (2003), 'The European Water Framework Directive and water management research', *Physics and Chemistry of the Earth*, **28** (12–13), 523–7.

Partzsch, L. (2007), *Global Governance in Partnerschaft. Die EU-Initiative 'Water for Life'*, Baden-Baden: Nomos.

Rechkemmer, A. and F. Schmidt (2006), *Neue globale Umweltpolitik. Die Bedeutung der UN-Reform für eine nachhaltige Wasser- und Waldpolitik*, Berlin: Erich Schmidt.

Sabatier, P.A. (1993), 'Policy change over a decade or more', in P.A. Sabatier and H.C. Jenkins-Smith (eds), *Policy Change and Learning: An Advocacy Coalition Approach*, Boulder, CO, pp. 13–39.

Shaw, M., S.J. MacLean and M. Nzomo (2000), 'Going beyond states and markets to civil societies', in T.C. Lawton, J.N. Rosenau and A.C. Verdun (eds), *Strange Power: Shaping the Parameters of International Relations and International Political Economy*, Aldershot: Ashgate, pp. 391–406.

SRU (Rates von Sachverständigen für Umweltfragen) (2004), *Umweltgutachten 2004*, Baden-Baden: Nomos.

Staudenmann, J. (2005), 'Linking IWRM and WSS. UNDP's Kazakhstan experience World Water Week', 24 August, EUWI Special Session, Stockholm.

UN Development Programme (UNDP) (2006),'Beyond scarcity: power, poverty and the global water crisis', Human Development Report, http://hdr.undp.org/en/media/hdr06-complete.pdf, accessed 15 May 2008.

UNDP Kazakhstan (2006), 'Integrated water resource management', www.voda.kz, accessed 15 March 2006.

UN Educational, Scientific and Cultural Organization (UNESCO) (2006), *Water a Shared Responsibility. The UN World Water Development Report 2*, World Water Assessment Programme, New York, USA and Paris, France: Berghahn Books.

UN Environment Programme (UNEP) (2006), *Challenges to International Waters: Regional Assessments in a Global Perspective*, Nairobi: UNEP.

van Ast, J.A. and S.P. Boot (2003), 'Participation in European water policy', *Physics and Chemistry of the Earth*, **28** (12–13), 555–62.

Walk, H. and A. Brunnengräber (2000), *Die Globalisierungswächter. NGOs und ihre transnationalen Netze im Konfliktfeld Klima*, Münster: Westfälisches Dampfboot.

WaterAid and Tearfund (2005), 'An empty glass: the EU Water Initiative's contribution to the water and sanitation millennium targets', www.devinit.org/PDF%20downloads/An%20empty%20glass.pdf, accessed 15 May 2008.

14 The introduction of floodplain rehabilitation and rural development into the water policy for the Tisza River in Hungary

Saskia E. Werners, Zsuzsanna Flachner and Piotr Matczak

> The whole world had already been created when the Tisza was standing alone before the Lord's throne. Then Jesus took a golden plough, harnessed a donkey to it and told the Tisza to follow. Thus he set the plough against the soil and ploughed the bed for the river, which followed faithfully everywhere. However thistles were scattered all around. The donkey that was feeling hungry, reached after one and then another, leaving the straight path. This is why the Tisza is so unpredictable, so winding and meandering. (Hungarian folk tale)

14.1 Introduction

This chapter analyses the early twenty-first-century transition in the water policy for the Tisza River in northeastern Hungary. In the spring of 2003, the Hungarian government issued a decree that marked a substantial shift in water management. The new water policy for the Tisza River recognized rural development and nature conservation as important objectives side-by-side with flood protection. Floodplain rehabilitation and land use change were introduced as water management measures to replace or complement the prevailing engineering approaches that primarily favoured flood levee construction. From an external perspective, this was surprising given a 150-year history of water management mainly through heavy river regulation, embankments and drainage, where water policy had mostly served the interests of large-scale agriculture.

We attempt to explain what happened in the period leading up to the breakthrough year of 2003 and in the subsequent years, when actors had to realize the direction taken in the new water policy. The development of the new water policy (called the New Vásárhelyi Plan) between 1998 and 2006 is taken as a main object of investigation. The changes in water policy for the Tisza have been described before, for example, from an institutional point of view (Matczak et al., 2008; Werners et al., 2009). This chapter takes a new angle, specifically assessing the role of individual actors and the strategies that they used – deliberately or not – in bringing about policy change.

We build on the research carried out in a series of Hungarian and international research projects including the ADAM project[1] and the NeWater project.[2] Data were collected in three ways: through 17 interviews with stakeholders representing national and regional organizations (ministries, water authorities, planners, academic institutions, non-governmental organizations (NGOs), the water board, municipalities and farmers); from a series of workshops with local and national stakeholders; and by analysis of policy and project planning documents and background studies.

We begin with an account of what changes in the water policy we call a transition and then describe the transition from the perspective of the five strategies introduced by Meijerink and Huitema (Chapter 2, theoretical introduction to this volume). A discussion follows of the importance of each strategy and a review of the individuals behind it. Our analysis shows the importance in particular of recognition of a new policy at a conceptual level by the responsible civil servants and of advocacy of the policy concept by a credible regional coalition. The international attention and domestic political focus following the 2000 cyanide disaster on the Tisza River, the 2001 floods and the 2002 national elections provided a window of opportunity for adoption of the new water policy. Ambiguity about the practical application of new policy concepts and the responsibility of different actors facilitated consensus on the new water policy, but has hampered its implementation in recent years.

14.2 Background to the Tisza River and the policy transition

A short historical overview of water management in the Tisza River basin helps provide the context for and reasons why the changes introduced into Tisza water policy can be called a transition.

The Tisza River is the largest tributary of the Danube, receiving water from the Carpathian Mountains in Romania, Slovakia and Ukraine. The Tisza River basin holds almost 50 per cent of Hungarian territory. Until the eighteenth century, river management was mainly organized around the operation of a system of small streams and channels regulating the water flow between the main river bed and the floodplain (Balogh, 2002). The inundation frequency determined land use. Mosaic floodplain production systems combined ploughed land, forest, floodplain orchards, meadows, fish and cattle. The floodplain provided a secure income for the communities along the river (Andrásfalvy, 1973). Since the 1750s, the Tisza River has been heavily modified. To cater for large-scale mono-agriculture and river transport, the river was canalized and straightened, and the floodplains drained. The major changes to the river were introduced by the first Vásárhelyi Plan, which was initiated in the nineteenth century. Dyke-building, regulation of one-third of the river's length, and floodplain drainage decreased the total naturally flooded area by 84 per cent (see Map 14.1)

These changes put an end to the traditional water management system and the related production systems (Bellon, 2004). The reoccurrence and high visibility of floods caused resources to be funnelled into an extensive flood defence system (Vári, 2001). Deforestation and river regulation made the river flow more extreme and, together with population growth in the low-lying reclaimed floodplain, added to flood risks (Csekö and Hayde, 2004). Besides its connection with flooding, water management has been associated with problems such as drought, water stagnation, soil salinization, and the degradation of peatlands and wetlands (Barta et al., 2000; Vámosi, 2002). The communist era after the Second World War supported the large-scale intensive agriculture in the region. Privatization at the beginning of the 1990s led to a drop in the operation and maintenance of the large irrigation systems and of agricultural output. At present, the region is socio-economically challenged by a high unemployment rate, ageing and migration (Sendzimir et al., 2004; Linnerooth-Bayer et al., 2006). Large areas have an unclear property status and unresolved responsibility for water system maintenance and taxation (Matczak et al., 2008). At the same time, the region has great potential for recreation and nature conservation (Vári et al., 2003).

The annual floods that returned in 1998 after 20 years of drought were a driving force behind the development of a new water policy for the Tisza River in Hungary known

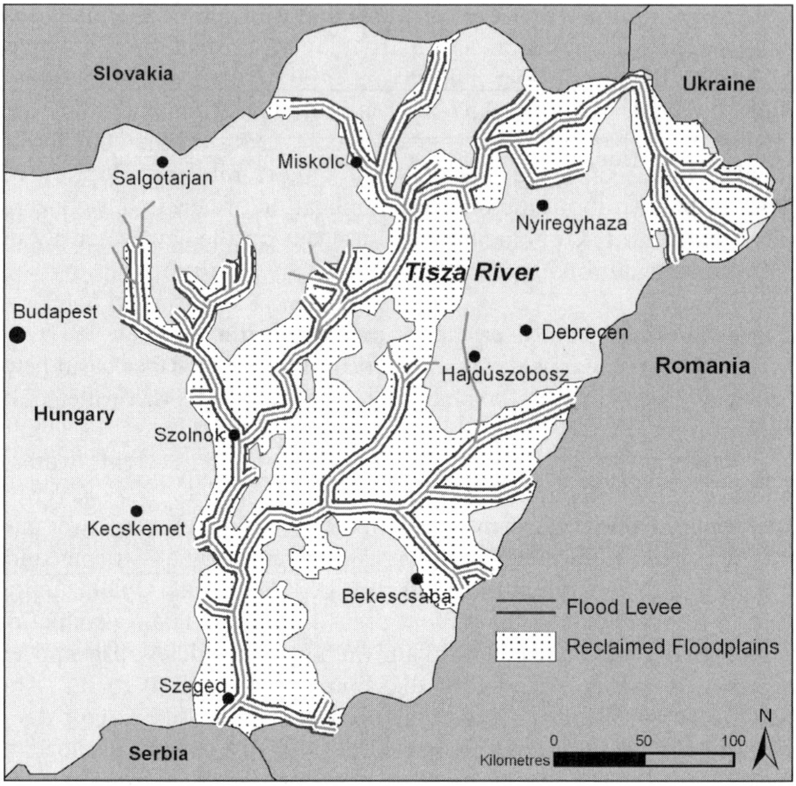

Source: Based on Szlávik and Ijjas (2003).

Map 14.1 Reduced river length in the Tisza

as the New Vásárhelyi Plan (VTT). Actors identify three stages in the development and implementation of the VTT. These are illustrated in Table 14.1.

Why do we call the development and implementation of the new water policy a transition? Meijerink and Huitema (Chapter 2 in this volume) define that a transition should become visible in the substance of policy or in the governance paradigm of water management. In the development of water policy for the Tisza, we observe a significant change in the substance of the policy and in the procedure applied in its design. The water policy endorsed in 2003 explicitly recognizes rural development and nature conservation as important objectives alongside flood protection. Floodplain rehabilitation and land use change were introduced to replace or complement flood levees that had been the preferred solution in water management for 150 years. The hegemony of the water authority was broken and the implementation plan for the policy was prepared through an intense collaboration of a large number of regional and national actors, facilitated by the national planning agency, VÁTI. Major decisions were taken by an interministerial committee, where regional and civic organizations were invited. Actors involved talk about a paradigm shift or change in philosophy. The government brochure for the new

Table 14.1 Stages in the development and implementation of the new Vásárhelyi water policy

Description	Background
1 1999–2001: announcement and first design of policy by water authority and Ministry of Transport, Communication and Water	Mainly work of water authorities. The first plan consisted mostly of classical flood protection measures (dykes, drainages, mono-purpose retention reservoirs). A few national NGOs, like WWF Hungary, have been invited to comment on the plan. Local NGOs protest against the plan, especially the new NGO Bokartisz. Decision to alter the plan.
2 2002–03: alteration of the policy in collaboration of ministries,* NGOs, research community and other stakeholders, facilitated by national planning bureau VÁTI	Government Decree (1022/2003 of February 2003) recognizes principles of nature conservation and rural development next to flood protection. It orders development of an implementation plan for the first six retention reservoirs, with floodplain rehabilitation where local stakeholders agree, including a rural development programme. The total costs of implementation until 2007 are estimated at €520 million (endorsed by Government Decree, October 2003).
3 2004 – present: implementation of the new policy	Approval of project sites is delayed. Implementation focuses on building emergency retention reservoirs. The associated rural development lags behind.

Note: * In 2002, the Ministry of Transport, Communication and Water was reorganized. Water affairs were transferred to the new Ministry of Environment and Water. This ministry led the interministerial committee for the water policy with the Ministry of Agriculture and Rural Development, the Regional Development Agency and other ministries (social, labour).

water policy captures this well and introduces typical examples of the rural development and nature conservation component of the policy:

> The government has adopted on the 15 of October, 2003 a decision on the most ambitious rural development program of past decades . . . The program reflects a new government philosophy, in that it takes as far as possible into consideration the interests of environmental protection and nature conservation . . . The complex project, the basic aim of which is to raise the living standards of the people in the Tisza Region, while ensuring a higher level of flood safety, would be accompanied by a number of important infrastructural developments. These include land drainage and sewerage, afforestation, construction of cycling lanes, as well as environment management schemes, like creation of a mosaic landscape pattern by water routing, streamlet rehabilitation and conveying the water down the full length of the flood bed. (VITUKI, 2004)

This new philosophy is also expressed in the logo for the water policy in Figure 14.1. In the words of a representative of the Ministry of Agriculture:[3] 'the design of the VTT

Figure 14.1 Logo and motto for new Vásárhelyi water policy in government brochure

meant a significant paradigm shift as the program was looking for alternative solutions instead of the traditional approach of increasing the height of the dams'.

14.3 Individuals and the transition

14.3.1 Introduction

This section describes the role of key individuals in the development and implementation of the new Vásárhelyi water policy (VTT) from the perspective of the strategies introduced in the theoretical introduction to this volume.

14.3.2 Develop new ideas

Transitions are related to the emergence of new ideas, which challenge an existing paradigm, ideas that are frequently developed deliberately as a policy change strategy. In this section we investigate the origin of the main new idea of floodplain rehabilitation, rural development and nature conservation as introduced into Tisza water policy in 2003.

After a decade of individual experiments and theoretical studies by individuals and separate groups, the NGO Bokartisz was founded in 2001. The founding group consisted of the municipal councils of 12 villages in the Bodrogköz (the area between the rivers Tisza and Bodrog) and three non-profit organizations. Bokartisz's first project was to coordinate a pilot study in the Bodrogköz area. Based on this experience, the coalition developed its concept of integrated floodplain management and floodplain rehabilitation for the whole Tisza area, the Bodrogköz in particular (Figures 14.2, 14.3). The floodplain rehabilitation aims to recreate the mosaic landscape structure that allows for regular flooding and sustainable rural development. The organization began advocating its concept under the name 'Last Straw' in 2001 (Botos et al., 2002). Bokartisz's goal is to realize sustainable economic development through floodplain rehabilitation. It encourages residents to become active participants in the shaping of their region's society, economy and ecology (www.bokartisz.hu).

We began by asking different parties about the origin of the new idea and whether one person or group of people could be identified as a clear source. VTT (Vásárhelyi-Terv Továsbbfejlesztése – New Vásárhelyi Plan) members from the Bokartisz NGO coalition

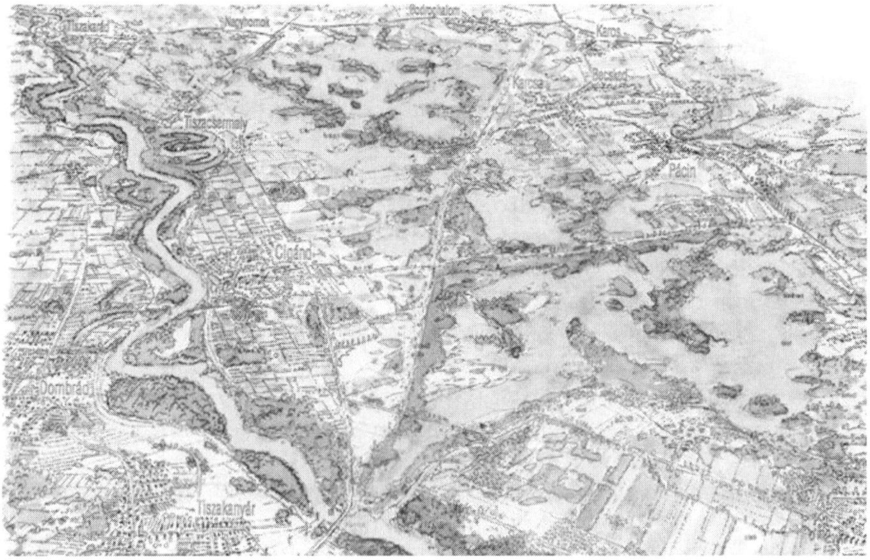

Figure 14.2 *Vision of Bodrogköz floodplain prepared for water policy with local and national partners (new elements – rural development and floodplain rehabilitation)*

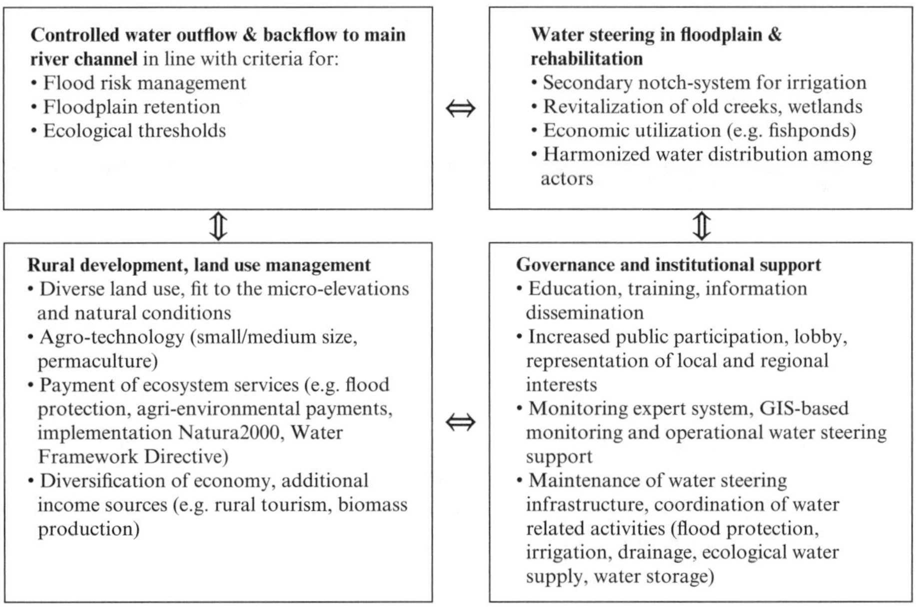

Source: After Botos et al. (2002).

Figure 14.3 *Integrated floodplain management*

point to their own organization, and specifically its leader, Géza Molnár, as the key individual behind the idea of floodplain rehabilitation in the Tisza region. They argue that their work and the publication of 'Last Straw' – inspired by a number of experiments in the field – initiated the discussion in the region and brought the idea to the national level (Botos et al., 2002). It is important to stress that Bokartisz does not present floodplain rehabilitation as something new, but rather refers back to floodplain utilization before river regulation. Its concepts derive from the shallow flooding, mosaic land use and community management of small-scale water infrastructure that was common in the Tisza region in the period 1500–1700. Due to the utilization of small sluices (*fok* in Hungarian) and creeks (*notch*), they call the traditional water management the '*fok-notch* system'. Since the 1980s, Géza Molnár has studied the *fok-notch* system and the socio-economic system in the Tisza valley from old documents (Andrásfalvy, 1973; Bellon, 1991; Molnár, 2002). Together with a small group of farmers and landowners he restored and experimented with traditional water steering systems on a small scale at various locations along the Tisza.

We distinguish between the concept of floodplain rehabilitation as advocated by Bokartisz (Botos et al., 2002) and the more detailed implementation plans for particular locations along the Tisza and in particular for the Bodrogköz area. Working both on the more abstract concept and the practical implementation at particular locations, Bokartisz discussed the feasibility of the concept with local mayors, farmers, national park authorities and NGOs, and demonstrated its application for the national government. In the words of a representative of the Hungarian Environmental Economy Centre and Bokartisz:[4] 'after we founded Bokartisz, we began to elaborate a land use and flood protection plan for the Bodrogköz. By the end of 2002/start of 2003, the concept was put together. The engine of course was Géza Molnár.'

What is significant here is that the new idea came from a bottom-up process based on practical local experience. It was negotiated with local municipalities in cooperation with a small interdisciplinary group of researchers. These actors combined elements of the traditional *fok-notch* water steering system with floodplain rehabilitation and rural development. Thus the new idea was developed in response to local problems independent of the national policy process and the problems recognized by the national administration. Bokartisz's idea was out there to be picked up and carried by the national policy stream.

Bokartisz's knowledge of the local situation and of floodplain rehabilitation is impressive. It is critical, however, to ask how widely this explanation of the origin of the new idea and its development is supported. We found that mayors who are members of Bokartisz and farmers who collaborate with Bokartisz share the opinion that Bokartisz originated the new ideas. The planning agency, VÁTI, also attributes the role of originator to Bokartisz. Some stakeholders, however, hold an alternative notion that the changes that were introduced may have resulted in a less direct fashion from the debates among different actors looking at the development opportunities for the region. The water board director for the larger Bodrogköz area recalls:[5]

> I have no clear memory when new ideas were introduced. It was more a vision and an evolutionary process. Many people were asking: 'If there is an option to keep the water, what extra benefits could this provide?' It was quite obvious not to use the reservoir only once every 30 years, when there is a high flood . . . this area has low agricultural value. Nature restoration was

> mentioned as a potential income source. For example, in 2002 the environmentally sensitive zones were launched and gave a push to this thinking. Later, the agro-environmental payments were introduced. And we are still waiting for other financial opportunities that will be attached to the reservoir-polder scheme.

The position of the administrative agencies brings another perspective on the origin of new ideas incorporated in the water policy transition. The then ministerial head of department responsible for the new water policy does not recognize floodplain revitalization and a rural development programme as necessary new requests from the counties in the region. Instead, he argues that the first version of the new water policy already allowed for new elements to be included in the implementation.[6] As an accession country to the European Union (EU), the national government knew it would soon have to comply with European Directives like the Water Framework Directive (Commission of the European Communities, 2000a) and was keen to make use of European funds. Integrated water management and rural development were more likely to comply with requirements attached to European support than classical flood protection. Representatives from various ministries participated in European consultations and study tours where the prospects of regional support were presented. The head of the Hungarian office of the World Wide Fund for Nature (WWF) was a member of the body charged with preparation of the new water policy in 2000, bringing experience from the WWF Living Rivers project (Zöckler et al., 2001). Notwithstanding these international influences, integral and participatory planning was relatively new in Hungarian water management. Many people at the water and planning authority had been trained as civil engineers and believed strongly in hard river regulation. Under the socialist regime, the water authority had always been a strong hierarchical organization with significant financial resources (Csekö and Hayde, 2004).

In the international context, the degree to which the new Hungarian approach of reintroducing floodplain rehabilitation into water policy was new is debatable. The concept of river rehabilitation did not originate from Hungary. Hungarian actors, then, may have taken inspiration from other countries. For example, in 2000, Hungarian scientists participated in the Conference on River Restoration in Europe that concluded: 'River restoration is internationally popular. Many river restoration projects are being implemented. River restoration will even get a higher attention within the framework of the implementation of the European Water Framework Directive' (Fokkens, 2001). Many countries were considering non-structural or 'soft' measures from a sustainability perspective or within a system approach (Kundzewicz, 2002; Meijerink, 2005).

Are there signs of investigation of river restoration and rehabilitation in the Hungarian scientific community? In the 1970s, the traditional water management system was studied by a small number of scientists in the region. Publications focused on the operation of the traditional system without making a link to present-day water management (Andrásfalvy, 1973). In the 1980s and 1990s, scientists in Budapest engaged in similar studies (Lászlóffy, 1982). Karácsonyi (2001) is one of the first papers in English. Building on the Association for Local and Regional Development's study (1997) the paper highlights benefits of reintroducing the traditional system and floodplain rehabilitation. Karácsonyi illustrates the importance of the traditional system in light of a disastrous pollution spill, but does not make reference to a new water policy. More recently, national and international researchers (Vári, 2001; Linnerooth-Bayer and Vári, 2003; Sendzimir et al., 2004) started to publish on the traditional system and flood risks, a combined treatment that also became an important element of international research and development projects.

In sum, two main sources of the new ideas in water policy in the Tisza are identified. Those in the Bokartisz coalition insist that their group developed the new idea concept outside involvement with the government. Bokartisz members – and especially its leader – combined elements from historical analysis, theory and field experiments. Thus the origin of the new idea can be seen as mostly local. On the other hand, people within the local and national administration point out that some elements were already present in the first version of the water policy, and that the new idea evolved from the development of the plan. What emerges from these separate points of view is the possibility that acceptance of the idea by 'both sides' as their own, or at least not as something completely new, may have been instrumental in its acceptance in the water policy in 2003.

14.3.3 Build coalitions and sell ideas

This section explores the second key strategy of coalition-building in order to sell a new idea. The theoretical introduction to this volume (Chapter 2) defines coalitions as groups of actors with shared beliefs and explicit agreements on how to use resources to achieve common goals. The theoretical chapter proposes the hypothesis that new ideas for a transition must be carried by individuals or groups of individuals that prepare them until they are ready for consideration.

Various regional and national NGOs emerged in Hungary in the 1980s that focused on rural development and nature conservation. In the Tisza region, a series of independent, locally driven initiatives began, each with objectives related to improving the economic and ecological situation at a small site (Government of Hungary and UNDP-GEF, 2004). In general, these initiatives involve local government, local representatives, NGOs, private companies and/or local and national experts. Capitalizing on local experience and cooperation, Bokartisz was established in 2001 and produced a set of concepts that were discussed and endorsed by all its members. Beyond concept development, coalition members contributed in different ways. The participating municipalities gave moral support and support in kind, for example by offering office facilities. Scientific and technical support came from national research institutes. Together with Bokartisz members, these undertook detailed social and ecological research on how Bokartisz's concepts can lead to multiple benefits.

We propose that the linking of traditional local water management with contemporary notions of nature conservation and rural development prepared the ground for the new coalition to form. For the purpose of this chapter, we focus on Bokartisz as an example of a central coalition with a major role in a water policy transition.

We start by analysing the relationship between Bokartisz and the development of the new water policy (VTT). In the autumn of 2001, Bokartisz hosted a meeting in the region where the responsible ministry also presented its first version of the water policy. According to Bokartisz:[7] 'We had our concept ready and wanted to present it at this meeting'. This summarizes Bokartisz's belief that the new idea was developed first, independently from the new water policy. Only when the government made its initial presentation, Bokartisz considered that it went against the new concept they had developed.

They started their opposition, beginning in 2002 to formulate their critique against the VTT. According to the Hungarian Environmental Economy Centre and Bokartisz:[8]

> It was a great shock to see the VTT in that form with the 13 reservoirs . . . it was against our concept as Géza analyzed it. The good thing for our concept was that the plan could not be done between two dykes, and water had to be channeled into reservoirs. The bad thing was that the core of the plan was against the holistic floodplain concept because its only aim was to decrease the flood level between the dykes. There was not much attention to what happens in and between the reservoirs.

Bokartisz challenged the prevailing water management as the cause of many of the problems in the Tisza region. This opinion was supported by scientists who showed that the existing flood defence system had reached its limits, and engineering measures (eliminating the bottlenecks, constructing higher and stronger dykes) alone could not accommodate higher flood levels (Timár and Rácz, 2002). As the Hungarian Environmental Economy Centre and Bokartisz state:[9]

> In the beginning of 2003, we issued a statement that the VTT in this form cannot be accepted against rational land use. As NGOs do, we issued it to the press and raised our voice at all venues. As a result, the Ministry [Ministry of Transport, Communication and Water, represented by the responsible Head of Department and his Deputy] said: 'Let the civic organizations and NGOs tell what they would like.' . . . A meeting was organized in Budapest where Géza [Molnár] and Tamas Cseloszki presented the concept for the Bodrogköz and the whole Tisza . . . This point of view was channeled into the VTT, it seemed. For me, my impression is that the high level leaders responsible for water management understood that we did not only like to shout, but that we had a concept and could work it out if given the opportunity.

Shortly after, in February 2003, the Hungarian government endorsed the decree that included rehabilitation of the primary floodplain and rural development as objectives for the new water policy along with flood protection. This was still a way off from the main innovation that Bokartisz was calling for: to channel water into the landscape and to connect water bodies. Yet, at the same time, Bokartisz was given the opportunity to 'sell' its concept as part of the new water policy as it was one of the partners to cooperate in developing the detailed implementation.

If it is assumed that Bokartisz managed to convince other parties to consider its new concept as suggested above, the question arises as to when its members felt their ideas were taken seriously and the degree to which they were deliberately strategic in their efforts towards this result. In the words again of the Hungarian Environmental Economy Centre and Bokartisz:[10]

> The moment was maybe the first negotiations in spring 2003 and then when it became part of the planning in summer 2003. The proof was when we saw our work as an attachment to the government decree. We said then: 'It's real' . . . It came as a surprise. As an expert and NGO you get accustomed that you write papers, you send them in, you give your own statement, and nothing happens. Sometimes when it happens, when you open the revised plan and you find your own proposal there, it is a big surprise.

The above events and accounts suggest the Bokartisz coalition was important in uniting the voice of various organizations in the region that had not been heard before.

Bokartisz developed its concept in time for it to be considered in the new water policy, although it had not planned from the start for this to happen. Rather it entered the opposition as soon as it considered the new water policy was at odds with the new concept. Representing the larger Bokartisz coalition, a small group of two to three people communicated with the central administration. They presented decision-makers at the Ministry of Environment and Water with their critique together with their new ideas both in general and in detail for some key locations.

Is this enough to explain the transition? And why did it occur at that particular moment in time? Who allowed Bokartisz to become a partner at the negotiation table? We look into these questions in more detail in the next section by reviewing the transition and key actors from the perspective of 'windows of opportunity'.

14.3.4 Create and use windows of opportunity

The hypothesis from the theoretical introduction to this volume is that a window of opportunity is needed to launch new ideas in the public sphere. Before we ask who created, recognized or used windows of opportunity, we describe the three streams that must converge for such windows to be exploited succesfully: the problem stream (issue on public agenda); the political stream (issue on political agenda); and the policy stream (attention to official policy options related to the issue) (Kingdon, 1995).

Looking at the problem stream, events began in 1998 with the first major flood in the Tisza for 20 years. The annual floods that followed each produced a new record water level in at least one section of the river (Timár and Rácz, 2002). In 2000, the eyes of the world were on the Tisza when first, in January, a cyanide spill at a gold mine in nearby Romania wiped out aquatic life, leading to tonnes of dead fish being pulled from the Tisza. Then in March, floodwaters rose to a 100-year high. Summer followed with a record-breaking heatwave. More disaster was seen in 2001 when the embankment broke at two places and the Bereg region was flooded, seriously damaging 2000 houses. In Figure 14.4, we illustrate the problem stream by looking at the Tisza news coverage. We approximate news coverage by the number of occurrences per month of the words 'Tisza' together with 'water' in Google News between 1998 and 2007. News coverage shows clear peaks around the floods in November 1998, March 1999, and 2001. The Tisza River got most attention in international news around the flood and cyanide spill in 2000 and the floods in 2001 and 2006.

Figure 14.4 also illustrates the political stream by tracking the attention paid to the Tisza on websites of the coalition government parties. The figure shows the number of hits per year for Google searches of the parties' websites. The 2001 flood happened during the election campaigns and attracted a lot of political attention. Then 2006 shows another peak in interest in the Tisza from the political parties as assessed by website coverage of the topic. This attention coincided with the 2006 elections and another major flooding event. A second illustration of the political stream in Figure 14.4 is the level of attention paid by the Hungarian parliament to the Tisza, approximated by the number of occurrences of the word 'Tisza' per year on the parliament website (www.mkogy.hu). It shows how the Tisza gained importance on the parliamentarian agenda in 2000, peaking in 2004 when parliament approved the 'Tisza Law'. The figure also shows that attention has waned, and that the Tisza has never since attracted the same level of parliamentary attention.[11]

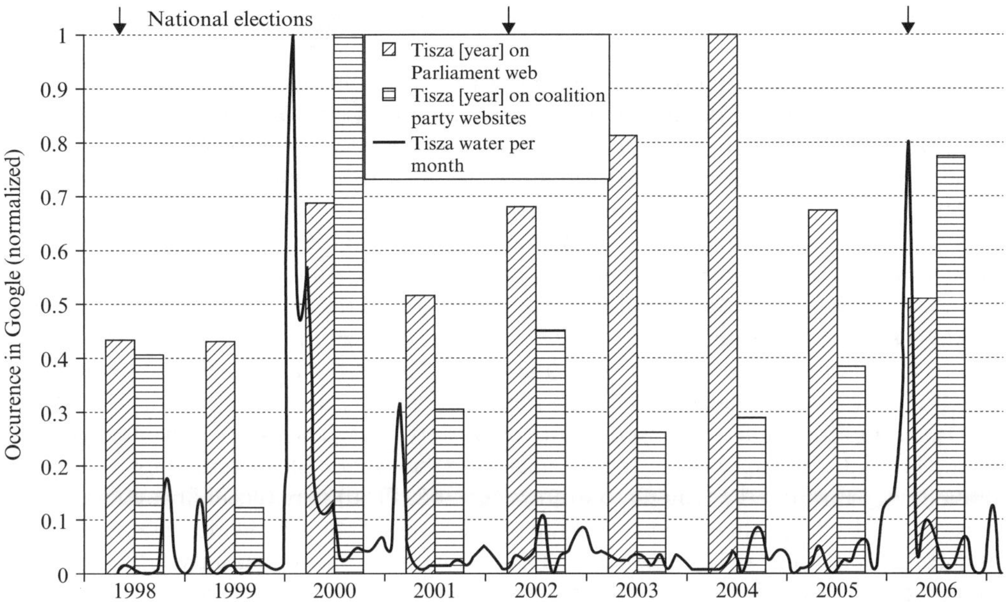

Figure 14.4 *Windows of opportunity illustrated by the normalized number of occurrences of specified terms arising from online search using Google*

The combined accounts of the problem, political and policy streams suggest that the sense of urgency at the political level to develop a new water policy for the Tisza was sparked by the 1998, 2000 and 2001 flood events. In June 2002, a new coalition of socialists and free democrats replaced the conservative, centre-right government. The coalition members wanted to put their mark on the ongoing policy process that they had inherited. From this perspective, the autumn of 2002 provided an excellent opportunity for the introduction of new ideas. As discussed in previous sections, it was by this time that the Bokartisz coalition had developed its new ideas and started actively to oppose the existing version of the VTT plan. Support for the existing policy options was waning. Municipalities and the Hortobágy National Park authority had rejected the location of a reservoir in their territory. The need for local support was becoming increasingly clear within the federal administration. At the same time, the government had to reduce its budget deficit as a condition for EU membership, making it imperative to find new sources of finance (Vári et al., 2003). European legislation, Directives and funding mechanisms, like the new agro-environmental schemes, favoured integrated policies and management approaches. In addition, the EU promoted participation and organized civil society as valuable support for a democratic system of government (Commission of the European Communities, 2000b, 2001).

Did the Bokartisz coalition use this window of opportunity by actually timing its interventions and local experiments to be ready for cooperation at the end of 2002? The accounts above of members of the Bokartisz coalition show that essentially they developed their ideas independently from the political stream. They first entered the political arena to present their ideas and then to oppose the existing plans when it became apparent that the plans conflicted with what the coalition wanted to achieve in the Bodrogköz region.

In contrast, the ministerial head of department responsible at the time for the water policy expressed a conscious and purposeful consideration of high-profile problems, national politics and policy options. He recalls using the sense of urgency after recurring floods, firstly to secure financial support for a new flood protection programme and, secondly, to turn it into a national programme.[12]

A review from the three-stream perspective shows a window of opportunity opening in 2000–2001. The cyanide spill and Bereg flood occurred as parties were campaigning for the 2002 elections. Whereas the water authority supported a hard engineering approach, Bokartisz was advocating its new ideas. The engineering approach was criticized by the NGOs and individual scientists for adding to the flood risk and causing unwanted side-effects in the region (Szlávik, 2001; Rakonczai, 2002; Szlávik and Ijjas, 2003; Bellon, 2004). In addition, after the elections it became associated with the preceding government. The window stood open. Individuals were then needed to take the opportunity to move policy through the window onto new ground. Producing this change required the recognition by individuals central to the administration, aware of the increasing local and European support for new ideas. For the window of opportunity to open in the first place, the high visibility of the problem stream and the negative attention for existing policy options may have been crucial. After the 2002 national elections, the water authorities were merged with the environmental authorities and their staff and budget reduced. This created a need to build new cooperation to ensure successful use of the window of opportunity. An interministerial committee was installed to develop the VTT plan further, both changing the role of the administration and adding a new venue in the transition process. We investigate the role of the new venues in the next section.

14.3.5 Play multiple venues

The previous sections have looked at the origin of the new ideas, the coalitions that were built to sell them, and the circumstances that favoured the introduction of the new ideas into the water policy at a particular moment in time. Here, we ask whether individuals or groups of individuals sought out alternative venues to promote the new ideas or undermine the existing paradigm. We focus in particular on the venues and actors noted in previous sections: the national government – particularly the ministry responsible for water policy; the local government, including the mayors and water boards; and the Bokartisz coalition.

The national government faced two main challenges to the implementation of any water policy: finances and local support. By 2000, the EU had opened up as a key new venue, providing new funding opportunities and policy requirements (Veres, 2004). Politicians and civil servants from the accession countries were actively briefed on European funding possibilities, such as structural funds, cohesion funds and agricultural subsidies. They were invited to participate in multinational projects and conferences in the EU. Although Europe was not a venue where civil servants deliberately lobbied for the new water policy, the restrictions and opportunities in the European regulations and financing conditions were recognized. Whereas national funding previously had supported engineering solutions, the European financing conditions offered opportunities for the new idea of river rehabilitation. Government decree 1022/2003 on the improvement of the Vásárhelyi water policy specifies that: 'financial resources should be determined with respect to the European Union's common joint financing conditions'. Text under '14.3.2 Develop new ideas' shows that the water authority already regarded nature restoration and agro-environmental schemes as potential new income sources.

The second main challenge for the national government was to get local support. Already two possible locations for a retention area had been rejected by local farmers, mayors and the national park authorities. Both Bokartisz and local mayors outside the Bokartisz coalition were aware of this situation. They offered the Bodrogköz area to the preparatory body for the new VTT water policy as an alternative location for a retention polder. In the words of Bokartisz:[13] 'Another key element in the success was that we could find a territory that we could offer. There was a territory always affected by inland water stagnation without high nature value: Cigánd.' Bokartisz reports that it was aware of this opportunity before the more detailed planning process started in 2003. It is less clear when and how the local mayors began to support Cigánd as a location for a VTT retention area. In general, they are keen to highlight the secondary benefits they ensured for their municipality by participating in the planning process. These range from national funding for a municipal sewerage system to city status for Cigánd.[14] This suggests that an analysis based on stakeholder interest could be a valuable addition to the idea-centred approach taken in this chapter.

To explore the idea of floodplain rehabilitation for the Tisza and for the particular location close to Cigánd in the Bodrogköz, at the instigation of the Ministry of Environment and Water, the national government created two major new venues. First, it established the interministerial committee, which represented a larger group of ministries than in the previous planning stage. The committee collaborated intensively with the planning agencies, research institutions, national park authorities, and local government and civic groups. The second venue consisted of five open public procurement tenders to support the drafting of the new water policy. The tenders were coordinated by VÁTI as the national planning agency for regional development, and VIZITERV (Vízügyi Tervező Vállalat – Designing Enterprise for Water Resources Management), the water resources design bureau. Established through government decree 1022/2003, the tenders allowed for increased stakeholder involvement and representation of local interests in the planning process. (We come back to the cooperation of regional representatives and researchers and the role of VÁTI in the next section.)

The discussion above touched on the role of the research community, which deserves notice as another new venue and group of actors in the transition. The transition had three significant research contributions. First, Bokartisz members – especially Géza Molnár – report that they were inspired partly by scientists to conduct the 1980s experiments in river revitalization. Studies of the traditional management of water and land use were of particular importance (Andrásfalvy, 1973; Bellon, 1991). The scientific community served a second time as a venue for the establishment in policy of floodplain rehabilitation during the broad cooperation that occurred between 2002 and 2003. The NGO members in the Bokartisz coalition sought cooperation for the development of the new ideas within the research community, especially at the national Water Resources Research Centre (VITUKI), which was responsible for studies related to water resources management and the collection of hydrological data. VITUKI is a fully state-owned organization, supervised by the ministry responsible for water management. Its accreditation gave water management schemes crucial national support and legitimacy. Bokartisz[15] explains: 'that VITUKI in 2002 was willing to test and support the idea by modeling to see whether the concept works gave a big push because it provided a more scientific foundation to the proposed water steering and shallow flooding'.

Individuals from the research community introduced a third venue in the form of international research and development projects. Following the cyanide spill in 2000, an increasing number of international research projects offered opportunities to experiment with floodplain management and river rehabilitation. In addition, these projects provided an international stage to discuss, test and promote these concepts. Three in particular were important for Tisza water policy: VITUKI won the coordination of the Tisza River Project on sustainable use of water resources and ecological values in the Tisza river basin, under the 5th Framework Programme of the European Commission in 2001 (www.tiszariver.com); a substantial United Nations Development Programme Global Environment Facility (UNDP-GEF) project for protecting biodiversity in the Tisza Basin to be co-financed by national and local government (www.elotisza.hu), a project that individuals from the research community helped secure, involving Bokartisz and other local organizations as partners; and INTERREG III project funding sought by members of the research community together with Bokartisz. We cannot say to what extent the emergence of international cooperation through research projects won new supporters for the concept of floodplain rehabilitation. We do know, however, that the national government was very keen to find new sources of funding and the projects above underlined that the new idea of floodplain rehabilitation attracted more international support than engineering-dominated solutions. Bokartisz and other Hungarian NGOs recognized this opportunity and used it to attract funding to develop their ideas further and present them on an international stage.

The Bokartisz coalition used additional venues to express its ideas or undermine the existing paradigm. As a coalition with members from the municipalities, NGOs and the scientific community, Bokartisz had in-house experience in playing different venues. As mentioned, Bokartisz issued protests and statements in the press, encouraged by the environmental NGO members in the coalition. The rural development NGOs and municipalities brought experience with training and information dissemination. Farmers in particular were approached to participate in pilot projects or public hearings and training sessions. Just like the national government, Bokartisz looked for new financial mechanisms, for example, helping farmers to apply for agro-environmental schemes and small pilot projects.

In sum, a number of new venues were used to sell and develop the new ideas. Most importantly, the EU offered new financial instruments and called for more participatory policy-making. Together with the national government's focus on European co-financing, this influenced the prioritization of policy options. The NGO members in the Bokartisz coalition used a range of venues to discredit existing policy options and to lobby locals to support Bokartisz's ideas. By entering the domestic and international research arenas, the coalition built up national credibility and drew international support for the new ideas. The transition in policy was supported by the national government's creation of the interministerial committee and a participatory planning process (supported by a series of open public procurement tenders) as two new venues.

How did the various actors cooperate in the venues associated with the policy transition? We take up this topic in the next section as a question of the emergence and orchestration of networks.

14.3.6 Orchestrate and manage networks

Turning to the last of the five strategies presented in the theoretical introduction (Chapter 2 in this volume), we ask what networks played a role in the transition and whether (groups of) individuals actively influenced the operation of networks. In particular, we analyse whether the breakthrough in the development of water policy for the Tisza was influenced by individuals breaking up or providing alternative policy communities.

Such analysis requires determination of what networks shaped water management before and during the water policy transition. In the 150 years before the transition, hard engineering solutions dominated water management. The water authority was extremely powerful, especially under the communist regime. Engineers were partly trained in the Soviet Union (Csekö and Hayde, 2004). A strong, informal network of water authorities and contractors developed from this period, with many individuals still in key positions today. Similarly, the environmental NGOs in Hungary form a strong network, where many individuals in key positions know each other from the communist environmental youth movement. In 2000, the environmentalists were becoming increasingly organized. They stressed that continuing levee construction and the resulting narrowing of the riverbed increased flood risks in Hungary. In a series of interviews and a public survey carried out in 2000–2001, the majority of the interviewees considered the strengthening and raising of the existing levees as inevitable. Opinions differed, however, on whether to increase the height of the whole levee system, a measure supported by most of the water authority experts, or to pursue alternative solutions like partial rehabilitation of the floodplains and removal of levees to create natural reservoirs, ideas supported by most of the mayors and NGOs (Vári et al., 2003).

Our analysis points to a number of important networks. The national level features the network of the water authority, the engineers and contractors. This can be viewed as a strategic alliance because it entails cooperation over several decades, but without the shared organization or resources fundamental to an advocacy coalition, such as Bokartisz.

At the regional level, the activities of the Bokartisz coalition built on the existing networks of key players in the region. The long-running collaboration (or animosity) between the mayors and other strong regional actors, however, extends well beyond Bokartisz in terms of both time and number of actors. At its heart are people who held central positions in state enterprises (for example, the agricultural cooperatives) and in the administration under the communist regime. In the turbulent times after 1989, well-informed landowners could profit from land consolidation. The consolidation in particular of land around the Cigánd reservoir serves to illustrate the importance of regional players. It has been suggested that some mayors and bigger landowners speculated on the location of the reservoir and acquired land shortly before the site was selected. These individuals could benefit from expropriation or from the compensation and the agro-environmental schemes proposed in association with the reservoir. Such benefits seem to have been more under consideration by these players than any particular water policy. In this way, they did not manage the transition, but were important in influencing the site-specific implementation. According to the Ministry of Agriculture and Rural Development:[16]

> Local mayors – who might also be members of Parliament – are among the most powerful actors. They have a voice in Ministries, since an important part of ministerial routine is to give an adequate response to initiatives coming from the local administrations; moreover, local mayors have strong capacity to influence the process of practical implementation of the plans and projects.

As described above, Bokartisz members engaged with researchers from national research institutes, thereby breaking up the engineering epistemic community that had designed most of the recent Hungarian water infrastructure, and initiating a more interdisciplinary network of researchers. The tenders following the 2003 government decree allowed this network to test the new ideas with facilitation by VÁTI, the national planning agency. VÁTI makes it clear that it recognizes the importance of strong regional cooperation and representation:[17]

> Another important element in this whole period is how local interests are incorporated. It is very difficult to request from them because these areas are the least developed and low capacity areas. They also have low lobbying power. The reason why Bokartisz has such a key role in the whole process is that immediately, from the beginning, they were articulating their ideas as a local stakeholder and had the scientific basis and support for them. In addition, they were accepted by the locals. We are coming from Budapest, and it would take us a long time to be accepted by the local people and to be part of negotiations. So throughout the work we did, we tried really to rely on Bokartisz and involve them in the whole process. They have this kind of special knowledge about the local situation that we from Budapest would never have. If there were more of this high quality organized local representation, then the whole VTT would be much farther ahead.

To summarize, we found strong existing networks at both the national and the regional level. New interactions between these levels occurred around the time of the transition. VÁTI became a major actor due to its facilitation of discussions between the national government, scientists and local representatives. It chose Bokartisz as a counterpart and NGO representative throughout the entire planning process. The tenders administered by VÁTI for the Ministry of Environment and Water broke open the coalition of the water authority, its engineers and contractors.

14.4 Discussion and conclusions

What can we learn from the Tisza case about transitions in water policy? It is too early to determine precisely who used what strategy for what reason. We may never be able to untangle the intricate web of actors and their interactions during the period of transformation of Tisza water policy. However, viewing the transition from the perspectives offered by the five strategies outlined in Chapter 2 yields a number of new insights relating to that turbulent time, with each perspective pointing to different key actors and events.

The recurring major floods and the cyanide spill on the Tisza River in 2000 were obviously significant, but in what way? The floods and resulting damage were severe, but these two factors were probably not sufficient in themselves to trigger a transition in the water policy. They did, however, highlight the problem stream and support the strategy of local NGOs to define the prevailing water management as unsustainable and as requiring a different approach. This proved instrumental in opening up the debate about an alternative water policy, helping to align all three issue streams – problem, political and policy.

The abstract notion of water retention, floodplain rehabilitation and integrated river basin management already had supporters in the national government and major NGOs like WWF, as well as abroad. It could change the water management dominated by hard engineering solutions. This was at least partly so because European Directives advocated such approaches and offered indispensable financial support. What was needed in addition was local support and a concrete application of the abstract approaches to the regional situation. The new Bokartisz coalition of municipalities in the Tisza region, NGOs and researchers offered both. Bokartisz's recently finalized concept of floodplain rehabilitation and shallow flooding – inspired by traditional water management and ongoing local experiments – was developed for the whole Tisza region and contextualized for specific locations. In the Bodrogköz area particularly, support from the water board and mayors inside and outside the Bokartisz coalition was high and opposition minimal. With strong awareness and advocacy of the new idea in place, the next alignment needed was that of political will.

The 2001 flood and dyke break highlighted the problem. The 2002 national elections brought to power a new coalition determined to prove itself different from the previous government. Water affairs were transferred to a new Ministry of Environment and Water, setting the scene for the appearance of the transition in new procedures. The then upcoming accession to the European Union favoured a shift towards participatory and integral planning. In this way, 2002 brought convergence of the issue streams, opening the window of opportunity for coupling the new idea and a relevant problem, and providing impetus for a few years thereafter. It required a policy entrepreneur to open and use this window. This key player came in the shape of the head of department assigned with the development of the plan at the new Ministry of Environment and Water.

The main advocates ready to sell the new idea at the regional and national level were the members of the Bokartisz coalition. Steps consciously undertaken during the transition by Bokartisz to support the concept of floodplain rehabilitation include:

1. Small pilot studies.
2. Development of the concept on paper outside government and based on practical experience and scientific research.
3. Establishment of a regional coalition in support of the concept and negotiation of site-specific implementation with local partners.
4. Solicitation of scientific support for ideas.
5. Issue of press releases.
6. Presentation of ideas to administrative authorities.
7. Cooperation in designing implementation plan.

These steps provide insight into the substrategies within the major strategies of idea development and the selling of ideas by a coalition.

Researchers at the Hungarian Academy of Science and the national Water Resources Research Centre lent authority to the ideas promoted by Bokartisz, helping them become integrated as policy. The new ministry's department head swiftly took the lead in strategy to ensure advancement of the policy, taking the coalition of the water authority, its engineers and contractors by surprise. A government decree was passed in February 2003 that acknowledged the concept of floodplain rehabilitation and rural development and created two new venues for the further development of the water policy: a new interministerial committee and a series of tenders to deliver support studies and a regional implementation plan.

Strong network development and management as a transition strategy followed and helped to achieve well-supported and well-developed new ideas and priority on the political agenda. During further development of the water policy, VÁTI as the national planning agency successfully managed the network of research partners and NGO and local representatives. Respecting regional representation and experience as well as national competencies, VÁTI used the tenders for policy development work to support a new interdisciplinary epistemic community in the creation of a new body of evidence on floodplain rehabilitation in the Tisza region. In close cooperation with the interministerial committee, VÁTI delivered the spatial plan for implementing the water policy. The 2004 Tisza Law and the October 2003 government decree provided endorsement of implementation by the national government. Endorsement notwithstanding, however, since then only one of the first six retention areas has been built and the related floodplain rehabilitation and rural development has either not been attained, or attained only after many delays.

Reflecting on the policy–science perspective presented in the theoretical introduction to this volume (Chapter 2), we conclude that the five strategies and the focus on policy entrepreneurs offer a simple and attractive frame for analysis of the transition in water policy. Important lessons include the way the new ideas allowed for issues to be linked through a new coalition. Furthermore, while this coalition elaborated its ideas at the regional level, national policy-makers recognized a window of opportunity to link regional support to the policy change being advocated and supported financially at the international level. Development of the policy saw new venues and networks arise that proved influential during the transition.

In terms of development and promotion of ideas, we suggest that extra attention could be paid to the small-scale experiments begun by members in the central coalition. These pilot projects built trust and physical proof of cooperation. With respect to managing networks in the region, legacy effects appear to be strong, with prior social networks and the historic legitimacy of actors determining the nature of the game. In regard to implementation, the management of interests and (financial) resources, either to facilitate or to slow the transition, emerged as another important strategy. It was suggested that: 'Those opposing the plan thought it would never pass parliament. When it did they were shocked. The easiest and most efficient way to block it is though the budget'.[18]

Framing analysis that focuses on policy change should be careful not to miss the activities of the private sector and so omit a potentially important actor and partner. Whereas the origin, advocacy and management related to the new policy ideas in the Tisza do not lie with bigger private agents, their cooperation becomes crucial in implementation for two important reasons. First, the cooperation of landowners was and still is required. Second, implementation of water policy in Hungary can no longer rely solely on national government support. New partnerships have to be built. The new water policy requires the cooperation of many partners. It calls for in-depth examination of multistakeholder organizations and institutions that were not well understood at the time of the transition and that represent an emerging and complex factor in many countries around the world.

With regard to the two elements of a transition – change seen in policy or in governance arrangements – the Tisza case exhibits little formal organizational change, partly as a result of the use of ambiguity to secure consensus in policy development. In fact, actors identify coordination and clarity of the organizational structure as components that have been sorely lacking in the implementation phase. This is one of the challenges that the policy entrepreneurs in the Tisza presently face as the implementation of the policy has been slow and could still fail. Explicit analysis of the role of individuals in implementation may uncover strategies complementary to those discussed in the theoretical chapter (Chapter 2).

Our record of the transition highlights interactions among key individuals and groups of individuals involved in the five key strategies discussed in Chapter 2. It attempts to provide a convincing account of the principal factors involved and does not argue that the result of the interactions was inevitable or that the policy is 'correct', only that the prevailing conditions and interactions made its adoption feasible. The narrative is necessarily compressed and does not do full justice to the work of the many people who cooperated in the Tisza, but we hope to have conveyed the main thrust of affairs. Beyond the desirability of proving the viability of an idea, the Tisza case clearly shows the need for policy entrepreneurs to initiate the transition, as well as those that can take it safely through a period of confrontation, change and reorganization.

Acknowledgement

Work on this chapter has been supported by a grant from the European Commission through the European research project ADAM (Project no. 018476-GOCE) and from the Dutch Ministry of Agriculture, Nature Conservation and Food Quality. The authors would like to thank research partners Géza Molnár, Péter Balogh, Istvan Láng, Zsolt Harnos and Márton Jolánkai for their support in the Tisza region. We thank all interviewees and participants of the Tisza regional workshops for sharing their experience. Special thanks go out to Joel Aberbach and Tom Christensen whose work in New Zealand has been a clear source of inspiration for this chapter. We are grateful for the valuable comments and suggestions of research partners, especially Louis Lebel, Maria Gordon, Dave Huitema, Sander Meijerink and Rik Leemans.

Notes

1. Adaptation and Mitigation Strategies: Supporting European climate policy, www.adamproject.eu (accessed November 2008).
2. New approaches to adaptive water management under uncertainty, www.newater.info (accessed November 2008).
3. Interview: 25 October 2006, Hortobágy.
4. Interview: 9 August 2007, Budapest.
5. Interview: 22 August 2007, Sárospatak.
6. Interview: 30 October 2006, Budapest.
7. Interview: 22 August 2007, Bodrogköz.
8. Interview: 9 August 2007, Budapest.
9. Interview: 9 August 2007, Budapest.
10. Interview: 9 August 2007, Budapest.
11. The incorporation of Internet-based figures comes with the recognition that the medium has become increasingly used by political parties and that the number of issues covered on their websites and the

number of pages itself have increased. Thus the figure should be read only as an indication of the changing attention of party websites to the Tisza and not as an absolute assessment of the political agenda of the parties.

12. Interview: 30 October 2006, Budapest.
13. Interview: 22 August 2007, Bodrogköz.
14. City status extends the legal status, functions and powers, financial resources, basic rules for asset management and status in parliament of the local government representatives. Obligatory functions and powers of local governments are determined by parliament, which simultaneously must ensure the financial means necessary for the fulfilment of such tasks as enumerated by the Act on Local Governments (Horváth and Kiss, 2000; Temesi, 2000).
15. Interview: 22 August 2007, Bodrogköz.
16. Interview: 25 October 2006, Hortobágy.
17. Interview: 30 October 2006, Budapest.
18. Interview: 23 October 2007, Bodrogköz.

References

Andrásfalvy, B. (1973), 'Ancient floodplain and water management at Sarkoz and the surrounding area before the river regulations', *Vízügyi történeti füzetek*, **6** (in Hungarian).

Association for Local and Regional Development (ALRD) (1997), 'Study on the rehabilitation of the 'notch'-system', Case study, Debrecen (in Hungarian).

Balogh, P. (2002), *Basics and Method of Floodplain Management on Middle-Tisza Valley*, Budapest: VATI Kht.

Barta, K., K. Bódis, T.L. Boga, T. Kiss, R. Kiss, G. Mezősi, I. Pálfai, J. Rakonczai, L. Szlávik and I.G. Török (2000), *A Tisza Vidék fejlesztését befolyásoló vízrajzi kockázatok, erőforrások és lehetőségek* (*The Hydrological Risks Influencing the Development of the Tisza Country; Resources and Opportunities*), Szeged, Hungary: Natural Geography Department, University of Szeged (in Hungarian).

Bellon, T. (1991), 'Floodplain husbandry along the Tisza in the XVIII–XIX century', *Alföldi társadalom II*, Békéscsaba, Hungary (in Hungarian).

Bellon, T. (2004), 'Living together with nature: farming on the river flats in the valley of the Tisza', *Acta Ethnographica Hungarica*, **49** (3), 243–56.

Botos, C., P. Kajner, G. Molnár and G. Ungvári (2002), 'Floodplain landscape management: concept and concrete steps at the Bodrogkös area. The "Last Straw" program's ecological historical foundation and the synthesis of the environmental economics analysis, Budapest–Karcsa–Nyíregyháza: Bokartisz, www.bokartisz.hu (in Hungarian).

Commission of the European Communities (2000a), *Directive 2000/60/EC of the European Parliament and of the Council Establishing a Framework for the Community Action in the Field of Water Policy* (EU Water Framework Directive).

Commission of the European Communities (2000b), 'The Commission and non-governmental organisations: building a stronger partnership', 18.1.2000, COM(2000)11 final, discussion paper, Brussels: European Commission.

Commission of the European Communities (2001), *European Governance: A White Paper*, COM(2001) 428 final. Brussels.

Csekö, G. and L. Hayde (eds) (2004), *Danube Valley: History of Irrigation, Drainage and Flood Control*, New Delhi: ICID.

Fokkens, B. (2001), 'Preface', in H.J. Nijland and M.J.R. Cals (eds), 'Proceedings of the Conference on River Restoration in Europe 2000', Wageningen, the Netherlands: Institute for Inland Water Management and Waste Water Treatment RIZA, Ministry of Transport, Public Works and Water Management, RIZA report no. 2001.023.

Government of Hungary and UNDP-GEF (2004), 'Conservation and restoration of the globally significant biodiversity of the Tisza River floodplain through integrated floodplain management (Tisza-biodiversity project)', UNDP project document, Government of Hungary and United Nations Development Programme – Global Environment Facility.

Horváth, T.M. and J. Kiss (2000), 'Politico-administrative relations in local governments in the case of Hungary', in 8th NISPAcee Annual Conference 'Ten Years of Transition: Prospects and Challenges of the Future for Public Administration', Working Group on Politico-Administrative Relations in Central and Eastern Europe, Budapest.

Karácsonyi, Z. (2001), 'Rehabilitation of the 'notch'-system as a tool for multipurpose floodplain management on the Upper-Tisza River', in H.J. Nijland and M.J.R. Cals (eds), *Proceedings of the Conference on River Restoration in Europe 2000*, Wageningen, the Netherlands: Institute for Inland Water Management and Waste Water Treatment (RIZA), RIZA report no. 2001.023.

Kingdon, J.W. (1995), *Agendas, Alternatives and Public Policies*, New York: HarperCollins.

Kundzewicz, Z.W. (2002), 'Non-structural flood protection and sustainability, International Water Resources Association', *Water International*, **27** (1), 3–12.

Lászlóffy, W. (1982), *Tisza River: Construction and Water Management in the Tisza Water Regime*, Budapest: Akadémiai Kiadó (in Hungarian).

Linnerooth-Bayer, J. and A. Vári (2003), *A Model-based Stakeholder Approach for Designing Disaster Insurance*, Laxenburg: International Institute for Applied Systems Analysis.

Linnerooth-Bayer, J., A. Vári and M. Thompson (2006), 'Floods and fairness and Hungary', in M. Verweij and M. Thompson (eds), *Clumsy Solutions for a Complex World: Governance, Politics and Plural Perceptions*, Houndmills: Palgrave Macmillan, pp. 181–204.

Matczak, P., Z. Flachner and S.E. Werners (2008), 'Institutions for adapting to climate change in the Tisza river basin', *'Klimá – 21' Füzetek* (English Special Edition), **55**, 87–100.

Meijerink, S. (2005), 'Understanding policy stability and change. The interplay of advocacy coalitions and epistemic communities, windows of opportunity, and Dutch coastal flooding policy 1945–2003', *Journal of European Public Policy*, **12** (6), 1060–77.

Molnár, G. (2002), 'Merlin mester üzenete', Budapest, Hungarian Environmental Economics Centre (MAKK).

Rakonczai, J. (2002), 'Current questions concerning the problem of flood-risk in the Tisza catchment basin', Third International Conference of Critical Geography, Békéscsaba, Hungary.

Sendzimir, J., P. Balogh, A. Vári and T. Lantos (2004), 'The Tisza River basin: slow change leads to sudden crisis', in S. Light (ed.), *The Role of Biodiversity Conservation in the Transition to Rural Sustainability*, Amsterdam: ISO Press, pp. 261–90.

Szlávik, L. (2001), 'Flood protection development in the Tisza Valley', Földrajzi Konferencia (Geographical Conference), Szeged, Hungary (in Hungarian).

Szlávik, L. and I. Ijjas (2003), 'Action plan on flood prevention for Tisza River', Precautionary Flood Protection in Europe, International Workshop, 5–6 February, Bonn.

Temesi, I. (2000), 'Local government in Hungary', in T.M. Horváth (ed.), *Local Governments in Central and Eastern Europe. Decentralization: Experiments and Reforms*, Budapest: Local Government and Public Reform Initiative, pp. 343–83.

Timár, G. and T. Rácz (2002), 'The effects of neotectonic and hydrological processes on the flood hazard of the Tisza region (East Hungary)', *EGU Stephan Mueller Special Publication Series*, **3**, 267–75, available at: http://www.stephan-mueller-spec-publ-ser.net/title_and_author_search.html.

Vámosi, S. (2002), *The Effect of Excess Waters on the Development of the Great Plain Regions*, Debrecen: University of Debrecen.

Vári, A. (2001), 'Flood risk management in the Upper Tisza region: results of stakeholder interviews', Draft Report, Laxenburg: IIASA.

Vári, A., J. Linnerooth-Bayer and Z. Ferencz (2003), 'Stakeholder views on flood risk management in Hungary's Upper Tisza Basin', *Risk Analysis*, **23** (3), 537–627.

Veres, J. (2004), 'Hungary in an enlarged Europe: fiscal and financial relations', *EIB Conference: Strategies and Frameworks for Hungary's Growth and Convergence after EU Accession*, European Investment Bank, Budapest.

VITUKI (2004), 'Rebirth of the River Tisza – The new Vásárhelyi Plan', Vásárhelyi Plan Intersectorial Committee Bulletin (www.vizugy.hu). Budapest, National Environment, Conservation and Water Authority; Water Resources Research Institute (VITUKI).

Werners, S.E., Z. Flachner, P. Matczak, M. Falaleeva and R. Leemans (2009), 'Exploring earth system governance: a case study of floodplain management along the Tisza River in Hungary', *Global Environmental Change*, **19** (4), 503–11.

Zöckler, C., E. Wenger and J. Madgwick (2001), 'Assessment of WWF river and floodplain restoration projects in Europe', in H.J. Nijland and M.J.R. Cals (eds), *Proceedings of the Conference on River Restoration in Europe 2000*, Wageningen, the Netherlands: Institute for Inland Water Management and Waste Water Treatment (RIZA), RIZA report no. 2001.023.

15 Spanish water management in transition: transition management watered down?

Nuria Font and Joan Subirats

15.1 Introduction

Over the last 20 years water management policy in Spain has shown signs of a gradual transition. The traditional supply-based approach to water policy promoting the construction of state-subsidized hydraulic projects, while not completely abandoned, has been complemented by the adoption of tentative steps towards a concept based more on sustainability of the resource. Many factors have facilitated this gradual transition, the most important of which has been the increasing erosion of the traditional balance in which water policy has long built up. In short, the growing competition for water combined with the emergence of policy entrepreneurs promoting sustainability-related issues has helped to fracture the traditional approach to water management and open new avenues for change. This chapter explores the strategies employed by policy entrepreneurs (Kingdon, 1995) to mobilize new policy ideas, generate knowledge and make use of multiple venues for influencing domestic water politics moving through several stages in a transition period.

15.2 Policy stability and policy change

Spain's traditional problem-solving approach to water policy has consisted of the regulation of the water supply by means of state-subsidized construction of large-scale infrastructure. In recent years this type of policy has been partly replaced by alternative supply-based approaches taken in policy involving more sustainable options. The 2001 National Water Plan in particular, adopted by the Popular Party administration, was revoked three months after the Socialist Party won the national elections in March 2004.[1] The national government adopted the AGUA[2] programme shortly afterwards, introducing desalinization as a policy aim intended as part of a shift to alternative, supply-based regulation methods.

In order to account for changes in water policy, the literature provides several perspectives on cultural, historical or agent-based approaches. From a cultural perspective, Tàbara and Ilhan (2008) argue that the role played by the production and use of cultural constructions, including worldviews, beliefs and values, has brought about transition in water management policies and power structures in Spain. For the authors, the so-called New Water Culture (Nueva Cultura del Agua) movement exemplifies such cultural triggers in adaptation toward sustainability. This interpretation would indeed apply to a change in the institutionalized policy image, as suggested by Baumgartner and Jones (2002), who propose that opponents to the dominant policy frames develop new ideas challenging the status quo. While such interpretation provides interesting insights on the transformation of ideas underlying policy changes, it draws less attention to the long-term processes that actually drive change and that appear to be of crucial importance in lengthy and incremental transitions.

Historical institutionalism (North, 1990; Thelen, 1999; Pierson, 2000) might account for why some policies become very stable. This approach emphasizes concepts of path-dependency discussed in the theoretical introduction to this volume (Chapter 2): previous steps in a given policy direction are seen to guide future movement in the same direction. Policies are thus self-reinforcing as a result of an increasing returns process (Pierson, 2000), which means that the costs of changing the policy direction become progressively higher. Policy inertia is perhaps one of the overall trends most inherent to water policies, which tend to perpetuate their constituent trends. One core premise of path-development processes is therefore that policies change incrementally. As this book's introduction makes clear, change may occur in either the substantive content of policy or in the dominant governance paradigm. A window can open for a change in the direction of policy with the advent of a critical juncture or a fundamental crisis. In order to account for both policy change and policy stability, historical institutionalism suggests adopting a long-term perspective. This is particularly appropriate when studying water policy, as it is an area of public initiatives based on long-standing core policy beliefs and practices, where changes, when they take place, are of an incremental nature. Costejà et al. (2004) and Bukowski (2007) in their analysis of water policy change in Spain share a methodological assumption of the need to take a long-term perspective. This chapter too, and in line with historical institutionalism, assumes that trends toward policy and institutional inertia are intrinsic to water policy, making it necessary to understand indicators of change in a historical context and as part of a cumulative succession of events.

However the focus on the chain of historical events is not sufficient to explain policy change and needs to be complemented by a review of actors' intentionality (see the theoretical introduction to this volume, Chapter 2 by Meijerink and Huitema). According to Pierson (2000), historical institutionalism is not at odds with this agent-based approach. Policy and institutional inertia may restrict or even lock in the course of possible actions, but it does allow room for strategic behaviour. In other words, action matters. Actors may be able to mobilize material and ideational resources strategically in order to change the course of a policy. In this sense, Bukowski (2007), in analysing Spanish water policy, adopts an advocacy coalition framework and identifies two coalitions – one of environmentalists, the other of marketizers – sharing core beliefs, that account for policy continuity and change. Possibly of greater interest are the modes in which individuals and groups develop entrepreneurial activities, including building such coalitions, raising alternative policy images and exploiting multiple venues in order to prompt policy change. As Richardson (2000) points out, these venues constitute resources actors are likely to exploit. Significantly, the European Union (EU) provides new venues, namely the complaints system and the Water Framework Directive (WFD), regularly used by policy entrepreneurs in the water sector.

Taking all these considerations into account, this chapter produces a three-stage model of water policy transitions based on the behaviour of policy entrepreneurs and their impact on policy. The three stages observed cover a two-decade period and follow a sequence: emerging protest; polarization followed by a fundamental crisis; and then a change of policy options. More specifically, the chapter focuses on the role played by policy entrepreneurs as proactive agents of change throughout the transition period in Spain. Policy entrepreneurs in this case include environmental organizations, the scientific community and the Socialist Party, all of which have developed entrepreneurial strategies and taken advantage of political windows of opportunity and used venues, particularly those in the EU, in order to promote alternative approaches to water management.

15.3 Water management in transition: a three-stage model

15.3.1 Introduction

Water management in Spain was institutionalized in the late nineteenth century by the enactment of the first and second Water Acts of 1866 and 1879, respectively, which were later modified by the provisions on water contained in the Civil Code in 1889. This legal framework lacked a fully developed policy dimension and gave the state administration few powers (Costejà et al., 2004). This situation started to change at the end of the nineteenth century partly because of the influential intellectual 'regeneracionist' movement's promotion of the idea of state investment in hydraulic infrastructures as a national strategy for modernizing the country (Costejà et al., 2004). State regulation of water resources, mostly for irrigation, was heavily used by the Franco regime.

The advent of democracy in the 1970s completely transformed the political system and opened windows for a transformation of the water management regime. The adoption of the Spanish Constitution in 1978 in particular helped to bring about change to water management by changing the larger institutional context in which it was embedded. The creation of a decentralized political structure, the so-called State of Autonomous Communities, led to the distribution of water management powers between national and regional governments. The former hold exclusive powers over the declaration of Continental waters, covering both surface and ground renewable waters as public domain as well as state control over access. The national administration is also responsible for: legislation, planning and granting of government concessions and authorization when waters flow through more than one autonomous community; the projection of hydraulic works when such are declared to be of public interest or to affect more than one autonomous community; and environmental protection legislation and planning. The Ministry of the Environment, created in 1996, has responsibility for water policy. Through the Dirección General del Agua (Directorate General of Water), it is responsible for the production, supervision and review of the National Water Plan, the definition of systematic criteria for the review of river basin hydrological plans, and the coordination of sectoral or regional plans affecting hydrological planning. Hydrographical confederations are the main water management bodies that, in the case of intercommunity river basins, are attached to the Ministry of the Environment. Autonomous communities are responsible for managing intra-community river basins, specifically for: the projection, construction and exploitation of hydraulic resources, channels and irrigation infrastructure of regional interest; legislation and management of mineral waters, thermal springs, shell fishing, aquaculture and fluvial fishing; and, in some autonomous communities, the introduction of legislation on environmental protection. There are cross-regional variations as not all autonomous communities have the same type and level of institutional development in regard to water management. In addition to the national and regional administrations, local administrations, including about 8000 municipalities, are responsible for water supply and sanitation.

Together with the erosion of powers entailed in the decentralization of the national water administration, sectoral legislation on water has modified the parameters of policy. The Water Act (29/1985) placed an emphasis on state intervention by extending the idea of the public domain and focusing on planning. It also moved towards a more integrated approach in its conceptualization of water as a unitary resource and by introducing the concepts of sanitation, water resource quality preservation, civil protection, environmental values and territorial planning (Costejà et al., 2004). The new legal framework attempted to adapt water policy to a political, socio-economic and technological context that had been thoroughly transformed over the previous decades in terms of the increase in the number of uses and users.

Following a mandate of the 1985 Water Act, the government presented a draft version of the National Water Plan in 1993. The plan perpetuated the traditional hydraulic paradigm consisting of state regulation of water resources through the construction of large-scale infrastructure, namely dams and interbasin transfers, with the ultimate objective of supplying water for everybody at zero cost whenever and wherever needed (Sauri and Del Moral, 2001). However the attempts by the government to adopt a water planning instrument, with successive drafts released between 1993 and 1995, were unsuccessful. The projected plan was heavily criticized, mainly by the donor regions – principally Aragón – environmentalists and practitioners. This marked a turning point because for the first time a water plan provoked enough negative reaction to make the government withdraw it. At this point the first signs of transition can be found. Thus the transition of water policy had its origins in the early 1990s and has evolved through a succession of stages. However the most evident changes of policy change towards a less traditional approach have taken place only recently.

The three stages identified in the water policy transition correspond to the terms in office of the three national single-party governments in office since the early 1990s (see Table 15.1). The first stage, emerging protest, covers the minority government of the Socialist Party between 1993 and 1996. In this period, protest by general public actors and practitioners prevented the government from adopting a planning instrument that promoted the construction of huge water infrastructure. The second stage, polarization, includes two Popular Party governments covering the periods 1996–2000 (minority government) and 2000–2004 (majority government). During the first minority government, the executive successfully reformed the existing legal framework. Yet it faced large-scale protests against the philosophy and content of the 2001 National Water Plan during the subsequent majority government. Finally, the third stage, the change of policy options, started with the change of government, by then controlled by the Socialist Party following the national elections of 2004. During the period of the third stage, the government envisioned abandoning the traditional water development approach and defining alternative supply mode types of intervention, mostly limited to the construction of desalinization plants. At the same time the EU WFD began to be implemented by means of a complex participation process that opened up the scenario to an even greater extent.

As discussed above, from the standpoint of historical institutionalism, the water transition in Spain is characterized as long term rather than related to specific circumstances. As noted, Spain's successive stages of transition are marked by changes of government. This does not mean, however, that change in the nation's water transition can be accounted for in terms of party politics. Similarly, while the dynamics of government

Table 15.1 Three-stage model of the water policy transition

	Stage 1 1993–96	Stage 2 1996–2004	Stage 3 2004 onwards
	Emerging protest	Polarization	Change of policy options
	Protest at the traditional supply approach	Market efficiency accepted but protest at the traditional supply approach	Alternative supply approach
Main policy output	1993–95 Draft Project of National Water Plan	2001 National Water Plan	2004 AGUA programme WFD adaptation
Main policy interventions	Large hydrological infrastructures: interbasin transfers and about 150 dams	Large hydrological infrastructures: Inter-basin transfers (Ebro) and about 120 dams	105 interventions, mostly desalinization plants
Policy approach	Supply-based	Market- and supply-based	Alternative regulation of water supply
Policy discourse	Economic development and modernization Territorial and social solidarity	Efficiency Economic development Territorial solidarity	Sustainability Rational use of water

and opposition must not be downplayed, they have not been the driving force of change. They have, though, opened windows of opportunity for policy entrepreneurs to act.

15.3.2 Stage one (1993–96): emerging protest

Following the 1985 Water Act mandate, the Ministry of Public Works, Transport and Environment in 1993 elaborated a project ahead of the National Water Plan. The so-called Plan Borrell, named informally after the minister, perpetuated the traditional problem-solving approach to water scarcity. Based on the territorial solidarity principle, it argued that Spain had enough supplies of water to solve both the unequal geographic distribution of resources and the periodic recurrence of severe droughts (Sauri and Del Moral, 2001). It aimed to satisfy the increasing demand for water and correct territorial imbalance through the construction of about 150 dams. After successive drafts and intense public debate, the plan was defeated in 1995 both in Congress, which made the adoption of a plan dependent on the production of a National Irrigation Plan, and in the Senate, which added the condition that each of the river basin plans should be presented before the plan could be adopted (Bakker, 2002). Furthermore the National Water Council also rejected the government's plan and submitted an alternative proposal calling for greater efficiency and rationality (Sauri and Del Moral, 2001).

The underlying factor leading to the 1995 defeat of the National Water Plan was possibly the weak position of the minority Socialist Party government during the turbulent 1993–96 legislature. This political situation stimulated the growing plurality of conflict-

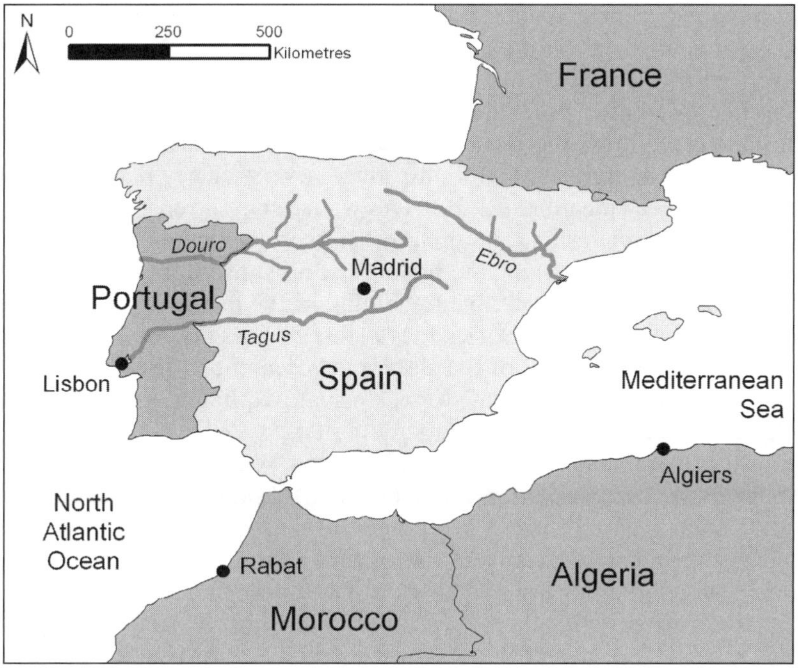

Map 15.1 Spain and its main rivers, including the Ebro

ing interests surrounding water management. On the one hand, the plan was welcomed by agricultural associations and the regions affected by water restrictions that saw it as a solution to the water scarcity problem. On the other hand, the traditional consensus that had dominated water policy was broken by alternative definitions of the policy problem promoted by different types of actors, including economists, environmentalists and the donor regions actors. Not surprisingly, by invoking the principles of efficiency and conservation, economists and environmentalists formed strategic alliances in favour of a demand-led water policy against the traditional supply-based engineering approach (Bakker, 2002). With a different perspective, the donor regions rejected the idea that water was a national asset and argued that water transfers would enhance economic activities in some Mediterranean regions, like Murcia, to the detriment of the more depressed ones, namely Aragón (Sauri and Del Moral, 2001). In this respect, Aragón was the first region to break with the traditional idea of territorial justice. In fact all political forces in the region attained the so-called Aragón Water Pact in 1992 by virtue of which the region would not consent to transfers from the Ebro River (Map 15.1) while regional needs were not satisfied. In addition, unrest crossed national frontiers. Portuguese opposition to water diversions from Spanish–Portuguese rivers added a new front of dissent.

15.3.3 Stage two (1996–2004): polarization

Following the failure of the projected 1993 plan, subsequent reforms of water policy were conducted in the late 1990s. The Popular Party government newly elected in 1996 reformed the 1985 Water Act in 1999 by means of Act 46/99, which opened the door to the adoption of the National Water Plan through Act10/2001. The national government, which was critical of the previous government's inability to define a national strategy to cope with the drought problem, urged that water policy strategy combine market efficiency and supply-based principles. As regards the former, the government created water banks and seven public corporations known as Sociedades Estatales. Water banks for the exchange of use rights aimed to introduce new efficiency parameters by making the water concession regime more flexible. This was seen in the creation of centres for the exchange of use rights for further cession to other users. Sociedades Estatales in turn are public–private partnerships designed to finance the most profitable hydraulic infrastructure projects, with those less profitable remaining in the hands of the state. As Bakker (2002) points out, the creation of Sociedades Estatales does not imply the abandonment of water policy as an instrument of the state, but represents a restructuring of funding mechanisms that would allow the state to seek private capital for profitable projects and EU funding for projects eligible under different criteria.

The Spanish government promoted the National Water Plan, which was adopted by parliament once the National Water Council had expressed its support. The plan contained two main aspects. First, it promoted the construction of large-scale hydraulic infrastructures transferring water from the surplus to the deficit river basins. By doing so, it aimed to satisfy the increasing demand for water arising from the combination of cyclical climate droughts with the increase in the demand for water due to industrial, tourism and urban development in the southeastern regions over the previous decade. The key project contained in the plan consisted of the transfer of 1050 hm^3 of water from the Ebro River, a central provider of the Spanish hydrological system, south along the Mediterranean coast to the Jucar, Segura and South Almeria basins and north to the internal basins of Catalonia. It also included the construction of about 120 water dams and river channels in Spanish territory.

The national government's strategy on water policy provoked mixed reactions. On the one hand environmentalists and the neoliberal sectors welcomed the adoption of more flexible management instruments (Bukowski, 2007). On the other hand the National Water Plan was widely criticized by social, territorial and political actors. Significantly, a wide range of environmental and social organizations, political parties and the scientific community mobilized a wide range of political and cognitive resources to influence public opinion and exercise pressure on the national government to suppress the plan. However in spite of intense protest against the National Water Plan, the national government, counting on a large majority in parliament as well as support from some autonomous communities, maintained its position. Protest intensified and extended as a result. In other words, the plan generated an adverse response from large parts of society, which became engaged in intense informational campaigns, mobilized scientific knowledge and ideas and increased pressure on the government.

15.3.4 Stage three (2004 onwards): change of policy options

The change of government in March 2004 generated a political opportunity for actors protesting against the construction of large water infrastructure. Shortly afterwards, the newly elected Socialist Party government, fulfilling an electoral promise, modified the 2001 National Water Plan,[3] including cancelling the Ebro River transfer, and then quickly adopting the AGUA[4] programme (2004–08) under the leadership of the Minister for the Environment, Cristina Narbona. By doing so, it took a step towards the abandonment of the traditional supply-based approach to water policy and several steps towards redefining the type of supply instruments. The AGUA programme included more than 100 initiatives, including the construction of water desalinization plants and the creation of public water banks. Proposals existed in the programme for alternative methods of regulating water supply, but it barely regulated water demand, although it included this measure as an option.

Partly to address this, the Ministry of the Environment launched a voluntary process assessing the environmental sustainability of the urgent initiatives contained in the AGUA programme for the Mediterranean basins. The document presented by the government received mixed reactions from the environmental community. Environmental groups, namely Greenpeace and WWF Spain, welcomed the assessment's inclusion of the 'water restriction' option following the consultation process launched by the Ministry of the Environment. However they objected to the document excluding from consideration some of the projects contained in the AGUA programme, such as the Júcar-Vinalopó interbasin transfer, where the 'water restriction' option was not being recommended because of the strong public criticism it would create. The two groups also criticized the fact that the programme was based on the consumption of water and energy resources.[5]

The Ministry of the Environment planned to create a public water bank in each hydrographical basin, which would allow historical water resources to be reallocated according to criteria of equity, efficiency and sustainability.[6] In October 2004, the government approved the implementation of centres for the exchange of water rights in the Segura, Júcar and Guadiana river basins. Environmental groups reacted positively to the creation of such a policy instrument by considering that it might hamper the impact of drought, but as WWF Spain relates, they rejected the implementation of the banks in those river basins with long-standing problems caused by illegal wells and markets.[7]

In addition to the changing orientation of water management schemes, the Water Framework Directive (WFD) has established a new outlook in water policy. The WFD was transposed to Spanish legislation by means of Law 62/2003 of 30 December 2003, amending the Revised Text of the Water Law (RTWL). The WFD establishes a specific public participation process in order to embark on the measures necessary to improve the environmental quality of water, with a deadline of 2009. In specific terms it points to the need to provide access to information relating to the entire planning process, the dissemination of environmental information, and public participation in the production of the improvement plan. Initiatives have been undertaken to comply with this directive since 2005 and starting in 2008 participation activities in various hydrographical basins have been implemented, which are changing the traditional roles and power balance in the sector.

15.4 Policy entrepreneurs

The first signs of crisis in the traditional model of water management appeared in the early 1990s. Intense public debate led to the involvement of experts and academics, environmentalists, consumer groups, trade unions and economists in a strategic alliance between free-market economists and environmentalists that questioned the supply-led engineering approach of the traditional paradigm (Bukowski, 2007). This alliance challenged the traditional policy image based on hydrotechnology for surface waters predicated on the basis of ever-increasing demand for water. Launched in the midst of a severe drought, the debate also exacerbated the 'water wars' between wet and dry regions. The lack of consensus on the National Water Plan prevented the project from being adopted in parliament.

Alternative policy ideas raised by opponents to the traditional schemes gradually gained room in the political debate and were included on the public agenda. The incipient mobilization of policy entrepreneurs opposing the institutionalized policy image in the early 1990s was the seed of a large-scale movement in the following years. During the second stage of the transition, a wide range of actors from environmental groups, the scientific community and the Socialist Party developed active entrepreneurial activities and played a leading role in questioning the traditional approach.

Environmental groups encompass diverse organizations, the most proactive of which include stable and long-standing groups – for whom work on the National Water Plan was one of many actions – and more focused smaller and ad hoc organizations. The former type comprises Greenpeace, WWF Spain, SEO/BirdLife and the Confederation of Organizations for Environmental Protection (CODA), and a variety of smaller organizations. CODA was formed by the New Water Culture Foundation (Fundación Nueva Cultura del Agua), COAGRET (the Association of People Affected by Large Dams) and smaller groups, such as the Association for the Defence of the Ebro River (Plataforma per la Defensa de l'Ebre). Beyond their differences, they shared common definitions of the water problem, adopted similar strategies and pooled resources in order to maximize their efforts. Coalition-building to enhance shared beliefs, ideas and action has thus been a key aspect in their entrepreneurial strategies.

Environmental groups were particularly active not only in producing ideas, but also in selling them in order to instigate policy change. They based this strategy on the launch of joint campaigns against the 2001 National Water Plan with three main objectives: first putting pressure on the government to withdraw the plan; second, and closely related to the first, creating a climate of opinion among Spanish society; and finally, based on the production of technical studies, proposing alternatives. To fulfil these objectives, environmental groups orchestrated ambitious campaigns combining various types of actions.

Major efforts were made by environmental groups to fulfil the first of the three objectives. In this respect, the largest environmental organizations, including Greenpeace, WWF Spain and SEO/BirdLife, launched supra-national and national campaigns condemning the social, economic and environmental consequences of the plan. Further, these groups actively used multiple venues in an attempt to promote policy change. Key actions included frequent provision of information and the lodging of complaints to the European Commission and the presentation of petitions and questions to the European Parliament, all of which alleged that the National Water Plan, which was expected to be co-financed with EU funds, went against several EU environmental Directives.

Environmental groups considered that the large-scale infrastructure proposed in the plan, in particular the Ebro diversion, had negative effects on Special Areas of Conservation and Special Protection Areas constituting the Natura 2000 Network, and that it would have severe consequences for most water ecosystems. The same groups reported that the plan did not comply with the objectives contained in the Water Framework Directive and that the government had not produced a strategic environmental impact assessment of the plan, but instead merely partial assessments.[8] The adoption by the Spanish government of both the Water Framework Directive and the Strategic Impact Assessment Directive reinforced the arguments of the environmental groups, since no legal requirement to enforce either Directive existed when the plan was adopted. Adoption of the Directives in practice opened a window of opportunity that the groups managed to exploit. Apart from these actions, many environmental organizations, including Greenpeace and WWF Spain, launched an intensive networking campaign in Brussels in order to sell ideas opposed to EU financing of the work projected in the National Water Plan.

The strategies employed in Brussels produced an institutional reaction. In response to the numerous complaints and petitions, the European Commission initiated contact with Spain's Ministry of the Environment in order to obtain information to prepare its own position. In October 2000, when the plan was still in its draft phase, the European Commission sent a letter to the Spanish government expressing its concern at compliance with Directive 79/409/EEC on the conservation of wild birds, Directive 92/43/EEC on the Conservation of Natural Habitats and Wild Fauna and Flora, and Directive 97/11 amending Directive 85/337/EEC on the Assessment of the Effects of Certain Public and Private Projects on the Environment. It also stated a reminder of the consequences of non-compliance in respect of the co-financing of projects. In July 2001, the European Commission also reminded the Spanish government to take the 2000/60/EC Water Framework Directive into account and recommended the production of a Strategic Impact Assessment pursuant to the 2001/42/EC Directive.[9] In response to EU pressure, in January 2002 following adoption of the National Water Plan, the government sent a Strategic Impact Assessment to the European Commission. Further technical dialogue took place between the Commission and the Ministry of the Environment until 2004. That year, the European Directorate-General for the Environment issued a report expressing doubts about the viability of the plan, its environmental impact and its eligibility to receive Cohesion Fund money.[10]

At the domestic level, environmental groups created a network of territory-based organizations, above all in the Ebro Delta and the Mediterranean regions. Their main initiatives included active involvement in the National Water Council, whose opinion was required before parliamentary adoption of the plan. The groups themselves also issued opinions and allegations, sending them to members of parliament and political parties, and condemned the expected consequences of the plan in the media. In addition to the major groups, those with an ad hoc focus managed to mobilize extensive resources.

The organization COAGRET was a case in point. It embraces representatives of the affected municipalities, environmental organizations and the scientific community as well as individuals and opinion leaders. One of its members, the Association for the Defence of the Ebro River, was particularly dynamic in activating political, cognitive and social resources against the adoption of the plan. Counting on the support of the association, COAGRET organized a wide range of entrepreneurial initiatives, including the publication of manifestos, participation in workshops and seminars and the organization of massive demonstrations.

The large and smaller ad hoc environmental groups succeeded also in maintaining constant media attention, both before and after the adoption of the plan. Needless to say, the production of science-based information was a crucial aspect of their entrepreneurial strategy. They were very active in both technically assessing the social, environmental and economic consequences of the National Water Plan and in proposing alternatives. For instance, COAGRET, the Association for the Defence of the Ebro River and SEO/BirdLife participated in the so-called Blue March in Brussels, where they lodged complaints about the National Water Plan to the European Commission and the European Parliament.

The New Water Culture Foundation has also been a strong policy entrepreneur, producing several technical studies evaluating the Strategic Impact Assessment of the National Water Plan made by the Ministry of the Environment. In this respect it is worth mentioning the leading role played by the president of the New Water Culture Foundation, Pedro Arrojo, a professor of economics at the University of Zaragoza. With the support of the foundation as well as a wide range of environmental organizations, he produced a diverse set of scientific studies questioning the 2001 National Hydrological Plan. He also gave advice to the European Commission on the Plan, including a strategic impact assessment and a socio-economic assessment of the transfer projects included in the Plan.[11] He was one of the leading voices against the institutionalized problem-solving approach as he sought out the most favourable venues for transforming the dominant policy image.

Other environmental organizations also behaved proactively in the mustering of knowledge-based resources. WWF Spain, for instance, produced an environmental assessment of the impacts of the projected dams and water transfer affecting sites contained in Natura 2000.[12] At the EU level, the main environmental players and the New Water Culture Foundation participated in the technical meeting hosted by the European Commission in October 2003, which was also attended by representatives of Spain's Ministry of the Environment and the Brussels representative of the Aragón Region. The meeting addressed two technical questions: the projected future water flow in the Ebro River, and the flow regime in the lower Ebro required to ensure the protection of the ecological and chemical status of the river and its delta.[13]

In the second stage, the involvement of the scientific community continued at all stages of policy. The Ministry of the Environment commissioned about 100 studies, some of which were favourable and others unfavourable.[14] Among the latter the technical reports issued by the New Water Culture Foundation and, as mentioned above, by its president Pedro Arrojo were vital in introducing changes in the dominant governance paradigm. The foundation was created in 1998 with the aim of stimulating a multidisciplinary debate on water and sustainability. Every two years since then, the foundation has organized the Iberian Congress for Water Management and Planning, bringing together representatives of the scientific community with the support of more than 70 universities.[15] In addition the foundation has organized various conferences and workshops, published articles and reports and organized other related activities. The proactive involvement of the New Water Culture Foundation has contributed greatly to the questioning of core policy beliefs and the promotion of alternative ideas on water management.

In contrast to environmental groups and the scientific community, the Socialist Party has not exhibited a coherent position since the early stages of water policy transition in the early 1990s. In a relatively short period of time, it shifted its position on water policy from a traditional approach, as adopted in the so-called 'Borrell Plan', to an alternative supply-based approach. The Socialist Party had initially faced poor internal cohesion around the long-standing distributive problem of water. At one extreme, the regional President of Aragón, Marcelino Iglesias, actively opposed the plan and the Ebro transfer by arguing that they would have negative social, economic and environmental consequences for the region, which was one of the least developed areas. At the other extreme, the long-standing regional Presidents of the Extremadura and Castilla-La Mancha, Juan Carlos Rodríguez Ibarra and José Bono, supported the National Water Plan in 2001 in order to protect territorial interests. They only changed their position when the Socialist Party produced an alternative plan (Conejero Paz, 2005). At that point the Socialist Party plan included the construction of a large number of desalinization plants and the elaboration of an irrigation plan, among other features (Conejero Paz, 2005). Since then, with a more coherent position, the Socialist Party promised to cancel the National Water Plan. Indeed, the reversal of the National Water Plan was one of the most outstanding electoral promises by the Socialist candidate to the Chief of the Executive at the 2004 elections, José Luís Rodríguez Zapatero. Following the Socialist Party's victory in the elections of March 2004, President Rodríguez Zapatero demonstrated his commitment to alternative schemes of water policy by means of appointing Cristina Narbona to head the Ministry of the Environment. Narbona had long held the respect of the environmental community, and was deeply committed to amending the 2001 National Water Plan. Undoubtedly the change of government resulting from the political scenario itself arising from the 2004 elections opened a political opportunity window for the Socialist President to put forward an alternative policy option. In short, the political stream derived from the changing government was coupled with the policy and problem agendas.

15.5 The implementation of the WFD: new opportunities in old scenarios

The implementation of the WFD in Spain has provided an obvious window of opportunity for those who had traditionally been excluded from the hard core of water policy formulation in Spain. Farmers, hydroelectricity producers, construction companies and engineers have seen their consensus and balance of relations changed in recent years. Their recent participation and discussions with very different actors on factors as crucial as the shared diagnosis of the environmental quality of water, determination of the ecological volume and the priorities for intervention in basins, have all had schedules that are very specifically stipulated in the WFD. In fact, as mentioned above, the WFD and policy with regard to water management within the European Union was an important resource for those opposed to the National Water Plan. The agents opposed to the National Water Plan have turned out to be the strongest backers of the participation processes mentioned in the WFD, while the agents who were most reluctant to accept the withdrawal of the National Water Plan are now those who are most sceptical in terms of their involvement in the participatory experiences that must be undertaken.

Many participatory experiences have recently taken place in order to implement the stipulations contained in the WFD. Without providing a complete overview, some significant studies show underlying trends (Espluga and Subirats, 2007). Generally, increasing debate on water policies has recently been observed. The usual participants include representatives of autonomous regional and state governments and administrations with responsibilities in the field, industrial users, energy users, urban supply users, irrigation users, users for leisure and recreation, environmentalism and ecological associations, universities and research centres, trade unions and other social movements, including residents' and cultural associations. This composition, which has been made possible by the outlook adopted by the WFD, opens up the field of debate a great deal and makes necessary deliberation processes and consensus strategies that have not been at all common in this type of policy. There has been variation in the type of participatory processes, although they generally include informational schemes, debates in workshops and plenary sessions for agreements to be reached. In some cases, the process remains in a purely deliberation phase, while in others agreements are reached and specific decisions are taken. In regard to the subjects, and regardless of the diversity and specific features of each basin, questions related to maintenance supplies, as well as industrial, urban and agricultural pollution of the water, and so on, have become dominant.

Until now the perception of the extent and significance of the participatory processes created by the WFD has been limited. It is unclear how participatory processes that are highly focused on particular sub-basins will be integrated in the dynamics of change in an entire basin. In addition these experiences may end up being very isolated and, although functional, may not ever change the correlation of forces between those involved and the agendas in each basin. Further, the long-standing nature of some of the conflicts has reduced the plurality of those involved in these participatory experiences. Be that as it may, what is significant is the opening up of these spaces for participation in a scenario which had previously been conditioned by the presence of very few and very specific interests that had not included the logic of sustainability among their strategies.

15.6 Conclusions

Water management is perhaps one of the most critical issues in Spain. The combination of historically cyclical droughts with a rapid increase in the number and types of water uses in recent decades makes water policy a focal point of public intervention and a subject of political, socio-economic and territorial controversy. In addition, the geographical and climatic diversity of the territory, in which a wet northwest and a dry southeast Spain cohabit, has accentuated the distributive conflict. Who has the right to use water, who bears the environmental costs and who pays for the real cost of water are just some of the core issues at stake. In recent years new considerations have been introduced into the water policy agenda, some of which emphasize sustainability and efficiency. Yet the prevalence of conflicting views on how to meet both of these guiding principles in dealing with the water scarcity has prevented the introduction of far-reaching changes in water management.

The institutionalized policy image of water management based on the promotion of state-subsidized large-scale supply infrastructure has been prevalent in recent decades, following long-standing traditions that date back to the 1950s and earlier. More recent water management has appeared to inherit some of the problem-solving approaches of the past. However some signals of transition towards a more sustainable management model have gradually emerged. Deviations from the traditional path consist of the gradual abandonment of the focus on large hydraulic infrastructures to alternative regulations on supply. This does not mean in turn that the new policy approach emphasizes the regulation of water demand. Instead it redefines the modes of intervention in water supply in apparently more sustainable terms.

For obvious reasons it is too early to assess the outcomes of the new policy approach and, even more so, whether this constitutes a consolidated change in the water management paradigm. The traditional modes of water governance have come to seem outdated. However the changes introduced in 2004 are too recent and might be episodic, not necessarily constituent of a long-term fundamental change. The need to comply with the requirements of the WFD combined with increasing supply problems may encourage more enduring processes of change.

In order to account for the outcome of escalated transition identified in the three-step model of transition, this chapter has focused on the ways in which policy entrepreneurs have succeeded in using strategies that question the institutionalized policy image (see Chapter 2, the theoretical chapter by Meijerink and Huitema). By means of social mobilization campaigns, generating and disseminating knowledge and exploiting political opportunities and EU venues, environmental organizations, the scientific community and some political actors have managed to develop a new policy image by which the dam approach is no longer associated with the idea of progress and territorial solidarity, but rather with unsustainability and territorial imbalance.

Which strategies have been used by the policy entrepreneurs? Five types of interrelated entrepreneurial strategy stand out. First and foremost, they have succeeded in developing and selling alternative policy ideas on water management, robustly founded on scientific research. Second, building coalitions of environmental groups and large sectors of the scientific community, with the support of most political parties in order to produce knowledge, enhance shared beliefs and pool resources to contest the government policy option and provide high visibility to the alternative policy image. Third, use of the political window of opportunity stemming from the combination of increasing attention to the issue on the political agenda, policy-makers' concern over water management options and the political scenario in the aftermath of the March 2004 elections. Fourth, shopping among multiple institutional venues, in particular at the EU level, as a way to amplify their concerns and find institutional alliances. Finally, joint efforts: network creation efforts that helped to break up the long-standing policy community around water management. In brief, water policy entrepreneurs in Spain, by opposing a deep-rooted policy paradigm, partially succeeded in their change strategies, managing to weaken the traditional policy image, alter the policy dynamics and redefine policy content.

Notes

1. Royal Government Decree 2/2004 of 18 June, amending Law 10/2001 of 5 July, concerning the National Water Plan.
2. Actuaciones para la Gestión y la Utilización del Agua (Actions for Water Use and Management).
3. RDL 2/2004 Modifying the National Water Plan and 11/2005 Act Modifying 10/2001 Act on the National Water Plan.
4. http://www.mma.es/secciones/agua/programa/quees.htm, accessed 11 January 2008.
5. http://www.wwf.es/aguas_continentales/gestion_agua.php; http://www.greenpeace.org/espana/campaigns/aguas, accessed 16 January 2008.
6. http://www.mma.es/secciones/agua/programa/como.htm, accessed 16 January 2008.
7. http://www.wwf.es/aguas_continentales/gestion_agua.php, accessed 16 January 2008.
8. http://www.greenpeace.org/espana/reports/esto-no-es-plan-la-visi-n-de, accessed 14 January 2008.
9. http://ec.europa.eu/environment/water/spanish_hydrological_plan.html.
10. http://www.mma.es/secciones/agua/informes.htm.
11. http://www.unizar.es/premio_goldman/index.htm, accessed 14 January 2008.
12. http://www.wwf.es/aguas_politica_phn.php, accessed 14 January 2008.
13. http://ec.europa.eu/environment/water/spanish_hydrological_plan.html, accessed 14 January 2008.
14. See reports at the website of the Ministry of the Environment, http://www.mma.es/secciones/agua/informes.htm, accessed 14 January 2008.

15. http://www.unizar.es/fnca/index3.php?id=1&pag=16, accessed 15 January 2008.

References

Bakker, K. (2002), 'From state to market? water *mercantilization* in Spain', *Environment and Planning A*, **34** (5), 767–90.

Baumgartner, F.R. and B.D. Jones (eds) (2002), *Policy Dynamics*, Chicago, IL: University of Chicago Press.

Bukowski, J. (2007), 'Spanish water policy and the NHP: an advocacy coalition approach to policy change', *South European Society and Politics*, **12** (1), 39–57.

Conejero Paz, E. (2005), 'El Plan Hidrológico Nacional. Un enfoque neopluralista', VII Congreso Español de Ciencia Política, Madrid, unpublished.

Costejà, Meritxell, Nuria Font and J. Subirats (2004), 'The evolution of water regime in Spain', in I. Kissling-Näf and S. Kuks (eds), *The Evolution of National Water Regimes in Europe. Transitions in Water Rights and Water Policies*, Dordrecht: Kluwer Academic Publishers, pp. 235–63.

Espluga, J.L. and J. Subirats (2007), 'Participación ciudadana en las políticas de agua en España', Fundación Nueva Cultura del Agua Workshop, Seville, December, mimeo.

Kingdon, J.W. (1995), *Agendas, Alternatives and Public Policies*, New York: HarperCollins.

North, D. (1990), *Institutions, Institutional Change and Economic Performance*, Cambridge: Cambridge University Press.

Pierson, P. (2000), 'Increasing returns, path dependence, and the study of politics', *American Political Science Review*, **94** (2), 251–67.

Richardson, J. (2000), 'Government, interest groups and policy change', *Political Studies*, **48** (5), 1006–25.

Sauri, D. and L. Del Moral (2001), 'Recent developments in Spanish water policy: alternatives and conflicts at the end of the hydraulic age', *Geoforum*, **32** (3), 351–62.

Tàbara, J.D. and A. Ilhan (2008), 'Culture as trigger for sustainability transition in the water domain', *Regional Environmental Change*, **8** (2), 59–71, http://www.springerlink.com/content/103880/?k=Ebro.

Thelen, K. (1999), 'Historical institutionalism in comparative politics', *Annual Review of Political Science*, **2**, 369–404.

16 Neoliberal transitions in hydropower and irrigation water management in Turkey: main actors and opposition groups

Aysegül Kibaroglu, Argun Baskan and Sezin Alp

16.1 Introduction

The main goals of state water allocation and planning policy in Turkey comprise the following: independence from imported energy sources; increased agricultural production; satisfaction of increasing industrial, urban and rural demands for water; and the resolution of regional economic and social imbalances in order to raise the living standards of the population. From the 1950s the inclusion of such social aims led to water resources planning and development being carried out by government agencies through public investment.

Since the early 1980s, however, a neoliberal[1] transformation of the Turkish political economy has resulted in significant changes in water policy and management. The privatization of irrigation water management in the early 1990s with the guidance and partial financing of the World Bank serves as illustration. Within the framework of an accelerated programme of management transfer, irrigation associations (IAs) were established to operate and maintain almost all the irrigation systems in the country. Liberalization of the hydroelectricity sector has been under way since the 1980s, reinforced by important legislation adopted in 2005. Hence this chapter focuses on the water policy and management transitions in the hydroelectricity and irrigation sectors[2] and outlines the role of the key actors fostering change and that of the opposing coalition.

The neoliberal economic transitions seen in Turkey since the 1980s have typically been initiated by high-level politicians, in particular Turgut Ozal during his premiership (1983–89) and presidency (1989–93). Yet policy change specifically in the water sector was supported and carried out by bureaucrats from the ministries concerned and their affiliated institutions. Additional key advocates of the policy change included the World Bank and other international actors such as the Organisation for Economic Co-operation and Development (OECD) and European Union (EU) and private national and international corporations. We describe and examine below the strategies used by these actors in implementing the neoliberal irrigation and hydroelectricity policies.

In the context of criticisms of and actions against the water policy transition by the coalition of opponents, we examine the position and activities of the Union of Chambers of Turkish Engineers and Architects (TMMOB in Turkish) as the actor that has developed the major critiques.

The TMMOB is a corporate body and a professional organization defined as a public institution as stated in Article 135 of the Turkish Constitution. An umbrella organization, it has 23 chambers and about 300 000 members. The Chambers of Electrical, Mechanical, Civil and Agricultural Engineers under the TMMOB have all filed lawsuits, produced numerous declarations and organized public meetings, briefings and national

Map 16.1 Turkey and its main rivers

water congresses with the main objective of blocking the shift to neoliberal water policy and management. The chambers in doing so have allied with parliamentarians from the opposition parties and academics from Turkish universities. The National Association of Industrialists and Businessmen (USIAD in Turkish) has also supported the general stance of the TMMOB, and produced reports in 2007 and 2008 critical of national water policy. In this chapter, we explore the key strategies of the opposing coalition, setting out their discourses and practices aimed at preventing or limiting the policy transition in the hydropower and irrigation subsectors. Map 16.1 depicts Turkey's main rivers.

16.2 The rise of neoliberalism in Turkey and the political economy of water

Until the beginning of neoliberal restructuring of the Turkish economy in the early 1980s, Turkey had long followed the Keynesian development model. The state had actively participated in the nation's economic life. As a result of low levels of private capital within the country and the effects of the Great Depression of 1929, the state acted from the 1930s on as the principal investor and producer in the main industrial sectors. It did so through the operation of State Economic Enterprises (SEEs).

State intervention in the economy aimed to provide domestic production of consumer goods in order to decrease dependency on imports. High quotas and tariffs were employed to protect the national manufacturing sector (Yeldan, 2001, p. 38). State economic intervention was institutionalized in 1960 with the establishment of the State Planning Organization. From then on, all public investment was arranged according to national five-year development plans.

The protectionist development model of import substitution was applied until the end of the 1970s when the system went into crisis. Managed with the aim of creating an independent national manufacturing sector, the system ironically created a high level of import dependency. Industry mainly based on small-scale assembly production geared to the domestic market became reliant on foreign technology, unable to meet demand filled through imports and unable to appeal to foreign markets (Boratav, 2006, p. 155). The current account deficit resulting from high imports and low exports was compensated for by short- and long-term foreign credits and the injection of foreign exchange by Turkish nationals working abroad as guest workers (ibid.). The 1973 oil crisis changed this situation. That year the Organization of Petroleum Exporting Countries (OPEC) quadrupled oil prices and caused a worldwide crisis. This produced a foreign exchange crisis in Turkey as foreign credit dried up (Yeldan, 2005, p. 15). Inflation skyrocketed.

Turkey's engagement with world markets started in this crisis environment with the government's economic stability programme known as the 'January 24 Decisions'. Announced on 24 January 1980, the programme was prepared by Turgut Ozal who became Deputy Prime Minister in the last days of 1979 (Boratav, 2006, p. 146). Undersecretary for the State Planning Organization, a World Bank adviser and top manager at the Sabanci Holding (one of the largest industrial groups in Turkey), Ozal led decision-making on the national economy. Although presented as a stabilization package, the January 24 Decisions featured the characteristics of a structural adjustment programme designed to achieve a neoliberal transformation of the Turkish economy. It incorporated all the elements of the International Monetary Fund (IMF)'s standard stabilization policy packages presented to underdeveloped countries in crisis and was along the same lines as the World Bank's structural adjustment programmes (Kepenek, 1995, p. 148).

The main purposes of the package were to decrease inflation, increase foreign exchange supply and relieve the state Treasury of the burden of SEEs. Devaluation of Turkish currency and wage suppression were the tools of an export promotion policy to boost foreign exchange. However the government in power did not have efficient means to implement this 'austerity programme'. The country experienced severe social and economic turmoil that led to a military coup on 12 September 1980 and the interruption of Turkish democracy. The military regime contended that the country's main economic problem was inflation stemming from state intervention and the related budget deficit. Hence the new government established under the control of the generals became a supporter of the January 24 Decisions. As a result, Turgut Ozal was reappointed Deputy Prime Minister with responsibility for the economy and the structural adjustment programme was applied under the control of the military regime (Tanor, 1997, p. 30). The shift in Turkey from Keynesian economics to neoliberalism was in this way almost concomitant with a military *coup d'état* (Baysoy, 2006, p. 48).

Ozal's Motherland Party (ANAP) won the successive general elections of 1983 and 1987 leading to his appointment as President (Head of State) in 1989. As the main architect of the neoliberal transformation of the Turkish economy, Ozal followed the paths of Margaret Thatcher and Ronald Reagan and touted the transformation with the slogan; 'There is no alternative'. From 1981 to 1988, under Ozal's economic leadership, the government worked to decrease state intervention and facilitated the shift to a free market economy. The country's foreign trade regime was liberalized with the abolition of quotas and decreased customs tariffs in addition to decreased levels of domestic industry protection. Suggesting that inclusion of free enterprise and privatization would produce economic efficiency, Law Number 3291, enacted in 1986, authorized the Council of Ministers to privatize the SEEs, entities seen as the cause of persistent budget deficits.

Privatization extended also to the deregulation of public services and concessions given to private entities for the provision of public goods and services that had been supplied by the state. The introduction of a build–operate–transfer (to private ownership) (BOT) model to the energy sector in 1984 with Law No. 3096 enabled the private sector to buy the right to generate, transmit and distribute electricity under a public need provision (Imre, 2001). This system was later modified and began to be applied in other sectors including water, where it was extended to the provision of public water and sanitation services in municipalities and to the construction, operation and management of infrastructure, such as dams, hydropower plants and irrigation systems.

Systematic water resources management in Turkey started with the establishment of the State Hydraulic Works (DSI in Turkish) in 1954. Comprehensive planning activities have been undertaken in Turkey since the 1950s, relating to the construction of physical structures to meet the energy and food demands of the growing population. Turkey since then has made considerable progress in augmenting its water supply. However the political and economic crises that have intermittently erupted since the 1970s have hindered the completion of water supply investment projects. Moreover demand for water for drinking, agriculture, industrial uses and energy generation increased at a very high rate in the second half of the twentieth century. In addition the physical structures built to increase water supply caused degradation and depletion of water and land resources (Kibaroglu et al., 2005).

Turkey has a subtropical, semi-arid climate with extremes in temperature. In hydrological terms, Turkey's territory features 25 river basins that exhibit large variations in average annual precipitation, evaporation and surface run-off parameters. In total, average annual run-off totals approximately 186 billion cubic metres (bcm) of which 112 bcm could be exploited at reasonable cost. At present Turkey utilizes 39.3 bcm of its overall capacity, which corresponds to 36 per cent of the total water supply.

At the beginning of the 1960s only 1.2 million hectares out of the 8.5 million hectares of the total irrigable land were irrigated. Among other factors, this prompted the government to pursue socio-economic development goals (under the 1961 Constitution) that aimed to deal with the relative backwardness of the Southeastern Anatolia region. This strategy included development of water and land resources in the region mainly through public investment. The Southeastern Anatolia Project (GAP in Turkish) was designed to increase agricultural productivity and hydroelectric power generation using the existing potential of water, land and human resources. The major objective was to irrigate the region's fertile lands, which make up one-fifth of the irrigable land in Turkey, by exploiting the Tigris and Euphrates rivers. The rate of addition to the country's hydroelectric capacity was increased at the same time, due among other things to the oil crises of the 1970s. Development of hydroelectric power plants and lignite schemes gained prominence as a way to decrease dependency on imported oil. Despite all such efforts, the mismatches between demand and supply continued as a result of increasing population, rapid urbanization and agricultural development.

According to the DSI statistics, annual per capita water availability is 1430 m^3 with a 2008 population of about 72 million. Availability is estimated to decline to 1000 m^3 per capita/year by 2030 with an expected population of 100 million. Turkey's major focus continues to be the development of water resources because of their potential economic and social benefits. However water pollution control and general protection of water-based ecosystems incorporating rivers, lakes and deltas, while increasingly acknowledged as important, have yet to reach satisfactory levels (Demirayak and Dıvrak, forthcoming). Both Turkey's National Environmental Action Plan and the Ninth Five Year Development Plan give top priority to these issues (DPT, 2007).

16.3 Liberalization of the hydroelectricity sector

16.3.1 Introduction

In line with the reasoning of the historical institutionalists (North, 1990), the fundamental water policy change occurred in Turkey during the major domestic political and economic crisis of the late 1970s and the regime change in its aftermath. The 1980 military *coup d'état* and the 1982 Constitution created a political window of opportunity for actors like Turgut Ozal to introduce economic liberalization policies that extended to social provisions including those of the water sector. These policies brought dramatic changes in the 1980s to political and economic thinking in Turkey. Liberalization and deregulation of the national energy sector and its subsectors, including hydroelectricity production, marked a clear departure from earlier decades. Public investment had characterized the period between 1950 and 1980 and particularly the 1970s, which had seen the construction of large dams with no participation by the private sector (Tutus, 2006).

Established in 1970 the Turkish Electricity Authority (TEK) formed a statutory monopoly until 1984. Authorization under the 1984 Law No. 3096 of private sector generation, transmission and distribution of electricity included not just BOT, but also build–own–operate (BOO) and transfer of operating rights (TOOR). In 1993 the TEK was split into two new bodies, the Turkish Electricity Generation-Transmission Corporation (TEAS) and the Turkish Electricity Distribution Corporation (TEDAS). After the introduction of the Electricity Market Law No. 4628[3] in 2001, TEAS was further unbundled into three companies: generation (EUAS); transmission (TEIAS) and wholesale (TETAS; all abbreviations in Turkish). However this structure has not yet taken its final form (International Energy Agency, 2005).

The Law on the Utilization of Renewable Energy Resources for the Purpose of Generating Electrical Energy (No. 5346), which entered into force in May 2005, is regarded as a 'revolutionary' step in the hydroelectricity business. The Law entails a 'guarantee of purchase' by the government as a strong incentive for private investment. Before the adoption of Law No. 5346, legal aspects of the private sector's involvement in hydroelectricity production business tended to be disputed.

16.3.2 Actors and their arguments opposing the transition to liberalized water policy

The core of the debate over liberalization lies in the objections of the Chamber of Electrical Engineers (CEE) and Chamber of Mechanical Engineers under TMMOB as well as the USIAD. These organizations formed a coalition of actors that tried to block change and demanded the protection of the common good and public interest through the maintenance of the public monopoly. Lawsuits by the CEE in the Council of State interrupted the technical and financial process designed to enable the entry of private companies into the hydroelectricity business. The Council of State's ruling to suspend the liberalization process was a legal victory for the CEE. The major objection to the privatization process was based on the claim that the extension to the private sector of the right to use public natural resources through agreements between government bodies and private companies was simply in contradiction with the public interest.

Additionally the CEE filed another suit questioning the competence of the Energy Market Regulatory Authority (EMRA) regarding the issuance of production licenses to private companies. The Council of State ruled for a second time in favour of the CEE's arguments. Again the privatization process was halted. The Law on Renewable Energy (No. 5346) was adopted in 2005 to overcome the legal obstacles. It made a critical change in the official management of the process and appointed the General Directorate of DSI as the main government body with the legal competence to implement the initial steps for the privatization process and to organize tenders. The result was a partial change in assignment of legal competence from the EMRA as an autonomous agency to another competent government agency, the DSI, which has undertaken regulatory functions in implementing the change. Currently private companies do not face any legal obstacles to construct and run private hydroelectric plants, but the CEE may take further legal action.

Neoliberal policies and globalization starting in the 1980s have clearly contributed to macroeconomic policies in Turkey and specifically to the liberalization of the electricity market. Privatization of electricity markets was on the agenda of the General Agreement on Tariffs and Trade (GATT) talks since 2000 (Thomas and Hall, 2006). In addition Turkey's relations with the World Bank have shaped the neoliberal thinking of Turkish governments for more than two decades. The close connection of national electricity policies with international financing institutions like the World Bank often receives harsh 'anti-imperialist' criticism in public debates and publications (Chamber of Mechanical Engineers, 2006). The OECD is another major international organization that has contributed to the neoliberal economic mentality. Relations with the EU had less direct impact except in regard to the environmental dimension of liberalization processes. Although the EU is accelerating the process of private sector involvement in the hydroelectricity business, and despite the fact that the Law on Renewable Energy represents the Turkish government's answer to findings against its liberalization policies, it is nevertheless a genuine example of the EU's effect on the environmental aspects of Turkey's national energy policy planning and its privatization of the hydroelectricity production sector.[4] Adoption of the Law on Renewable Energy is one of the major steps which Turkey has undertaken to bring its legal framework into line with the EU energy Directives.

There have also been calls using strong 'nationalist' and 'anti-imperialist' rhetoric by the members of the coalition advocating the development of a more 'national' energy policy and restriction of the role of foreign players. The USIAD published a report in 2006 on the need for a national electricity policy that aimed at influencing government as well as private sector actors (USIAD, 2006). This can be understood as a call for protectionism and economic nationalism by avoidance of the challenge of foreign market participants in terms of ownership, capital transfers and employment conditions. The main argument has been related mainly to macroeconomic policies where water policy is presented as a case in point. Entry of foreign capital into almost all national sectors, especially energy and telecommunications, is accepted as carrying risk not only for the national economy, but also for the independence and security of the country. Problems with the EU accession process[5] also feed suspicion of foreign capital. Other objections question not only the management of the liberalization process, but the whole notion of private sector involvement in the business of hydroelectricity production. Behind this kind of objection lies the belief that renewable energy sources should be developed by public bodies and strictly only in the public interest. An accompanying argument points to the failure of similar privatization experiences in European countries, especially the United Kingdom.[6]

It is too early to make an overall assessment of the ongoing, but still young privatization process enabling the construction of private hydroelectricity stations.[7] A view, though, of early experiences in the sector produces a somewhat pessimistic outlook. The inherited national energy policies that failed to provide sufficient public investment and that neglected production capacity and infrastructure, especially in the field of electricity generation, have been stalled. The 1980s BOT model that opened the way for private investments provided purchase guarantees from the Treasury to participating companies. Some BOT power stations are still operating today. BOT-type power stations in general, however, have not lived up to expectations. The coalition opposing the transition in the water irrigation and hydroelectricity subsectors asserts that management of BOT power stations raised energy prices above those prior to privatization. Hence, from the late 1980s, the TMMOB organized several public meetings, and issued numerous public statements and reports to draw the attention of the public to the matter of increasing energy prices, caused largely by the improvident energy policies of successive governments. Since its establishment in the late 1990s, the USIAD has contributed to the efforts of the TMMOB, by producing research reports and through mass media appearances. End users suffered the increases and, due to the Treasury's purchase guarantees, public spending increased so much that even investigation reports prepared by several state organs recorded this failure.[8] From the 1980s to the mid-1990s the successive governments tried to overcome the legal, practical and fiscal problems in the liberalization of the electricity market. Supervision of the overall system, however, was weak.[9] The experience of the process led to the introduction of the current model under the 2001 Electricity Market Law.

The coalition against the transition has advanced a series of arguments concerning several persistent technical and bureaucratic ambiguities of the liberalization process. The transition opponents propose in-depth examination of the technical competence and financial background of applicant companies. Under the official procedures companies first apply to the DSI, then present their projects to the EMRA, which is responsible for financial assessment. The coalition suggests that a unified mechanism for application assessment would achieve more comprehensive and thorough evaluations. On the technical side, transition opponents point out that gaps can develop between production targets and actual electricity power output since companies have still to reach satisfactory production performance levels. There are indeed indications today of poor realization of some private investment projects (Keloglu, 2006, p. 23), where only about half of the targeted 10 000 megawatts of electricity supply to end consumers might be achieved.[10] The coalition also argues that the private company investment plans need to incorporate environmental sustainability measures for the 25 river basins identified by the DSI.

Transition proponents (that is, high-level bureaucrats from the government institutions), counter that high entrance costs to the market and high initial costs for infrastructure construction make it reasonable to grant state support through various schemes to encourage investment financing by private companies (Turkyilmaz, 2006). Similarly it is suggested that an acceleration of the bureaucratic procedure for application assessment and approvals would facilitate the liberalization process and an increase in production. Articulation of comprehensive data on Turkey's hydropower production potential would also contribute to the management of the planning and investment initiatives. Rather than moves to slow or end liberalization, supporters hold that greater speed of applicant review and more precise data on the current situation could help to narrow the potential gap between today's technical forecasts and actual future energy production levels.[11]

The opposition retorts that the present small number of companies limits the range of customer choice. A more competitive business environment could be a necessity for a fully productive and efficient hydroelectricity production system. Increasing competition in the national electricity sector as a whole would require a well-designed preparation process to precede the introduction and auditing of privatized systems. The recent legal challenges do not seem to have derailed the process for the moment, but they could discourage future investments (Economist Intelligence Unit, 2005). The Chamber of Construction Engineers, as part of the opposition coalition, raised doubts that Turkey's bureaucracy is ready to undertake all the related tasks successfully. The administrative system needs restructuring to include more transparent procedures and more sophisticated technical competence (Committee of Water Resources, 2006).

Key policy entrepreneurs include high-level bureaucrats from the DSI, particularly the former Director General, Dr Veysel Eroglu, who is the Minister of Environment and Forest under the current Justice and Development Party (AKP, in Turkish) government. Such entrepreneurs have promoted the liberalization process in the hydroelectricity sector, arguing that the future of the hydroelectricity liberalization process is related to the national energy policy[12] and to changes in the overall national economy. Positive developments would facilitate Turkey's integration with the EU in the energy field and good relations with the new EU-backed Energy Community in the Balkans. In this way the liberalization of the hydroelectricity sector could be considered a part of larger energy policy and other national concerns.

While the production side of the hydroelectricity business will probably remain the domain of national investors, foreign investors are expected to participate in hydroelectricity delivery systems. Turkey has the potential to become an electricity exporter or even a regional leader in the South European electricity production and transmission system. EU–Turkey rapprochement on energy policy issues could contribute to general political relations between the EU and Turkey (Isik, 2004). Official interest in opening the hydroelectricity market to private participation has put the subsector on the larger agenda relating to global climate change. Under increasing public, practical and moral pressure to transform its current carbon-dominated energy production sector, the ruling government has come to see the private hydroelectricity production as a relatively small, but still promising tool to diversify the nation's energy production structure.

The TMMOB has built and led the opposition coalition of professional chambers, business associations, labour unions and the opposition parties. In doing so, the TMMOB has used multiple venues, including: the judiciary – in order to file lawsuits; the legislative arena – where it advised opposition party delegates and parliamentary research commissions; public meetings and conferences; media briefings and interviews.

16.4 Devolution in irrigation management

16.4.1 Introduction

The agricultural sector plays a key role in the social and economic development of Turkey. Yet the sector resembles those of countries with low economic production that produce in alignment with the 'laws of nature', unlike the agricultural sector of highly developed countries, where production is typically protected, buffered or separated from natural conditions (Gunaydin, 2006). Turkish agriculture depends heavily on climatic conditions. Average annual precipitation is 643 mm, ranging from 250 mm in the southeastern part of the country to over 3000 mm in the northeastern Black Sea coastal area. Water is a limiting factor for agriculture throughout much of the country. Irrigation has gradually been expanded since the late 1950s to the point where it now accounts for the largest allocation of water by sector at 74 per cent. Facility construction and other projects related to irrigation constitute 60 per cent of the budget of the DSI, the main public authority in the national water sector (Yildiz, 2007).

Development of irrigation infrastructure in Turkey has been carried out by the DSI and the General Directorate of Rural Services (GDRS), both public sector institutions. The DSI is responsible for planning, design, construction, operation and management of dams, pumping stations and canals for large-scale irrigation systems. Before its dissolution in 2005, the GDRS worked simultaneously with the DSI on large projects and sought to consolidate and even out the capacity of minor irrigation projects (Republic of Turkey, 2003). Itself lacking enough operation and management capacity, the GDRS had been transferring small-scale surface groundwater irrigation schemes to relevant authorities like village organizations and irrigation cooperatives. The transfers made by the GDRS were informal, involving the direct devolution of construction schemes to appopriate bodies without any legal agreement being signed, in contrast to transfers administered by the DSI.

The DSI has the primary task of management and operation of irrigation systems, including the collection of water fees and delivery of water to farms. However Establishment Law Code 6200 (1954) entitles the DSI to transfer the operation and maintenance (O&M) of irrigation systems to irrigation management organizations (IMOs), such as village administrations, municipalities, cooperatives, irrigation associations (IAs)[13] and other private legal entities. From the early 1960s the DSI had a programme for such transfers relating to secondary and tertiary canals. Until 1993, however, the DSI was able to transfer O&M irrigation systems amounting only to approximately 70 000 ha to various types of IMOs. The process has gained momentum since 1993 and from 1993 to 2009 nearly 2 million ha of irrigation schemes have been given over to local administrations or to IAs.

16.4.2 Accelerated transfer programme: the DSI, irrigation associations and the World Bank as policy entrepreneurs

Since the World Bank offered a loan in 1993 to address problematic irrigation practices in Turkey, irrigation management in the country has undergone a rapid transformation. The loan was conditional on the full transfer of O&M responsibilities of irrigation facilities to water user groups. The transfer was presented as the main 'cure' for irrigation problems in Turkey. As a result, the government started the irrigation management transfer (IMT), a national accelerated programme of O&M transfers in the irrigation sector starting in 1993, whereby almost 90 per cent of irrigation schemes were transferred to IMOs, IAs in particular.

Key background conditions leading to the IMT include: adoption of neoliberal policies in overall macroeconomic decisions since the mid-1980s, which ended up decreasing levels of public investment and public sector borrowing in many sectors including irrigation; a national budgetary crisis that led to severe limitations on financial allocations to the DSI in general and to the O&M department in particular; and progressive deterioration of the irrigation infrastructure due to deferred maintenance (Svendsen and Nott, 1999, p. 20). Moreover the country was experiencing escalating labour costs nationwide. By implementing the O&M transfers, the government aimed to decrease the number of DSI personnel responsible for O&M services. This in turn was intended to release O&M staff for priority design and construction work (DSI, 2004). It was expected also that O&M costs undertaken by the state would decrease as a result of devolving authority to the water users. By making users directly responsible for costs, the transfers would end the previous poorly functioning cost recovery system, where fee collection levels were very low and only realized two years after the costs were incurred. Although water fee rates collected from farmers should have covered 100 per cent of O&M service costs, in practice only a fraction of real costs were recovered owing to high inflation rates. What is more, as water fees were collected and retained by the Ministry of Finance, the collected funds made no contribution to the DSI budget.

Conditions at the national level combined with the motivation of the World Bank for the transfers became an important factor among those leading to transition in irrigation management. From the mid-1980s World Bank authorities had been pushing the Turkish government to take measures to reduce operation, management and investment costs of irrigation facilities (World Bank, 1992). A World Bank supervisory mission visited the DSI in 1992 to facilitate the transfer process. The World Bank staff emphasized that if the DSI transferred the responsibility for O&M of irrigation systems as well as the collection of fees to the IAs, the overall cost recovery system would function properly. As part of the loan agreement the World Bank offered to provide international experience on transfer of irrigation and drainage facilities, which could be adopted as models for the participatory irrigation management practices in Turkey. Hence the World Bank team organized Bank-sponsored study trips to Mexico and the United States where the DSI staff accumulated knowledge pertaining to the irrigation management transfer to locally controlled organizations (see also Wilder, Chapter 5 in this volume).

We conclude that the key policy entrepreneurs in the devolution of irrigation management in Turkey were the national- and international-level bureaucracies, namely those of the DSI and the World Bank as the donor agency. With the technical and financial support of the World Bank, the DSI played the critical role of establishing the IAs using a top-down approach. The IAs have since taken up the role of policy entrepreneur at lower administrative levels. Thus the DSI, the World Bank and the IAs formed a coalition of actors advocating and implementing the irrigation policy transition. Meanwhile the opposition group composed of the members of the Chamber of the Agricultural Engineers under the TMMOB, some bureaucrats from government agencies, such as the GDRS and the Southeastern Anatolia Project Regional Development Administration, and academics critical of the process, constituted a coalition against the accelerated IMT process.

The coalition against the transition asserted that it would be imperative in implementing the accelerated IMT programme to consider first and foremost socio-economic and socio-cultural peculiarities, by paying particular attention to regional disparities in economic development. In order to involve farmers fully in the transfers, members of the coalition called for the transfers to be more gradual and to include targeted timetables. While critics agree that farmers should be involved in irrigation management, they believe that participation will not reach predicted levels, particularly in regions like Southeastern Anatolia where a tribal social structure still dominates social and economic life. This leads them to suggest that public agencies like the DSI should develop support programmes to strengthen the democratic, administrative and technical formation of the IAs (Kibaroglu, 2002).

16.4.3 Opposition coalition criticisms of the accelerated transfer programme: more privatization, less participation?

The IAs in Turkey have been established through existing local government structures. The IA as a concept is not a product of any grass-roots movement or organization, but came largely from DSI staff as the major initiators and executors of the transfer programme. Yet there were intensive negotiations between the DSI as the relevant agency and the local communities, particularly the representatives from the local governments, concerning the terms of transfer.

As a result three main types of organizations are presently involved in the management, operation and maintenance of the irrigation schemes, extending from the primary water source down to the farm level:

1. DSI – managing bulk water supplies and the main supply canals and controlling the distribution of water to the IAs.
2. IAs – managing the secondary systems and controlling the distribution of water at the head of the tertiary hydraulic units.
3. Informally organized groups of irrigators – controlling water distribution to individual farmers within the tertiary hydraulic units.

Under this structure, O&M responsibility for irrigation schemes is shared between the DSI and the relevant bodies that it authorizes.[14] Based on the aim of providing irrigation for all needy areas in the country, the IMT was restricted to the O&M service provision, leaving the irrigation schemes and equipment in the possession of the state.

If an irrigation facility is used by only one local administration, for example a village or a municipality, the DSI transfers O&M of the facility to this specific local unit. If the facility provides the irrigation service for more than one local unit, say two municipalities and villages, then the transfer is made either to an irrigation cooperative formed for each pair and as based on the Cooperatives Law No. 1163 (adopted in 1969), or to the irrigation associations established according to the Municipality Law No. 1580 (DPT, 2007). A new law on Local Government Associations, Law No. 5355, was enacted in 2005 with a section devoted to the establishment of water user associations. Although the new law makes some improvements on the previous municipal law, considerable difficulties still arise in registering essentially private entities under a public administration law (World Bank, 2006).

The establishment, membership, management, rights and obligations of IAs are governed by three principal legal instruments: the Municipality Law, the Transfer Agreement between the DSI and each individual IA, and the Statute of each IA. The use of the Municipality Law, under which the IAs were established, appears to have been dictated by administrative convenience rather than appropriateness for the management of relatively complex irrigation and drainage systems. As IAs are based on existing local administration, it was possible to form a large number of them quickly, but there has been no direct involvement of irrigators in the transfer process. IAs have had a very short period to establish themselves before starting operations and they have so far lacked the necessary technical and managerial skills (Halcrow–Dolsar–RWC Joint Venture, 1994).

The IA Statute is the document that establishes an IA as a corporate body. IAs operate directly under the statute and have not developed formal internal regulations or written procedures for management and conduct of their meetings or sanctions, despite the fact that the statute requires that such regulations are prepared and approved by the Governor's office.[15] Model regulations have not been available from the DSI.

The Transfer Agreement, prepared by DSI, sets out the rights and responsibilities of the IA and the DSI. It specifies the parties to the agreement, the irrigation facilities and the O&M instructions for the IA. The agreement does not specify details of irrigation, drainage and service facilities transferred, and no inventory or condition report is made at the transfer. IAs have not been provided with system specifications or performance, detailed drawings, maps or O&M manuals. As a result, IAs cannot carry full responsibility for managing irrigation systems in an efficient and equitable way. They cannot, for example, be considered to have a responsibility for drainage. O&M costs and cost-sharing for systems providing services in common are not defined. The transfer is also silent on water entitlement and the conditions under which the DSI might reduce or withdraw supply.

An IA is administered under municipal laws by two main governing bodies: a general assembly and a board. A general assembly, typically with 30 to 70 members, consists of mayors and village administrators (*muhtars*, in Turkish) as 'natural members' and a number of 'selected members', who are chosen by the natural members. A DSI representative serves as an observer member. An IA board typically has seven members. The general secretary and accountant are board members and either the assembly or the board members elect the remainder from general assembly members. The chair is regarded as an administrative post of the IA and entails specific duties, which include representing the IA, preparing budgets, implementing general assembly and board decisions, acting as paymaster and conducting board meetings. The IA must employ a general secretary, who must be an agricultural engineering graduate and who has no specific duties other than to take on tasks of the IA under the direction of the chair. Some IAs employ an O&M technician to supervise fieldwork. Maintenance is undertaken by daily hired labour, if at all. Recruitment of staff is not transparent and depends on the decision of the chair, who often favours relatives (Erdogan, 2004). Tribal culture affects village- and farmer-dominated IAs, particularly in eastern and southeastern regions of the country where tribal chiefs maintain their power through land ownership and political power. IA membership of landowners who belong to different ethnic groups can cause divisions within the general assembly and board. The dominant group will occupy the post of chair and all the seats on the board, excluding other groups from IA governance (see Chapter 12 on Tanzania by Goldin and Kibassa in this volume).

The IA general assembly, made up of *muhtars*, does not represent the irrigator, whose only recourse is elections for the post of *muhtar* held every five years. As a result the IA typically does not look after the interests of the majority of its customers. The chair and board members, who are usually large landowners, can and do favour themselves and other individuals with exemptions from paying water charges or extra water supplies, and tend to try to reduce water charges to below sustainable levels. Some training support has been given to IA accountants, through seminars run by the DSI, but there is no adequate system to identify training needs and support for IA staff. This is an ongoing need as accountancy staff are periodically replaced. The DSI provides some support to general secretaries through annual workshops and seminars, but there are no formal training programmes, working manuals or sets of procedures available for new staff. This leads to a wide variation in IA practices, and standards of management range from good to very poor. Without standard procedures it is difficult to identify and correct poor practices. The persistence of a lack of clearly defined operational guidelines and insufficient numbers of adequately trained staff has led to poor coordination and inefficient management and implementation of operations and maintenance at all levels in the irrigation systems. The result has been a poor and unreliable delivery of services to the end user, the farmer.

The current process of establishing and handing over an IA also contributes to poor IA performance and very considerable long-term risks and economic costs (Kibaroglu, 2002). The potential benefits of introducing good O&M into new irrigation schemes at an early stage are great, and the costs of doing so relatively small. IA staff need to be properly trained before taking over the system, and require adequate support through O&M guidelines, systems, manuals, full supervision during initial seasons and performance monitoring.

Successive governments have seemed quite content with the 'reform process' in irrigation management, and the World Bank has pronounced Turkey a 'case of success' (see also Chapter 5 on Mexico by Wilder in this volume). The coalition against the accelerated irrigation management transfer process, however, has levelled considerable criticism. The Chamber of Agricultural Engineers has led the objections, focusing on a number of issues (Soylu, 2006). The participatory aspect of the transfers in particular has been questioned owing to the exclusion of irrigators from IA general assemblies and boards. In addition, the top-down approach, adopted instead of a grass-roots approach generated by farmer interest and involvement, has caused fierce debate over the characterization of the associations as democratic. According to the Chamber, irrigation cooperatives that were established based on the Cooperatives Law and that have been functioning since the 1960s provide sufficient ground for democratic, participatory management. The implementation of transfers based on IAs, whose formation has not been governed by any specific law, is seen as problematic. The coalition against the policy transition contends that transfers should be made to irrigation cooperatives rather than IAs typically lacking in technical, administrative and legal capacity (Suicmez, 2008).

Critics stress that maintenance, rehabilitation and modernization of the irrigation canals, some of which are 40 years old, cannot be accomplished due to these kinds of capacity deficiency. The Chamber of Agricultural Engineers is of the opinion that Turkey will not completely implement the new irrigation management policy unless public institutions are assigned full responsibility for building new irrigation systems concomitantly with the rehabilitation of old systems (Gunaydin, 2007). As the major opponent of the transition, the TMMOB in general asserts that irrigation management should be constituted as a public service. The group argues that because irrigation uses the largest share of water supply in the country and comprises an important element of socio-economic development, it should not be left to the technically inefficient and organizationally inequitable IAs.[16]

The coalition against the transition has used strategies such as organizing public events attended not only by the opposition bloc, but also by key agents of change, including high-level bureaucrats, experts and international agency representatives. Events of this kind include two successive national conferences on water policy organized in 2006 and 2008 by the TMMOB and public meetings organized on World Water Day almost every year since the late 1990s. Other mechanisms like public reports, declarations and briefings have also served as attempts to block the change using venues such as press conferences and the Internet. However the efforts of the coalition protesting against the new policy have proved quite ineffective since more than 90 per cent of the public irrigation systems have now been transferred to the IMOs in the form of IAs under the accelerated transfer programme.

16.5 Conclusion

A general overview of the Turkish water sector focused particularly on the bureaucracy of water management displays some changes, such as: the realignment of the DSI as the major water agency within the Ministry of the Environment and Forestry (formerly the Ministry of Energy and Natural Resources); the abolishment of the GDRS without assignment of its main functions of rural soil and water conservation to a relevant public authority; and the adoption of a series of laws relating to water quality with the aim of harmonizing with EU legislation in the same sector. In our view these changes, in contrast to their consequence of deterioration in O&M policy, are changes in governance rather than substantial changes. The associated bureaucracy in recent years has made attempts to review the institutional and legal structure of the water sector as part of EU harmonization efforts. No real change in the form of new practices has emerged thanks to a lack of political will, bureaucratic inertia and the exclusion of water users at all levels from the review process. Some changes in participation can be observed, however, in terms of the involvement in the water sector of non-governmental organizations, such as environmental foundations and associations. Such groups have been more visible since the mid-1990s through activities like public awareness campaigns and pilot implementation projects aimed at integrated water resources management. Yet such involvement in the sector has not generated significant changes in the habits of either water suppliers (public and private) or users.

Changes in water management are often the result of drastic changes in government policy and regulations. In Turkey the major domestic political and economic crisis in the 1980s paved the way for the neoliberal transformation of the water sector. In this chapter we have traced the water policy and management transitions in the hydroelectricity and irrigation subsectors, describing the work of the coalition of actors who carried out the water policy change and of the coalition who tried to block the transitions. Our analysis shows that the interactions and negotiations between the key policy entrepreneurs and the opposing coalition have been very weak.

The transitions were imposed largely by successive national governments and the bureaucracies of the agencies involved including those of supportive international agencies. These effective policy entrepreneurs took advantage of the national and international window of neoliberal political opportunity. Most of the experts, academics and intellectuals (that is, epistemic communities) at the national level were not consulted on or included in the process of formulation and planning of the new water policy. More important, water users at urban and rural levels had barely any chance to deliberate on the new policies; nor were they informed about the changes that resulted, changes that would have serious consequences for their social and economic status. The TMMOB has led the coalition against the transition, which has tried to accomplish its mission of blocking the change by using multiple venues, including several for public communication, the judiciary and legislative mechanisms. The impact of the coalition, however, has been quite limited. Liberalization in the hydroelectricity sector has been under way since the mid-1980s and devolution of irrigation management is almost complete. It seems likely, then, that the members of the coalition against the transition stand a chance of seeing their preferred policies and practices realized only in the case of victory by the sympathetic major opposition party in parliamentary elections. Such a shift in governing power seems essential to open a window of opportunity onto a fresh political landscape. Given sufficient alignment of political, epistemic and community factors, it might then be possible for the present opposing group to transform into an advocacy coalition with the strength and configuration needed to implement a new transition in water policy in Turkey.

Notes

1. Neoliberalism is an ideology and political strategy that has aimed to improve conditions of capital accumulation on a world scale since the late 1970s, on the basis of financial and trade liberalizations as well as privatizations accompanied by political, economic and legal restructurings in many countries. For further discussion on neoliberalism see Bedirhanoglu (2008).
2. Even though commercialization of water services (drinking water and sewerage) represents another case of liberalization of the water sector, this dimension of the subject matter is not addressed in the chapter due to space limitations.
3. 'The purpose of this law is to ensure the development of a financially sound and transparent electricity market operating in a competitive environment under provisions of civil law and the delivery of sufficient, good quality, low cost and environment-friendly electricity to consumers and to ensure the autonomous regulation and supervision of this market.' See the Energy Market Regulatory Authority website: www.emra.org.tr (accessed in September 2008).
4. Interview with Hamza Ozguler, General Directorate of State Hydraulic Works, December 2007.
5. These problems are related to Turkey's prolonged EU membership negotiation process and to growing EU scepticism in the country, which originates from the nature of transformation (that is, privatization policies) imposed on Turkey by the political and economic criteria set by the the EU Commission for EU membership.
6. Interview with Kemal Ulusaler, former chairman of the Chamber of Electrical Engineers, December 2007.
7. Interview with Dursun Yildiz, senior expert and manager (retired), General Directorate of State Hydraulic Works, June 2008.
8. Interview with Kemal Ulusaler, December 2007.
9. Interview with Dursun Yildiz, June 2008.
10. Interview with Dursun Yildiz, June 2008.
11. Interview with Dursun Yildiz, June 2008.
12. Interview with Hamza Ozguler, December 2007.
13. A form of transfer considered innovative, where the irrigation scheme covers more than one local administrative unit (for example, village or municipality) (Kibaroglu, 1999).
14. Article 2 of Law No. 6200.
15. Turkey is divided into 81 administrative provinces. Each province has a governor serving as head of the provincial government, appointed by the Council of Ministers with the approval of the President. As chief executive of the province and principal agent of the central government, each governor supervises other government officials assigned to carry out ministerial functions in his or her province.

16. Interview with Erkan Alemdaroglu, former Regional Director, GAP Regional Development Administration, member of the Chamber of Agricultural Engineers, Ankara, June 2008.

References

Baysoy, E. (2006), 'Political economy of development in a historical context: international and Turkish experiences', unpublished Masters thesis, Middle East Technical University, Ankara.

Bedirhanoglu, P. (2008), 'Methodological Investigations on the State in Neoliberal Globalisation: State Restructuring in Turkey', paper presented at the Fifth Historical Materialism Annual Conference, Many Marxisms, University of London, School of Oriental and African Studies, London, 7–9 November.

Boratav, K. (2006), *Türkiye Iktisat Tarihi: 1908–2005* (The economic history of Turkey: 1908–2005), Ankara: Imge Yayinevi.

Chamber of Mechanical Engineers (2006), 'Turkiye'nin Dogal Gaz Temin ve Tuketim Politikalarinin Degerlendirilmesi Raporu' ('Assessment report on Turkey's policies of natural gas obtainment and consumption'), Ankara: Makine Muhendisleri Odasi (Chamber of Mechanical Engineers).

Committee of Water Resources Planning and Electricity Power of the Ankara Branch of the Chamber of Construction Engineers (2006), 'Su Kaynaklarimizdan Enerji Uretimi Politikamiz ve Gelecek Projeksiyonlari' ('Our hydroelectricity policy and projections for the future'), in *TMMOB Su Politikalari Kongesi – Bildiriler Kitabi 1. Cilt* (*Union of Chambers of Turkish Engineers and Architects – Proceedings of Congress of Water Policies, Volume 1*), Ankara, 21–23 March, Ankara: Insaat Muhendisleri Odasi, pp. 308–9.

Demirayak, F.and B.B. Dıvrak (forthcoming), 'Growing role of NGOs in water resources management in Turkey, case study: Konya closed basin', in K. Ayşegül, A. Klaphake, A. Kramer and W. Scheumann (eds), *Turkey's Water Policy, National Frameworks and International Cooperation*, Berlin: Springer-Verlag.

DPT (State Planning Organization) (2007), '9th Development Plan, Land and Water Special Committee Report', Ankara.

DSI (2004), 'Irrigation management transfer', Capacity Building Symposium on Integrated Water Resource Management, www.emwis.org/documents, accessed in May 2008.

Economist Intelligence Unit (2005), 'Turkey country profile', www.economist.com, accessed 13 September 2006.

Erdogan, G. (2004), 'Sustainable irrigation in Harran Plain, Turkey: the roles of stakeholders', unpublished MSc thesis, Tema V, Faculty of Arts and Sciences, Linköping University, Sweden.

Gunaydin, G. (2006), 'Türkiye Tarım Sektörü', ('Turkish agricultural sector'), *Tarim ve Muhendislik* (*Agriculture and Engineering*), 76–77.

Gunaydin, G. (2007), Chair, Chamber of Agricultural Engineers, 'Opening Speech', at the symposium on Climate Change, Drought and Water Management, Ankara, 21 March.

Halcrow–Dolsar–RWC Joint Venture (1994), '*Identification report*', Volume 1, Main Text, GAP-MOM Study, Management Operation and Maintenance of GAP Irrigation Systems 3, Sanliurfa, Turkey.

Imre, E. (2001), 'Build–operate–transfer model in Turkey, legal structure and application', www.ydk.gov.tr/seminerler/turkiyede_yid_modeli.htm#t5, accessed in April 2008.

International Energy Agency (2005), *Energy Policies of IEA Countries: Turkey 2005 Review*, Paris: IEA Publications Service.

Isik, Yusuf (2004), 'Turkey's energy prospects in the EU–Turkey context', EU–Turkey Working Papers No. 9, Centre for European Policy Studies, www.ceps.be, accessed 4 February 2007.

Keloglu, N. (2006), 'Turkiye Su Kaynaklarinin Gelecekteki Kullanimi ve Hidroelektrik Enerji Uretimi' ('Water use and hydroelectric energy production of Turkey in the future'), in *TMMOB Su Politikalari Kongesi – Bildiriler Kitabi 1. Cilt* (*Union of Chambers of Turkish Engineers and Architects Proceedings of Congress of Water Policies, Volume 1*), Ankara, 21–23 March, Ankara: Insaat Muhendisleri Odasi, pp. 13–24.

Kepenek, Y. (1995), *Turkiye Ekonomisi* (*Turkish Economy*), Istanbul: Remzi Kitabevi.

Kibaroglu, A. (1999), 'Turkish experience in irrigation management transfer', paper presented at the *Conference on the Irrigation Reform Policy Process*, Irrigation and Water Engineering Group in the Department of Environmental Sciences at the University of Wageningen, World Bank and the Ministry of Water Resources of India, Hyderabat, India, 11–14 December.

Kibaroglu, A. (2002), 'Building bridges between key stakeholders in the irrigation sector: GAP-RDA's management operation and maintenance model', in I.H. Olcay Ünver and R.K. Gupta (eds), *Water Resources Management: Crosscutting Issues*, Ankara: METU Press, pp. 172–99.

Kibaroglu, A., W. Scheumann, A. Klaphake, A. Kramer and A. Carius (2005), 'Cooperation on Turkey's Transboundary Waters', Report prepared for the Federal Ministry for Environment, Nature Conservation and Nuclear Safety and Berlin Technical University, Berlin.

North, D. (1990), *Institutions, Institutional Change and Economic Performance*, Cambridge: Cambridge University Press.

Republic of Turkey (2003), 'Turkey country report', prepared for the Third World Water Forum, www.

worldwatercouncil.org/fileadmin/wwc/Library/Publications_and_reports/country_reports/report_Turkey.pdf, accessed in May 2008.

Soylu, N. (2006), 'Turkiye Su Kaynaklari ve Sulama Hizmetleri Yapilanmasi' ('Water resources in Turkey and structuring of irrigation services'), in *TMMOB Su Politikalari Kongesi – Bildiriler Kitabi 2. Cilt* (*Union of Chambers of Turkish Engineers and Architects – Proceedings of Congress of Water Policies, Volume 2*), Ankara, 21–23 March, Ankara: Insaat Muhendisleri Odasi, pp. 339–48.

Suicmez, B.R. (2008), 'Turkiye'deki Sulama Isletmeciligi' ('Irrigation management in Turkey'), www.zmo.org.tr/etkinlikler/usy03/13.pdf, accessed in April 2008.

Svendsen, M. and G. Nott (1999), *Irrigation Management Transfer in Turkey: Process and Outcomes*, Economic Development Institute (EDI) Participatory Irrigation Management Case Studies Series, Washington, DC: World Bank.

Tanor, B. (1997), 'Siyasal Tarih: 1980–1995' ('Political history: 1980–1995'), in S. Aksin (ed.), *Türkiye Tarihi Cilt 5: Bugunku Turkiye: 1980–1995* (*The history of Turkey, Volume 5: Today's Turkey: 1980–1995*), Istanbul: Cem Yayinevi, p. 30.

Thomas, S. and D. Hall (2006), 'GATS and the electricity and water sectors', PSIRU, Business School, University of Greenwich, London, http://www.psiru.org/reports/2006-03-WE-GATS.doc accessed in November 2008.

Turkyilmaz, O. (2006), 'Enerji Politikalari – Yerli, Yeni ve Yenilenebilir Enerji Kaynaklari Raporu' ('Energy policies – report on national, new and renewable energy sources'), Ankara: Makine Muhendisleri Odasi (Chamber of Mechanical Engineers).

Tutus, A. (2006), 'Turkiye'de Elektrik Enerjisinin Tarihsel Gelisimi ve Yeni Piyasa Duzeni Icerisinde Hidroelektrik Enerjinin Yeri' ('New market model and hydroelectric projects in Turkey'), in *TMMOB Su Politikalari Kongesi – Bildiriler Kitabi 1. Cilt* (*Union of Chambers of Turkish Engineers and Architects – Proceedings of Congress of Water Policies, Volume 1*), Ankara, 21–23 March, Ankara: Insaat Muhendisleri Odasi, pp. 318–20.

USIAD (2006), 'Elektrik Enerjisinde Ulusal Politika Ihtiyacimiz: Elektrik Enerjisinde Bugun ve Gelecegimiz Raporu' ('Our need for a national electricity policy: report on temporary and future prospects on electricity power'), USIAD (National Association of Industrialists and Businessmen), www.usiad.net, accessed 13 September 2006, pp. 24–6.

World Bank (1992), 'Turkey irrigation management and investment review', Washington, DC, Report No 11589–TU.

World Bank (2006), 'Economic sector work (ESW) irrigation and water resources with a focus on irrigation prioritisation and management', Working Paper, Water Resources and Institutions.

Yeldan, E. (2001), *Kuresellesme Surecinde Turkiye Ekonomisi: Bolusum, Birikim ve Buyume* (*Economy in Turkey in the process of globalization: distribution, accumulation and development*), Istanbul: Iletisim Yayinlari.

Yeldan, E. (2005), 'Assessing the privatization experience in Turkey: implementation, politics and performance results', Report Submitted to the Economic Policy Institute, Washington, DC.

Yildiz, D. (2007), 'Su Raporu: Ulusal Su Politikası Ihtiyacimiz' ('Water report: our need for national water policy'), USIAD, Ankara: National Association of Industrialists and Businessmen.

17 Transitions to adaptive approaches to water management and governance in Sweden

Per Olsson and Victor Galaz

17.1 Introduction

Human well-being and societal development depend on ecosystem services such as food, timber, medicines, water and air purification, carbon storage, pollination, soil formation, and the provision of aesthetic and cultural benefits (Millennium Ecosystem Assessment, 2005). Fresh water – the 'bloodstream of the biosphere' (Falkenmark, 1999; Ripl, 2003) – is crucial in this respect as it drives critical processes and functions in ecosystems like forests, woodlands, wetlands, grasslands, croplands and other terrestrial systems. This stream of interconnected social–ecological systems, however, is becoming increasingly complex to manage. This is due to human-induced environmental changes, from the local to the global scale, that have serious impacts on water flows and on ecosystems. Some of these changes are incremental and possible to prepare for with integrated planning and monitoring (for example Bates et al., 2008). Others, however, can unfold as surprises and trigger biophysical processes with irreversible ecological repercussions (Scheffer et al., 2001; Millennium Ecosystem Assessment, 2005). Such shifts can erode the capacity of ecosystems to generate essential services and make them more vulnerable to perturbations. Impaired water-related ecosytems, for example, can become less resilient to sudden flooding, nutrient or chemical leakage and algal bloom or high levels of toxic pollutants. Since such perturbations are an inherent or typical part of social–ecological systems, the challenge is to safeguard or restore the capacity of life-supporting ecosystems to respond to change without losing important structures and functions.

Many scholars emphasize the need for new flexible, integrated, holistic forms of water management and governance that can deal with the complexity, uncertainty and surprise entailed in social–ecological systems and their associated freshwater resources (for example Falkenmark and Folke, 2000; Pahl-Wostl et al., 2008; Galaz et al., 2008). Adaptive management, ecosystem management and integrated water resource management are all promising approaches in this context. Two key questions arise. First: can the suggested approaches enhance the resilience of complex freshwater resources? Second: is it possible to facilitate and steer transitions from rigid and centralized governance modes to adaptive governance approaches? (see Pahl-Wostl et al., 2008).

In the first part of this chapter, we use the lens of 'resilience' to highlight some of the major challenges facing conventional management approaches. We focus on the capacity of institutions and broader governance mechanisms to deal with uncertainty and abrupt change in social–ecological systems. The second part of this chapter looks at social–ecological transitions, specifically the social and ecological dynamics that lead to shifts in policy to adaptive governance approaches. We review the mental and institutional barriers to change, and strategies for moving towards new flexible forms of governance that secure the capacity of ecosystems to generate essential services for human well-being.

This scope includes the triggers for change and the role of agency to overcome barriers and to navigate transitions. We concentrate in particular on three aspects of such shifts: improving the capacity for dealing with abrupt change and uncertainty; enhancing learning and experimentation; and supporting participation and collective action. We draw on the insights from our earlier studies in Kristianstad Vattenrike Biosphere Reserve in southern Sweden as well as a wider literature.

Lastly we elaborate on the main differences between conventional water governance approaches and those based on an understanding of complex social–ecological systems. We do so by using the ongoing implementation of the European Union (EU) Water Framework Directive (WFD) in Sweden as a case study (see also Chapters 13, 14 and 15 on the EU, Hungary and Spain, respectively, in this volume). Implementation involves one of the most fundamental reforms of water governance ever seen in Sweden, and the start of a major transition. The question is: to what? Is Sweden moving towards new adaptive approaches? We use an adaptive governance framework (Folke et al., 2005; Olsson et al., 2006) to analyse the degree to which the transition in Swedish water policy incorporates social dimensions of ecosystem management. We examine interactions between individuals, organizations and institutions at multiple levels, and the factors for responding to crisis, shaping change and building resilience.

The answers to the questions posed in the chapter are of relevance not only for those interested in Swedish water policy and current European water policy reform. This chapter's findings relate to one of the major challenges in the management and governance of water resources and ecosystems: can transitions toward new, adaptive approaches be centrally steered? And if so, what are the major pitfalls and possibilities?

17.2 Governance and fresh water as complex adaptive social–ecological systems

17.2.1 The problem of fit between ecosystems and governance systems

Humans are part of ecosystems and ultimately dependent on the capacity of ecosystems to generate services. Ecosystems are complex and adaptive, characterized by historical dependency, non-linear dynamics, threshold effects, multiple basins of attraction and limited predictability (Levin, 1999). Gordon et al. (2008) show, for example, how agricultural modifications of hydrological flows can lead to a variety of ecological regime shifts – changes in the characteristic conditions under which processes occur – that operate across a range of spatial and temporal scales ranging from soil structure to salinity and vegetation cover. Such shifts can have severe implications for food production, the quality and quantity of irrigation, industrial and drinking water and other ecosystem services such as climate regulation and coastal replenishment (see also Resilience Alliance and Santa Fe Institute, 2004). Hence water flows have to be viewed and managed as an intrinsic part of interconnected social–ecological systems, rather than as an easily bounded biophysical resource.

Institutions, planning processes and policy prescriptions that fail to acknowledge this tight interconnection are likely not only to provide ill-founded guidelines, but also to steer societies onto undesirable pathways. An environmental policy or regime cannot be effective unless it incorporates an understanding of the larger social, economic and political context and its dynamics. But neither can a social system succeed, no matter how adaptive, if it is formed out of ecological illiteracy. Current approaches for managing ecosystems often fail to match social and ecological structures and processes operating at different spatial and temporal scales, and including threshold and cascading dynamics (Berkes and Folke, 1998; Carpenter and Gunderson, 2001; Berkes et al., 2003; Cumming et al., 2006; Folke et al., 2007). The factors behind this governance failure lie not only in weak environmental legislation, lack of enforcement power or poor monitoring systems (United Nations Environment Programme, 2007), but also in attempts to control a few selected ecosystem variables in their efforts to deliver efficiency, reliability and optimization of ecosystem goods and services (Holling and Meffe, 1996). Such governance amounts to a command-and-control approach and can bring considerable benefits to humans in the short term. However, treating a set of desirable ecosystem goods and services as stable can create mismatches between institutions and ecosystems that can in turn introduce or increase vulnerability into the systems affected, and also lead to undesirable regime shifts and ecological surprises (Gunderson and Holling, 2002; Folke et al., 2003).

The mismatch between ecological and institutional dynamics is often referred to as 'the problem of fit' (Young, 2002; Folke et al., 2007; Galaz et al., 2008). A major challenge concerning the problem of fit lies in addressing the governance dimension of ecosystem management and the social factors that enable such management. This includes factors that stimulate the development of institutions that respond to both predicted and unexpected environmental changes and that help to maintain the capacity of ecosystems to generate services for human well-being (Folke et al., 2007). It also includes recognizing the importance not only of scientific monitoring and analysis, but also of social processes involved in monitoring ecosystem changes and in generating, accumulating and transferring ecological knowledge and understanding.

17.2.2 Enhancing the fit between ecosystems and governance systems

The fit between biophysical systems and environmental and resource regimes can in principle be enhanced, but not without addressing the fundamental need for governance solutions that build the capacity to harness the dynamics of a highly interconnected social, political and ecological world. There is increased interest from the scientific community ranging from researchers of international environmental regimes and global governance (Biermann, 2007; Young et al., 2008), to natural resource management scholars elaborating the foundations of more adaptive approaches to the governance and management of social–ecological systems (Armitage et al., 2007). Scholars of coupled social–ecological systems often emphasize the importance of social learning, the robustness provided by polycentric institutions and institutional diversity as features needed to deal with complex adaptive systems and to overcome misfits (for example, Berkes and Folke, 1998; Gunderson and Holling, 2002; Westley, 2002; Berkes et al., 2003; Dietz et al., 2003; Folke et al., 2003; Olsson et al., 2004a; Pahl-Wostl et al., 2008; Armitage et al., 2008).

Social learning in this case refers to the development of a common framework of understanding, the creation of a joint basis for action, and the joint analysis of system dynamics in the form of identifying feedbacks, driving forces, thresholds, possible regime shifts and major uncertainties (Walker et al., 2002; Gallopín, 2002; Schusler et al., 2003). These are ongoing processes of trial and error and learning by doing that typically involve the consideration of a range of future outcomes, the weighing of probabilities, small-scale pilot projects, actions designed to be useful across a range of potential futures, reversible actions favoured over irreversible ones, results monitoring and, accordingly, policy modification. This way institutional and organizational arrangements and ecological knowledge can be tested and revised in a dynamic process that includes multiple sources of knowledge, different competencies and distributed decision-making (Gunderson, 1999; Carlsson and Berkes, 2005).

Such learning builds on collaboration and partnerships among multiple stakeholders, like governmental agencies, local communities and resource users, local governments and non-governmental organizations (NGOs). Stakeholders negotiate and share the responsibility and power related to managing a specific area or a set of resources (Berkes, 2002; Schusler et al., 2003; Borrini-Feyerbend, 2004; Carlsson and Berkes, 2005). These processes involve trust-building, resolving conflict and dialogue. In the case of the Northern Highlands Lake District of Wisconsin, for example, the use of scenario planning encompassing social and ecological driving forces served as the basis of multistakeholder dialogue (Peterson et al., 2003a).

A polycentric institutional configuration includes redundant institutions and nested layers of institutions (McGinnis, 2000). Arguments in favour of redundancy focus on increased system reliability in the face of environmental or operational uncertainty (Streeter, 1992). Low et al. (2003) suggest that redundancy and overlapping functions may play a central role in absorbing disturbance and in spreading risk. As observed by Ostrom (2005), polycentric institutions have the capacity to compensate for failure at different scales. Corruption at a local level (for example, in an area trying to cope with illegal logging) may be compensated for by action at higher levels, in the form, say, of national government or international intervention (see Berkes, 2002). Institutions and the interplay among them can enable or stifle self-organization and learning. If institutions are nested, they can create enabling conditions and regulation that encourage learning and self-organization at the levels needed for adaptive governance approaches, potentially accommodating, say, both centralized and decentralized modes.

Institutional diversity – a heterogeneity of institutional types, such as hierarchies, markets or self-governance – is often argued to be far more than inefficient deadweight. Becker and Ostrom (1995) and Dietz et al. (2003), for instance, discuss the risks associated with 'one-size-fits-all' solutions for common-pool resources in a constantly changing world. Norberg and Cumming (2008) highlight the important role of institutions in providing a greater range of options for responding to environmental change (see also Folke et al., 2005). A similar argument has been put forward for the emergence of different trading emission credits for carbon dioxide under the framework provided by the Kyoto Protocol (Victor et al., 2005).

Others have emphasized the role of organizational diversity in addition to institutional diversity. Hong and Page (2004) show how individual diversity within a group can enhance problem-solving. Further, Reidsma and Ewert (2008) show how regional farm diversity can reduce the vulnerability of food production to climate change. Imperial (1999, p. 459) highlights the links between institutional and organizational diversity and argues that polycentric governance creates a rich environment that can 'encourage innovation and experimentation by allowing individuals and organizations to explore different ideas about solving [complex] problems'. This has several implications for governance and management. Independent planning teams may develop alternative management plans based on complementary observations and knowledge, enhancing the range of response options. Redundancy and diversity in environmental and resource regimes can become a major source of stability and strength by providing multiple ways of coping with or reorganizing after change and unexpected events (Low et al., 2003).

Several studies have looked at the role of social networks in interorganizational collaboration and collective action in relation to natural resource management (see, for example, Agranoff and McGuire, 1999, 2001; Carlsson, 2000; Mandell and Steelman, 2003; Imperial, 2005; Manring, 2007; Pahl-Wostl et al., 2008). Although these studies address the linking of network members, there is a need also to increase understanding of the role of networks in dealing with uncertainty and abrupt change in social–ecological systems (Scheffer et al., 2001; Bodin and Norberg, 2005; Janssen et al., 2006; Ernstson, 2008). Westley (2002) holds that the capacity to deal with the interactive dynamics of social and ecological systems requires learning environments and networks of interacting individuals and organizations at different levels in order to create the right links at the right time around the right issues. A number of scientific papers offer an overview and explore typologies of learning and various avenues for learning in collaborative environmental management (for example, Garaway and Arthur, 2004; Cook et al., 2004; Fazey et al., 2005; Armitage et al., 2008).

It should be noted that all aspects presented need further elaboration. However the core features of adaptive governance approaches allow for: interactions across organizational levels; experimentation; new policies for ecosystem management, novelty regarding cooperation and relationships within and among agencies and stakeholders; new ways to promote flexibility; and new institutional and organizational arrangements (Dietz et al., 2003; Folke et al., 2005; Ostrom, 2005; Voß et al., 2006).

Even though the science community has made important advances in understanding the features of adaptive governance approaches, few have explored how to understand and facilitate that transition (Folke et al., 2005; Olsson et al., 2006; Pahl-Wostl et al., 2008). Put bluntly, it is one thing to know where to go. It is another to know how to get there. As we discuss next, moving towards adaptive governance approaches seems to call for skilful stewardship.

17.3 Navigating transitions in social–ecological systems

17.3.1 Introduction

New frameworks are emerging for investigating what occurs when long periods of stability are followed by abrupt change in social–ecological systems (Gunderson and Holling, 2002; Chapin et al., 2006). Researchers in the social sciences and humanities have long recognized that rigidity, lock-in traps and path-dependence are common characteristics of institutional development and public policy-making. They have also focused on understanding sudden change and 'punctuated equilibrium' where long periods of stability and incremental change are interrupted by abrupt, non-incremental, large-scale change (for example, Baumgartner and Jones, 1991; True et al., 1999; Repetto, 2006). Ingram and Fraser (2006) use a 'punctuated equilibrium' framework to analyse policy innovations in water management in California, where water management and policy were locked into a highly engineered infrastructure that reinforced one policy and excluded others. This resulted in rigidity in policy-making that stifled innovation and the capacity to deal with crises. With new awareness among stakeholders, California policy and water management have shifted and broadened to incorporate a wider array of state and federal agencies and private and public organizations (see Chapter 4 by Ingram and Lejano, this volume). Understanding such shifts and the role of innovation has been the focus of studies of transitions in socio-technological systems and transition management (for example Martens and Rotmans, 2005; Loorbach, 2007; Geels and Schot, 2007; Smith and Stirling, 2008).

17.3.2 The role of agency in overcoming barriers to change

Research on social–ecological systems has started to identify barriers for shifting to new forms of water management and governance (Sendzimir et al., 2007; Olsson et al., 2007; Pahl-Wostl et al., 2008), especially in regard to incorporating uncertainty and surprise, enhancing learning and supporting experimentation, and facilitating participation and collective action. This work also focuses on understanding the role of agency and leadership strategies (Westley, 1995, 2002; Olsson et al., 2006). There are at least three aspects for which agency seems particularly important: changing peoples' perceptions and mental models; developing new institutional and organizational structures; and developing learning networks. In the following sections we test the validity of this view using insights from the case of the Kristianstads Vattenrike Biosphere Reserve (KVBR) in southern Sweden (Map 17.1), to understand further the dynamic interactions among individuals, networks and institutions at multiple levels and shifts to new forms of management and governance of water resources.

The unique wetlands of the KVBR and their surrounding agricultural landscape generate a variety of essential ecosystem services for the region and beyond, including flood control and the maintenance of species diversity, as well as cultural, recreational and educational services. In the late 1980s and early 1990s the management of the area went through a transition and a new approach emerged through local initiatives, as a response to ecosystem changes and uncoordinated management efforts that threatened the cultural and natural values of the lower parts of a river catchment. Since 1989 a flexible collaborative governance approach has been in use, which promotes management that treats humans as part of ecosystems. KVBR management today is based on collaborative processes that involve international organizations; national, regional and local authorities; corporations; researchers; non-profit associations and farmers as well as other landowners. The approach developed in the following way (Olsson et al., 2004b):

1. Scope of management widened from a particular issue (floods) to a broad set of issues related to freshwater flows and ecological processes across scales.
2. Management expanded from individual actors to groups of actors.
3. Organizational and institutional structures evolved as a response to deal with the broader set of water and ecosystem issues.
4. Knowledge of ecosystem dynamics developed as a collaborative effort and became part of the organizational and institutional structures at multiple levels.
5. Social networks developed to connect institutions and organizations at multiple levels and facilitate information flows, identify knowledge gaps and create nodes of expertise of significance for flexible and collaborative management of the catchment.

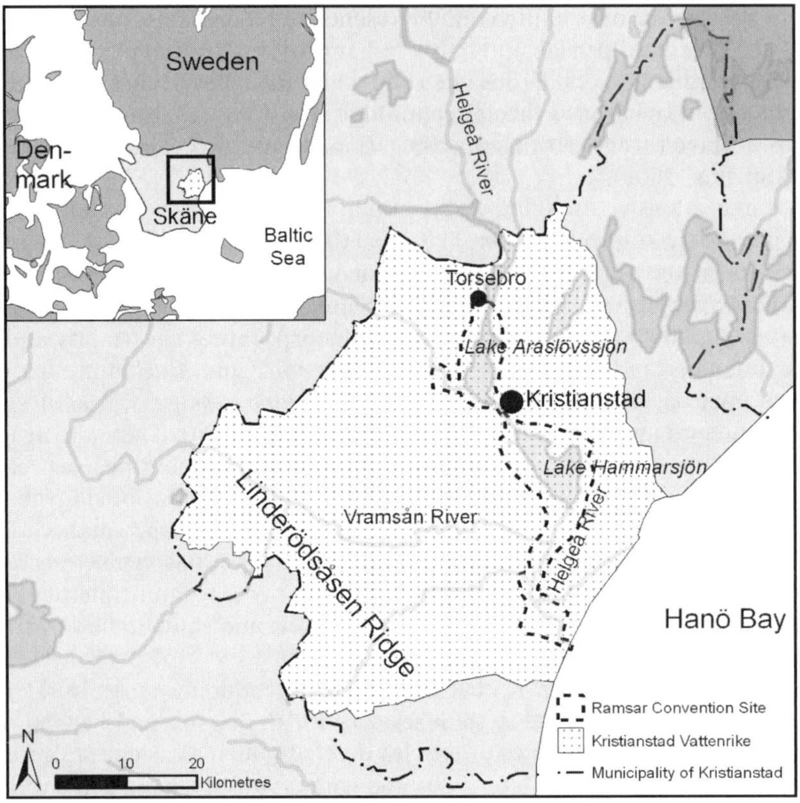

Map 17.1 Sweden and the case study region

6. The social network mobilized knowledge for management, which complemented and refined local practice and improved the capacity to deal with future uncertainties and surprises.

17.3.3 Changing people's perceptions and mental models

Simplified, linear, narrow mental models of social–ecological systems that do not consider the complexity of a system, including uncertainty and change, can lead to inappropriate management, unsustainable resource use and poor learning patterns (Holling and Meffe, 1996; Gunderson and Holling, 2002). Several studies have shown that changing people's perceptions is critical in altering the trajectory of natural resource management (for example Trosper, 2003; Huitema and Kuks, 2004). In the Netherlands a recent shift to more integrated forms of water management demonstrates that a change in people's mental models, from 'fighting the water' to 'living with the water', was critical for adaptive management (Van der Brugge et al., 2005; see Huitema and Meijerink, Chapter 19 in this volume). In Australia, the Great Barrier Reef Marine Park Authority used a number of communication strategies to change public perception of the Great Barrier Reef from a well-protected, pristine coral reef ecosystem to a vulnerable and complex seascape requiring active stewardship (Olsson et al., 2008).

In the KVBR, agency was important in altering the perception of both the public and local politicians, which proved critical in the reserve's transition to ecosystem-based management (Olsson et al., 2004b). The original view of the wetlands around the city of Kristianstad as a problem and the area as 'water sick' was transformed into one of the wetlands as a potential resource that generates a variety of ecosystem services and the area as 'water rich'.

Activities like extensive dredging and draining of the wetlands and making them the site of the city dump in the 1960s have stopped, due to the shift in public and political perception. Today the wetlands are a part of the identity of the city of Kristianstad and used to promote the region. A key individual behind this change coordinated a network to develop a new, ecosystem-based approach for integrated landscape-level solutions to environmental problems in KVBR. With support from a wide range of groups, this actor seized a window of opportunity to bring the idea to two municipal politicians and make them aware of emerging problems in the area and the need for action. Using a vision of ecosystem-based management, he linked the proposal to other goals such as regional development and managed to change the perception of these politicians. The politicians in turn convinced the Municipality Executive Board to support the idea. Increased political interest helped to tip governance in the new management direction with broad stakeholder engagement.

Changing people's values and perceptions continues to be the focus of the work in the KVBR. The reserve also provides a continuous role for agency in the new flexible and reflexive governance regime (Hahn et al., 2006; Schultz et al., 2007). A number of projects aim at raising public awareness of freshwater ecosystem services, how these contribute to human well-being and how they are threatened. These projects emphasize humans as part of ecosystems, human dependence on ecosystem services, and the importance of maintaining critical functions and interactions in nature to maintain the capacity to generate these services. The KVBR projects aim to build an understanding of ecosystem dynamics, including change, uncertainty and surprise, into governance systems. They also promote a landscape perspective that addresses the connectivity between social and ecological components in the KVBR, and emphasize the benefits of partnerships, working with specific actor groups to view other actor groups as part of the solution, not the problem.

17.3.4 Changing institutional and organizational structures

In KVBR, a bridging organization (BO) has been created to serve various actors and interests, including local actors and governmental bodies. The development and establishment of the BO was part of the reserve's transition to collaborative management. A key actor was the 'architect' behind the design and implementation of the BO as a component of existing organizational and polycentric institutional structures. The BO has a staff of five people and is part of the municipality's organization; it reports directly to the municipality board, like a municipality administration. However it is quite unique in that it is not an authority and has no power to make or enforce rules, but nevertheless has strong legitimacy and trust among stakeholders (Hahn et al., 2006). It relies on several funding sources, including the Municipality of Kristianstad, the County Administrative Board and the Swedish Environmental Protection Agency.

It plays a key role as a facilitator and coordinator in the collaborative processes to maintain the ecosystem services of the area. More specifically the BO is involved in developing policy, designing projects, coordinating and administering conservation and restoration efforts, and developing goals for the KVBR, as well as producing management plans, agreements, follow-up reports and updates for specific areas (Olsson et al., 2004b; Hahn et al., 2006; Schultz et al., 2007). The BO functions as an umbrella association and has played a key role in responding to environmental feedback and in developing new knowledge and understanding necessary for managing the area. Issues include: managing floods; dealing with the crop damage problem caused by increasing numbers of cranes and geese; protecting and restoring tributaries of the major river in the area; and the creation of social structures and processes to secure the continued cultivation of the flooded meadows.

Besides establishing the BO, the agency has also developed three distinct forms of KVBR management organizations for managing the KVBR: a consultancy group, theme groups and 'adhocracy' groups. These groups provide diversity useful for dealing with social and ecological problems at different scales and of different types, including uncertainty and surprise. The consultancy group operates at the scale of the KVBR and was formed to build trust, mitigate conflict, produce mechanisms for conflict management, identify common interests and discuss differences of opinion in a constructive way. The theme groups are formed to work on specific projects within the KVBR like flooded meadows or groundwater. 'Adhocracy' groups are organizations that emerge in response to a surprise, exist as long as the particular problem persists, and subsequently dissolve (Hahn et al., 2006). This latent, active phasing relies on a dormant set of connections in a network of actors involved in the management across organizational levels of the KVBR. These connections have developed around the BO over the years and can be seen as 'sleeping links' that are triggered by exogenous events, such as extreme floods. At such critical times an adhocracy group helps tune social and ecological dynamics by monitoring, combining knowledge, and developing management practices and responses to environmental change and impending conflicts.

17.3.5 Developing learning networks

We argue that ecosystem management is an information-intensive endeavour that requires knowledge of complex social–ecological interactions and related uncertainty and abrupt change in order to monitor, interpret and respond to ecosystem feedback at multiple scales (Folke et al., 2003). Such knowledge is dispersed among individuals and organizations in society (Berkes, 2002; Brown, 2003; Gadgil et al., 2003; Olsson et al., 2004a; Carlsson and Berkes, 2005). It requires social networks that span and draw on multiple domains and levels, and institutional arrangements that enable integration and mobilization of knowledge at critical times (Imperial, 1999; Olsson et al., 2006). Such networks have proven crucial for their capacity to deal with environmental change and crises such as sudden flooding, unexpected high levels of water pollutants, or lake systems that 'flip' into a new undesirable state (for example Folke et al., 2003; Tompkins and Adger, 2004; Moberg and Galaz, 2005; Galaz et al., 2008).

The adaptive governance approach used in the KVBR relies on a social network of actors and actor groups of which the BO is the key node (Hahn et al., 2006; Olsson et al., 2007). The BO plays a central role in creating a learning platform by eliciting common goals, creating an atmosphere of trust, brokering organizational and individual contributions and deploying skills necessary to implement strategic plans. These actions are important for stimulating, facilitating and sustaining adaptive approaches to water management and governance, including the emergence of governance networks (see Hajer and Wagenaar, 2003; Sorensen and Torfing, 2005) for dealing with uncertainty and change. The BO plays the role of a broker in developing new networks and connecting existing networks. They can create cross-scale links, mobilize a loosely connected network of actors at critical times, and draw on multiple external sources of information and knowledge, such as scientists and practitioners, to deal with abrupt change and crisis. Bridging different networks and creating opportunities for new interactions is important for dealing with uncertainty and change, and is a critical factor for learning and nurturing integrated adaptive responses to change (Stubbs and Lemon, 2001).

Agency is needed to initiate social networks with a wide scope of actors to connect institutions and organizations across scales and build trust, facilitate information flows, identify knowledge gaps and create nodes of expertise for adaptive water management. In the KVBR networks are 'shadow networks', formed alongside formal arrangements, and they played the important role of steering the emerging new governance regime away from a developing crisis. Shadow networks can explore new approaches and experiment with social responses to uncertainty and change, and thereby generate innovations that could trigger the emergence of new forms of governance and management (Gunderson, 1999; Olsson et al., 2006).

17.3.6 Steering networks of networks: the role of central coordination

Although the KVBR case study provides some important insights about where agency is needed for making transitions, it involves only a subcatchment and a part of a larger water district. The question is how to deal with problems and emerging crises of a larger scale, like the acidification crisis in Sweden in the late 1960s and early 1970s (Lundgren, 1998). There is a need for coordination of institutional interplay among existing social and policy networks in various policy arenas, such as water, security, land and health, to provide fast joint response to abrupt changes in biophysical systems that cascade through social–ecological systems and across time and spatial scales (Galaz et al., 2008). The aim here is not necessarily the creation of new bureaucratic organizations, but rather the development of a capacity to utilize existing social networks and institutions in diverse policy fields or to compensate where they are non-existent or maladaptive.

The Swedish state and the recently established water district authorities have an important role to play in coordination and capacity-building, but this requires a change in how central managers think about their own role and the development of new steering mechanisms. The state's role could change, for example, from authoritative allocation 'from above' to one of 'activator' of various institutional arrangements (Eising and Kohler-Koch, 1999). Different types of misfits between governance and biophysical systems might require a plethora of organizational options and different patterns of interaction among actors at multiple levels. This creates the challenge of defining the boundary of participation and requires the mobilization of actors in relation to the misfit type to be addressed. The activator must have the capacity to facilitate the emergence of such governance networks.

A further challenge arises in the need to develop new steering mechanisms for network governance. Researchers analysing the features of network-based governance have identified a number of network management strategies (Kickert et al., 1997) ranging from promoting mutual adjustment by negotiation and consultation, to more direct interventions, such as restructuring relations or the 'selective activation' of networks (Koppenjan and Klijn, 2004). These management approaches are worth exploring in trying to match institutions and wider governance systems with biophysical systems containing the risk of devastating cascading effects.

The coordination and management of networks requires skilful leadership. Network leadership and guidance is very different from the command-and-control of hierarchical management (Agranoff and McGuire, 2001). It requires steering to hold the network together (Bardach, 1998), and balancing social forces and interests that enable self-organization (Kooiman, 1993). It follows that instead of superimposing ready-to-use plans for ecosystem management on local contexts, the role of central authorities and agencies could be to legislate to enable self-organization processes, provide funding and create arenas for collaborative learning (Berkes, 2002; Olsson et al., 2004a; Hahn et al., 2006). Folke et al. (2003) refer to such an activator role as 'framed creativity' of self-organization processes. Such learning processes require mechanisms for aggregating knowledge claims and interests among multiple actors.

The interplay between individual actors, organizations and institutions at multiple levels is central to transitions in social–ecological systems (Folke et al., 2005). In the governance systems of the KVRB and also the Everglades in Florida in the USA (Olsson et al., 2006), successful transitions occurred because of the ability of leaders to:

1. Reconceptualize key issues.
2. Generate and integrate a diversity of ideas, viewpoints and solutions.
3. Communicate and engage with key individuals in different sectors.
4. Span scales; that is, to move across levels of governance and politics.
5. Promote and steward experimentation at smaller scales.
6. Recognize or create windows of opportunity and promote novelty by combining different networks, experiences and social memories.

Leaders who navigate transitions are able to understand and communicate a wide set of technical, social and political perspectives regarding the particular resource stewardship issues at hand. Visionary leaders fabricate new and vital meanings, overcome contradictions, create new syntheses and forge new alliances between knowledge and action.

17.4 Closing the window of opportunity? The EU WFD in Sweden

The primary purpose of the Water Framework Directive (WFD) is to improve and manage the quality of water by identifying and controlling all pollutants and activities that affect the status of water. It also aims to manage the quantity of surface and groundwaters and to protect aquatic ecosystems and wetlands. Specific measures of the WFD include (see also Partzsch, Chapter 13 in this volume):

1. Expanding the scope of water protection to all waters, surface waters and groundwater.
2. Achieving 'good status' for all waters by a set deadline.

3. Water management based on river basins.
4. A combined approach of emission limit values and quality standards.
5. Getting the prices right.
6. Getting the citizen involved more closely.
7. Streamlining legislation.

It is expected that by 2027 each body of water should achieve EU 'good water' status from both a chemical and an ecological perspective. This period includes two six-year cycles of planning. The Directive and its timeline apply to the quality, quantity and pollution levels, unless there are grounds for derogation by member states, and more specifically river basin administrations, on financial and safety grounds. There is also a general provision that there should be 'no deterioration' in status, which requires the management of the quality, quantity and structure of aquatic environments.

The WFD also requires the reduction to below set quality standards and the ultimate elimination of priority hazardous substances. The Directive maintains existing European water policy commitments and introduces a number of new areas into legislation, but perhaps most importantly, it creates a new administrative instrument for accomplishing its aims (Howe and White, 2002). The importance of the WFD should not be underestimated. Its common legislative framework will have a long-lasting impact on all EU members and carries the implication that it will secure the use of water by hundreds of millions of inhabitants.

With its vision and goals, the WFD provides an opportunity for shifting to more collaborative, flexible forms of management and governance of water resources and ecosystems in Europe. However the WFD does not in itself guarantee such change, since much depends on how the framework is implemented by individual states. We focus here on the implementation of the WFD in Sweden and its potential for the reform of existing freshwater management approaches and governance regimes.

According to the report 'Klart som vatten' (Statens Offentliga Utredningar, 2002, p. 105), Sweden has been divided into five water districts based on the connection of geographical areas with the sea basins of the Bay of Bothnia, North Baltic Sea, South Baltic Sea and North Sea. In each of the five districts, a water authority has been established with the purpose of ensuring fulfilment of the national government's environmental objectives for water. The Swedish government appoints a chief director to each of the five water authorities, where they oversee an official delegation with management responsibility as assigned to one of the district's county administrative boards. The authorities are to be responsible for environmental objectives, action plans and administrative plans for their respective water districts, and will ensure that watercourses are analysed and monitored. An additional layer of institutions includes Water Delegations linked to each water authority. These delegations are decision-making bodies consisting of representatives from, for example, regional administration, academia and NGOs. Lastly, and more locally, a number of voluntary water councils (Vattenråd in Swedish) are embedded in this multilevel institutional landscape. These councils are highly heterogeneous, made up of diverse actors, from individual businesses and farmers' associations to local councils.

The number of acknowledged stakeholders, the increase in legislation and recommendations, as well as the need to coordinate better the work of agencies across administrative boundaries both vertically and horizontally, together pose a number of new challenges for actors in Swedish water administration. The processes involved are coordinated around a water planning cycle, which includes the following steps:

1. Characterization of river basin.
4. Environmental monitoring and evaluation.
2. Setting of environmental objectives.
3. Creation of river basin management plans with water improvement measures.
5. Reporting and preparing for new characterization.

17.5 Does the WFD increase resilience?

17.5.1 Introduction

In the following section we investigate the opportunity for change provided by the implementation of the WFD in Sweden. We build on an analysis by Galaz (2005a) of the implementation and the strategies key water policy-makers are likely to apply in the near future in regards to incorporating uncertainty and surprise, enhancing learning and supporting experimentation, and facilitating participation and collective action. Galaz used key characteristics of an adaptive management approach to analyse the current status of the WFD in Sweden, and the results are summarized in Table 17.1. The study raises serious concerns about the role of central authorities in steering transitions and the possibilities of shifting to adaptive freshwater management and governance regimes that support such management.

17.5.2 Incorporating uncertainty and surprise

As discussed in the previous section, the promotion of social learning and organizational and institutional diversity is important for dealing with the dynamics of freshwater resources, including uncertainty and change. There is no recognition by European authorities in present guidelines to member states (see Common Implementation Strategy, 2003b, p. 66) of the importance of incorporating understanding of ecosystem dynamics and complexity into governance systems. When dealing with uncertainties, Swedish water managers tend to use strategies that involve discussions with other central agencies, or to concentrate their work on areas where uncertainty is low, instead of analysing uncertainties. As interviews with key water administrators show, environmental uncertainty is tackled by relying on expert (mathematical) models, by using existing expertise within the organization, or by relying on previous experience and research. None of these strategies can be seen as systematic attempts to tackle an increased complexity in freshwater management.

The impacts of climate change and increased climate variability present a major challenge for freshwater management. Although some initial modelling attempts and case studies highlight possible impacts and scenarios (Andréasson et al., 2004; European Commission Joint Research Committee, 2005; Statens Offentliga Utredningar, 2002, p. 60), large uncertainties remain regarding the impacts of climate change on both the quantity and quality of freshwater resources. These uncertainties, however, are not mentioned in the Directive itself (European Commission Joint Research Committee, 2005, p. 137), nor in the guideline documents designed specifically to support the work of water managers in Sweden (Naturvårdsverket, 2002, 2003, 2004a, 2004b, 2004c). Neither

Table 17.1 Contrasting the EU Water Framework Directive in Sweden and adaptive freshwater management

Dominant WFD perspective	Adaptive freshwater management
Stakeholder participation is promoted to secure the legitimacy and efficiency of water management activities	Collective action and network-building is promoted to strengthen actors' joint capacity to tackle social and ecological uncertainty, and unexpected events
Social learning is limited or realized to create consensus around water management initiatives	Social learning is institutionalized to understand freshwater system dynamics and identify major uncertainties
Institutions are designed to achieve fixed quality and quantity targets	Institutions designed to allow for adaptation to environmental change and crises
Evaluation is unsystematic and evaluation is applied ad hoc	Policy is treated as a hypothesis and management as an experiment from which managers can learn
Strategies to deal with uncertainty are absent	Developing strategies and stakeholder-driven processes to tackle uncertainty and complexity are a fundamental aim
Emphasis on solutions to achieve fixed water quality and quantity targets	Emphasis on solutions that change structures in freshwater systems with the objective to reduce vulnerability and to strengthen the users' capacity to respond and adapt
High reliance on models to describe status of water resources, and as a base in river management plans	Models are important in collaborative processes aiming to define the dynamic behaviour of freshwater systems, and to identify critical thresholds
Institutional homogeneity is promoted to secure administrative equality across the country	Institutional diversity is encouraged to promote innovation and reduce vulnerability
Multilevel water governance is encouraged to secure legitimacy and efficiency of fixed targets	Multilevel governance is promoted to secure local ecological knowledge, reduced vulnerability and to strengthen the users' capacity to respond and adapt

Source: Galaz (2005a).

are they mentioned in recent analyses of Swedish freshwater resources (for example Naturvårdsverket, 2005a). Interviews with water directors in Sweden confirm that strategies for dealing with uncertainties and surprises are also missing (Galaz, 2005a).

17.5.3 Facilitating participation and collective action

The WFD creates a number of possibilities to include the ecological knowledge of local stakeholders in the multilevel governance structure. The importance of getting a wide span of social actors together is widely recognized at the EU level, both in the WFD and in Common Implementation Strategy documents (for example Common Implementation Strategy, 2003b). The importance of stakeholder participation is also acknowledged in the Swedish governmental reports (Statens Offentliga Utredningar,

2002, p. 105; Naturvårdsverket, 2005b). Formalized cooperation among stakeholders is expected to evolve at the catchment, local and the 'super local' levels (Statens Offentliga Utredningar, 2002, pp. 105, 169ff). This includes the Local Water Councils (LWCs) that are initiated by municipalities, existing water user associations, water-consuming industries, farmer organizations and other stakeholders themselves. The idea is to allow these councils to develop through a self-organizing process without involvement from central government (Statens Offentliga Utredningar, 2002, pp. 105, 161). LWC functions include supplying water authorities not only with advice, but also with water monitoring data, taking part in the formulation of precise ecological standards for the resource and, if needed, detailed river management plans (Statens Offentliga Utredningar, 2002, pp. 105, 159f). This procedure is expected to help Swedish water authorities to design efficient solutions sensitive to the needs of local and regional stakeholders (Svenskt Vatten, 2002, p. 27).

Even though a number of voluntary catchment-based organizations have evolved in Sweden since the 1950s, a number of studies point to deficiencies in this form of cooperation (Galaz, 2005b). The problem is well known in both academic circles (for example McCay and Acheson, 1987; Berkes, 1989; Kellogg, 1998; Wondolleck and Yaffee, 2000; Sproule-Jones, 2002; Ostrom et al., 2002; Peterson et al., 2003b; Olsson et al., 2004a; Pahl-Wostl, 2005) and public administration circles. However Swedish guideline documents fail to elaborate the strategies and incentives needed to support such collective action. Key implementation documents such as Statens Offentliga Utredningar (2002, p. 105) and Naturvårdsverket (2003, 2005a) all lack analysis of what factors actually drive and promote collective action and self-organization of networks in freshwater management. Although water directors acknowledge the need for participation and collective action for reaching the goals of the WFD, the skills and capacity-building, as well as organizational and institutional arrangements for achieving such cooperation, seem to be lacking (Galaz, 2005a).

17.5.4 Enhancing learning and supporting experimentation

Collective learning, trust and network-building are all recognized in the Common Implementation Strategy documents as key factors for sustainable freshwater management in the Common Implementation Strategy documents (Common Implementation Strategy, 2003a). Several tools to promote trust among stakeholders and the public are presented, such as demonstration projects and improved access to information for emerging networks of stakeholders. However learning is defined as mutual respect by central managers and stakeholders of each other's views and the diversity of stakes, rather than as a mutual understanding of system dynamics and of a way to deal with uncertainty and surprise.

While learning is mentioned in the EU documents, it appears to be missing in the key Swedish implementation documents (for example Statens Offentliga Utredningar, 2002, p. 105; Naturvårdsverket, 2003). The emphasis for stakeholder involvement seems to be on supplying data and providing input on the plans presented by water authorities. Although stakeholders are expected to be involved at various stages of the water planning cycle, none of the five water authorities has concrete plans to involve stakeholders in learning processes and the joint analysis of freshwater system dynamics. Instead the planning cycle relies heavily on existing and natural science-based models.

The Common Implementation Strategy documents also include an extensive discussion on how to evaluate public participation projects, thus treating attempts to achieve efficient stakeholder participation as experiments from which managers can learn and adapt (Common Implementation Strategy, 2003a, pp. 50–60). Even though environmental targets will be evaluated by the European water authorities, there seems to be no plan to evaluate local water improvement projects systematically to understand system dynamics, nor to assess the thresholds of freshwater systems to avoid sudden and unwanted state shifts. In the Swedish guideline documents, such as Naturvårdsverket (2002, 2003, 2004a, 2004b, 2004c) and Statens Offentliga Utredningar (2002, p. 105), there is no discussion on the role of experimentation to understand the dynamics of a social–ecological system, nor for testing of management practices and policies and organizational and institutional arrangements. On the contrary a tendency seems to exist to reduce institutional and organizational diversity in order to secure homogeneity in the five water districts (Galaz, 2005a).

17.6 Conclusions

The WFD provides an opportunity to shift from conventional to more adaptive water management and governance approaches. The Swedish government and the recently established water authorities have an important leadership role to play in facilitating such a transition. However in order to seize this opportunity there is a need for new thinking and behaviour among managers and governmental officials, and for innovative organizational and institutional arrangements that can enhance social learning. Central authorities could, for example, provide space and enabling conditions for learning networks to form, including financial, political and moral support, along the lines possibly of transition arenas for shifting water management in the Netherlands (Van der Brugge et al., 2005). Such learning platforms can generate a diversity of ideas and solutions as a resource to be drawn from at critical times. Such experimentation and diversity might nurture innovations for renewal and reorganization, and increase the capacity to deal with uncertainty and abrupt change. Central authorities could play the important roles of activator and coordinator in such governance networks, providing a bridging function that facilitates cross-level interactions and synthesizes lessons for incorporation into national policies and guidelines for water management.

New management and governance approaches should build on initiatives in play like those of the KVRB, and on their capacity to innovate. The two cases examined in the chapter, however, show the need not to rely too heavily on self-organization, and the necessity of redefining the role of central authorities to coordinate, to help diffuse new insights and to respond to events that go beyond the scope of the local initiatives.

References

Agranoff, R.I. and M. McGuire (1999), 'Managing in network settings', *Policy Studies Review*, **16** (1), 18–41.

Agranoff, R.I. and M. McGuire (2001), 'Big questions in public network management research', *Journal of Public Administration Research and Theory*, **11** (3), 295–326.

Andréasson, J., S. Bergström, B. Carlsson, L.P. Graham and G. Lindström (2004), 'Hydrological change: climate change impact simulations for Sweden', *Ambio*, **33** (4), 228–34.

Armitage, D., F. Berkes and N. Doubleday (eds) (2007), *Adaptive Co-Management: Collaboration, Learning, And Multi-Level Governance*, Vancouver: UBC Press.

Armitage, D., M. Marschke and R. Plummer (2008), 'Adaptive co-management and the paradox of learning', *Global Environmental Change*, **18** (1), 86–98.

Bardach, E. (1998), *Managerial Craftsmanship: Getting Agencies to Work Together*, Washington, DC: Brookings.

Bates, B.C., Z.W. Kundzewicz, S. Wu and J.P. Palutikof (eds) (2008), *Climate Change And Water: Technical Paper of the Intergovernmental Panel on Climate Change*, Geneva: IPCC Secretariat.

Baumgartner, F.R. and B.D. Jones (1991), 'Agenda dynamics and policy subsystems', *Journal of Politics*, **53** (4), 1044–74.

Becker, C.D. and E. Ostrom (1995),'Human ecology and resource sustainability: the importance of institutional diversity', *Annual Review of Ecology and Systematics*, **26**, 113–33.

Berkes, F. (ed.) (1989), *Common Property Resources: Ecology and Community-Based Sustainable Development*, London: Belhaven Press.

Berkes, F. (2002), 'Cross-scale institutional linkages: perspectives from the bottom up', in E. Ostrom, T. Dietz, N. Dolsak, P.C. Stern, S. Stonich and E.U. Weber (eds), *The Drama of the Commons*, Washington, DC: National Academy Press, pp. 293–321.

Berkes, F., J. Colding and C. Folke (eds) (2003), *Navigating Social–Ecological Systems: Building Resilience for Complexity and Change*, Cambridge: Cambridge University Press.

Berkes, F. and C. Folke (eds) (1998), *Linking Social and Ecological Systems: Management Practices and Social Mechanisms for Building Resilience*, Cambridge: Cambridge University Press.

Biermann, F. (2007),'Earth system governance as a crosscutting theme of global change research', *Global Environmental Change*, **17** (3–4), 326–37.

Bodin, Ö. and J. Norberg (2005), 'Information network topologies for enhanced local adaptive management', *Environmental Management*, **35** (2), 175–93.

Borrini-Feyerabend, G., M. Pimbert, M.T. Farvar, A. Kothari and Y. Renard (2004), *Sharing Power: Learning by Doing in Co-Management of Natural Resources Throughout the World*, Tehran, Iran: International Institute for Environmental Development/World Conservation Union/Committee on Environmental and Economic Policy/Collaborative Management Working Group/Center for Sustainable Development.

Brown, K. (2003), 'Integrating conservation and development: a case of institutional misfit', *Frontiers in Ecology and the Environment*, **1** (9), 479–87.

Carlsson, L. (2000), 'Policy networks as collective action', *Policy Studies Journal*, **28** (3), 502–20.

Carlsson, L. and F. Berkes (2005), 'Co-management: concepts and methodological implications', *Journal of Environmental Management*, **75** (1), 65–76.

Carpenter, S.R. and L.H. Gunderson (2001), 'Coping with collapse: ecological and social dynamics in ecosystem management', *Bioscience*, **51** (6), 451–7.

Chapin, F.S. III, A.L. Lovecraft, E.S. Zavaleta, J. Nelson, M.D. Robards, G.P. Kofinas, S.F. Trainor, G.D. Peterson, H.P. Huntington and R.L. Naylor (2006), 'Policy strategies to address sustainability of Alaskan boreal forests in response to a directionally changing climate', *PNAS*, **103** (45), 16637–43.

Common Implementation Strategy (2003a), *Handledning angående allmänhetens deltagande i förhållande till ramdirektivet för vatten*, Stockholm: Naturvårdsverket.

Common Implementation Strategy (2003b), 'Public participation in relation to the Water Framework Directive', *Guidance Document No. 8*, Brussels, European Commission.

Cook, W., D. Casagrande, D. Hope, P. Groffman and S. Collins (2004), 'Learning to roll with the punches: adaptive experimentation in human-dominated systems', *Frontiers in Ecology and Environment*, **2** (9), 467–74.

Cumming, G.S., D.H.M. Cumming and C.L. Redman (2006), 'Scale mismatches in social–ecological systems: causes, consequences, and solutions', *Ecology and Society*, **11** (1), 14, http://www.ecologyandsociety.org/vol11/iss1/art14/, accessed November 2008.

Dietz, T., E. Ostrom and P. Stern (2003), 'The struggle to govern the commons', *Science*, **302** (5652), 1907–12.

Eising, R. and B. Kohler-Koch (1999), 'Introduction: network governance in the European Union', in R. Eising and B. Kohler-Koch (eds), *The Transformation of Governance in the European Union*, London: Routledge, pp. 3–13.

Ernstson, H. (2008), 'In Rhizomia: actors, networks and resilience in urban landscapes', Doctoral thesis, Stockholm University.

European Commission Joint Research Committee (2005), 'Climate change and the European water dimension – a report to the European Water Directors', EU Report No. 21553, European Commission Joint Research Committee.

Falkenmark, M. (1999), 'Forward to the future: a conceptual framework for water dependence', *Ambio*, **28** (4), 356–61.

Falkenmark, M. and C. Folke (2000), 'How to bring ecological services into Integrated Water Resources Management?', *Ambio*, **29** (6), 351–2.

Fazey, I., J. Fazey and D.M. Fazey (2005), 'Learning more effectively from experience', *Ecology and Society*, **10** (2), 4, http://www.ecologyandsociety.org/vol10/iss2/art4, accessed November 2008.

Folke, C., J. Colding and F. Berkes (2003), 'Synthesis: building resilience and adaptive capacity in social-ecological systems', in F. Berkes, J. Colding and C. Folke (eds), *Navigating Social–Ecological Systems: Building Resilience for Complexity and Change*, Cambridge: Cambridge University Press, pp. 352–87.

Folke, C., T. Hahn, P. Olsson and J. Norberg (2005), 'Adaptive governance of social–ecological knowledge', *Annual Review of Environment and Resources*, **30**, 441–73.

Folke, C., L. Pritchard, F. Berkes, J. Colding and U. Svedin (2007), 'The problem of fit between ecosystems and institutions: ten years later', *Ecology and Society*, **12** (1), 30, http://www.ecologyandsociety.org/vol12/iss1/art30/, accessed November 2008.

Gadgil, M., P. Olsson, F. Berkes and C. Folke (2003), 'Exploring the role of local ecological knowledge in ecosystem management: three case studies', in F. Berkes, J. Colding and C. Folke (eds), *Navigating Social–Ecological Systems: Building Resilience for Complexity and Change*, Cambridge: Cambridge University Press, pp. 189–209.

Galaz, V. (2005a), 'Social–ecological resilience and social conflict: institutions and strategic adaptation in Swedish water management', *Ambio*, **34** (7), 567–72.

Galaz, V. (2005b), 'Does the EC water framework directive build resilience? Harnessing socio-ecological complexity in European water management', Policy paper no. 1. Swedish Water House.

Galaz, V., P. Olsson, T. Hahn, C. Folke and U. Svedin (2008), 'The problem of fit among biophysical systems, environmental and resource regimes, and broader governance systems: Insights and emerging challenges', in O.R. Young, L.A. King and H. Schröder (eds), *Institutions and Environmental Change: Principal Findings, Applications, and Research Frontiers*, Cambridge, MA: MIT Press, pp. 147–82.

Gallopin, G.C. (2002), 'Planning for resilience: scenarios, surprises and branch points', in L.H. Gunderson and C.S. Holling (eds), *Panarchy: Understanding Transformations in Social–Ecological Systems*, London: Island Press, pp. 361–92.

Garaway, C.J. and R.I. Arthur (2004), *Adaptive Learning: A Practical Framework for the Implementation of Adaptive Co-Management – Lessons from Selected Experiences in South and Southeast Asia*, London: MRAG.

Geels, F.W. and J. Schot (2007), 'Typology of socio-technical transition pathways', *Research Policy*, **36** (3), 399–417.

Gordon, L.J., G.D. Peterson and E.M. Bennett (2008), 'Agricultural modifications of hydrological flows create ecological surprises', *Trends in Ecology and Evolution*, **23** (4), 211–19.

Gunderson, L.H. (1999), 'Resilience, flexibility and adaptive management: antidotes for spurious certitude?' *Conservation Ecology*, **3** (1), 7, http://www.consecol.org/vol3/iss1/art7, accessed November 2008.

Gunderson, L.H. and C.S. Holling (eds) (2002), *Panarchy: Understanding Transformations in Social–Ecological Systems*, London: Island Press.

Hahn, T., P. Olsson, C. Folke and K. Johansson (2006), 'Trust-building, knowledge generation and organizational innovations: the role of a bridging organization for adaptive co-management of a wetland landscape around Kristianstad, Sweden', *Human Ecology*, **34**, 573–92.

Hajer, M.A. and H. Wagenaar (eds) (2003), *Deliberative Policy Analysis: Understanding Governance in the Network Society*, Cambridge: Cambridge University Press.

Holling, C.S. and G.K. Meffe (1996), 'Command and control and the pathology of natural resource management', *Conservation Biology*, **10** (2), 328–37.

Hong, L. and S.E. Page (2004), 'Groups of diverse problem solvers can outperform groups of high-ability problem solvers', *PNAS*, **101** (46), 16385–9.

Howe, J. and I. White (2002), 'The potential implications of the European Union water framework directive on domestic planning systems: a UK case study', *European Planning Studies*, **10** (8), 1027–38.

Huitema, D. and S.M.M. Kuks (2004), 'Harboring water in a crowded European delta', in J.T.A. Bressers and S.M.M. Kuks (eds), *Integrated Governance and Water Basin Management: Conditions for Regime Change towards Sustainability*, Dordrecht: Kluwer Academic Press, pp. 61–100.

Imperial, M.T. (1999), 'Institutional analysis and ecosystem-based management: the institutional analysis and development framework', *Environmental Management*, **24** (4), 449–65.

Imperial, M.T. (2005), 'Using collaboration as a governance strategy: lessons from six watershed management programs', *Administration and Society*, **37** (3), 281–320.

Ingram, H. and L. Fraser (2006), 'Path dependency and adroit innovation: the case of California water', in R. Repetto (ed.), *Punctuated Equilibrium and the Dynamics of US Environmental Policy*, New Haven, CT: Yale University Press, pp. 78–109.

Janssen, M.A., Ö. Bodin, J.M. Anderies, T. Elmqvist, H. Ernstson, R.R.J. McAllister, P. Olsson and P. Ryan (2006), 'A network perspective on the resilience of social–ecological systems', *Ecology and Society*, **11** (1), 15, http://www.ecologyandsociety.org/vol11/iss1/art15/, accessed November 2008.

Kellog, W.A. (1998), 'Adopting an ecosystem approach: local variability in remedial action planning', *Society and Natural Resources*, **11** (5), 465–84.

Kickert, W., E-H. Klijn and J. Koppenjan (1997), *Managing Complex Networks*, London: Sage Publications.

Kooiman, J. (ed.) (1993), *Modern Governance*, London: Sage Publications.

Koppenjan, J.F.M. and E.H. Klijn (2004), *Managing Uncertainties in Networks. A Network Approach to Problem Solving and Decision Making*, London: Routledge.

Levin, S. (1999), *Fragile Dominion: Complexity and the Commons*, Reading, PA: Perseus Publishing.

Loorbach, D.A. (2007), *Transition Management: New Mode of Governance for Sustainable Development*, Utrecht: International Books.

Low, B., E. Ostrom, C. Simon and J. Wilson (2003), 'Redundancy and diversity: do they influence optimal management', in F. Berkes, J. Colding and C. Folke (eds), *Navigating Social–Ecological Systems*, Cambridge: Cambridge University Press, pp. 83–114.

Lundgren, L.J. (1998), *Acid Rain on the Agenda: A Picture of a Chain of Events in Sweden, 1966–1968*, Lund, Sweden: Lund University Press.

Mandell, M.P. and T.A. Steelman (2003), 'Understanding what can be accomplished through interorganizational innovations', *Public Management Review*, **5** (2), 197–224.

Manring, S.L. (2007), 'Creating and managing interorganizational learning networks to achieve sustainable ecosystem management', *Organization and Environment*, **20** (3), 325–46.

Martens, P. and J. Rotmans (2005), 'Transitions in a globalising world', *Futures*, **37**, 1133–44.

McCay, B.M. and J.M. Acheson (eds) (1987), *The Question of the Commons: The Culture and Ecology of Communal Resources*, Tucson, AZ: University of Arizona Press.

McGinnis, M. (ed.) (2000), *Polycentric Governance and Development*, Ann Arbor, MI: University of Michigan Press.

Millennium Ecosystem Assessment (2005), *Ecosystems and Human Well-being: Synthesis*, Washington, DC: Island Press.

Moberg, F. and V. Galaz (2005), 'Resilience: going from conventional to adaptive freshwater management for human and ecosystem compatibility', Swedish Water House Policy Brief, 3, Stockholm, Sweden, SIWI.

Naturvårdsverket (2002), *Helhetssyn i vattenvården*, Stockholm, Sweden: Naturvårdsverket.

Naturvårdsverket (2003), *En basbok om ramdirektivet för vatten*, Stockholm, Sweden: Naturvårdsverket.

Naturvårdsverket (2004a), *Handbok för vatten – Att planera arbetet med vattenplanering*, Stockholm, Sweden: Naturvårdsverket.

Naturvårdsverket (2004b), *Handbok för vatten – Att göra konsekvensanalyser*, Stockholm, Sweden: Naturvårdsverket.

Naturvårdsverket (2004c), *Handbok för vatten – Förvaltningsplan för vattendistrikt*, Stockholm, Sweden: Naturvårdsverket.

Naturvårdsverket (2005a), 'Beskrivning, kartläggning och analys av Sveriges vatten – sammanfattande rapport, ytvatten', Report no. 721-3909-04, Stockholm, Sweden: Naturvårdsverket.

Naturvårdsverket (2005b), *Handbok för vatten – samverkan*, Stockholm, Sweden: Naturvårdsverket.

Norberg, J. and G. Cumming (eds) (2008), *Complexity Theory for a Sustainable Future*, New York: Columbia University Press.

Olsson, P., C. Folke and F. Berkes (2004a), 'Adaptive co-management for building social–ecological resilience', *Environmental Management*, **34** (1), 75–90.

Olsson, P., C. Folke and T. Hahn (2004b), 'Social-ecological transformation for ecosystem management: the development of adaptive co-management of a wetland landscape in southern Sweden', *Ecology and Society*, **9** (4), 2, http://www.ecologyandsociety.org/vol9/iss4/art2, accessed November 2008.

Olsson, P., L.H. Gunderson, S.R. Carpenter, P. Ryan, L. Lebel, C. Folke and C.S. Holling (2006), 'Shooting the rapids: navigating transitions to adaptive governance of social-ecological systems', *Ecology and Society*, **11** (1), 18, http://www.ecologyandsociety.org/vol11/iss1/art18/, accessed November 2008.

Olsson, P., C. Folke, V. Galaz, T. Hahn and L. Schultz (2007), 'Enhancing the fit through adaptive co-management: creating and maintaining bridging functions for matching scales in the Kristianstads Vattenrike Biosphere Reserve, Sweden', *Ecology and Society*, 12 (1), 28, available at: http://www.ecologyandsociety.org/vol12/iss1/art28.

Olsson, P., C. Folke and T.P. Hughes (2008), 'Navigating the transition to ecosystem-based management of the Great Barrier Reef, Australia', *PNAS*, **105** (28), 9489–94.

Ostrom, E. (2005), *Understanding Institutional Diversity*, Princeton, NJ: Princeton University Press.

Ostrom, E., T. Dietz, N. Dolsak, P.C. Stern, S. Stonich and E.U. Weber (eds) (2002), *The Drama of the Commons*, Washington DC: National Academy Press.

Pahl-Wostl, C. (2005), 'Actor based analysis and modeling approaches', *Integrated Assessment Journal*, **5** (1), 97–118.

Pahl-Wostl, C., E. Mostert and D. Tàbara (2008), 'The growing importance of social learning in water resources management and sustainability science', *Ecology and Society*, **13** (1), 24, http://www.ecologyandsociety.org/vol13/iss1/art24/, accessed November 2008.

Peterson, G.D., T.D. Beard Jr., B.E. Beisner, E.M. Bennett, S.R. Carpenter, G.S. Cumming, C.L. Dent and T.D. Havlicek (2003a), 'Assessing future ecosystem services: a case study of the Northern Highlands

Lake District, Wisconsin', *Conservation Ecology*, **7** (3), 1, http://www.consecol.org/vol7/iss3/art1/, accessed November 2008.
Peterson, G.D., G.S. Cumming and S.R. Carpenter (2003b), 'Scenario planning: a tool for conservation in an uncertain world', *Conservation Biology*, **17** (2), 358–66.
Reidsma, P. and F. Ewert (2008), 'Regional farm diversity can reduce vulnerability of food production to climate change', *Ecology and Society*, **13** (1), 38, http://www.ecologyandsociety.org/vol13/iss1/art38/, accessed November 2008.
Repetto, R. (ed.) (2006), *Punctuated Equilibrium and the Dynamics of US Environmental Policy*, New Haven, CT: Yale University Press.
Resilience Alliance and Santa Fe Institute (2004), 'Thresholds and alternate states in ecological and social–ecological systems', Resilience Alliance, http://www.resalliance.org/index.php?id=183, accessed November 2008.
Ripl, W. (2003), 'Water: the bloodstream of the biosphere', *Philosophical Transactions – Royal Society of London, Biological Sciences*, **358** (1440), 1921–34.
Scheffer, M., S.R. Carpenter, J.A. Foley, C. Folke and B. Walker (2001), 'Catastrophic shifts in ecosystems', *Nature*, **413** (6856), 591–696.
Schultz, L., C. Folke and P. Olsson (2007), 'Enhancing ecosystem management through social–ecological inventories: lessons from Kristianstads Vattenrike, Sweden', *Environmental Conservation*, **34** (2), 140–52.
Schusler, T.M., D. Decker and M.J. Pfeffer (2003), 'Social learning for collaborative natural resource management', *Society and Natural Resources*, **16** (4), 309–26.
Sendzimir, J., P. Magnuszewski, Z. Flachner, P. Balogh, G. Molnar, A. Sarvari and Z. Nagy (2007), 'Assessing the resilience of a river management regime: informal learning in a shadow network in the Tisza River Basin', *Ecology and Society*, **13** (1), 11, http://www.ecologyandsociety.org/vol13/iss1/art11/, accessed November 2008.
Smith, A. and A. Stirling (2008), 'Social–ecological resilience and socio-technical transitions: critical issues for sustainability governance', STEPS Centre Working Paper 8.
Sorensen, E. and J. Torfing (2005), 'The democratic anchorage of governance networks', *Scandinavian Political Studies*, **28** (3), 195–218.
Sproule-Jones, M. (2002), *Restoration of the Great Lakes: Promises, Practices and Performances*, East Lansing, MI: Michigan State University Press.
Statens Offentliga Utredningar (2002), 'Klart som vatten – Betänkande av Utredningen svensk vattenadminstration', ('Statens Offentliga Utredningar, Final Report from Government Inquiry Commission'), Stockholm: Ministry of the Environment (in Swedish).
Streeter, C.L. (1992), 'Redundancy in organizational systems', *Social Service Review*, **66** (1), 97–111.
Stubbs, M. and M. Lemon (2001), 'Learning to network and network to learn: facilitating the process of adaptive management in a local response to the UK's National Air Quality Strategy', *Environmental Management*, **27** (3), 321–34.
Svenskt Vatten (2002), *Ramdirektivet för vatten – en kommunal utmaning*, Report M126, Stockholm, Sweden: Svenskt Vatten.
Tompkins, E.L. and N. Adger (2004), 'Does adaptive management of natural resources enhance resilience to climate change?' *Ecology and Society*, **9** (2), 10, http://www.ecologyandsociety.org/vol9/iss2/art10/, accessed November 2008.
Trosper, R.L. (2003), 'Policy transformations in the US forest sector, 1970–2000: Implications for sustainable use and resilience', in F. Berkes, J. Colding and C. Folke (eds.), *Navigating Social–Ecological Systems: Building Resilience for Complexity and Change*, Cambridge: Cambridge University Press, pp. 328–51.
True, J.L., F.R. Baumgartner and D.J. Bryan (1999), 'Explaining stability and change in American policymaking: the punctuated equilibrium model', in P. Sabatier (ed.), *Theories of the Policy Process*, Boulder, CO: Westview Press, pp. 97–115.
United Nations Environment Programme (2007), *Global Environment Outlook: Environment for Development, Vol. 4*, Valletta, Malta: Progress Press.
Van der Brugge, R., J. Rotmans and D. Loorbach (2005), 'The transition in Dutch water management', *Regional Environmental Change*, **5** (4), 164–76.
Victor, D.G., J.C. House and S. Joy (2005), 'A Madisonian approach to climate policy', *Science*, **309** (5742), 1820–21.
Voß, J.-P., D. Bauknecht and R. Kemp (eds) (2006), *Reflexive Governance for Sustainable Development*, Cheltenham, UK and Northampton, MA, USA: Edward Elgar.
Walker, B., S.R. Carpenter, J. Anderies, N. Abel, G. Cumming, M. Janssen, L. Lebel, J. Norberg, G.D. Peterson and L. Pritchard (2002), 'Resilience management in social–ecological systems: a working hypothesis for a participatory approach', *Conservation Ecology*, **6** (1), 14, http://www.consecol.org/vol6/iss1/art14, accessed November 2008.
Westley, F. (1995), 'Governing design: the management of social systems and ecosystems management', in

L.H. Gunderson, C.S. Holling and S.S. Light (eds), *Barriers and Bridges to Renewal of Ecosystems and Institutions*, New York: Columbia University Press, pp. 391–427.

Westley, F. (2002), 'The devil in the dynamics: adaptive management on the front lines', in L.H. Gunderson and C.S. Holling (eds), *Panarchy: Understanding Transformations in Human and Natural Systems*, Washington, DC: Island Press, pp. 333–60.

Wondolleck, J.M. and S.L. Yaffee (2000), *Making Collaboration Work: Lessons from Innovation in Natural Resource Management*, Washington, DC: Island Press.

Young, O.R. (2002), 'Institutional interplay: the environmental consequences of cross-scale interactions', in E. Ostrom, T. Dietz, N. Dolsak, P. Stern, S. Stonich and E.U. Weber (eds), *The Drama of the Commons*, Washington, DC: National Academy Press, pp. 265–91.

Young, O.R., L.A. King and H. Schröder (eds) (2008), *Institutions and Environmental Change: Principal Findings, Applications, and Research Frontiers*, Cambridge, MA: MIT Press.

18 Germany: transitions in flood management in the Rhine basin

Gert Becker

18.1 Introduction

The River Rhine is the third-largest river (1320 km) in Europe and the region's busiest and economically most important international waterway. Its basin features major industrial and urban development (Overeem, 2005; Koordinierungskomitee Rhein, 2005). The Rhine offers a particularly interesting set of cases for examination of the transition in German flood management, since they incorporate several important aspects, including: renewed flood safety issues caused by continuous human intervention; a high degree of institutional complexity in the basin;[1] accompanying socio-ecological aspects; and the concurrent beginning of new environmental and democratic discourses in the 1960s and 1970s (Blatter, 1994; Tittizer and Krebs, 1996; Dieperink, 1997, 2000; Cioc, 2002; Frijters and Leentvaar, 2003; Dombrowsky and Holländer, 2004; Becker et al., 2007; Nienhuis, 2008; Mostert, 2009).

One hundred and fifty years of human intervention to adapt the water and river systems of the Upper Rhine to different user functions had serious, even critical consequences for the ecological and hydro-morphological resilience of the river system. The loss of 75 per cent of floodplains between 1817 and 1928 and, particularly, of another 130 km^2 (13 per cent) between 1957 and 1970 significantly increased flood risk[2] and the potential for related damage[3] downstream (Hochwasserschutzkonzept Köln, 1996). In 1968 the International Commission for Research on Floods of the River Rhine (HSK)[4] was established to investigate the history and background of flooding of the Upper Rhine. Based on the final report of the commission in 1978 and the recommendation to provide a storage volume of 220 million m^3 to re-establish the former, pre-1957, flood protection against a 200-year flood, the riparian countries, Germany and France, came to a formal agreement[5] in 1982 to restore the original safety standards (Disse and Engel, 2001). Although basically designed as a hydrological compensation project, the agreement proved to be the starting point for the transition to a new flood management policy in Germany (Bernhardt, 2002; Dister, 2002).

The German federal arrangement does not facilitate such transitions as it is characterized by a polycentric administrative structure, with a high degree of interdependence of federal and state (*Land*) governmental levels and of political and policy interconnection (*Politikverflechtung*) (Jänicke et al., 2001; Gunlicks, 2005). Legal framing power exists mainly at the national level, while the power of policy application and implementation in most areas, including water, agriculture and spatial development, lies with the federal states and at their lower administrative levels. This institutional setting fosters individual policy developments in the federal states (*Länder*) and promotes equifinality, or in other words, follows the principle that in open systems a given end state can be reached by many potential means. In effect many different perspectives, ideas and interests are

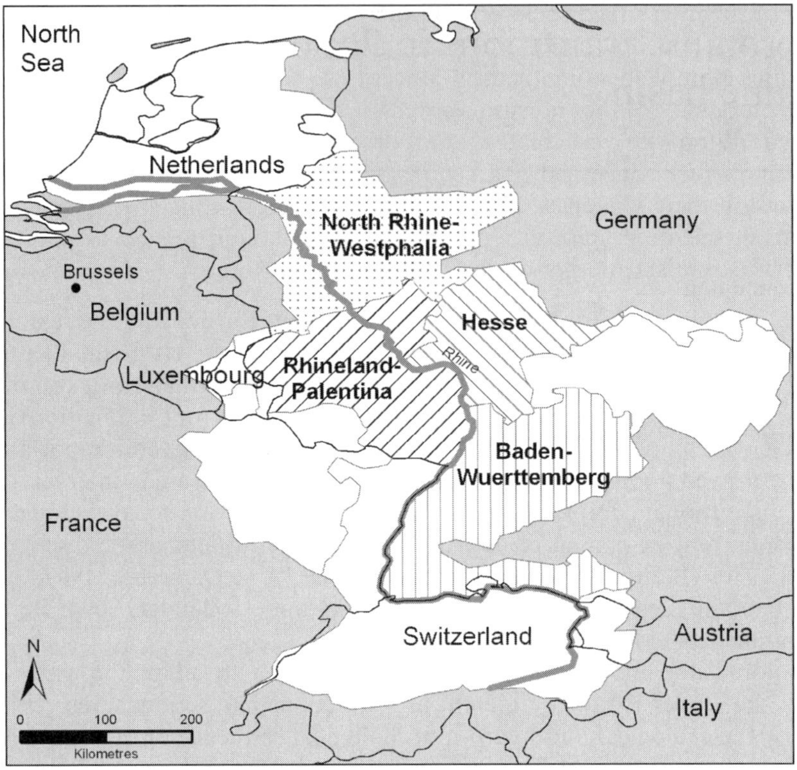

*Map 18.1 The Rhine basin and the main German riparian federal states (*Länder*)*

accommodated and produce similar outcomes conducive to consolidation (Broschek, 2007). The distribution of power and the frequent incongruence of decision structure and problem structure complicate coordination and require a considerable amount of both vertical and horizontal cooperation (Mayntz, 1999; Gunlicks, 2005). The interrelationship among federal, regional and local levels, and the strict legal system and bureaucracy, make political change difficult particularly in the case of interaction problems – problems that require joint planning and decisions (Wieland, 1998; Barzelay and Füchtner, 2003). The cases studied reflect such developments and interrelationships between levels.

In line with their devolved responsibility for flood protection, transition was initiated in the major *Länder* along the Rhine (Map 18.1). First in the Upper Rhine in Baden-Wuerttemberg (B-W), Rhineland-Palentina (R-Pf) and Hesse (HE) and at a later stage, independently, in North Rhine-Westphalia (NRW) in the Lower Rhine area. Policy changes were initiated and endorsed, first separately and at a later stage as a consolidated national and basin-wide regime, but with responsibility for implementation remaining at the *Länder* level.

Each of the *Länder* provides a case study for examination of the arrangements made for the change in Germany from a traditional, technocratic safety discourse in flood management to one based on an ecological risk perspective. We provide a description of the individual approaches, showing how they hold in common a more integrated consideration of the interlinked socio-ecological system and its non-linear processes full of uncertainty. We show how a new cross-boundary policy came to be institutionalized and the difficulties in implementing the new concept.

This chapter analyses the transition process – the path to policy adoption and implementation and how this was created – and describes the specific concepts developed in the different *Länder* concerned. I focus particularly on key actors who promoted the changes and on their strategies. The empirical data supporting the analysis have been collected from scientific literature, political documents and numerous interviews with flood experts, scientists and political officials along the Rhine.[6]

18.2 Preparing the system for change

18.2.1 Introduction

Three general and defining aspects had a major influence on the transition and its course. These comprised: the federal structure and the power distribution in German water management; the administrative authorities and their respective actors (Box 18.1); and the mandate of the HSK and the recommendations of the 1978 report.

In 1982, the governments of France and Germany concluded a binding agreement to re-establish flood safety standards for the Upper Rhine based on the findings of the HSK report. Under Germany's federal system, the German government had first to negotiate formal agreements with the three Upper Rhine *Länder* concerned, B-W, R-Pf and HE. This instituted a framework for the distribution of obligations that would arise from the bi-national accord. The individual *Länder* would be responsible for the design and planning of their respective part of the project. B-W and R-Pf committed to provide additional retention space as proposed by the HSK study, whereas HE would play a passive role, hiding behind the bilateral and national agreements that entailed only financial contributions to the project.[7]

In the Lower Rhine area, NRW developed its own new flood protection strategy during the mid-1980s, independently of the Upper Rhine activities although with similar objectives for improvement of the hydrological resilience of the river, rather than the previous emphasis on protection of human structures and activities at the expense of the health of the river and the attendant benefits to human society.

The different institutional structures in the autonomous states allowed a different framing of the problems in each *Land* according to the particular situation and local requirements. This produced individualized promotion of political change in each state, which added various elements to the general 'more room for the water' proposition.

In B-W the task was framed as an ecological problem, with restoration of floodplains and riparian forest as complementary components, and included an emphasis on hazard reduction. R-Pf and NRW took a more socio-economic standpoint, treating flood protection as a precautionary risk management task in the former *Land* and integrating flood prevention, navigation and ecological measures in the latter.

Major flood protection projects and concepts have their origin in the Upper Water Administrations of the *Länder*, which fall under the authority of the state ministry responsible for water management (Box 18.1). Their technical experts were the main initiators, transmitters of new knowledge and project managers. In all cases, members of the technical administration in each of the *Länder* could be identified as main drivers

BOX 18.1 INSTITUTIONAL ASPECTS OF GERMAN WATER MANAGEMENT

The responsibility of the main water resource management tasks, such as water supply, waste water disposal, flood management and so on, are located at the Federal State (*Länder*) level. At the national level the framework legislation is formulated and some specific overarching tasks are exercised, such as navigation, international environmental aspects and water-related international relations (Krämer, 1999, Jänicke et al., 2001, Winnegge and Maurer, 2002). The *Länder* in turn interpret and implement the framework legislation into State Water Acts (*Landeswassergesetz*) and formulate their policies and objectives accordingly.

In most Federal States, water resource management is integrated in the general administration:

1. The Supreme Water Authority, predominantly within the Ministry for the Environmental functions, sets the basic policies and makes the strategic decisions and is in charge of superior administrative procedures and control, supported by technical expert agencies (State Environment Agency, *Landesumweltämter*)
2. The Upper Water Authority, usually the Regional government (B-W: *Regierungspräsidium*, NRW: *Bezirksregierung*, R-Pf: *Struktur und Genehmigungsdirektionen*), is responsible for the planning and implementation of major procedures and projects
3. The Lower Water Authority, normally the District government (*Landratsamt*) or major cities including their technical authorities, executes the permitting and licensing procedures, monitoring local implementation and gives technical advice for local implementation
4. The municipalities (*Kommunen*) are responsible for water supply and wastewater discharge with a certain freedom of organization ranging from municipal agencies and river associations to private companies. In most *Länder* they are also in charge of the local flood protection whereas at the Lower Rhine (North Rhine-Westphalia) specific associations (dyke associations) assume this task.

The recent administrative reforms (*Verwaltungsreform*, R-Pf 2000, B-W 2005, NRW 2007) integrated regional license, control and expert functions into the Upper Water Authorities.

In 1956 the State Water Working Group (LAWA, *Länderarbeitsgruppe Wasser*) was established to harmonize policies and legislation across the Federal States and to coordinate their participation at the international level, for example, in the formulation and implementation of EU legislation.

Floods are seen as natural disasters and the government is usually not legally liable for the damage but requires and promotes private provision. However, after major floods, public pressure on governments may urge *ex post* (partial) compensation out of public budgets as happened after the 2002 Elbe flood disaster, just before national elections, totalling over 10 billion euros. The opportunity to insure against flood risk has not been used frequently, although such systems are available (Bouwer et al., 2007).

All major construction projects have to run through a planning approval procedure (and a required land use planning procedure and environmental impact assessment), during which communities, stakeholders and the general public have the opportunity to comment or raise objections. The planning approval decision can be challenged at court. This provides a strong power base for lower water authorities and municipalities.

of change, but given the individual context in each *Land* and the rigid overall regulatory system, these policy entrepreneurs needed specific characteristics to be successful. Fresh from university, placed in a newly established administrative office (created as a result of the 1982 agreement – B-W; specific departments – R-Pf; or a special agency – NRW), they took up and processed the new holistic environmental spirit of the time.[8] They were keen on multidisciplinary cooperation rather than afraid of it. Nor were they blinkered by traditional routines and old-hand experience. Open-minded, they incorporated and transferred broader and new scientific knowledge. Their enthusiasm enhanced the prospects for innovative alternatives, although initially their relatively low position in the hierarchy was an obstacle to their aim to persuade traditional-minded superiors or to engage political decision-makers.

A third and inviting aspect of the transition was inherent in the HSK report in the form of its focus on 'more room for the water'. The reputation and international weight of the epistemic community of the HSK helped to establish this new idea in the political arena. From the very beginning, however, the actual measures recommended by the HSK had contained a number of weaknesses that did not fully align with the concept of providing space for water. This made them impossible to implement as originally proposed, for technical as well as ecological and legal reasons, and also for societal ones:[9]

1. The HSK study was conducted as a compensation project with a merely technical approach. It proposed storing as much water as possible in the smallest possible space. The resulting maximum storage height of 9–10 metres in case of a major flood made the original water storage options technically infeasible and outside the requirements of (future) ecological legislation.
2. The hydrological focus overlooked the socio-ecologic system aspects. It prevented tighter links with related domains like agriculture, spatial planning and ecology, and it blocked a more integrated approach of flood protection and risk reduction. These limitations and poorly conceived components of the programme would not be acceptable to the local officials and the public.
3. The HSK study investigated only the effects and possible measures downstream as far as Worms rather than considering a consolidated basin-wide approach.[10] This precluded the integration of flood protection for the Lower Rhine (NRW) as well as

synergies in the fields of planning, monitoring, learning and implementation support among the *Länder*.

The policy entrepreneurs, however, interpreted the deficiencies as triggers of transition rather than barriers, seeing an opportunity to review traditional routines, present new ideas and consider individual approaches in view of local situations and conditions.

18.2.2 Baden-Wuerttemberg (B-W)

As the major origin of the problem and the *Land* with the largest workload under the 1982 agreement, B-W was the first to take on its obligation to create additional retention space and established in 1983 a new technical administration agency, the Neubauleitung Hochwasserschutz Oberrhein (Upper Rhine Flood Damage Supervisory Office, NHO), to implement the programme. The technocratic flood protection discourse based on hydrology was still widespread within the water administrations of the *Länder*, while the general attitude of the political decision-makers was quite passive, mindful of potential opposition from municipalities and other stakeholders. As a result, the unstated mandate became: 'Do something, but not too much.' The NHO, however, with its young, amenable water engineers, was keen to accept the new challenges.[11]

Meanwhile, the new environmental discourse had arrived, marked and moved forward by the EU's introduction of the Environmental Impact Assessment (EIA) Directive[12] for application to all major construction. The NHO anticipated additional ecological demands during the multiyear implementation phase of the 1982 agreement and decided to carry out EIAs (the first in Germany) at three planned retention sites in order to avoid any future surprises. The EIAs and experience with test flooding of the retention area at Altenheim confirmed the necessity for additional storage space if they were to meet the technical and ecological targets. However, additional locations were a major deviation from the original planning and required new political decisions.

Faced with this dilemma, the NHO started building a coalition by looking simultaneously for external complementary environmental expertise and a highly respected partner to help it gain societal and political support in strongly pro-environmental times in Germany. The search for both ended at the World Wide Fund for Nature (WWF) and with Professor Emil Dister, the Founding Director of the Auen Institute on Floodplains Ecology.[13] He became the independent academic adviser to the NHO, which asked him to co-develop a new, revised planning concept that would re-establish the original flood protection standard of 1957 between Iffezheim and Mannheim and complement the technical programme with ecological objectives, including the restoration of alluvial floodplains and habitats, and the incorporation of the new concept of ecological flooding.[14]

The building of the partnership between academia and the NHO also signalled as in traditional practice by the technical administrators in the form of shared leadership of the project. Rather than staying locked into the technocratic safety discourse, the multidisciplinary partnership became able to initiate and promote the major shift towards an integrated 'system-ecology' discourse. In addition, Professor Dister's special relations with high-level politics served as a major driving force for change. He used the international successes and reputation of the WWF and public respect for the Auen Institute as the door-opener to the political decision-making level. Most significantly, Dister could take on the task of proposing the new concept with the necessary emphasis to the relevant minister.[15] Such an opportunity would not have arisen to the same degree for the NHO alone due to traditional, hierarchical ties and the lack of political weight. The minister in turn could use the independent scientific expertise and reputation of academia and recognition by the environmental movement of the WWF as an upstanding and effective non-governmental organization (NGO) as powerful arguments to promote the ecological aspects of the NHO programme to the general public and to support its extension at the local level.

18.2.3 Rhineland-Palentina (R-Pf)

R-Pf approached the flood protection issue from a more socio-economic and participatory angle. A first, 'conventional', plan to provide the main part of the retention volume at two sites (Hördt, Neupotz) failed due to strong local opposition, negative environmental expert reports (Hördt: 1986, 1990) and rejection by the district government (Neupotz). The planning process had to restart in order to regain the goodwill of the local people, which had been lost as a consequence of the first 'top-down' initiatives by the authorities. In 1988, the R-Pf technical administration commissioned a multidisciplinary expert group that included environmental NGOs to conduct a spatial assessment study of potential retention sites. Based on the final report (1992) and after intensive dialogue (within an 'open planning' process) with the related communities and associations, the state government decided to designate ten retention sites[16] – partly as recommended by the study and partly to achieve political acceptability (Meuser, 2006).

An increasing appreciation of environmental and cultural values by the general public together with scepticism regarding major construction projects, however, prompted the responsible technical experts in the R-Pf Ministry for Environment and Forestry to adopt a more comprehensive approach to flood protection and a fundamental reconsideration of its tasks and responsibilities.[17] The starting point came with discussion about the correct application of cost–benefit calculations for flood protection measures: 'Which damage data should be used – the potential damage in front of the dykes or this combined with the potential damage behind the dykes?' This question immediately introduced probabilities of disaster and the uncertainty of natural events, the involvement of multiple causes and the possibility of multiple countermeasures. It raised further questions on a range of topics: resilience and damage minimization; damage values – their objectivity and their dependence on precautionary measures; utilization; conflicts of scale; responsibilities; divergent perceptions between the strategic and local levels; and of solidarity issues between the up- and downstream partners. The provision of retention polders or floodplain restoration no longer seemed like the final social and sustainable solution. Flood protection – traditionally the task of hydrological engineers – had to be redefined, and the socio-economic dimension had to be added to address the multicause nature of flood issues. Within the R-Pf technical administration, the idea of risk (defined as the probability of an event multiplied by its consequences) was articulated, including the social aspect, where a certain share of responsibility for risk limitation is assigned to society as cause as well as victim of floods.[18]

Dr. Karl Heinz Rother, head of the Department of Water Management in the R-Pf Ministry of the Environment and Forestry, authored a concept paper that did not stop at hazard (as did the flood protection measures in B-W). The paper introduced the idea of precaution and covered the issues of flooding consequences and damage minimization at different spatial scales. It included flood management and restoration of the rural water net (later implemented as the 'Action Blue' programme[19]), technical flood protection measures along the Rhine and at the local level, and further precautionary measures like land use restriction, emergency services and risk mapping as well as provisions for individual precautionary behaviour. The new approach represented a paradigm change from the former 'public safety promise' to a socio-ecological system perspective as a key focus of a new risk discourse.

Thus the case of R-Pf shows how the adoption of the new ideas was accelerated by a fortunate constellation of circumstances, including: an assertive, expert team within the administration, capable of identifying the important elements of flood risk management and addressing them in consistent and conclusive proposals; and a receptive minister[20] open to convincing and well-argued propositions. Both the experts and the minister viewed the new approach as cutting-edge and in alignment with the R-Pf running retention programme. To the technical administration team, the ecological risk management discourse represented a positive summary of their expertise and achievements. For the minister, it vindicated her political action and foresight.

18.2.4 Hesse (HE)

In the *Land* of Hesse also a proposal was made to integrate flood protection and nature restoration with a major retention polder near Trebur.[21] However HE serves as an example of transition hampered by political change and individual ambition, disregarded hierarchies, poor process coordination and poor communication.[22]

After the state election of 1991 the two new ministers[23] responsible for the environment (including water) and agriculture (including nature) respectively were concerned to build their political reputation. The political situation at the time was delicate. Both ministers hesitated to go public with a firm retention, protection and restoration proposal. An accidental and uncoordinated publication of the plans provoked vehement resistance from local municipalities, industry and citizens, who were taken by surprise and felt offended by the violation of the institutional information and consultation rules. It became very easy for a well-organized and experienced (through successful campaigns against major construction projects) local opposition to find sufficient counterarguments, such as the fact that the Upper Rhine flood problem did not originate in HE, that political pressure to protect specific sites of major economic value was low, and that an acceptable alternative was available of increasing the height of existing dykes. The responsible minister (Joschka Fischer, Green Party) followed the local mood and turned down the application for a polder at Trebur. The defensive attitude seen in HE was actually in accordance with the 1982 convention, which demanded only financial contributions from HE. Nevertheless it can also be interpreted as a 'Not in my back yard' (NIMBY) stance that rejected the option of solidarity with the other Upper Rhine *Länder*, even at the risk of flooding.

18.2.5 North Rhine-Westphalia (NRW)

NRW was not part of the Upper Rhine programme or the French–German agreement, although subsequent commissions[24] confirmed negative impacts of the Upper Rhine spreading down to Köln and the regional loss of some 80 per cent of floodplains[25] increased flood risk to the Lower Rhine area as well. And indeed a similar transformation away from technocratic flood protection began in the mid-1980s, although triggered in a different way – by the risk of failing dykes. An extensive inspection at this time confirmed the necessary rehabilitation from scratch of some 150 km[26] (50 per cent) of the dykes. The responsible technical administration in NRW[27] took the opportunity to expand the constructive maintenance part of the programme and to incorporate hydrological (provision of more room for floodwater) ecological and socio-economic elements.[28]

The NRW guidelines for natural river training and maintenance of water flow[29] already (by 1980) suggested renaturalization and use of retention areas to avoid protective construction. What was new was to apply these measures on a large scale on waterways like the Rhine, where dykes traditionally guaranteed flood protection. Appropriate alternatives to traditional measures were elaborated by the NWR technical administration under Robert Kolf, department head of Water Management in the NRW Ministry of Environment, Spatial Planning and Agriculture, who actively championed the new concept at the political level. He convinced the responsible minister[30] and secured his support, using 'imminent threat to public safety' as the most persuasive argument for political action.

The result was the *Generalpläne* (General Plans) for Flood Protection for the Lower Rhine (1990/1991)[31] that not only defined priorities for the dyke rehabilitation programme, but also streamlined the organization of the dyke associations and proposed feasible sites for dyke relocation and floodplain restoration to create more space for the river. To gain broader political support for the proposals, other major actors and players from political domains affected in NRW, like navigation and environmental administration, became involved and confirmed their agreement.[32] This led to the publication in 1992 of an integrated concept, the Master Plan for Flood Protection, Ecology, and Navigation[33] that considered and assessed also the ecological and navigational impacts of the planned measures. The spatial elements of the *Generalpläne* and the Master Plan still required detailed design and governmental approval, but both documents served as working papers for the responsible NRW authorities.

Each of the *Länder* along the Rhine developed (or, in the case of HE, avoided) a new flood protection position with the support of academia and multidisciplinary study groups or experts from related sectors, but mainly led by the policy entrepreneurs from the respective technical flood administrations. The following paragraphs discuss the strategies used to get the new concepts endorsed as official policies in the individual *Länder* and later at the federal level.

18.3 Political endorsement

18.3.1 Introduction

Meijerink and Huitema (see the theoretical introduction to this volume, Chapter 2) discuss the aspects of political stability and administrative inertia, and the importance of external perturbations, new political positions, calamities and changes of public mood as drivers of fundamental political change.

The 1980s and 1990s saw such a concurrence in Germany as a whole of circumstances favourable for a policy transition in the area of river management: the 1982 bi-national agreement, the new environmental and democratic discourses of the era, and the major flood events of 1993 and 1995, all of which were sources of change and opened 'windows of opportunity' that allowed policy entrepreneurs to promote their preferred solutions (Kingdon, 1984 [1995]).

18.3.2 Baden-Wuerttemberg (B-W)

In B-W, the 1982 accord, in combination with the practical impossibility of executing the original proposal of the HSK study, became an external perturbation, smoothed in turn by the new environmental discourse and pushed forward by the new European Environmental Assessment legislation. The power of the coalition with the Auen Institute, the weight of academic advice and the reputation of the WWF were powerful enough to gain political acceptance for the new concept of ecological flood protection.

In 1988 the government of B-W agreed to develop the Integrated Rhine Programme (IRP),[34] confirming this, after internal discussions and amendments, by Cabinet decision in 1996. The term 'integrated' was used to convey the aim to combine technical flood protection and ecological restoration. The programme is designed to relocate dykes and two weirs to produce a retention volume of 1673 million m^3 at 13 sites with controlled polders originally to be finalized around 2010 (latest estimates revised to between 2015 and 2020). The investment costs are estimated at some €775 million. In 2008 three of the polders had been completed, and one was under construction, bringing the available retention volume to some 40 per cent of the projected total.[35]

In R-Pf and NRW as the two other *Länder* involved, it took the external shock of the major floods of 1993 and 1995 to push forward the transition to integrated flood management policy.

18.3.3 Rhineland-Palentina (R-Pf)

The 1993 flood was judged by the state technical administration as the appropriate time to introduce risk management and the precautionary flood protection concept into the political arena, with an explanation of the work entailed. The responsible minister took the opportunity to demonstrate forward-looking political action and presented the new policy in a Governmental Declaration to the R-Pf parliament in December 1994. The initial political reaction was limited, but became enthusiastic after the next major flood in 1995. Suddenly flood protection came into political focus, opening a wide window for the new ideas and proposed solutions. The upper political echelon could react instantly to a critical event because the technical administration had comprehensive plans and solutions already prepared.[36] By 2008 almost 40 per cent of the planned retention was operational (28 million m^3) and completion was planned for 2012 (Meuser, 2006).

18.3.4 North Rhine-Westphalia (NRW)

In NRW as well, the 1993 floods helped to prime the political scene, which saw dramatic change after the 1995 flood when the General Plans and the Master Plan of 1992 became of major political value. An agitated public and the media aggressively pushed the responsible Environment Minister for practical solutions. Based on the above plans, Kolf, as head of the NRW technical flood administration, was able to prepare at short notice a consolidated flood management concept, enabling the responsible minister to arrange an instant press conference to present the proposed retention and restoration measures as the most modern and effective means to reduce flood risks in the future.

After the 1995 general election, the new Minister for Environment in NRW, a member of the Green Party[37] elaborated the concept (dyke relocations, controlled flood polders and embankments) and strengthened the ecological elements, such as restoration of floodplains and rural water management to ensure high safety levels by international standards.[38] The NRW Concept of 'Sustainable Flood Protection for the Lower Rhine' (NRW, 1996) was updated by Kolf in 2006 to include a firm timing and more emphasis on risk management and damage prevention. In 2008 four dyke relocations provided a retention volume of 68 million m^3 (44 per cent of the planned volume), while three additional relocations and three controlled polders were scheduled for completion by 2015.[39]

18.3.5 Towards a new, broader equilibrium

Three new flood policies had been institutionalized: the Integrated Rhine Plan in B-W (1988, 1996); the precautionary risk concept paper in R-Pf (Government Declaration 1994); and the *Generalpläne* and Masterplan in NRW (1990, 1991, 1992). However the new policies stopped at the borders of the *Länder*.

Yet the consecutive flood events of 1993 and 1995 created the necessary pressure for political decision makers to review flood policies also at the national and basin-wide level. This offered the opportunity for the policy entrepreneurs of the *Länder* to present and consolidate their new concepts and to instigate upwards scale change.

On the national level, flood prevention was discussed in the State Water Working Group (LAWA)[40] with Dr Rother as chairman. Based on the precautionary risk concept adopted in R-Pf, a guidance paper[41] for German rivers in general was formalized in 1995, summarizing strategies and plans to minimize damage and to emphasize as well the individual responsibility to take appropriate action to minimize the impact of floods. The guidelines served as a basis for the individual flood protection programmes in the *Länder*.[42]

An additional focusing event was required to propel national implementation of the new concept and the guidelines.[43] The catastrophic floods in August 2002 of the Elbe and Donau rivers brought to light shortcomings of cross-border coordination and flaws in the application of the guidelines. Triggered by the scale of the damage and pushed by public attention in an election season, the German government took action and presented a 'Five Points Programme' at a national conference, outlining five main work stages to improve preventive flood management.[44] The programme basically reconfirmed the LAWA guidelines on precaution and was institutionalized in 2005 by the Flood Control Act (Gesetz für den vorbeugenden Hochwasserschutz).[45] As a composite Act, it covers the amendment of related legislation, sets rules for disaster management and spatial development, and aims to promote *Länder* interplay.

On the international level, and again influenced by the 1993 flood, the 11th Conference of Rhine Ministers in 1994 agreed to add 'water discharge and quantity' to the task list of the International Commission for the Protection of the Rhine (IKSR).[46] Further pressure due to the subsequent 1995 flood led to the Declaration of Arles (February 1995) in which the national ministers in charge of water protection from the contracting countries expressed the need for a better-integrated, basin-wide approach and urged the IKSR to develop a flood protection plan for the Rhine.

This enabled the policy entrepreneurs of the *Länder* to use the IKSR venue to upscale their ecological retention and risk concept proposals because much of the work had already been done in the *Länder*, particularly R-Pf, and as a follow-up by the LAWA. The German delegation, led by Dr Rother, presented the LAWA guidance document to the IKSR as a useful basis for discussion. Supported particularly by Switzerland and the Netherlands, and further fine-tuned in the IKSR working group, chaired by Dr Rother, the Flood Action Plan (FAP) of the IKSR[47] was finally agreed in 1998.

The FAP combined the existing projects of the *Länder* and riparian countries and expanded them into a consolidated basin-wide programme. In the IKSR discussions, Dr Rother pushed for ambitious, political and quantifiable objectives, and the principle of 'risk reduction' was agreed. The FAP institutionalized a basin-wide regime with common, quantified targets, budgets, timings and a strict monitoring scheme (IKSR, 1995, 1998).[48]

The IKSR as an additional actor in Rhine flood management can be seen as a political focus and major boost for more active interplay between the countries, *Länder* and sectors to synchronize and coordinate the individual programmes. The IKSR, however, has no supra-national status or decisive power and can only function as an adviser, political broker and provider of a negotiation platform (Dieperink, 1997, 2000; Kampa et al., 2003; Dombrowsky and Holländer, 2004). To gain the full benefit of efficient collaboration and synergy, stronger guidance with a broader political mandate and including all water management domains may well be required.

The 14th Conference of Rhine Ministers in October 2007 might serve to support this point of view. The intermediate monitoring report for the year 2005 (IKSR, 2006, 2007a, 2007b) suggests that the FAP might not achieve its 2020 targets, a prospect that prompted political action (IKSR, 2007c): the IKSR was given a mandate for further adaptive measures and extended coordination. A project was agreed to identify additional potential retention space and also to include potential climate change impacts, an issue that had until then been considered only regionally for the German part of the Rhine (for example, the KLIWA programme[49] of B-W, Bavaria, and R-Pf).

18.4 Implementing change

The status of implementation as the final phase of transition requires assessment of a variety of factors. Using the retention measures as an indicator (Table 18.1), it can be concluded that the transition at present has not yet reached the micro, or local, level. After some 20 years, just 40 to 60 per cent of the retention measures are operational.

The delays or fading focus in the implementation phase have numerous reasons that can be attributed to several and often related political, organizational or psychological factors that include:

1. The general split in interlinked political domains like water–environment and space–agriculture, which reduced the possibility of reframing flood protection as a spatial planning problem.
2. The numerous administrative reorganizations and reforms in the different *Länder*,[50] which caused a reallocation of responsibilities, and a disintegration of mutual trust and cooperative routines.
3. The various governmental elections held during the long implementation phase, producing changes in political direction, priorities and local configurations, which

Table 18.1 Retention volumes

Retention volume (millions of m³)	Original concept 1982	Concept FAP 2020 (estimate)	Operational 2005
France	56.0	51.6	58.4
B-W	126.0	167.3	96.0
R-Pf	40.0	64.9	28.0
NRW	–	155.0	68.0
Retention			
Upper Rhine	222	291	144
Lower Rhine	–	155	68
Total	–	445	213

Source: IKSR (2007a).

caused some dilution of the original concepts (B-W), or altered decision-making constellations emerging as a result of political competition (HE).

4. Inadequate interplay at the federal level that became evident in the fragmentation of resources, hampered flexibility and institutional barriers seen during the 2002 flood catastrophe.
5. Reduced risk perception and the lack of a sense of urgency in both the public and political arena during long periods with no major flood events. A scientific report (Lammersen, 2004) that considers the impacts of climate change on future flood risks at the Lower Rhine and anticipates colossal flooding of the Bonn–Köln region did not make any deep impression: it was distributed in 2005 to all members of the NRW parliament without any political reaction.
6. Low interest in personal precautionary measures like flood-proof building and flood insurance, a factor obviously related to complacency about risk, but possibly also to economic conditions and a lack of related regulation;
7. The lax handling of spatial development restrictions leading to numerous exemptions in various communities.

Most hindrance, however, stemmed from the increased local opposition that began in the 1990s. In dominant 'top-down' management style, the technical administration defined the main elements of the programmes without any early involvement of local communities or of the general (but, at the time, basically uninterested) public. Most often, and mainly at the later, more concrete stage, local resistance built up within the legal frame of land use planning approval and environmental assessment procedures.

In this phase, limited risk perception and disbelief by the public met with limited experience of the technical administration of the *Länder* in transferring new knowledge, understanding local interests and achieving comprehensive cooperation at the operational level, all necessary for successful implementation (Ufit Umweltforschungsinstitut, 2006).

It became obvious that a new mode of governance was required, one with more emphasis on competencies in process management, more transparency and open communication across disciplines and across administrative and social boundaries to involve all parties concerned down to the operational level. Two examples may illustrate how the technical authorities as the traditional leading actor tried to adapt the traditional top-down project management style.

In R-Pf, the Supreme and Upper Water Authorities attempted to apply a policy of open dialogue including intensive moderation procedures with local stakeholders (Hördt and Ingelheim; see examples below). In B-W and R-Pf, so-called 'flood partnerships' (*Hochwasser-Partnerschaften*) have been established between spatial planners, water managers and municipalities to promote interests bottom-up and secure involvement at an appropriate, early stage. These approaches have shown initial signs of success. Polder Ingelheim was the first polder to be agreed without court proceedings. Inauguration of the polder came in 2006 after consensus was reached among agriculture, ecology and water management stakeholders, and a construction period of less than two years. At Hördt, earlier proposals for a retention area had been rejected as a result of a controversial ecological assessment. The request of the state parliament of R-Pf in 1995 to provide additional retention space prompted the federal authorities to reconsider the region around Hördt as an emergency polder for extreme flood events. This time responsibilities for the development of an integrated usage concept were delegated downwards to the local stakeholders involved in the planning phase. During 2006 the R-Pf technical administration organized a moderated discussion process between the authorities and local stakeholders covering alternatives in order to isolate and neutralize issues and to agree a final regulatory framework, including compensation regulation, safety standards and operational procedures. The Water Management Department of the R-Pf Ministry of the Environment, Forestry and Consumer protection was elaborating the details in 2008 and the project should be completed by 2020.

The Hördt case also indicates certain limitations to participation and the risk of reduction to the lowest common denominator. Experts have suggested that because additional retention space will be required to achieve the set FAP targets, a better option might be a controlled retention polder of the kind already operating in medium flood events, or a dyke relocation rather than the agreed emergency polder.[51] However no stakeholder has found this proposal sufficiently in their particular interest and a potentially more adaptive solution has fallen by the wayside.

In NRW, dyke associations are a main actor as they perform dyke maintenance, for which they raise a tax;[52] they possess the necessary technical and local knowledge and are the main representatives of farmers, who in turn are the main users of most of the land involved. In this way dyke associations have a key function in 'delivering' the required common consent, space and budget. Here the technical administration intensified the building of local alliances with the communities and dyke associations as a general way to foster bottom-up involvement at the early phase of the project and to facilitate implementation of retention measures. Two supporting instruments motivated the affected dyke associations to cooperate more actively and transcend individual interests: the identification of win–win projects for communities or industrial sector stakeholders; and generous financial arrangements, such as attractive compensation schemes to enable a socially acceptable consolidation and reallocation of agricultural land holdings (*Flurbereinigung*) – this without expropriations and with primary financing by the *Land* to cover more than 80 per cent of retention and infrastructure expenses.

As the above examples demonstrate, implementation as the final step gained momentum with the introduction of a more collaborative management style featuring certain key elements, like: open dialogue and partnerships to introduce more transparency and to create awareness during preparation and planning; delegation of responsibilities downwards to promote shared leadership and involvement; the agreement and use of clear political instruments and incentives to build trust. But, as the retention volumes listed above indicate, these approaches have still to be applied to the extent needed to ensure full implementation.

18.5 Discussion: key aspects of German flood management transition

18.5.1 Introduction

Research (van der Brugge and Rotmans, 2007) shows that since the 1970s transition involving significant political change has occurred in European flood management. Different phases during the transition can be distinguished and are presented by Olsson et al. (2006), starting with the development of innovation and new structures (preparation phase), a second phase of transition (including exploiting the window of opportunity), and finally the stabilization and resilience build-up of the new system. Van der Brugge and Rotmans (2007) distinguish similarly between four phases: pre-development, take-off (innovations, networks), acceleration (new structures) and stabilization.

Various factors relating to policy entrepreneurs came into play during the different phases of transition, starting from the initial and acceleration phase (individual push of new ideas), followed by endorsement (building coalitions and relations, preparing windows of opportunity) and implementation.

At the strategic, macro level, we can observe a stabilized political transition occurring from the technocratic and controlling perspective of safety and of water as something to fight against, to a holistic approach of accommodating water to minimize risk. Despite the institutionalization, through legislation or accord, of the precautionary socio-ecological, system-oriented concept in the form of a national and bi-national flood regime, transition at the lower administrative level has still to pass the acceleration phase. This is indicated by the continuing lack of full coordination among the *Länder*, the absence of harmonized standards, and the individual interpretation of rules and future requirements in the *Länder*. The recent monitoring results of the FAP and the political reactions at the 14th Conference of Rhine Ministers (2007) also suggest a number of challenges and opportunities for a more robust and united approach.

Given that the transition process across scales faces a number of issues, I conclude that the design of German institutions for river management has a number of implications relating to policy entrepreneur strategies and mechanisms.

18.5.2 Policy entrepreneurs

In most cases policy entrepreneurs are seldom identified as individuals managing a transition on their own. Typically, organizations and coalitions are the key drivers, working in changing configurations where different actors dominate different phases of the transition. In the cases studied, academia and NGOs, politicians and the national government play a significant role. Nevertheless I find that the transition in flood management in the *Länder* and at the different venues was very much driven by a number of individual actors.

The main contributions in the different contexts along the Rhine came from experts in the technical administration of the *Länder* and, in the B-W case, also from academia. Young, flexible and innovative, these individuals served as policy entrepreneurs within the traditional hierarchy by recognizing and signalling the problems, promoting new ideas and alternatives (technical, ecological, strategic) and by involving other parties as required. By introducing new procedures and routines, they played a prominent role in opening the window of opportunity (Kingdon, 1984) for radical policy change. They were able to create or exploit supportive departmental constellations. This allowed them to build alliances of appropriate capacity to push their ideas to the political level and to prepare non-traditional alternatives in time to be launched at the right moment aided by academic support or focusing events.

These technical administration experts were also able to learn. It is interesting to note that the key policy entrepreneurs and initiators (as I define them) were present over a long period in leading administrative or academic (B-W) positions as they actively manoeuvered the new discourses upwards through the levels of administration. This suggests that they had time to develop the necessary management skills and characteristics to lead the transition through the different phases of initiation, if probably to a lesser extent and from a distance in terms of implementation. In this respect, the experts acting as policy entrepreneurs provided a stabilizing element over time in regard to political rearrangements and changing decision-maker constellations.

18.5.3 Strategies for transition

A range of actors has to be involved and specific political circumstances must prevail to institutionalize fundamental policy change in the highly regulated German structure with its fixed responsibilities. Such circumstances are described in Kingdon's (1984) metaphor of three different independent 'streams' to tie into a package and push through a policy window. Focusing events like natural disasters or political change can influence the problem and political streams and open a window, which then offers an opportunity for policy entrepreneurs to push their preferred solutions. Olsson et al. (2006) confirm this view, finding that new discourses, focusing events and careful preparation like knowledge production are instrumental in creating the opportunity, but that leadership and networking are essential to exploit an 'open window'.

We can observe that all three streams – the problem stream (flood events, perspectives), the political stream (public mood, political pressure, venues and changes) and the policy stream (instruments, possible solutions and concepts) – contributed to the establishment of the ecological risk discourse and the success of the transition in flood management in the Rhine Basin. The political stream, which included the environmental movement of the 1970s as a general background of change, supported the genesis of new ideas and helped to form the working routines and postmodern perceptions of the new administrative generation and their strategies.

To gain the necessary knowledge and to sell new ideas and concepts, the technical administration experts had to find collaborators. Despite a long tradition of public involvement in the Upper Rhine region, active public participation was not the most promising option (Blatter, 1994; Löser, 2002). Civil initiatives and NGOs were engaged in the anti-atomic power movement or specific local 'nature' problems to save specific habitats and, in the case of flood management, unable to form a regional or basin-wide network. For this reason the technical administration in B-W remained dominant in the initiating phase.

Given the 1982 accord and the ecological requirements as the frame, the building of a powerful coalition in B-W with environmental experts from academia and the connection with an international NGO (WWF) with a high social reputation proved a more viable and successful strategy to influence political decision-makers and to institutionalize the IRP. 'Knowledge production' and the acquisition of experience of more sustainable flood management solutions were supporting drivers.

Learning from initial failures, R-Pf addressed the 'more room for the river' proposal by organizing appropriate alliances and commissioning multidisciplinary study groups, including NGOs, as well as more 'open planning' and dialogue with local communities and key stakeholders.

In NRW, circulation of the new concept of ecological flood protection stayed even more within the realm of managers. Pushing safety remained the main strategy to win political agreement, but the master plan was prepared in cooperation with players from the linked political domains of navigation and environment as a way to increase the weight of arguments and to pre-empt unnecessary discussions in the political decision stage.

In HE, coalition-building between administrative experts and academia did not succeed. The elections and governmental change opened a different kind of window of opportunity – a political window for local opposition. The exogenous pressure to act was too low; locally acceptable alternatives were available, and the political competition together with a poorly prepared process led to the rejection of the new flood management concept.

Essential triggers for transition in R-Pf and NRW (and later for scaling up) were the floods of 1993 and 1995. Here the value of preparatory work became most evident. The policy entrepreneurs had produced sufficient new concepts and alternatives to float around in the 'primeval soup' of the policy stream ready to be picked up and linked as solutions to problems. The floods provided a 'reality test' that undermined confidence within expert circles, and even more importantly of the public, in engineered protection and confuted the idea of safety as technically controllable. The opportunity was used by the policy entrepreneurs not only to keep control in order to steer the transition in their respective *Länder*, but also to use venues at higher administrative and political levels to establish the cross-boundary regime.

In R-Pf, Dr Rother had prepared the precautionary risk concept, which became the basis of the government declaration in 1994. As a follow-up he used the venues of the LAWA and IKSR to constitute the new risk discourse at the national and international levels.

In NRW, the quick succession of the floods raised the public attention and political pressure. Again, concepts of integrated ecological flood protection were ready for implementation – gratefully picked up by the political decision-maker and further assisted by the election of a new Green Party minister.

In the implementation phase, top-down planning remains dominant, but in all cases new strategies have increasingly been applied to foster local implementation and bottom-up involvement. Such strategies include open dialogue and delegation of responsibilities to improve transparency and local acceptability (B-W, R-Pf) or the creation of win–win situations and attractive financial arrangements to improve vertical interplay between authorities (NRW).

18.5.4 Institutional impact

The German federal structure has a number of general implications for the mechanism and route of the transition. Its strength is its network, where multiple centres of expertise can produce ideas and solutions as appropriate for specific regional and functional contexts. This produces competitive innovation and institutional redundancy as observed in the cases studied, where different elements of the new flood management policy emerged in different regions (see Chapter 17 by Olsson and Galaz in this volume). Ideas and solutions were advanced by different actors with different intentions and objectives that could finally be consolidated at the national and international level. The contributions of each of the *Länder* combined in the form of a more complete and robust regime.

The transition as a whole also exhibits frequently criticized elements: a lack of basin-wide thinking, ecological misfit, coordination problems and lost opportunities for synergies or learning (Moss, 2003, 2005). The binding frame of the German–French agreement of 1982 and the individual treaties between the federal government and the *Länder* had a number of limiting effects: reduced potential for flexibility through mutual planning; prevention of a more integrated programme in the initial stage; restricted cooperation; constraints on both learning and a continuous review of assumptions or strategies, as shown by the HE case and the basically unchanged design of the FAP measures.

As a major financial contributor, the federal government intended, through the national Flood Act, to gain more coordinating influence on the flood protection programmes of the *Länder*. The impact of this corrective attempt and the potential added value of the Act have still to be realized. Criticism of the Act comes on the one hand because it contains too many opportunities for local interpretation and does not set a clear example of ecological flood prevention.[53] On the other hand, more detailed and far-reaching regulation could discount the spirit of risk management and prevention, which does not consist of fixed, uniform and standardized solutions, but takes a sustainable 'best solution' approach, adaptable to local situations and needs. Yet, as the first national law on flood protection, the Flood Act confirmed the new regime. The extended discussion and the rejection of initial, stricter proposals in the Federal Council (Bundesrat), however, indicate the firm intent of the *Länder* to defend their individual flood management powers.

The *Länder* are not only responsible for water management, but also for the linked political domains, like spatial development, agriculture and environmental conservation. This distribution of competencies across different ministries and their local execution among different administrative levels creates horizontal and vertical institutional barriers (Krämer, 1999; Mayntz, 1999). For example, the missed chance to frame flood protection as a spatial development problem presents a major issue. Despite rules for settlement and spatial restrictions in the new concepts and the Flood Act, various exemptions and interpretations form one reason for delayed implementation.

In addition the administrative structure defines a sensible 'tension triangle' of political decision power, concept generation and local implementation. Most major projects and new concepts have their origin in the upper-level technical administration. At the same time, these experts are squeezed into a complex hierarchical power constellation that sits between the political decision and local operational levels – depending upwards on the support of the deciding ministerial authority, and downwards on the cooperation of the powerful lower authorities, stakeholders and local political configurations.

18.6 Completing the transition

The cases studied show that the construct of the transition is set; the question remains as to what needs to be done to achieve completion. A number of factors and possibilities come into play:

1. The potential impact of climate change will demand additional major flood management decisions with significant societal consequences.
2. A mutual and operational vision for flood management in the Rhine basin could provide a sound basis for future challenges shared by national and international decision-makers and local shareholders alike.
3. Rather than determination by sovereignty and federal or hierarchical considerations, a holistic frame is required with harmonized standards, interpretations and mutually agreed assumptions, with structured monitoring and coordinated experimental schemes to learn and profit fully from current implementation experience and the numerous case studies.

However, not only is knowledge required about what needs to be done, but it must also be transferred and embedded into society. Expert knowledge and social acceptance both form part of the task, and both should follow the principles of transparency and rationality.

Transparency of the process is needed rather than evaluation alone of progress. As a first step, indicator systems can help both to assess the status and to define the deficits openly and clearly of precautionary flood protection in a municipality, for example. Follow-up of certification or benchmarking systems, goal-setting, presentation of options for improvement, active participation of stakeholders and monitoring are essential.

Objective information, clear social and ecological goals and a sound, transparent approach are needed rather than feelings alone or political conformism as a basis for decision-making to produce the most effective result.

18.7 Concluding remarks

We can assume that the basin-wide and international venue will gain in importance. The recent decisions of the Conference of Rhine Ministers to give the IKSR a more distinct coordination mandate for adaptive strategies are a promising corrective signal. In addition the new European regulations, such as the Flood Directive,[54] will create new networks, relations and responsibilities to stabilize the regime and will also build up the pressure to follow the principles of solidarity and subsidiarity.

Solidarity is needed to increase cooperation and interplay up- and downstream, to develop common climate scenarios and hydro-atmospheric models, and to develop common solutions.

Subsidiarity provides an avenue to acceptance of responsibility for action and improvement together with the freedom to apply the most appropriate measures. This would require a better integration of the political domains and better cross-border coordination. This could lead to a more integrated and encompassing management concept and combination of responsibilities, including possibly a 'water manager' responsible for all related management aspects of large rivers.

Given the above, implementation of the transition in flood management is not just a question of additional legislation or political instruments, nor one of new measures and alternatives, many of which have been brought forward already by the LAWA, IKSR or NGOs. Nor is there a lack of knowledge or capability. What is needed is the political will to take action and to follow a common vision. It remains to be seen whether sufficient political leadership will emerge to finalize the transition process. Even greater cooperation among a range of actors and levels of governance is needed than that seen so far in the Rhine basin. In the absence of strong political drive, the possibility exists not of measured, collaborative implementation steered by human policy entrepreneurs through a current window of opportunity, but instead of urgent, stringent political action forced by a gaping hole of necessity caused by the next major disaster.

Acknowledgements

The major part of the research reported has been carried out under the framework of the projects NeWater (Contract no. 511179 (GOCE) under the 6th Framework Programme) and ACER (developing Adaptive Capacity to Extreme events in the Rhine basin, www.adaptation.nl). I would like to thank the European Commission, the Dutch BSIK '*Klimaat voor Ruimte*' programme for the financial support received. The personal interviews provided essential information and insight and I would like to thank all participants for their contributions and their cooperation. Particularly, the useful details and constructive comments of Dr Karl-Heinz Rother, Professor Emil Dister, Dr Ulrike Pfarr and Mr Robert Kolf are gratefully acknowledged.

Notes

1. The basin comprises nine countries: Switzerland, Austria, Germany, France, the Netherlands, Luxembourg, Belgium, Italy, Liechtenstein and eight German federal states: Baden Wuerttemberg (B-W), Rhineland-Palatinate (R-Pf), Hesse (HE), North Rhine-Westphalia (NRW), Bavaria (BA), Lower Saxony (SAX), Thuringia (THU) and Saarland (SA).
2. The probability of peak floods increased from once every 200 years to once every 50–60 years and the running time of the flood wave from Bale to Karlsruhe reduced from 64 hours to 23 hours.
3. The consequences of a 200-year flood in the three *Länder* (B-W, R-Pf, HE) would be disastrous: between Iffezheim and Bingen some 2000 km^3 with 95 cities and communities and a population of 700 000 would be at risk; the potential economic damage has been estimated above €6.5 billion (IKSR).
4. The Hochwasser-Studien-Kommission (HSK) was established in 1968 on the initiative of B-W with experts from Germany, Austria, France and Switzerland to quantify the increased flood risk downstream as far as Worms. In the final version of the report (1978), the HSK recommended measures to reclaim more flood space for the river: special operation of the power stations, using the former Rhine bed for discharge; three adjustable retention weirs; and controlled retention polders.
5. In a number of conventions and supplementary agreements, France and Germany regulated rights and future water management developments along the Upper Rhine. For further details see www.untreaty.un.org, accessed June 2008.
6. A more detailed list and minutes of the interviews are available from the author.
7. One argument was that the origin of the flood problem was not caused within Hessian borders and that the adjacent position of HE to the Rhine above Worms is minor. The current spending amounts to some €40 million.
8. Interview with the Director of the Auen Institute and the former President of the R-Pf State Environment Agency.
9. Interview with the Director of the Auen Institute.
10. Subsequent expert groups quantified the flood conditions further down the Rhine: HSG Worms-Kaub

1985 (Dister, 2002); HSG Kaub-Rolandswerth 1993 (Hochwasserschutzkonzept Köln, 1996; Meuser, 2006); HSG NRW 2002 (www.lanuv.nrw.de); DE-NL HW AG (Lammersen, 2004).

11. Interview with the Director of the Auen Institute.
12. The European Environmental Impact Assessment Directive was introduced in 1985 and amended in Germany in 1997; www.ec.europa.eu/environment/eia/home.htm, accessed November 2008.
13. The WWF, and Professor Dister as an independent academic adviser, were already involved in the early 1980s in several international activities to protect or restore endangered alluvial forests and riparian wetlands also at the Upper Rhine.The successes and the recognized importance of the river line landscape prompted the WWF to hasten the progress of the foundation of a specific research institution that was established in 1985 at the Auen Institute (Institute for Floodplains Ecology) in Rastatt with Professor Dister as its founding director. See www.auen.uni-karlsruhe.de, accessed November 2008.
14. Controlled flood storage areas (retention polders), which are only operated in flood situations, must be regularly flooded to restore and maintain dynamic habitats (ecological flooding). See, for example, www.forwara.org, www.auen-uni-karlsuhe.de, accessed November 2008.
15. Gerhardt Weiser, 1976–96: Minister of Agriculture, Forestry and Environment.
16. R-Pf agreed originally to provide a floodwater storage volume of 44 million m^3 plus reserves of 7 million m^3 to account for future decision processes in the course of more detailed planning. The actual planned volume amounts to almost 65 million m^3. Also in R-Pf a special construction office (NGO R-PF, Neubaugruppe Hochwasserschutz Oberrhein) was in 1990 given the task to design, plan and implement the retention programme.
17. Interview with the former President of the R-Pf State Environment Agency and with the Director of the Auen Institute.
18. Interview with the former President of the R-Pf State Environment Agency and with a departmental head of the R-Pf Environmental ministry.
19. Action Blue (Aktion Blau) was launched in 1994 as a programme to renaturalize home waters in terms of quality, habitat and natural flood protection. As a communication tool to interest the public and particularly young people, so-called 'good parenthood' (*Patenschaften*) for local river segments were offered to schools, clubs and so on, www.wasser.rlp.de, accessed November 2008.
20. Klaudia Martini, 1991–2001: Minister for Environment and Forestry in R-Pf.
21. Following the B-W experience, a coalition of technical administration and academia (Professor Dister, Auen Institut and Technical University of Darmstadt) was established that recommended a portfolio of spatial assignments with the retention polder near Trebur as the most important piece of the jigsaw.
22. Interview with the Director of the Auen Institute.
23. Joschka Fischer, 1991–94: Minister for Environment and Energy, Jorg Jordan, 1991–95: Minister for Spatial Development, Agriculture, Forestry and Nature Protection.
24. See note 10.
25. From the original floodplain area of 1800 km^3 only some 300 km^3 are left.
26. The dyke rehabilitation requirements were revised over time. Current estimates increased to 261 km to be rehabilitated, of which 115 km have been completed. Forty-five km have still to be inspected (http://www.umwelt.nrw.de/umweltinformationen/umweltbericht/ub-2006/php, accessed May 2008).
27. Environmental Agency of NRW, Chief Dyke Inspector H.G. Schmitz and his colleague R. Kolf.
28. Interview with a former department head for water management in the NRW Ministry for Environment, Nature Protection and Agriculture.
29. Richtlinie für den naturnahen Ausbau und Unterhalt der Fliessgewässer in NRW (Guidelines for natural river training and maintenance of water flow in NRW) 1980 (new draft, 1990), http://home.landtag.nrw.de/mdl/johannes.remmel/f11.html.
30. Klaus Mathiesen, 1983–95: Minister for Environment, Spatial Planning and Agriculture in NRW.
31. 'Generalpläne Hochwasserschutz am Niederrhein' (1990, compiled by R. Kolf) and 'Hochwasserschutz am Rhein im Regierungsbezirk Köln' (1991). The documents reviewed also set out the ownership and responsibilities of dyke maintenance and the consolidation of dyke associations.
32. Interview with a former departmental head for water management in the NRW Ministry for Environment, Nature Protection and Agriculture, 15 April 2008.
33. Gesamtkonzept Hochwasserschutz, Schifffahrt, Ökologie am Rhein in NRW, published by the Ministry for Environment, Spatial Planning and Agriculture in NRW (1992).
34. www.rp.baden-wuerttemberg.de/servlet/PB/menu/1191692/index.html, www.rp-freiburg.de, accessed November 2008.
35. Interview with the departmental head 'Integrated Rhine Programme' of the B-W Regional Council, 12 March 2008.
36. Minister Margit Conrad confirmed the anticipatory risk concept of 1994 again in a Governmental Declaration, March 2005.
37. Bärbel Höhn, 1995–2005: Minister for Environment, Spatial Planning and Agriculture in NRW.

38. At the Lower Rhine, flood safety levels are: at Köln 1/200 years, between Düsseldorf and Duisburg 1/300 years, and between Duisburg and the Dutch border 1/500 years (Lammersen, 2004).
39. The measures will reduce the flood waves at the Dutch border by more than 10 cm and delay the arrival of the flood wave at the Dutch border by more than 12 hours.
40. A special institution, the LAWA (Länderarbeitsgruppe Wasser, State Water Working Group, established 1956) has the task of harmonizing policies and legislation across the federal states; www.lawa.de, accessed November 2008.
41. Leitlinien für einen zukunftsweisenden Hochwasserschutz 1995 (Guidelines for forward-looking flood protection, www.lawa.de). In addition to traditional construction-based protection measures (dykes, retention basins), key elements include natural retention and rural water management as well as precautionary actions like restrictive land use, flood-proof building and preventive action against risk (public provision, individual provision, insurances, awareness programmes).
42. B-W: www.um.baden-wuerttemberg.de/servolet/is/975; www.ikone-online.de; R-Pf: www.wasser.rlp.de/servlet/is482.
    NRW:www.regionale2010.de/fileadmin/user_shares/2010_bildwelten/Rheinkonferenz_2005/Rheinkonferenz_Vortraege/hochwasser/Kolf.pdf; www.nr-feldmann.de/Downloads/HWSK-Text.pdf; HE:www.hmulv.hessen.de/irj/HMULV_Internet?cid=c42172c88d8a9e75b704e201958c01cf; all accessed November 2008.
43. The LAWA guidelines were confirmed in 2003 and supplemented with practical advice for their implementation (Instruments and Guidelines for the Implementation of the Guidelines), www.lawa.de.
44. www.bdu.de/gewaesserschutz/doc/3114.php, Press Statement 8 August 2003, J. Trittin, Federal Minister for the Environment, Nature Conservation and Nuclear Safety, www.bdu.de, accessed November 2008.
45. www.bmu.de/files/pdfs/allgemein/applications/pdf/Hochwasserschutzgesetz.pdf.
46. The Internationale Kommission zum Schutz des Rhein (International Commission for the Protection of the Rhine, IKSR) was founded in 1950 and formalized by the convention of Bern 1963 (strengthened 2003) as a basis for international law for future cooperation. Due to a Dutch initiative, the bordering countries France, Switzerland, Luxembourg, Germany and the Netherlands (and additionally the EU, 1976) created a common forum to discuss pollution issues of the Rhine and to seek and agree solutions accordingly; www.iksr.org, accessed November 2008.
47. The Flood Action Plan (FAP) includes the planned measures of the riparian countries Switzerland, France, Germany and the Netherlands.
48. The four action targets are: to reduce damage risk 10 per cent by 2005 and 25 per cent by 2020; to reduce flood levels up to 30 cm by the year 2005 and 70 cm by 2020; to increase public awareness of flooding by drafting risk maps by the year 2005; and to improve flood forecasting systems and prolong the forecasting period by 100 per cent by the year 2005.
49. www.kliwa.de, accessed November 2008.
50. Administrative reforms (*Verwaltungsreform*): R-Pf 2000, B-W 2005, NRW 2007.
51. Interview with the Director of the Auen Institute, 13 March 2008.
52. Dyke maintenance is financed by the Dyke Associations; new constructions, like the retention programme, are mainly (up to 80 per cent) financed by the *Land*.
53. The NOG of BUND (Bund für Umwelt und Naturschutz) crticized the missed chances to integrate political domains: 'Hochwasserschutz – aus den Fehlern nichts gelernt', Bilanz, August 2007, www.bund.net, accessed December 2008.
54. EU Flood Directive 2007/60/EC on the assessment and management of flood risks, Water Framework Directive 2000/60/EC; www.ec.europa.eu, accessed November 2008.

References

Barzelay, M. and N. Füchtner (2003), 'Explaining public management policy change: Germany in comparative perspective', *Journal of Comparative Policy Analysis: Research and Practice*, **5**, 7–23.

Becker, G., J. Aerts and D. Huitema (2007), 'Transboundary flood management in the Rhine basin: challenges for improved cooperation', *Water Science and Technology*, **56** (4), 125–35.

Bernhardt, C. (2002), 'PCCP processes in history: the model of the Upper Rhine region', contribution to the UNESCO–Green Cross International project From Potential Conflict to Cooperation Potential (PCCP).

Blatter, J. (1994), 'Erfolgsbedingungen grenzüberschreitender Zusammenarbeit im Umweltschutz. Das Beispiel Gewässer- und Auenschutz am Oberrhein', EURES Discussion Paper 43, www.eures.de/de/publikationen/download/dp-43_ocr.pdf, accessed 15 July 2006.

Bouwer, L., D. Huitema and J. Aerts (2007), 'Adaptive flood management: the role of insurance and compensation in Europe', Internal Report W07/08, Department of Spatial Analysis and Decision Support, Free University Amsterdam, the Netherlands.

Broschek, J. (2007) 'Federalism and political change: Germany and Canada in historical–institutionalist perspective', paper presented at the annual meeting of the Canadian Political Science Association, 29 May – 1 June, University of Saskatchewan, Saskatoon, Canada.

Cioc, M. (2002), *The Rhine: an Eco-Biography, 1815–2000*, Seattle, WA: University of Washington Press.

Dieperink, C. (1997), 'International regime development: lessons from the Rhine catchment area', *TDRI Quarterly Review*, **12** (3), 27–35.

Dieperink, C. (2000), 'Successes in the international cooperation in the Rhine catchment area', *Physics and Chemistry of the Earth*, **25**, 341–7.

Disse, M. and H. Engel (2001), 'Flood events in the Rhine basin: genesis, influences and mitigation', *Natural Hazards*, **23** (2–3), 271–90.

Dister, E. (2002), 'Gefährdete Landschaften: Vom falschen und richtigen Umgang mit Flüssen und Auen', in Jörg Kachelmann (ed.), *Die grosse Flut*, Reinbeck, Germany: Rohwohlt Verlag, pp. 112–43.

Dombrowsky, I. and R. Holländer (2004), 'Erfahrungen im integrierten Management grenzüeberschreitender Flüsse in Europa: Das Beispiel des Rheins', *Zeitschrift für angewandte Umweltforschung (ZAU)*, **15–16**, 443–59.

Frijters, I.D. and J. Leentvaar (2003), 'Rhine case study', UNESCO–Green Cross project PCCP, http://webworld.unesco.org, accessed 15 January 2006.

Gunlicks, A.B. (2005), 'German federalism and recent reform efforts', *German Law Journal*, **6** (10), 1283–96.

Hochwasserschutzkonzept Köln (1996), 'Abschrift des Auszuges aus dem Beschlussbuch 1 February 1996', www.steb-koeln.de/hskkln.html, accessed 19 June 2007.

IKSR (1995), 'Grundlagen und Strategie zum Aktionsplan Hochwasser', Internationale Kommission zum Schutze des Rheins, Koblenz, Germany.

IKSR (1998), 'Aktionsplan Hochwasser', Internationale Kommission zum Schutze des Rheins, Technisch-wissenschaftliches Sekretariat, Koblenz, Germany.

IKSR (2006), 'Nachweis der Wirksamkeit von Massnahmen zur Minderung der Hochwasserstände im Rhein infolge der Umsetzung des Aktionsplans Hochwasser bis 2005', Report 153d, www.iksr.org, accessed 12 February 2008.

IKSR (2007a), 'Umsetzung des Aktionsplans Hochwasser, Bericht 2005', Report 157d, www.iksr.org, accessed 12 February 2008.

IKSR (2007b), 'Nachweisinstrumente für die Reduzierung von Schadensrisiken: Wirksamkeitsnachweis der Maßnahmen zur Minderung der Schadensrisiken infolge der Umsetzung der IKSR Aktionsplan Hochwasser bis 2005', Report 157d, www.iksr.org, accessed 12 February 2008.

IKSR (2007c), 'Living and linking Rhine – common challenge of a watershed', Conference of Rhine Ministers, 18 October, Bonn, Germany, www.iksr.org, accessed 20 November 2007.

Jänicke, M., H. Jörgens, K. Jörgensen and R. Nordbeck (2001), 'Governance for sustainable development in Germany: institutions and policy making', Forschungsstelle für Umweltpolitik, www.oecd.org/dataoecd/27/32/1828117.pdf, accessed 27 July 2008.

Kampa, E., N. Kranz and W. Hansen (2003), 'Public participation in river basin management in Germany, "From borders to natural boundaries"', Work Package 4 of the HarmonoCOP project, Ecologic – Institute for International and European Environmental Policy, Berlin, Germany.

Kingdon, J.W. (1984), *Agendas, Alternatives and Public Policies*, (2nd edn 1995), New York: HarperCollins College Publishers.

Koordinierungskomitee Rhein (2005), 'Internationale Flussgebietseinheit Rhein', Bericht an die EU Kommission nach Richtlinie 200/60/EG, Teil A., www.iksr.org/index.php?id=26, accessed 23 August 2007.

Krämer, A.R. (1999), 'Water management and policy in Germany', www.ufrgs.br/iph/kraemer_water_management_and_policy_in_Germany.pdf, accessed 22 August 2005.

Lammersen, Rita (2004), 'Grensoverschrijdende effecten van extreem hoogwater op de Niederrhein', Final Report, Ministerie van Verkeer en Waterstaat, Directie Oost-Nederland, Arnhem, the Netherlands.

Löser, G. (2002), 'Die Badisch-Elsässischen Bürgerinitiativen', www.badischelsaessische.net/geschichtekurz.htm, accessed 6 July 2007.

Mayntz, Renate (1999), 'Multi-level governance: German federalism and the European Union', in Carl Lankowski (ed.), *Governing beyond the Nation-State*, Report No. 11, Washington, DC: American Institute for Contemporary German Studies, pp. 101–14.

Meuser, A. (2006), 'Bau der rheinland-pfälzischen Rückhaltungen am Oberrhein – von der Version zur Realität', Bundesanstalt für Gewässerkunde (BfG) Kolloquium, 17 January, Koblenz, Germany.

Moss, T. (2003), 'Regional governance and the EU Water Framework Directive: a study of institutional fit, scale and interplay', paper presented at the workshop of the EU Thematic network project REGIONET, 29–31 January, Lillehammer.

Moss, T. (2005), 'Dissecting Institutions: Bestandteile einer institutionellen Konfiguration am Beispiel der Wasserwirtschaft', ICAR Diskussionspapier 7/2005.

Mostert, E. (2009), 'International co-operation on Rhine water quality 1945–2008: an example to follow?' *Physics and Chemistry of the Earth*, **34** (3), 142–9.

Nienhuis, P.H. (2008), *Environmental History of the Rhine–Meuse Delta: An Ecological Story on Evolving Human and Environmental Relations Coping with Climate Change and Sea-Level Rise*, Dordrecht: Springer Science and Business Media.

Olsson, P., L.H. Gunderson, S.R. Carpenter, P. Ryan, L. Lebel, C. Folke and C.S. Holling (2006), 'Shooting the rapids: navigating transitions to adaptive governance of socio-ecological systems', *Ecology and Society*, **11** (1), 18, www.ecologyandsociety.org/vol11/iss1/art18/.

Overeem, A. (2005), 'Description of the River Rhine basin', www.knmi.nl/~overeem/rhinr.pdf, accessed 8 March 2007.

Tittizer, T. and F. Krebs (eds) (1996), *Ökosystemforschung: der Rhein und seine Auen, eine Bilanz* (Ecosystem research: the Rhine and its floodplains), Berlin: Springer.

Ufit Umweltforschungsinstitut (2006), 'Das integrierte Rheinprogram IRP in Baden-Württemberg: Kommunikation, Beteiligung, Akzeptanz, Component B: Entwicklung einer gemeinsamen Planungsmethode', www.freudeamfluss.eu/docs/FAF%20Component20B6%20ufit%DE.pdf, accessed 5 October 2007.

van der Brugge, R. and J. Rotmans (2007), 'Towards the transition management of European water resources', *Journal of Water Resource Management*, **21** (7), 249–67.

Wieland, M. (1998), 'Wie aktuell ist die Politikverflechtungsfalle', www.mwieland.mw.ohost.de/mathiaswieland/pol/pv-falle.htm, accessed 17 October 2008.

Winnegge, R. and T. Maurer (2002), 'Water resource management, country profile Germany', Global Runoff Data Centre (GRDC), Federal Institute for Hydrology (BfG), Koblenz, Germany.

19 Policy dynamics in Dutch water management: analysing the contribution of policy entrepreneurs to policy change

Dave Huitema and Sander Meijerink

> The eternal challenge of the sea dominates Dutch history, and the response to that challenge tells us much about human capability to manipulate an adverse environment. (Reuss, 2002, p. 465)

19.1 Introduction

The Netherlands is a relatively small and crowded country located adjacent to the North Sea and in the delta of four European river basins, those of the Rhine, the Meuse, the Scheldt and the Ems (see Map 19.1). With a population of about 16 million, the Netherlands has one of the highest population densities in the world (about 470 inhabitants per km^2). The largest population concentrations are in the low-lying areas in the west of the country, which is the urbanized region including cities like Amsterdam, Rotterdam, The Hague and Utrecht.

This chapter considers the way water policies have evolved in the Netherlands since 1970. The change we describe is one towards the 'greening of water management', a term that points to the realization within the Netherlands that the long-standing efforts to control water have brought safety and prosperity, but also degradation of ecosystems and the landscape, and that this degradation should end. In a country where 15 per cent of the land is below sea level and half of its area would disappear without dams and dykes, it is understandable that taming the 'water wolf', as this natural enemy is known, has been an important societal goal (Reuss, 2002). It is also logical that success in achieving this goal in the period between the storm surge of 1953 and the present time has instilled a measure of faith in the technologies and institutions used so far to accomplish it. This may be one of the reasons why the Netherlands is a country where a high level of flood risks is coupled with a low awareness of such risks, especially in those areas most threatened (De Boer et al., 2003). It is also an important background factor shaping the context for the activities reviewed in this chapter: the work of policy entrepreneurs trying to take water management in a greener direction.

What do we mean by the greening of water management? Two temporally and spatially divergent answers apply. In the 1970s the practice meant ameliorating the negative environmental effects of new water infrastructure. In the form of dams and dykes or the large canalization projects of this time, infrastructure was often large in scale and normally designed to achieve maximum efficiency in realizing societal goals. The most important goal for water managers was and essentially continues to be flood safety. A major 'focusing project' at the national level, for instance, consisted of the South Sea Works (Zuiderzeewerken) whose history goes back to the nineteenth century. The South Sea (Zuiderzee), an inland arm of the North Sea, was a key traditional fishing ground,

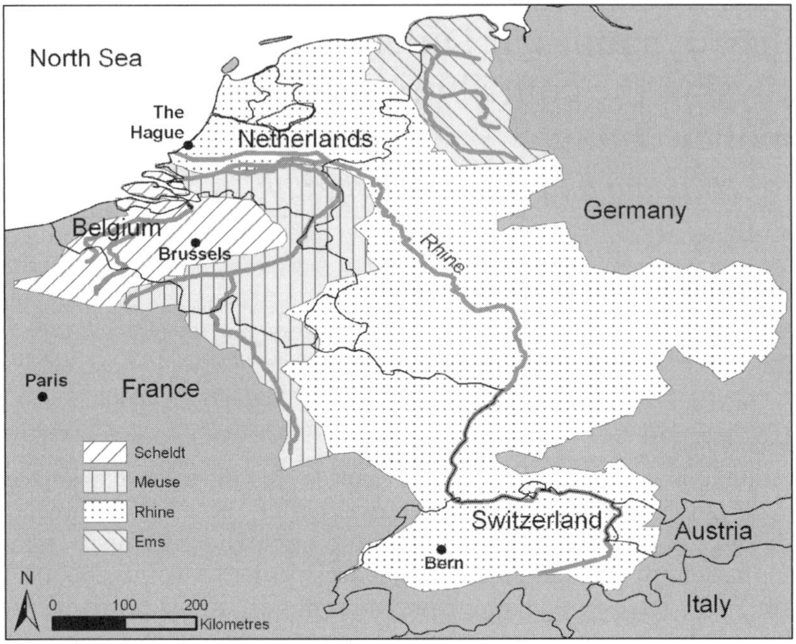

Map 19.1 Position of the Netherlands at the northwestern end of Europe and the major river basins located in the Netherlands

a sailing route for boats from the harbour of Amsterdam and a practice area (shooting range) for the army. When a major flood occurred in 1916, however, plans held for decades to close off the inland sea were taken down from the shelf and a 30-km dam was built between the South Sea and the North Sea in order to guarantee the safety of the hinterland. The South Sea Works took the better part of the twentieth century to complete and greatly increased the prestige of the National Public Works Agency (the Rijkswaterstaat). The Works were accompanied by the digging of a canal that gave Amsterdam harbour a direct link to the North Sea. Meanwhile the importance of the army reduced considerably, and the fishing interests were too small to make a united effort to prevent flood safety from becoming a dominant concern in the newly created freshwater lake called the IJsselmeer (Huitema, 2002; Huitema and Kuks, 2004).

The main opponent of the National Public Works Agency in this project and in many others was the Ministry of Finance, which was suspicious of the amounts of money invested. The Ministry of Agriculture has traditionally lined up as an ally of the Public Works Agency in this intergovernmental battle where, as in the South Sea Works, the reclamation of land has formed an integral part of many projects, partly to shorten the coastline, but mainly to increase agricultural production. Massive new polders (stretches of dried-out land on the former sea floor) were created and prepared for intensive agricultural use. Other interests were not taken much into account, although recreation lobbyists in the 1930s succeeded in convincing water managers that the dykes around the new polders should not all be straight and that the new lake would be more appealing with various mini islands. The flood safety–agriculture coalition held sway until the end of the 1960s and 1970s when food shortages became a distant memory due to agricultural overproduction in Europe, and new linkages became necessary to justify the massive infrastructural projects being proposed. These linkages were made with the housing sector, which was producing massive numbers of new and better-quality houses at the time. Construction of the last polder of the South Sea Works, completed in the mid-1970s, was therefore mainly defended for its importance for housing, although the polder was also mentioned in connection with a second national airport (see Huitema, 2002).

In the same period the Netherlands saw the advent of the environmental movement. Groups formed that advocated a more natural appearance for the landscape and a softening of boundaries between ecosystems outside and inside the water infrastructure. They started protesting against several planned projects, including elements of the South Sea Works and various local-level 'improvement' projects. The paradigmatic case of this era, however, is the Eastern Scheldt Dam. Here the extant paradigm of reducing the coastline by dam construction became contested, and a major concession that spoke to the public imagination was achieved in the direction of greener water management (see Meijerink, 2005). This is the first case study analysed here.

The second concept informing the greening of water management arose in the 1990s. Those advocating the notion began to promote the even more radical ideas of restoring the water system to a much more natural state and of 'living with water' rather than stopping it at every point. Living with water implies enhancing the operation of the natural dynamics of the water system and in various circumstances the dismantling of the existing water infrastructure – especially the summer dykes – to make 'space for water'. This has moved the geographical focus of the debate from the seacoast and more to the basins of the several large European rivers that flow into the Netherlands (see Map 19.1). The decision process on flood measures in the main river area provides the second case study of this chapter.

Before discussion of the case studies, some preliminary remarks are in order. First, we have minimized explanations of formal organization and policies. We instead refer the reader to other works, such as those of Bressers et al. (1995), Reuss (2002), Van Slobbe (2002), Kuks (2004), Huitema and Kuks (2004) and Kissling-Näf and Kuks (2004). Second, the transition we have chosen to analyse is very much one of policy substance rather than of the process of governance. As a result we focus on one specific aspect of the transition as described by Van Ast (2000) and Van der Brugge et al. (2005). These authors suggest that Dutch water management has evolved from a 'technocratic scientific style towards an integral and participatory style' (Van der Brugge et al., 2005, p. 164). We are interested in the substantive change towards a more integral approach to water issues and leave aside the move towards greater public participation in Dutch water management, although it is clear that the presumed transition towards greater participation has affected the venues for policy debates (Huitema and Kuks, 2004).

Third, we note that the question itself of the greening of water management in the Netherlands can be seen as the subject of a transition. The definition of transitions as instances of 'radical policy change' (see the theoretical introduction to this volume, Chapter 2) implies a contextual element where what is seen as radical may change over time and place (see also Chapter 5 by Wilder in this volume). The ideas proposed by environmentalists in the 1970s and 1980s were perceived as radical at the time, but have now become mainstream. They can also be seen as 'add-ons' to the then dominant engineering approach. The ideas about living with water that emerged in the 1990s remain quite radical and not (yet) completely accepted in policy circles. We are aware that by grouping policy changes in the 1970s and 1980s under 'the greening of water management' with the changes of the 1990s, we imply that these changes are different expressions of the same development. We think that this is defensible as both ideas brought forward in the respective eras emerged from the environmental movement, and because the radicalism of the ideas of the 1990s reflects the acceleration of environmental thinking since the earlier period.

Fourth, in our focus on national policy we have attempted to select cases that reflect the discussion in the 1970s and 1980s, and the debates of the 1990s. We contend that the two cases we describe are paradigmatic in that they were (and are) the focusing projects through which new policy development was negotiated. We are aware, however, that in leaving out many other stories, for instance those that played out at the regional level, we fail to describe the work of many policy entrepreneurs (but see Huitema and Kuks, 2004). Whether or not, then, our two case studies are also representative is open to argument. One factor that suggests our cases are less representative is the fact that 'large infrastructure projects' – which is what they are – tend to have their own dynamic, which is often more complex than the dynamic of smaller projects (see Flyvberg et al., 2003). In addition, generalization from our findings may be complicated given the factor of space. Ideas about making space for water are potentially more easily realized – and transitions thus more easily achieved – in the context of river systems where more options for finding space may exist. Along the Dutch seashore, with its massively fortified chains of dykes, which are only as strong as the weakest link, and where the potential for policy lock-in is great, such ideas have more difficulty taking hold.

With the above four important qualifications to our analysis, we present the two case studies in the following two sections and the lessons to be learned from both in the final section.

19.2 The Eastern Scheldt Storm Surge Barrier: a focusing project

19.2.1 Introduction

Decision-making on the construction of a storm surge barrier in the Eastern Scheldt is often considered a turning point in the history of Dutch water resources management (Leemans and Geers, 1983; Westerheijden, 1988; Bijker, 2002; Disco, 2002; Meijerink, 2005). After a lengthy stalemate and societal disagreement on the necessity of damming the Eastern Scheldt estuary, the government in 1976 under Prime Minister Den Uyl decided to construct a half-open (semi-permeable) storm surge barrier on piers. This unique masterpiece of coastal engineering made flood safety possible along the estuary, while at least partly maintaining the estuarine dynamics. Under normal weather conditions the dam is open, and sea and river water are mixed. During extreme wet and/or windy weather the dam is closed.

The decision to construct the Eastern Scheldt Barrier broke with centuries of flood policies designed to achieve flood safety and to serve only a few economic interests, such as water provision for agriculture. Ecological values featured significantly in decision-making on the new dam, the first time this had occurred in relation to large-scale flood defence infrastructure in the Netherlands. Why was the Dutch government prepared to spend billions of euros to preserve the Eastern Scheldt ecosystem? The answer comes from several related questions: which parties opposed the closure of the Eastern Scheldt estuary, and why were they ultimately successful? What strategies did they use to challenge successfully the traditional engineering paradigm and to influence the decision-making process? And why had the flood management community no choice but to accept a compromise that they had opposed for a long time? The history of decision-making on the Eastern Scheldt Storm Surge Barrier, as sketched briefly below, provides useful context for an examination of these questions.

The last sea flooding in the Netherlands took place in 1953. On 1 February that year a storm surge raised water levels along the Dutch coast to extreme heights, with dramatic consequences: 1836 people died and about 200 000 hectares were inundated. This is a classic example of a shock or focusing event that raises awareness. The Netherlands' vulnerability to flooding became dramatically visible. The Dutch government, which (perhaps understandably) had devoted more attention to rebuilding the country after the Second World War than to flood safety, responded immediately by establishing a Delta Committee to advise on how to prevent future sea flooding and guarantee flood safety. The Delta Committee continued the work and activities of the Storm Water Committee, which before the Second World War had already formulated various recommendations on how to improve flood safety and had issued the so-called Delta Plan in 1954. A part of the plan entailed the construction of a series of large dams to close the Rhine, Scheldt and Meuse estuaries in the southwestern part of the Netherlands. Map 19.2 indicates the various proposed closures.

The Delta Act, which was approved almost unanimously by the Dutch parliament in 1957, made the Delta Plan official government policy. Because of the sense of urgency, however, a newly established Delta Project Department had already begun preparation for various projects in 1954, taking over responsibilities previously held by the local water boards. Although the centralization of responsibility for water might be seen as an important governance transition, the substance of the Delta Plan definitely did not break radically with the past. In essence the plan continued the policies developed around the time of the South Sea Works (Disco, 2002; Huitema and Kuks, 2004; Meijerink, 2005). The construction of the series of large dams to close the estuaries formed the main substantive strategy of the Delta Act, following exactly the solution applied by the Rijkswaterstaat engineers in the construction of the Closure Dam (*Afsluitdijk*) and the damming of the Zuiderzee in 1932. The importance of the Delta Works for flood safety in the Netherlands cannot be denied, but the resulting dams did not reflect a new paradigm in water management. New discourses and fresh paradigms began to make an impact only when the Delta Project Department began planning the closure of the last estuary, the Eastern Scheldt, setting the scene for a breakthrough in Dutch flood management.

19.2.2 The development of new ideas

The Dutch flood policy domain has been long dominated by the Rijkswaterstaat as the central water management agency, and the epistemic community of Delft civil engineers. These water management experts usually frame water issues as technical problems for solution by technical means and measures, such as the construction of dams and dykes. Because of the relative success of these strategies, the Rijkswaterstaat for a long time had a very good reputation in the Netherlands. The Rijkswaterstaat employees were even

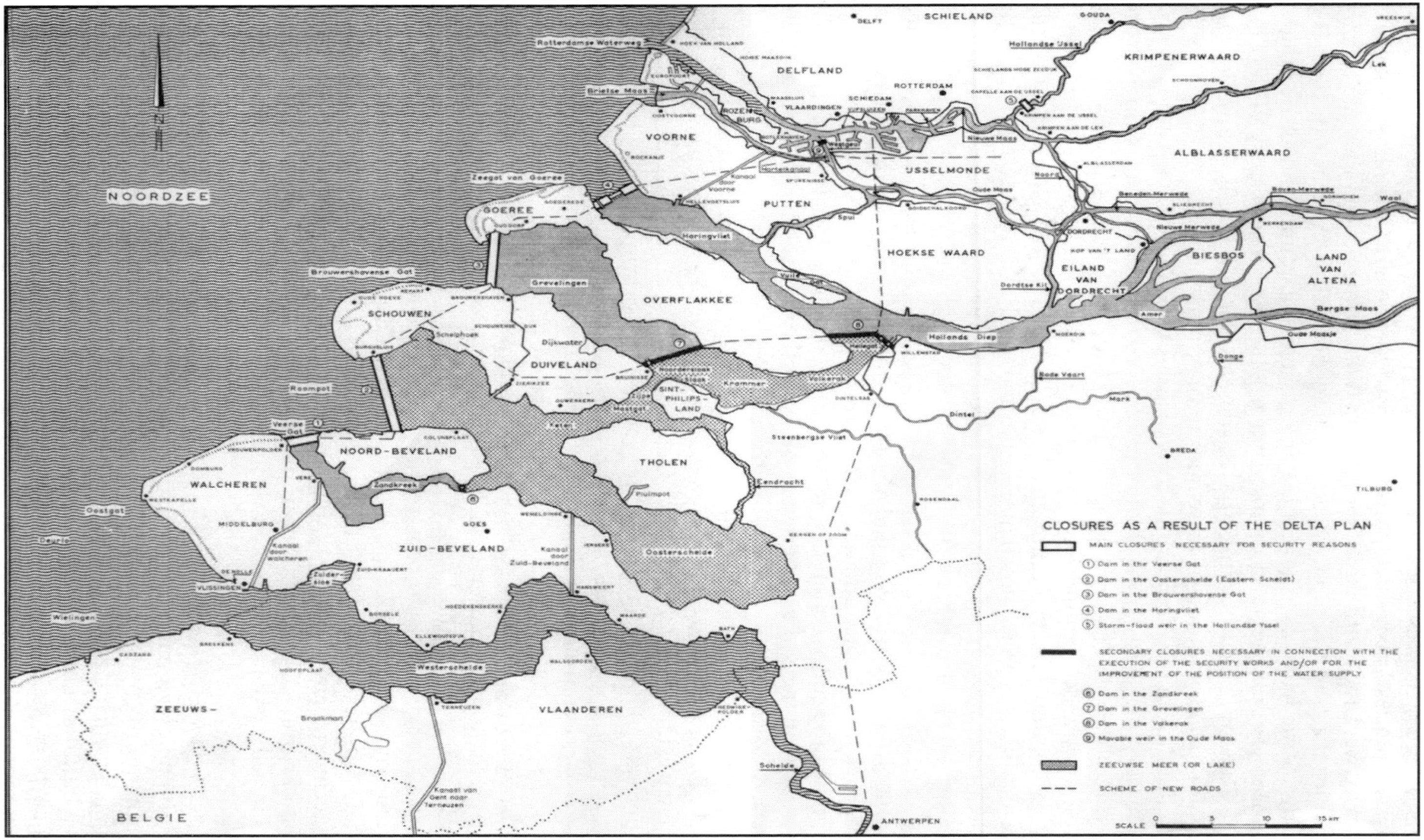

Note: The plan proposed the closing of all Rhine, Scheldt and Meuse estuaries, with the single major exception of the Western Scheldt. This estuary is shared with Belgium and forms the navigation channel to the port of Antwerp.

Map 19.2 Closures resulting from the Delta Plan

called the Dutch 'water wizards' (Den Doolaard, 1948). In the 1960s and 1970s, however, this positive reputation would change markedly. Two main developments put pressure on the Rijkswaterstaat: the growth of the environmental movement and the democratization of Dutch society. The Rijkswaterstaat faced huge difficulties in incorporating the new environmental discourse, which gained dominance rapidly after the publication of the Club of Rome report *Limits to Growth* (Meadows, 1972). Due to its hierarchical nature and its military origins, the Rijkswaterstaat was hardly responsive to the new demands for openness and transparency.

While the first dams in the Rhine and Meuse delta, such as the Haringvliet dam, could be planned and built without any serious opposition, the closure of the Eastern Scheldt became a major focusing project as it exposed the limited adaptive capacity of the Rijkswaterstaat (Lowry, 2006). The environmentalists pointed to the unique ecological values of the Eastern Scheldt estuary and framed the issue as a matter not only of safety, but also of ecology. The new ideas of ecosystem-based water management were developed by an advocacy coalition of scientists, environmental non-governmental organizations (NGOs) and representatives of government agencies. The Delta Institute, an independent research institute of the Dutch Royal Academy of Sciences, for example, made inventories of the flora and fauna in the estuary, and forecast the ecological losses that would occur if the estuary were closed (Saeijs, 1982; Leemans and Geers, 1983). The Eastern Scheldt issue was discussed at a series of meetings and conferences that started with a conference organized in 1967 by the Royal Zeeland Society of Sciences (Koninklijk Zeeuws Genootschap) (Westerheijden, 1988).

Although the Rijkswaterstaat clearly faced difficulties in adapting to the new circumstances, it did recognize the urgent need for developing new knowledge and expertise in the field of ecology and biology within the Rijkswaterstaat itself. In 1971 the Delta Project Department gained a specialized environmental section, which was headed by Henk Saeijs, the Rijkswaterstaat's first university-educated biologist (Disco, 2002). This section was charged with carrying out research on the ecology of the Delta waters, and to formulate recommendations on how to preserve the ecosystems.

Saeijs was an important policy entrepreneur who used an arsenal of strategies to influence water policies. He appointed various other ecologists and biologists within the Delta Project Department, and although faced with a lot of resistance inside the Rijkswaterstaat he managed gradually to reform the organization from within. Following his appointment the external pressures on the bastion of engineers became paralleled and strengthened by increasing internal pressure. Saeijs managed to create a network of experts who disseminated ecologically inspired conceptions of water management throughout the agency. Saeijs later worked at the central ministry in The Hague. This new position offered him more opportunities to influence directly the direction of national water policies.

He was one of the authors of the influential policy report 'Dealing with Water', issued in 1985 by the Institute of Inland Water Management and Waste Water Treatment (Ministerie van Verkeer en Waterstraat, 1985) and considered by many to be the founding statement of the concept of integrated water management. Saeijs and others introduced the concept to emphasize the need for water management based on ecosystem preservation. At a later stage in his career, as a professor of sustainable water management at the Erasmus University of Rotterdam, he often referred to the disturbed processes of erosion and sedimentation and other negative impacts of the Delta Works, and argued that nature is the best engineer in the end (see for example Saeijs et al., 2004). This statement probably summarizes best the ideas and visions that he has promoted within the Dutch water management community during his career.

19.2.3 Build coalitions and sell new ideas

The parties that wanted to include ecological values in decision-making, and questioned the need to close the Eastern Scheldt, gradually learned how to improve coordination of their activities and build a powerful advocacy coalition. Scientists, environmental NGOs and politicians worked together in the Eastern Scheldt Study Group and the Action Group for an Open Eastern Scheldt. These platforms later joined forces in the Committee for an Open Eastern Scheldt (Ten Dam and Hillenius, 1982). The parties involved shared key policy beliefs and their important coalition challenged the beliefs underlying policy of the Delta Project Department (Meijerink, 2005). The environmentalists also cooperated intensively with another group opposing the closure of the Eastern Scheldt: a coalition of fishermen who were dependent on shellfisheries in the tidal salt and brackish waters, and yacht owners. This coalition did not share deep core beliefs, but found common ground in short-term economic interests (ibid.). Each coalition based its opposition to the plans to close the Eastern Scheldt estuary on very different reasons. We have called this type of cooperation a strategic alliance (ibid.). The Eastern Scheldt case demonstrates in this way that advocates of policy change may join forces with parties of different beliefs and hence different problem frames, as long as they can reach agreement on specific policy options or preferences.

The opposition to the closure of the Eastern Scheldt is a clear example of collective entrepreneurship, where parties in the policy network pooled resources despite holding different positions on the issues. While scientists improved scientific evidence regarding the ecological disaster that a closure of the estuary would cause, NGOs organized societal protest and played the media to attract attention to their alternative policy ideas.

19.2.4 The recognition, manipulation and exploitation of windows of opportunity and venues

As described above, the 1953 flood served as a focusing event. As such it opened a window of opportunity, one that highlighted the problem at the root of the issue. This problem window was exploited effectively by the Rijkswaterstaat and its epistemic community of civil engineers. Plans developed before the Second World War were further elaborated and brought to fruition, which resulted in the famous Delta Works. However the Delta Works, also as indicated above, continued the technical water management discourse and accelerated the implementation of existing policy rather than contributing to a policy transition. Decision-making on the damming of the Eastern Scheldt instead served as a turning point as it moved through a political window, namely the change of the Dutch government in 1973.

In that year Prime Minister Den Uyl formed the most progressive Cabinet in Dutch history. One of the political parties participating in the Den Uyl government was D'66, a liberal democratic party with an ambitious environmental programme. In addition the Den Uyl Cabinet adopted the election programme of the cooperating left-wing parties that together formed 'Keerpunt 1972' (Turning Point 1972). A couple of new ministers in the Den Uyl Cabinet, who had strong ties with the environmental movement, used strategies both of venue change and venue manipulation to influence the course of decision-making on the Eastern Scheldt. As the closure of the Eastern Scheldt was recognized as a crucial environmental and political issue, decision-making authority on the Eastern Scheldt was transferred upwards from ministry level to the Dutch Cabinet, which is an interesting example of venue change. Moreover the Cabinet were successful in their attempt to instal an independent Eastern Scheldt Committee, a new venue, to advise on possible solutions for the controversies over the closure of the estuary. They ensured that the committee was formed on multidisciplinary lines that included ecological and spatial planning expertise in decision-making. The words of one researcher sum up the situation well:

> While the installation of the Den Uyl government marked the end of the policy monopoly of the Ministry of Transport, Public Works and Water Management and its 'safety coalition', the installation of a multidisciplinary Eastern Scheldt Committee marked the end of the hegemony of Delft civil engineers in the coastal flooding policy domain. (Meijerink, 2005, p. 1072)

19.2.5 Orchestrate and manage networks

The strategies of policy entrepreneurs discussed so far were all aimed at strengthening a specific position, in this case the position advocated by environmentalists. Environmental players collected scientific evidence for their ideas, formed coalitions and exploited a political window by contacting newly elected politicians, who in turn exploited and manipulated venues: a clear example of collective entrepreneurship. In spite of the relative success of these strategies, the coalition did not gain complete dominance over the flood policy domain: they remained largely dependent on the cooperation of the ministry, the Rijkswaterstaat and the Delta Project Department. The various accounts of the Eastern Scheldt crisis demonstrate that neither the ministry nor the coalition of environmentalists made serious attempts to manage interdependencies actively and to bridge differences. Because of the high level of conflict a basic condition for cross-coalition learning was absent. As a result the 'discussions' can best be characterized as a 'dialogue of the deaf' (Jenkins-Smith and Sabatier, 1993).

The conflict had to be settled at the highest political level, that is, within the Dutch government. In this respect Westerheijden's account of political decision-making is highly informative (Westerheijden, 1988). The Minister of Transport, Public Works, and Water Management within the Den Uyl administration, Tjerk Westerterp, was a Christian Democrat who at first supported the ministry's policy to continue the implementation of the Delta Plan and to close the Eastern Scheldt estuary. Within the Cabinet, however, some liberal-democratic and social-democratic ministers who sympathized with the environmentalist advocacy coalition successfully tried to influence decision-making. The social-democratic Minister of Finance, Willem Duijsenberg, from the start opposed the new plans for the construction of a semi-permeable dam, as the project would entail too much government expenditure. Prime Minister Den Uyl wanted to prevent a crisis in his Cabinet. He managed, during a boat trip on the Eastern Scheldt, to convince Duijsenberg of the need to meet the demands of the environmentalists, in this way playing an important role as policy broker. When Westerterp heard about Duijsenberg's change of opinion, he was pragmatic enough to give in and to cooperate further on the new Eastern Scheldt policy (Westerheijden, 1988).

This case illustrates, then, the major impact individuals, here politicians, may have on the course of a particular decision-making process. Although it could be argued that, due to the change of national mood, a development toward more ecologically oriented water management policies would have been inevitable anyway, a change of opinion by the Prime Minister or Minister of Finance could have resulted in a different outcome and an irreversible closure of the Eastern Scheldt estuary. This also illustrates the limits to the influence of policy entrepreneurs. Although in this case collective entrepreneurs have successfully influenced and manipulated decision-making, they were not able fully to control the outcome of the decision-making process. Entrepreneurship is a condition *sine qua non*, but one not sufficient alone for policy change.

19.3 Management of the main rivers

19.3.1 Introduction

Map 19.3 shows the main rivers in the Netherlands. The change in Dutch policy on management of these rivers involves complex dynamics and circumstances that have demanded considerable strategizing on the part of policy entrepreneurs. The greening of water management in this instance refers to a set of interventions designed to give more space for the rivers to flow freely or, simply, to give 'space for water'. This is to be achieved through a set of related measures, notably: the removal of the 'summer dykes' (the lower dykes closer to the river); the digging out of clay and sand deposits in old river beds; the creation of water-carrying trenches close to the river; the allowing of alluvial forest development; and finally the reintroduction of large plant-eating animal species. Rather than add-ons to existing or proposed water infrastructure projects, these measures form an alternative approach to river management, one premised on the aim to create a self-maintaining and self-regulating river system. The 'inventors' of these ideas suggested that the long-standing approach to river management was self-defeating in that it stimulated the deposit of sand and clay in the riverbed, thereby perpetuating the need to increase the height of the dykes. The new approach is intended to bring safety

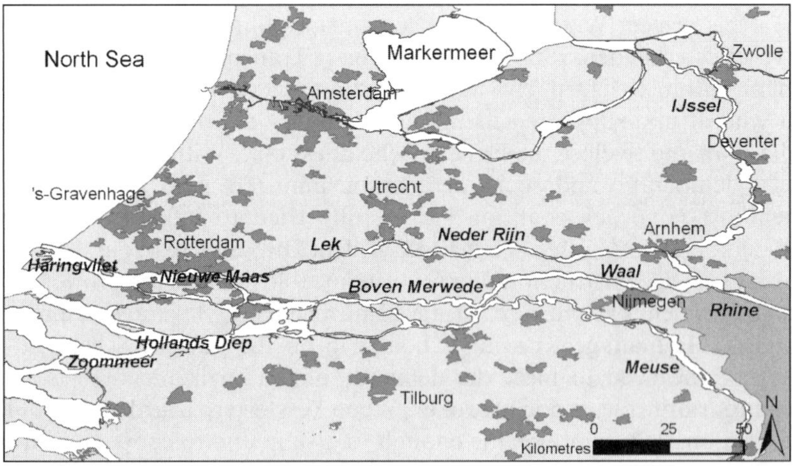

Map 19.3 The main river area in the Netherlands

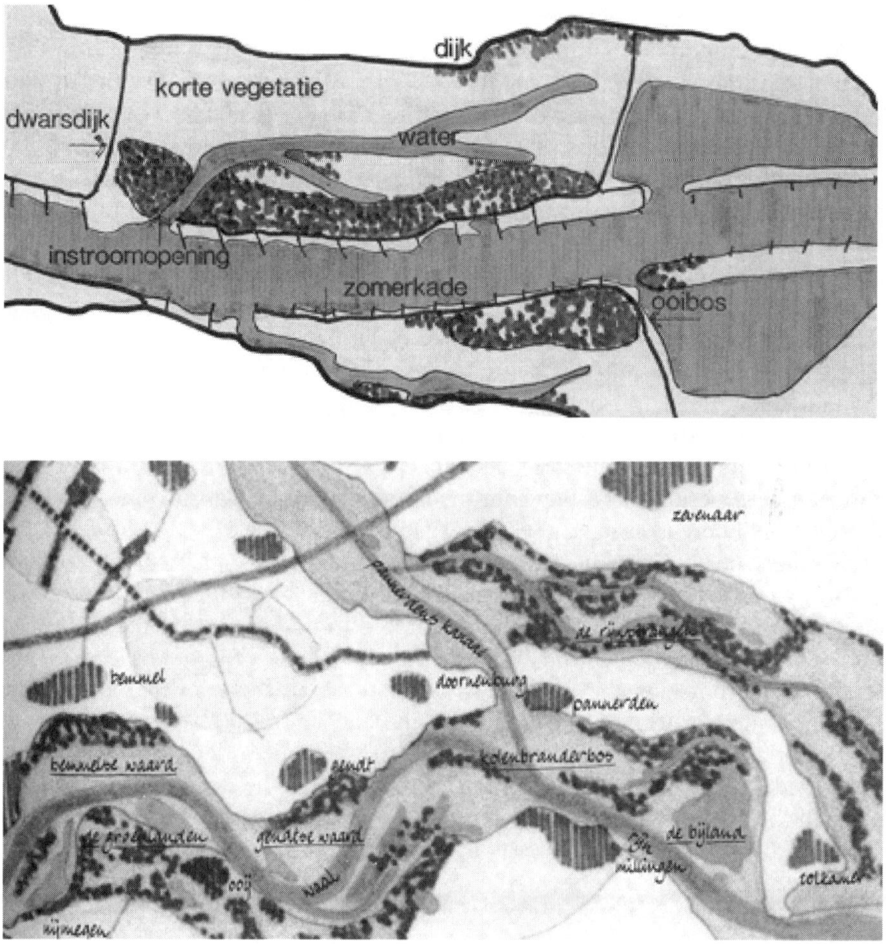

Note: The river is de-canalized by reintroducing natural bends and the creation of overflow areas that are sometimes flooded, sometimes dry. There is room for spontaneous vegetation and therefore ecosystem development next to the river, which enhances the attractiveness of the area for tourists, and the river's relatively shallow tributaries provide excellent habitat for a range of species. The idea is to facilitate a system that requires very little human intervention.

Source: Website of the Ministry of Agriculture, Nature Management and Food Safety, www2.minlnv.nl/thema/groen/ruimte/ols/algemeen/vb/inftgroav47.shtml, accessed 29 April 2008.

Figure 19.1 New water management concepts of the 1980s and 1990s

by excavating clay and sand deposits, to create an attractive riverside area for tourists and recreation, and greatly to increase biodiversity. The projects are planned not only to maintain themselves through the restoration of natural dynamics, but also to be self-financing because the clay and sand deposits can be sold on the construction market. Figure 19.1 illustrates the new ideas.

Several aspects of this case of river management stand out. First, the new ideas for management have never fully supplanted the 'old' approach, even though the new approach has been formally endorsed and integrated in government plans. The official government policy documents thus describe the new approach as one option for river management, not necessarily 'the' option. The underlying reason for this is the fact that the new approach has never won unanimous support from the expert community. Second, the case clearly demonstrates the importance of individuals playing various roles, such as decision-maker, expert or campaigner, that involve various activities, but also shows that these individuals actually operated as a group in the form of a shadow network rather than on their own. In this way the case highlights collective rather than individual policy entrepreneurship. Third, the case demonstrates how the (near-) floods of 1993 and 1995 did not provide a window of opportunity for proponents of a new approach. In fact almost the opposite occurred: the floods acted as a window for supporters of traditional river management. Those advocating new ways experienced great difficulty in keeping their approach alive in policy terms. Finally, it is important to note that although the transition of interest was formalized in official policy documents only late in the 1990s, its conceptual roots go back to a much earlier period, indicating the importance of perseverance as a factor in the management of transitions.

The story we tell here is based on secondary analysis. Many other authors have extensively researched the case, and here we build on their work. We particularly make use of Van der Brugge et al. (2005), Roth et al. (2006), De Jonge and Van der Windt (2007) and Ploegmakers (2008), and have essentially attempted to find clues that reveal the various policy entrepreneurs involved and their strategies. Most of the authors mentioned suggest, however, that it is not just the work of these entrepreneurs that explains the policy change under scrutiny. De Jonge and Van der Windt (2007), for instance, suggest that it is rather a combination of developments at the societal level and the innovative work of individuals at the micro level that put pressure on the water policy system. This system, as all the authors observe, has exhibited a degree of resistance to change, as the old approach to river management was firmly embedded in legislation, organizations, and so on. According to De Jonge and Van der Windt (2007), novel ('transformative') concepts played a key role in the transition, but advocates of these concepts have had to work hard to achieve acceptance. Their activities to raise awareness and secure a positive reception include facilitation of an active debate over, and a selection process for, alternative ideas and the promotion of new concepts as connected to a desire for change in society, to ongoing policy work and to the actual management of the water infrastructure through pilot projects (see also Olsson and Galaz, Chapter 17 in this volume).

19.3.2 The development of new ideas

The new approach to river management, as outlined above, is rooted in the larger realm of environmental management. It can be seen, therefore, to stem from a transition that originated outside the domain of water management. Ploegmakers (2008) connects the new ideas on river management to the change in discourse seen in Dutch 'nature management' circles in the 1980s and 1990s. Some saw the traditional discourse of nature conservation as too defensive. These managers propagated an agenda of nature 'development' as a more assertive strategy in several senses. It proposes the expansion of space available for nature and the establishment of new corridors to connect natural areas to assist exchange in the gene pool. The concept also includes a non-traditional underlying approach to nature as preferable when left wild and unregulated, rather than when controlled to create essentially man-made landscapes where wild flora and fauna must coexist with agriculture. Proponents of nature development seek the separation of areas intended for agriculture and the use of them to their full potential for that function. Such separation would leave no place for agriculture in 'natural' areas. The idea of nature development quickly gained legitimacy in the Netherlands and became one of the cornerstones of the 1990s National Nature Policy Plan.

The main rivers were an early testing ground for these new ideas. Nature development gained credibility due to the fact that it could successfully translate into proposals for attractive landscaping. This helped take the new ideas further on their way to formal endorsement. De Jonge and Van der Windt (2007) highlight the importance during this process of the policy community involved in landscaping. Several members of this community were disappointed with what they could achieve through agricultural land reallocation projects (*ruilverkaveling*) at the provincial level. Seeing very limited possibilities for establishing meaningful ecological values within such projects, policy community members began to favour the separation of ecology and agriculture, which they referred to as the 'cascading' of functions.

The river area was considered a potentially good option for application of the new ideas as agricultural activity between the summer and winter dykes along the rivers was economically marginal. Inspiration on how to fill this area if vacated by agriculture came from abroad when one entrepreneur behind the new policy, Willem Overmars, vacationed in the Loire Valley in 1985 (Ploegmakers, 2008). Based on his observations of the alluvial forests of the Loire, Overmars began to imagine a similar river landscape in the Netherlands. His ideas were further elaborated by a small circle of public officials, one that by then included an official from the Ministry of Water Management. The work of this group culminated in the 1987 publication of a detailed plan called 'Plan Stork' (Plan Ooievaar) – a name chosen to reflect the ambition to increase biodiversity with a central focus on rehabilitation of the habitat of the black stork, which had once lived in abundance in the Dutch large river area.

One interesting aspect of the plan is that the writers present it as a living document. They openly suggest that it should be seen as a container full of ideas that have yet to be applied (Ploegmakers, 2008, p. 108). In the process of application the authors propose that many ideas will be improved, changed or rejected, thus recognizing the need for collective entrepreneurship. Like support, criticism came from various sources, notably from the water management sector, where people doubted whether hydrological aspects had been well enough integrated, and from aquatic biologists, who saw the plan as a good starting point, but believed it contained too few opportunities for the restoration of aquatic life. Plan Stork subsequently inspired a range of similar proposals that better addressed these concerns and did indeed contribute to collective entrepreneurship, as almost all those involved in its design became active proponents of a new approach to water management. Private enterprise played a significant role in this collective attempt to influence policy: the landscape consultancy firm Stroming (Flow) and the related management company Association Ark (as in Noah's Ark), which uses the profits of the consultancy firm to fund projects on the ground.

Plan Stork came also to have an impact beyond Dutch borders in the World Wide Fund for Nature (WWF) Living Rivers plan (as mentioned in the chapters on China and Hungary in this volume). The Dutch branch of the WWF had long been concerned with environmental conservation outside the Netherlands, but under a new chairman, former Netherlands Minister of the Environment Ed Nijpels, the organization decided to engage also with conservation in the Netherlands. The organization was keen to minimize disruption of the activities of colleagues and decided therefore to focus on nature development along the rivers, which was then still an open niche. The developers of Plan Stork were invited in 1992 to help the WWF to draft its Living Rivers plan. This interaction led to the novel idea of creating trenches in the riverbed and established a strong connection between the Plan Stork developers and the WWF. This turned out to be very important for the impact of the new approach on policy and practice, because WWF started to provide some of the funds needed to buy land where the new approach could be implemented, and because WWF chairman Ed Nijpels became an extremely effective contributor to the collective entrepreneurship at work in this case.

19.3.3 Build coalitions and sell ideas

De Jonge and Van der Windt (2007) suggest that Plan Stork succeeded in bringing together what they refer to as a 'blue–green coalition', connecting the policy fields of environmental management and water management. This is true only to a certain extent in terms of a coalition between the Ministry of Agriculture, Nature Management and Fisheries, and the Ministry of Traffic and Water Management. Full cooperation was blocked because the ideas of Plan Stork were embraced by the political leadership of the Ministry of Traffic and Water Management, but were the subject of a great deal of doubt among the non-elected public officials within the ministry. Interestingly, almost the opposite situation existed in the Ministry of Agriculture, Nature Management and Fisheries, where many non-elected officials were enthusiastic, but the political leadership was less so. In this ministry the politicians were primarily concerned with agriculture and did not find appealing the idea of giving away agricultural land to develop nature. All participants in the collective entrepreneurship within the ministry had to be careful in choosing the venues for presenting their work, which brought the need to operate as a shadow network. This contributed to the decision by several to leave the ministry after Plan Stork was presented.

The situation in the Ministry of Traffic and Water Management needs a little more explanation. Support from the political leadership was largely due to the personal interactions between the minister and Dick de Bruin, lead author of Plan Stork and employee of the ministry. De Bruin accompanied the Minister of Traffic and Water Management, Neelie Smit-Kroes, on several trips abroad and managed to get the ear of the minister. About six months after the plan was published the minister went on national television to announce her decision to puncture all river dykes in the Netherlands, but without very clearly indicating that this involved only the interior dykes. This attracted a lot of media attention and surprise at both the national and international level. It did, however, also signal her commitment to Plan Stork-like measures, which subsequently became the basis for a relatively ambitious – and fairly green – national river management policy document under the fourth National Land Use Plan, referred to as the NURG ('Nadere Uitwerking Rivierengebied', further elaboration river area), and a vision for the Meuse River.

The take-up of ideas by the Minister of Traffic and Water Management was unexpected for many, including the creators of the Stork Plan. One of them even suggested that the plan had received attention in the 'wrong' policy sector (water management), since it was intended as input to the debate on nature management and landscaping (De Jonge and Van der Windt, 2007). The majority of the public officials within the ministry were also taken somewhat by surprise. The issue of flood safety in the main river area had been under consideration for quite some time. Various government committees were studying the issue and trying to navigate out of the stalemate between those who sought radical fortifications of dykes (essentially most water managers) and those who were blocking such proposals in order to spare the landscape and the houses that were built on or around the dykes (essentially local inhabitants and environmental NGOs). Those advocating dyke fortification were far from convinced that the ideas of Plan Stork could guarantee flood safety. In essence the ensuing technical debate has still to wane. The debate covers a range of issues and arguments too wide to relate here, but the effect of clay and sand excavation and the impact of alluvial forests were and are especially hotly contested. Many 'traditional' experts suggest that excavation has few beneficial effects for flood levels and that forests actually worsen floods, while the 'new water builders' insist that these are solid measures. These positions relate to a degree to where the experts involved were educated: the engineers from Delft generally prefer dyke fortification, and ecologists and biologists from places like Wageningen favour 'softer measures'. Ploegmakers (2008) concludes that the arrival of biologists in the Rijkswaterstaat and the Ministry of Traffic and Water Management has affected the outcome to a high degree because it has led to the advocacy of the new ideas also within these bodies.

Following the success of the coalition of policy entrepreneurs in influencing the management of rivers, the current debate on flood management seems to be tilting quite strongly towards the new, 'greener' ideas. The chronology of the case of river management shows, however, that this was far from guaranteed. The government committee that analysed the issue in 1992 (Boertien I) was rather critical and expressed serious doubt that ideas would be self-financing (Ploegmakers, 2008, pp. 88–9). Following its initial experiences with public opposition to soil excavation in the Meuse basin in the first half of the 1990s, government was leaning toward a far more concentrated approach that emphasized large local water retention ponds (Roth et al., 2006). The lobbying of WWF chair Ed Nijpels helped especially to keep the alternative ideas alive. Ploegmakers (2008) describes how Nijpels realized the importance of coalitions with industry (in this case stone fabrication from river clay) so that local partners would propagate the approach, and how he managed to achieve advocacy of the new ideas by becoming a member of various government committees or influencing their composition. Ploegmakers (2008) also relates how the relatively quick start of several pilot projects by Association Ark was instrumental in selling the ideas. Participants in the collective entrepreneurship used marketing savvy, organizing countless excursions to and presentations about these projects and assigning intriguing names and terms to their initiatives, including Plan Stork, Green Rivers, The Genius of Place, and so on. These kinds of entrepreneurial tactics seem to have helped to keep the debate on water management moving in a green direction.

19.3.4 Recognize and exploit windows of opportunity

Various windows of opportunity have appeared in the story of policy change in Dutch river management. Some of these windows were political as when, for instance, the various under-ministers of water management took somewhat different positions in the debate, but with the common aim of using the main river flood safety situation to draw attention to themselves. In this political stream flood safety does not normally take top priority for ministers who also deal with far more politically sensitive issues related to traffic. Perhaps in line with this general observation De Jonge and Van der Windt (2007) mention that Plan Stork was published in a period with relatively little policy activity, possibly allowing policy-makers and politicians time to engage with the new plan and to find politicians seeking recognition and therefore willing to listen to new concepts.

Flood issues do not normally rank high on the general public's priority list. The situation, however, in 1993 and 1995, with floods along the Meuse and a near dyke break along the Rhine and the associated evacuation of up to 250 000 inhabitants, changed this dramatically. The floods also threatened to swing the pendulum back. The Ministry of Traffic and Water Management, with strong support from parliament, came out very determined to improve the weaker points in the dykes rapidly, even if at the expense of attractive landscapes. Alternative ideas received relatively short shrift as parliament adopted emergency legislation that allowed for quick and centralized decision-making. Opponents of dyke fortification experienced strong community pressure. The proponents of the alternative approach to flood management, headed by WWF chair Nijpels, faced a difficult situation and could only look on grudgingly as many summer dykes along the Meuse were rapidly fortified.

The advocates of greening, however, did not waver. Just two days after the return of the last evacuees, Nijpels was suggesting the new approach as the best way to deal with flooding issues (Ploegmakers, 2008, p. 91). Two weeks later WWF presented a list of locations where safety enhancement and nature development could go hand in hand. In addition, WWF suggested that the response plan under consideration and to be embedded in emergency legislation (referred to as a new 'Delta Act' and as a new 'Delta Plan') should not focus solely on dykes, but encompass the river area in its entirety. This strategy was effective in part because WWF arranged for a visit by the Dutch Prime Minister to one of the locations featuring the alternative approach. The Prime Minister indicated he was impressed and suggested that the approach should be part of the new Delta Plan under the condition that safety would not be compromised and no delays caused (Ploegmakers, 2008, p. 92). This was an important victory, as nature development projects could now be implemented under the same forceful legal regime as the rest of the Delta Project. Those involved in this phase of policy entrepreneurship suggest that this success is largely attributable to the positive framing of their agenda. Rather than resist the plans of others they chose to highlight the positive attributes of their own approach (ibid.).

Although the alternative plans have 'survived' a major policy window for opponents, more recent policy discussions show that the battle will be long. Discussions particularly on emergency retention polders and the National Spatial Plan 'Room for the River' (see Meijerink, 2004; Roth et al., 2006) reveal the main factors constraining the discourse. In the 'Room for the River' plan, for example, nature development is formally part of the design task, but stringent safety regulations, if not the low budget, prevents the occurrence of much serious related work.

19.3.5 Recognize, exploit, create and/or manipulate the multiple venues in modern societies

Returning to the origins of the new ideas in Plan Stork, it is important to reiterate that the collective entrepreneurship began with a shadow network. Ploegmakers (2008) finds that the immediate bosses of the public officials involved were aware of their activities and allowed them to continue, but that such support had to remain quiet and invisible to higher leadership. The initial efforts of some participants in the collective entrepreneurship focused on the development of alternative landscaping plans and on trying to ensure a role in decision processes on land reallocation (*ruilverkaveling*). These attempts were resisted by the higher ranks in the Ministry of Agriculture, Nature Management and Fisheries (De Jonge and Van der Windt, 2007), and another venue had to be found.

This occurred when an association related to the National Spatial Planning Bureau opened a competition to find the best spatial plan for the river area. Plan Stork proved the winner, a feat that achieved several things at once. An important point here is that the jury initially doubted the practicability of the plan and consulted the German expert institute on alluvial forests to check whether the ideas made sense (Ploegmakers, 2008). The response was affirmative, thus enhancing the credibility of the plan as a potential solution. Probably even more important is the fact that the prize generated momentum for Plan Stork, leading to a steep rise in interest in what the plan entailed. The plan developers responded by producing a slide show, which was used to communicate the plan to a range of audiences, including schoolchildren. De Jonge and Van der Windt (2007) conclude that this strategy was very important in eliciting the public support needed to create external pressure on the policy system for reconsideration of the new ideas.

Another principal element in the strategy of proponents of the new water strategy was the creation of pilot projects or prototype sites as noted above. The new ideas could be implemented at these sites in collaboration with land-owning nature conservation organizations. These projects became venues in themselves, as they helped to show the public how attractive such locations could be and to increase knowledge about the effects of implementation. Such knowledge was key in scaling up the new approach.

19.3.6 Orchestrate and manage networks

The case of the main rivers features several examples of network orchestration, including the activities to build a constituency for the new approach and connections to influential politicians as described above. De Jonge and Van der Windt (2007) make additional observations. They note how ideas began to 'cascade' through networks of experts as a result of sustained interaction among landscape designers across the country. These designers shared experiences and developed a common understanding of what was wrong with the then current approach. This prepared the way for the emergence of the shadow network and its collective vision.

In addition De Jonge and Van der Windt (2007) suggest that the network as it grew up around Plan Stork qualifies as a 'small relational, large cognitive distance' group since it involved people who saw each other often and shared concepts, but who also contributed different pieces of knowledge. What is more, the network included individuals who could play a role in the collective entrepreneurship in different fora, such as: good communicators of the vision; internal critics who could improve the plans; and players who understood how to connect to ongoing policy processes. Even though both aspects described by De Jonge and Van der Windt are not networking strategies per se, they do present points of attention for future transition managers.

19.4 Conclusions

The above two case studies exemplify important changes in water management practices: the 1970s incorporation of ecological values in decision-making on a storm surge barrier for the Eastern Scheldt, and the practices of nature development along the Dutch main rivers in the 1990s that started with Plan Stork. Both examples illustrate the ongoing transition from a traditional technocratic (engineering) paradigm towards what we have labelled a greener approach to water management in the Netherlands.

What do these cases reveal about agency, specifically that of policy entrepreneurs and their strategies? We first need to acknowledge that the transition described is dependent on a range of factors that cannot be managed by policy entrepreneurs. Policy entrepreneurs do not cause floods or (typically) select new ministers and did not give rise to the emergence of the environmental movement. In this way De Jonge and Van der Windt (2007) are accurate in their finding that transitions consist of an interplay of micro-, meso- and macro-level developments. Yet a naturalistic explanation of policy change, assigning causation of the greening of water management, for instance, to the emergence of the environmental movement, is too simplistic. Equally, the connection between focusing events and radical policy change is not at all straightforward. While floods served as focusing events in the cases studied, the response was not a change in policy paradigm, but a push to implement existing policies more vigorously as evidenced by the drafting in both cases of a so-called Delta Plan (a rhetorical reference since the 1950s denoting an ambitious national project, a radical solution of issues).

We suggest that the work of policy entrepreneurs can perhaps best be seen as a necessary, but insufficient factor in policy change. In each case the most significant contribution of policy entrepreneurs was the new ideas and discourses, which were developed and conveyed by individuals and shadow networks. Some individuals have succeeded in making a difference in this way, using a variety of strategies to see 'their' ideas realized either deliberately or incidentally. They brought attention to neglected dimensions of an issue (in our cases the ecological and 'nature' dimensions), created shadow networks (the epistemic community of ecologists and biologists, and the authors of the Plan Stork, respectively), worked to alter the roster of participants in decision-making venues (such as the composition of the Eastern Scheldt committee), exploited policy windows (such as the river floods of 1993 and 1995) and/or demonstrated political leadership. A combination of various strategies rather than any single strategy appears to account for success. The strategies observed are largely political, necessarily as the management of water transitions is a political enterprise: it involves the politics of ideas and expertise, of influence, access to decision-makers and negotiations.

Likewise the case studies demonstrate that it does not take only an ideal-type leader or a champion, but a mixture of various key actors, each playing a specific role, which may trigger a transition. Biologist Hank Saeijs has undoubtedly played a crucial role in disseminating the ideas of ecosystem-based water management. Without the active involvement of ministers in the Cabinet of Den Uyl, however, the Eastern Scheldt estuary would now be closed, and the authors of Plan Stork would not have been able to see their plans brought to fruition through Ed Nijpels acting as a political champion. Individuals, however brilliant their ideas, skills and strategies, rely on the cooperation of others to put concepts into action. Policy entrepreneurship in water management is thus a multifaceted endeavour, one of a collective nature, a characteristic that both stems from and reflects the complexity of Dutch water policy dynamics.

Acknowledgements

The authors gratefully acknowledge the feedback on earlier drafts received from Anjali Bhat, Vishal Narain, Erik Mostert and Arwin van Buuren. Maria Gordon helped us greatly by editing a previous version of this chapter.

References

Bijker, W.E. (2002), 'The Oosterschelde storm surge barrier: a test case for Dutch water technology, management, and politics', *Technology and Culture*, **43** (3), 569–84.

Bressers, J.Th.A., D. Huitema and S.M.M. Kuks (1995), 'Policy networks in Dutch water policy', in J.Th.A. Bressers, L.J. O'Toole and J.J. Richardson (eds), *Networks for Water Policy: A Comparative Perspective*, London: Frank Cass, pp. 24–51.

De Boer, J., H. Goosen and D. Huitema (2003), 'Bewust werken aan waterbewustzijn: Studie naar de rol en relevantie van het begrip waterbewustzijn voor het waterbeleid', IVM rapport (E-03/09). Instituut voor Milieuvraagstukken, Vrije Universiteit, Amsterdam.

De Jonge, J. and N. van der Windt (2007), 'Doorbraken in het rivierengebied. De levensloop van transformerende concepten en hun netwerken in het centrale rivierengebied 1970–2005', Wageningen, Alterra Report 1502.

Den Doolaard, A. (1948), *Roll Back the Sea*, J.B. Mussey (trans.), New York: Simon & Schuster.

Disco, C. (2002), 'Remaking nature: the ecological turn in Dutch water management', *Science, Technology and Human Values*, **27** (2), 206–35.

Flyvbjerg, B., N. Bruzelius and W. Rothengatter (2003), *Megaprojects and Risk: An Anatomy of Ambition*, Cambridge: Cambridge University Press.

Huitema, D. (2002), 'Nurturing nature: how to make a lake and maintain it: the IJsselmeer case as an example of some modern dilemma's and tendencies in Dutch water management', Enschede, University of Twente, CSTM.

Huitema, D. and S.M.M. Kuks (2004), 'Harboring water in a crowded European delta', in J.Th.A. Bressers and S.M.M. Kuks (eds), *Integrated Governance and Water Basin Management: Conditions for Regime Change towards Sustainability*, Dordrecht: Kluwer Academic, pp. 59–98.

Jenkins-Smith, H.C. and P.A. Sabatier (1993), 'The dynamics of policy-oriented learning', in H.C. Jenkins-Smith and P.A. Sabatier (eds), *Policy Change and Learning: An Advocacy Coalition Approach*, Boulder, CO: Westview Press, pp. 41–56.

Kissling-Näf, I. and S.M.M. Kuks (eds) (2004), *The Evolution of National Water Regimes in Europe. Transitions in Water Rights and Water Policies towards Sustainability*, Dordrecht, Germany: Boston, MA, USA and London, UK: Springer (Kluwer Academic Publishers).

Kuks, S.M.M. (2004), *Water Governance and Institutional Change*, dissertation, Enschede: University of Twente.

Leemans, A.F. and K. Geers (1983), *Doorbraak in het Oosterscheldebeleid* (Breakthrough in Eastern Scheldt policy), Muiderberg: Coutinho.

Lowry, W. (2006), 'Potential focusing projects and policy change', *Policy Studies Journal*, **34** (3), 313–35.

Meadows, D. (1972), *De grenzen aan de groei, rapport van de Club van Rome*, Utrecht and Antwerp: Llitgeverij Het Spectrum NV.

Meijerink, S. (2004), 'Rivierbeheer als leerproces, Een tussentijdse evaluatie van het procesontwerp voor de PKB Ruimte voor de Rivier', *Bestuurswetenschappen*, **58** (5), 406–27.

Meijerink, S. (2005), 'Understanding policy stability and change: the interplay of advocacy coalitions and epistemic communities, windows of opportunity and Dutch coastal flooding policy 1945–2003', *Journal of European Public Policy*, **12** (6), 1060–77.

Ministerie van Verkeer en Waterstraat (1985), *Omgaan met Water. Naar een integraal waterbeleid*, 's-Gravenhage: Ministerie van Verkeer en Waterstraat.

Ploegmakers, H. (2008), *Doorbraken in beleid. Crises en hervorming in de ruimtelijke ordening en de rol van voorstanders van verandering in dit proces*, MA thesis, Nijmegen: Radboud University.

Reuss, M. (2002), 'Learning from the Dutch', *Technology, Management and Water Resources Development*, **43** (3), 465–72.

Roth, D., J. Warner and M. Winnubst (2006), *Een noodverband tegen hoog water. Waterkennis, beleid en politiek rond overloopgebieden*, Wageningen: Wageningen UR.

Saeijs, H.L.F. (1982), 'Changing estuaries: a review and new strategy for management and design in coastal engineering', Rijkswaterstaat communications, No. 32.

Saeijs, H.L.F., A.J.M. Smits, W. Overmars, D. Willems, J.A. van Ast and J.J. Bouma (2004), 'Changing estuaries, changing views', Internal report Nijmegen, Radboud University Nijmegen/ ESM.

Ten Dam, H. and H. Hillenius (1982), 'De afsluiting van de Oosterschelde: een netwerkanalyse', *M&O*, **1**, 19–39.

Van Ast, J.A. (2000), 'Interactive management of international river basins; experiences in Northern America and Western Europe', *Physics and Chemistry of the Earth*, **25** (3), 325–8.

Van der Brugge, R., J. Rotmans and D. Loorbach (2005), 'The transition in Dutch water management', *Regional Environmental Change*, **5**, 164–76.

Van Slobbe, E. (2002), *Waterbeheer tussen crisis en vernieuwing*, Wageningen: Wageningen University.

Westerheijden, D.F. (1988), 'Schuiven in de Oosterschelde, Besluitvorming rond the Oosterschelde 1973–1976' (Pushing in the Eastern Scheldt: decision-making around the Eastern Scheldt 1973–1976), PhD thesis, University of Twente: Enschede.

PART VI

CONCLUSIONS

20 Water transitions, policy entrepreneurs and change strategies: lessons learned

Sander Meijerink and Dave Huitema

20.1 Introduction

In collecting in-depth analyses of international water policies and of water transitions in 15 countries we had two objectives: to gain a better understanding of agency in water transitions; and to help the development of successful strategies by those who wish to influence transitions. This leads us in our final chapter to take up the research questions set out in the introductory chapter (Chapter 1).

We explain in our chapter on theory (Chapter 2) why we expect the literature on policy continuity and change to be helpful in answering these questions. Drawing on this literature we have developed a typology of strategies that advocates of change can use to challenge the status quo and bring about change. Here we aim to refine this typology further and to explore the possibility and potential limitations of a more generic theory of change strategies in water management. How much do the particular geographical, institutional and political landscapes rule out different types of strategy, or can we identify similarities among cases that indicate where certain types of policy entrepreneurs and strategies might be successful?

We first reflect on the concept of transition, which we have equated with major policy change (see Huitema and Meijerink, introduction to this volume, Chapter1) and on patterns of continuity and change in the various case studies. We ask to what extent it has been possible and useful to distinguish between major and minor policy change, in other words, between partial and complete change. Next we discuss the types of change agents who have played a role in the cases studied: were they individuals, organizations or coalitions? Are policy subsystems changed from within, that is, by governmental bureaucracies and/or elected politicians, or by 'outsiders', such as non-governmental organizations (NGOs) and scientists? We then address the central question of this chapter and this book: what strategies have policy entrepreneurs and defenders of the status quo used either to realize or to block change? Finally, acknowledging the structure–agency dilemma, we examine the interaction between the institutional context and the roles and strategies of policy entrepreneurs. How does the institutional context constrain or enable their activities? We conclude by summarizing the main lessons learned from the various examples of policy entrepreneurship featured in this volume, addressing both potential pitfalls and possibilities for improving the chances of a successful 'navigation' toward water transitions (Olsson et al., 2006).

The case studies in this book present examples of both substantive and governance transitions. Substantive water transitions include moving from structural flood defence to integrated flood risk management, and switching to more sustainable management of ground water mining and surface water capture and distribution. Typical examples of governance transitions are the liberalization and privatization of water services,

Table 20.1 Substantive and governance transitions analysed in the case studies presented

Substantive transitions		
Country/region	Chapter	Substantive transition analysed
United States	4	Integrated water management Multiple ways of knowing water
China	6	River restoration Ecosystem-based water river management
Thailand	8	Wet to dry (all-year-around irrigation) Farm to city (secure supplies to urban users) Good to service (manage multiple services)
Australia	10	Environmental water allocation Sustainable groundwater management
South Africa	11	Sustainable management of mining water
Hungary	14	River restoration Ecosystem-based water management
Spain	15	Sustainable alternatives to supply-based management
Germany	18	Space for the river Flood risk management
The Netherlands	19	Greening of water policies River restoration Space for the river
Governance transitions		
Country/region	Chapter	Governance transition analysed
Mexico	5	Decentralization, participatory governance (water user associations) Marketization (water markets, water pricing)
India	7	Decentralization, participatory governance (water user associations)
Indonesia	9	Decentralization, participatory governance (water user associations)
Tanzania	12	Liberalization, privatization Decentralization, participatory governance (water user associations)
European Union	13	Marketization (full cost recovery, water pricing) Public participation, stakeholder participation
Turkey	16	Liberalization of water services Decentralization, participatory governance (water user associations)
Sweden	17	Adaptive management

decentralization and the move to more participatory forms of governance. Although substantive and governance transitions are often related, most chapter authors have put the emphasis on either a substantive or a governance transition. Table 20.1 gives an overview of the main transitions analysed in the studies and based in various countries and the European Union.

The water policy transitions presented in this volume involve strongly diverging contexts, some arising in developed countries, some in developing countries, others in democracies, and others in semi-democracies or authoritarian regimes. The methodological perspective of this book aligns with the 'most different systems' approach, which assesses whether a certain theoretically expected connection holds across a range of settings. The cases were compiled to test the expectation that they would support the notion that policy change occurs because of the presence of policy entrepreneurs and their use of strategies. In synthesizing the results of the research, a drawback of the approach emerged in its lack of a systematic characterization of the societal, institutional or ideological context of the various national or European transitions described. This makes it difficult to suggest which strategies may work 'best' in which context. To address this shortcoming to some degree, we present certain patterns of transitions that can be inductively derived on the basis of the cases presented in this volume. Of particular note are the difference between 'top-down' and 'bottom-up' transitions, and the difference between countries that offer multiple venues ('institutionally rich environments') and countries that offer only limited venues for effecting change. These patterns will be described in the sections below entitled, 'The strategies of policy entrepreneurs' and 'How institutions matter'.

20.2 Patterns of continuity and change

Various theories of the policy process, such as the advocacy coalition framework (Sabatier, 1993), punctuated-equilibrium theory (Baumgartner and Jones, 1991) and historical institutionalism (for example Pierson, 2000), distinguish between major, radical or paradigmatic policy change and incremental or shallower forms of policy change. The theoretical approach of this volume equates transitions with major policy change – policy change that entails a change in a core policy belief, a policy image or a policy path.

The chapters in this book concern policy dynamics and offer clear examples of change embodied in either the content of water policies or in the overarching governance paradigm. It has been difficult, however, to make a clear distinction between major and minor policy change in most cases.

The complicating factor lies in the existence of new policies and policy practices alongside extant policies and policy practices. Font and Subirats, for example, in their description of the development of alternatives to traditional supply-based policies in Spain (Chapter 15), argue that: 'This does not mean in turn that the new policy approach emphasizes the regulation of water demand. Instead it redefines the modes of intervention in water supply in apparently more sustainable terms.' In the same vein, te Boekhorst et al., analysing the introduction of integrated river basin management (IRBM) in China (Chapter 6), conclude that: 'The introduction and implementation of IRBM as a new paradigm can be regarded as the emergence of resistance to the Chinese hydrological mission and its unintended consequences. However the ongoing execution of large hydro-engineering projects indicates a competition between two paradigms and an outcome that has still to be determined.' Ingram and Lejano, on US water policies (Chapter 4), similarly argue that: 'existing [ways of knowing] need not be considered totally bankrupt. In fact one of the lessons of this chapter is that [ways of knowing] continue and persist even as new ways of knowing are introduced.' In addition, the case studies of flood management in Hungary, Germany and the Netherlands (Chapters 14,

18 and 19) all demonstrate how new ideas of ecosystem-based water management and flood risk management put pressure on stable policy communities, but in none of these cases has the 'old' hydrologic paradigm of fighting the water by regulating rivers and constructing dykes disappeared completely. In some cases, water managers rather try to integrate, combine or balance the 'old' engineering and 'new' ecological approaches, while in others managers actively attempt to frustrate the change in direction.

Most of the transitions in governance practices described in this book are partial. Although many serious efforts are being made to involve stakeholders and the public in water management, many governance practices are still based on more traditional modes of governance, most notably the activities of governmental bureaucracies. The Thai and Tanzanian case studies (Chapters 8 and 12) show how new rules and procedures overlaid existing local institutions without replacing them and without formally addressing the way in which they were supposed to interact. As a consequence prior institutions persist and sometimes continue to dominate decision-making processes, for example in areas where local irrigation organizations are still active. In other cases, such as Indonesia (Chapter 9), the management and maintenance of irrigation systems was formally decentralized, but the central government continued to play a crucial role in funding and monitoring and hence continued to be a powerful actor in irrigation management (see also the case studies on Mexico, India, Tanzania and Turkey in Chapters 5, 7, 12 and 16).

It is possible, of course, that 'old' substantive and governance paradigms or discourses will be replaced entirely in the long run, but many of the transitions that are central to this volume have not yet reached that stage and may never reach it. Indeed in many cases, such as those from Thailand and Tanzania, it may not be desirable for former practices and concepts to disappear completely. Integration of valuable new and old components might well form an element of successful transitions.

Besides this complication in distinguishing between major and minor policy change, the need exists to address the classic distinction between policy formulation and policy implementation. This book focuses on policy change and its two-step process: changing policies on paper and changing policies on the ground. It is well known from the literature that even if formal policies change radically, implementation constitutes a new round in the policy game, where established routines are often less amenable to change. Several case studies in this volume illustrate this dynamic, since opponents of policy change have often successfully blocked the implementation of changed policy. The chapters on India and Indonesia probably provide the best examples in this respect, but the issue of policy implementation arises in many other contributions to this volume (see Chapters 5, 14, 18 and 19 on Mexico, Hungary, Germany, and the Netherlands).

20.3 Policy entrepreneurs

20.3.1 Introduction

All contributing authors were asked to use an agency perspective, that is, to find out who lies behind the changes observed and to focus on the roles and strategies of the individuals or organizations so identified. Individual policy entrepreneurs stood out as agents of change in some cases, but it proved difficult in others to pinpoint key players. In the latter cases analysis focused primarily on the roles and strategies of organizations. This occurred for various reasons. Firstly, the social science training of some of the authors focuses on organizations and collectives, leading to inexperience or even unease with analysing the role of individuals and presenting detailed accounts of their strategic behaviour. Secondly, some authors faced a substantive difficulty in that entrepreneurship is often a collective undertaking, where a small group of individuals navigates a transition, each representing different organizations and playing a different role. An additional complicating factor in identifying key change agents comes with 'the politics of claiming success'. As Lebel et al. (Chapter 8, this volume) suggest:

> Policy change may be more about writing history than making it. In the Upper Ping, upland watershed organizations and lowland irrigation groups existed long before river basin and sub-basin organizations were introduced . . . Industrial estates demanded priority access to water and got it before such policy became explicit. Housing estates raised land levels, redirected irrigation canals and had their own flood protection walls built. Government plans and efforts at coordination came later. Policy transitions at the national level have frequently followed changes in actual practices within the Upper Ping.

If a transition is generally perceived as successful, many parties and persons will claim responsibility for the success. Because the reputations of individuals and sometimes also powerful organizations are at stake, the claiming of success is a political game. Success may well be attributed to those most adept at playing the media. The contribution of those lacking the necessary promotional skills may be thought to be less important to a particular transition. For this reason most of the case studies presented in this book largely draw on document analysis and multiple interviews with individuals who played a key role in the transition process. Existing accounts of policy change, as available for the Netherlands for instance, proved instrumental in this way. The consideration of someone or some group as key in either realizing or blocking change receives support when both documents and a majority of respondents refer to the same person or organization, thus increasing the credibility of the findings.

20.3.2 Individual policy entrepreneurs

The various case studies provide ample evidence for the crucial role of key individuals and organizations in realizing transitions. Moreover the observations corroborate Kingdon's thesis that individual policy entrepreneurs can be found anywhere (Kingdon, 1995), including within governmental bureaucracies (for example Dutch case: Henk Saeijs), political parties (for example US case: Bruce Babbitt), NGOs (for example Chinese case: various representatives of the WWF) or expert communities (for example German case: Professor Dister). These individuals share a common willingness to invest their resources (time, reputation and/or knowledge) in a particular proposal for policy change, and possess good networking skills. What is more, most of these entrepreneurs demonstrated considerable perseverance: they often worked on particular transitions during significant parts of their career or they had to make changes in their career to achieve the transitions they sought.

Experts within the bureaucracy and the scientific community have played a major role in all case studies described in this volume. This is not surprising given the highly technical nature of the water sector. The Australian case study (Chapter 10) reveals how bureaucrats take action 'to steer science and research toward policy goals (ambitious flow targets and forestry regulation) quite independently and with great success'. Both the Dutch and German case studies refer explicitly to the potential for contribution to policy change by a 'new guard', often someone with a different scientific-disciplinary background. As Becker (Chapter 18, this volume) in his account of transitions in German flood management says: 'Fresh from university, placed in a newly established administrative office . . . they took up and processed the new holistic environmental spirit of the time. They were keen on multidisciplinary cooperation rather than afraid of it. Nor were they blinkered by traditional routines and old-hand experience.'

20.3.3 Collective policy entrepreneurship

Although the cases of water transition illustrate the role of key individuals as change agents or obstacles to change, most cases also demonstrate the importance of networks of individuals (coalitions) in successfully challenging the status quo. Such networks often consist of representatives of government agencies at various levels of government, thus creating connections among various scales and levels of decision-making, NGOs and/or research groups. Collective policy entrepreneurship holds two main advantages. First, people in different positions may draw on a different arsenal of strategies to influence a change trajectory. Where experts working at either governmental research institutes or universities have excellent possibilities to develop and test new ideas and approaches, senior policy advisors or politicians generally are in a better position to help to achieve the adoption of new policies. Second, people may have different capacities and skills independent of their positions. Some have charisma and excel at using the media. Others possess skills in developing new policy concepts or in finding common ground.

Our cases confirm, to a degree, that shadow networks, those operating outside the formal circuits of power, are important for effecting policy change. Gupta (Chapter 3, this volume) points to the importance of 'science dominated by the global North', and of private companies and NGOs with a worldwide reach in developing new ideas for water management at the global level. Huitema and Meijerink (Chapter 19, this volume) show that several officials working on alternative approaches to water management in the Netherlands had to do so with only silent support from their superiors and eventually had to take jobs outside the bureaucracy. A 'shadow position' achieved in this way gives them greater flexibility in advancing their ideas. Similarly, ideas about radically different approaches to water management were developed in shadow networks of academics and NGOs in Hungary and Spain, but also in India. Returning to our first observation of the importance of collectives in effecting transitions, we must conclude that shadow networks are especially important in the phase of idea development and, in several cases, also in showing the applicability of their ideas in principle through pilot projects. The actual uptake of their ideas requires interaction with the formal policy network and gives politicians, former politicians and high-ranked bureaucrats a key role with their ability to translate the innovations into new policy. This is consistent with the finding of Olsson et al. (2006) that members of such shadow networks develop and test new ideas in the shadow of formal decision-making arenas, but need to develop links to formal decision networks to challenge a dominant policy paradigm successfully.

20.3.4 Donor organizations

While entrepreneurial bureaucrats, politicians, experts or representatives of NGOs may play a role in cases in the developing world, they are joined by a key additional change agent in the form of the various international donor organizations (see the chapters on India, Indonesia, Mexico, Tanzania, Thailand and Turkey). Organizations such as the World Bank, the IMF, Inter-American Development Bank and the Asian Development Bank impose strict funding conditions on national governments (see also Gupta, Chapter 3 in this volume, on 'Aid as a driver of change'). Such conditions often entail fundamental changes in governance regimes, including privatization, decentralization and participatory governance (for example by the establishment of water user associations in irrigation management). These all constitute elements of integrated water resources management, but the way in which these elements are implemented varies greatly across the case studies. The EU is an influential donor organization both in Europe (see Werners et al. on Hungary, Chapter 14, this volume) and in the developing world (Partzsch on EU foreign policies, Chapter 13, this volume). Just like other international donor organizations, the EU tries to steer water transitions by formulating funding conditions.

20.4 The strategies of policy entrepreneurs

20.4.1 Introduction

How, then, did policy entrepreneurs realize change, and what strategies have they used? The authors contributing to this volume have described the strategies of the various change agents in detail. Not all these strategies were used purposefully to bring about change. Some appear to have resulted from a relatively chaotic and unconscious process. Strategic analysis always runs the risk of rationalization by hindsight of the actions of the players involved. Some cases yield convincing evidence that strategic actions were indeed deliberate. The Spanish change coalition, for example, used the EU venues to put pressure on the national government (Font and Subirats, Chapter 15, this volume). In the Hungarian case, however, the change agents had not developed a strategy to link their ideas of ecosystem-based water management to formal decision-making fora, and the adoption of their plans results from a more serendipitous process (Werners et al., Chapter 14, this volume).

20.4.2 Top-down and bottom-up strategies for developing and disseminating new ideas

Various types of change agents clearly have used different sets of strategies. Where donor organizations were the main change agents, change could be mandated through the formulation of strict funding conditions, including demands for new forms of governance. Gupta (Chapter 3, this volume), in describing the driving forces related to global freshwater governance, presents an overview of various organizations operating internationally. Donor organizations, such as the World Bank and the International Monetary Fund, have been particularly successful in disseminating the ideas of liberalization and decentralization of water management. It is not a coincidence that the case studies of Turkey, Mexico, India, Tanzania and Indonesia all present examples of efforts to privatize and decentralize water and irrigation policies.

It may be argued that in these cases the national governments actually had no choice but to change their governance paradigm. It should be noted, however, that government elites on the national and regional levels often gain personal benefits, in the form of both influence and new job opportunities, from funding by international donor organizations (for example Goldin and Kibassa, Chapter 12, this volume). It is mainly cases of externally (enforced by donor organizations) and centrally (enforced by the central government) imposed transitions where the phase of policy implementation offers a good second chance for vested interests to block change. This is important, since transitions are more often than not characterized by serious implementation problems. Various case studies presented in this volume (in particular the establishment of water user associations in India, Indonesia, Mexico, Tanzania and also Turkey, and water reforms in Thailand) lead to serious questions regarding the effectiveness or even appropriateness of externally and centrally enforced forms of top-down transitions.

The Tanzanian case study serves as an interesting example of how colonists introduced a new system of water management that destroyed the adaptive capacity of local networks. Cynics might argue that this colonialism is not over yet because the studies of transitions in India, Indonesia, Mexico and Tanzania and of EU external water policies demonstrate how Western countries continue to implement their Western-developed models of good governance in a non-Western context. The origins of the various international water and governance discourses reveal a general pattern of upload to the international level of ideas developed in the Western world and their subsequent download in developing countries (see also Gupta, Chapter 3, this volume).

Goldin and Kibassa (Chapter 12, this volume) offer an insightful quote from an interview with officials from the Ministry of Livestock Development in Tanzania:

> Donors come and talk about water in Zambia, water in Sweden. Well, I have never been to Sweden and I have never looked at water in Zambia, but they still used this to make suggestions about our policy. The World Bank will often bring the aide-mémoire already written and present it to me with no time to digest . . . some donors come with big money and this can skew priorities.

Gupta (Chapter 3, this volume) also refers to the problems created by 'one size fits all' solutions: 'Current science is dominated by the global North and focuses primarily on issues of the North; and where it focuses on issues of the South, it does not often take contextual aspects into account, presenting instead technological fixes or ready-made solutions' (see Narain, Chapter 7, this volume).

This strategy is entirely different from the strategies used by certain individual policy entrepreneurs and (shadow) networks in possession of very different resources to influence decision-making. What these agents have in common is that they carry a particular set of ideas, whether it is a belief in the merits of water markets, water pricing, participatory governance or of ecosystem-based river management, and undertake efforts to get these ideas realized. Donor organizations, however, are in a position to enforce policy change, making a crucial difference between them and individual or collective entrepreneurs who need to bring attention and attract supporters to their ideas. The use of small-scale pilot projects to demonstrate the feasibility and benefits of their newly proposed approaches stands out as one of the bottom-up strategies used by this type of change agent. Examples from this volume include the Bokartisz coalition and floodplain rehabilitation in the Bodrogköz area in Hungary (Werners et al., Chapter 14), the WWF and wetland restoration in Dongting Lake in China (te Boekhorst et al., Chapter 6), and the authors of the Plan Stork and nature development along the Dutch main rivers (Huitema and Meijerink, Chapter 19). Such actors use the media to communicate the results of such projects, to disseminate their ideas and to change people's perceptions and mental models (Olsson and Galaz, Chapter 17, this volume).

Our set of cases shows very clear similarities in the ways in which water management issues are framed and how transitions of a similar type are under way in many countries around the globe. International NGOs, such as the World Wide Fund for Nature (WWF), have contributed to the dissemination of the concepts of river restoration and river basin management. These observations bring to mind the analogy made by Richardson (2000) between policy ideas and viruses, which itself reflects the concept of 'viral marketing' and Richard Dawkins's earlier idea of the 'meme'. Just like viruses or memes, policy ideas and discourses are contagious and spread around the globe.

Based on the case studies presented in this book, we conclude that top-down strategies to achieve water policy transitions, as applied by many donor organizations, tend to cause more implementation problems than the transitions induced by bottom-up strategies that attract supporters to newly developed policy ideas and practices. This is not to say that hierarchy does not play any role in realizing bottom-up induced transitions. It is exactly a combination of bottom-up and top-down strategies that causes most transitions to occur. Their relative importance depends largely on the particular institutional context or opportunity structure. In the China case study, for example, the establishment of contacts at the highest decision-making levels was indispensable to the transition (see also the section below, 'How institutions matter').

20.4.3 Building coalitions: balancing between advocacy and brokerage

This brings us to the strategy of gaining support and building successful coalitions and alliances. Based on the case studies, at least three types of coalition and coalition-building emerge. First, a coalition can be made up of those who share the same or very similar ideas, beliefs and values. They share either a common disciplinary background, that is, they may be part of an epistemic community, or they share certain ideas, beliefs and convictions unrelated to any particular discipline. Typical examples include environmentalist or pro-development coalitions. The primary activity of such a coalition is advocacy of a particular set of ideas. Premised on a stance at once ontological and epistemological, similarities in belief systems or shared meanings, known as 'ways of knowing' (Ingram and Lejano, Chapter 4, this volume) are found to be the binding element of these coalitions.

A second type of coalition, which elsewhere we have called a strategic alliance (Meijerink, 2005), is a coalition between parties who do not share the same policy beliefs, value preferences or worldviews, but nevertheless share an interest in realizing a particular sort of policy change. Partzsch (Chapter 13, this volume), in her account of the development of the EU Water Framework Directive, describes how the water industry and environmental NGOs joined forces because they shared an interest in water pricing, although for entirely different reasons. Whereas the drinking water companies hoped to benefit from this governance transition financially, the environmentalists wanted to ameliorate surface water quality and to restore water ecosystems. The Dutch case quite similarly shows how a coalition of fishermen, who tried to safeguard their income, and environmentalists, who wanted to preserve the estuarine ecosystem, formed a successful alliance opposing plans to close an estuary (Huitema and Meijerink on the Netherlands, Chapter 19, this volume). Finally Font and Subirats (Chapter 15, this volume) show how economists, environmentalists and (water) donor regions, each with different beliefs, interests and ways of knowing water, jointly opposed the traditional supply-based engineering approach in Spain.

A third type of coalition includes parties who do not share beliefs, problem perceptions or policy preferences, but are simply dependent on each other for realizing their diverging objectives. Policy transition results as incidental to the successful achievement of the separate goals of the coalition members. While shared beliefs or meanings, and shared policy objectives, form the glue binding advocacy coalitions and strategic alliances respectively, resource dependence forms the rationale behind this third coalition type.

The role played by a policy entrepreneur varies not only according to the prevailing institutional and opportunity structures, but also to the coalition type involved. Olsson and Galaz (Chapter 17, this volume) in their Sweden case study explain how a key individual managed to change the perception of these politicians from seeing wetlands as a problem ('water sick') to seeing them as a valuable resource ('water rich'). The policy entrepreneur in question did so by developing and applying landscape-level solutions to environmental problems, solutions originated by networks, and by linking proposals to additional goals such as regional development. Successful change agents clearly have an eye for the preferences and desires of other parties and try to meet their demands, winning support for their proposals in the process. Needless to say, coalition-building among parties with different value priorities and policy objectives often entails processes of negotiation and compromise, similar to the formation of a coalition government in a multiparty system. Successful policy entrepreneurs, therefore, have to balance continuously on the continuum between advocacy and brokerage (Kingdon, 1995). On the one hand, they must be good advocates of specific concepts. They need to be able to communicate their ideas and message in an appealing and convincing way. On the other hand, they need the skills to manage the interactions within networks, that is, to negotiate and cooperate with those who have different ideas, worldviews or interests, but who possess crucial resources.

Expanding on our finding that successful entrepreneurship is often collective entrepreneurship, the cases presented show that different persons may play complementary roles: while some may excel in generating new ideas, others are particularly skilled in advocating these ideas, and still others have the capacity to broker or negotiate.

Policy entrepreneurs can also benefit by understanding the importance of building networks across different 'ways of knowing' water, as emphasized by Ingram and Lejano (Chapter 4, this volume). These researchers observe that boundary organizations or key individuals can play a crucial role in connecting various ways in which water issues are framed, thereby inserting new values and perspectives. 'New elements of policy design can be introduced that have appeal across several ways of knowing. Policy entrepreneurs can facilitate recognition of elements common to several perspectives. Through reflexive communication, adherents to different ways of knowing can tell that they have been understood' (Ingram and Lejano, Chapter 4, this volume, citing O'Leary and Bingham, 2007). Such processes of networking and shared 'meaning-making' are often a part of policy brokerage.

Where Ingram and Lejano point to the usefulness of brokering between (or connecting) various ways of knowing, Narain and Lebel et al. (Chapters 7 and 8, this volume) insightfully show how discursive strategies may also be used strategically by both advocates of change and those who try to block it. Such discursive strategies can be used to attract supporters to new policy proposals and to justify policy interventions. Lebel et al. (Chapter 8, this volume) distinguish between various powerful framings, such as the 'drought narrative', the 'forests make water narrative' and the 'living with the river narrative'. In their study of policy change in the Upper Ping Basin, Lebel et al. found that adherence to such standard narratives formed a key strategy for individuals wishing to promote and secure policy changes. Players in the transition would, for example, try to fit new policy ideas into existing narratives or, conversely, appeal to such narratives to oppose change. Their analysis also demonstrates how the very same narrative used to promote one particular water transition can be used as a counter-narrative to block another water transition.

20.4.4 Anticipating, manipulating and exploiting windows of opportunity

Coalition building and shared meaning making, however, are not sufficient for realizing change. Coalitions need opportunities to launch their ideas. The case studies repeatedly illustrate the importance of windows of opportunity in water policy transitions. Table 20.2 gives an overview of some of the key windows discussed in various chapters.

Several examples exist of policy entrepreneurs who successfully exploited crises. Media coverage of environmental emergencies, like sea or river floods, periods of extreme drought, or cases of accidental pollution, raises public awareness and thus also increases political attention to specific water problems. Both public and political attention to an issue offers opportunities for policy entrepreneurs to gain support for new insights, policy proposals and policy directions. The river floods in Hungary (1998, 2001), Germany (2002), Thailand (2005) and China (1998), and the sea (1953) and river floods in the Netherlands (1993 and 1995) all offered windows of opportunity to launch or articulate ideas of ecosystem-based water management and of creating more space for the water.

As can be easily determined from the literature (Birkland, 1997; Boin et al., 2009), the framing of such flood events is a crucial strategy here. In the Netherlands case, the river floods mentioned above could be framed either as a sign that insufficient resources had been allocated for realizing and maintaining structural flood defences, or as a sign that the policy paradigm of controlling the water with technical infrastructure had failed. The serious threat of river floods in 1993 and 1995 and the resulting large-scale evacuation of residents dramatically increased awareness of water issues and policy. Although plans for creating more space for the river had been developed in the early 1990s, the floods also threatened to bring a return to former policy directions. The Ministry of Traffic and Water Management, with strong support from the Dutch parliament, was very determined to make rapid improvements to the weaker points in the dykes, even if this meant sacrificing attractive landscapes. Alternative ideas received relatively short shrift as parliament adopted emergency legislation that allowed for quick and centralized decision-making (Huitema and Meijerink on Dutch water transitions, Chapter 19, this volume). The proponents of a greener approach to water management had to operate very carefully in this situation, where their ideas could easily have been thrown aside. By cautiously linking their ideas to the public's desire for safety and ensuring that their approach would not be eliminated from official policy documents, they were able to prevent the pendulum swing back to earlier policy models.

Table 20.2 Windows of opportunity arising in case studies presented

Problem windows		
Country	Chapter	Problem window analysed
Mexico	5	Asian financial crisis
China	6	Yangtze River flood 1998
Thailand	8	Financial crisis 1997–98
		Upper Ping River flood 2005
Indonesia	9	Financial crisis 1997–98
		Severe drought caused by El Niño
Australia	10	Severe drought 1990s
Hungary	14	Tisza River flood 1998
		Cyanide spill Tisza 2000
		Tisza River flood 2001
		Tisza River flood 2006
Turkey	16	Economic crisis of the 1980s
Germany	18	Rhine River flood 1993
		Rhine River flood 1995
		Elbe and Donau River floods 2002
The Netherlands	19	Sea flood 1953
		(Near) Rhine River flood 1993
		(Near) Rhine River flood 1995
		Extreme rainfall 1996
		Extreme rainfall 1998
Political windows		
Country	Chapter	Political window analysed
United States	4	Appointment of Secretary of Interior Bruce Babbitt 1993
Mexico	5	Regime change 1990s
Thailand	8	Regime changes 1991, 1997 and 2006
Indonesia	9	Regime change 1998
South Africa	11	Regime change 1994
Hungary	14	Elections 2002
		Elections 2006
Spain	15	Elections 1996
		Elections 2004
Turkey	16	Political crisis of the 1980s
The Netherlands	19	Elections 1973

Other examples of such 'framing contests' (Boin et al., 2009) can be found in the chapters on Thailand and Hungary. In 2005 Chiang Mai experienced a series of major floods. Lebel et al. (Chapter 8, this volume) describe how these 'events were turned into a crisis by public figures, business owners and the media, creating opportunities to get projects implemented', how different parties framed the causes of these events differently, and how:

> 'Debate continues over whether or not to remove several traditional weirs from near the centre of the city (as these also raise water levels and distribute flood waters), whether high walls should be built to protect the city from peak flows, or whether efforts should focus on restoring river channel width, banks and floodplains.

In the Hungary case study the framing contest also centred on the causality of the river floods (Werners et al., Chapter 14, this volume):

> The cyanide spill and Bereg flood occurred as parties were campaigning for the 2002 elections. Whereas the water authority supported a hard engineering approach, Bokartisz was advocating its new ideas. The engineering approach was criticized by the NGOs and individual scientists for adding to the flood risk and causing unwanted side-effects in the region.

The case studies of strategic framing of disasters show that policy entrepreneurs engaged in a framing contest over the causality of such disasters and the desirability of specific policy options, but did not turn so much to the framing of disasters as to lapses of leadership by responsible authorities. This may be explained partly by the dependence of the policy entrepreneurs on the responsible authorities for achieving change. The policy entrepreneurs rather saw problem windows primarily as opportunities to launch their policy alternatives, to stress the merits of these alternatives and to position themselves better to think in line with and cooperate with policy-makers. The development and marketing of attractive policy alternatives that take into account the interests of many stakeholders have taken precedence as the strategy employed by policy entrepreneurs presented with windows of opportunity.

In all cases of river floods analysed in this volume, the new policies, which were adopted afterwards, had been developed and implemented on a small scale before the focusing event took place. In the Netherlands, the regional 'Plan Stork' preceded the adoption of the generic 'space for the river' policies; the Hungarian 'Bokartisz coalition' had gained experience with floodplain restoration in the Bodrogköz area before this approach was adopted at the national level; and in Germany, floodplain restoration in Baden-Wuerttemberg and flood risk management in Rhineland-Palatinate were later taken up at the national level.

These occurrences fit perfectly with Kingdon's multiple streams model, in which policy alternatives in circulation can be attached to new problem definitions when a window opens (Kingdon, 1995). They also underscore Olsson et al.'s (2006) argument that systems need to be prepared in readiness for change.

Several case studies point to the relevance of crises outside the realm of nature or even the water management subsystem. Examples include the Asian financial crisis (for example Lebel et al. and Bhat and Mollinga, Chapters 8 and 9, this volume) or the economic crisis in Turkey (Kibaroglu et al., Chapter 16, this volume). Where floods or natural disasters offered opportunities to garner active assistance and encouragement for substantive policy changes, exploitation of financial and economic crises aimed primarily to realize change in water governance, most notably toward privatization and decentralization.

Political windows also have a role to play in most of the case studies. Such windows range from those created by regime change, as in South Africa, Indonesia, Mexico and Thailand, to the change of an elected government or a single minister, as in the election of the Den Uyl administration in the Netherlands in 1973, the appointment of the Indonesian pro-reform President Abdurrachmann Wahid, or of Bruce Babbitt in the USA. Some researchers characterize contemporary society as postmodern and typified by new modes of governance and forms of deliberative democracy that are taking over the central role of traditional party politics. Many case studies in this volume, however, demonstrate the crucial importance of traditional party politics, general elections and changes of national governments as factors in water policy transitions. The case study on hydro-politics in Spain offers a clear example: while the conservatives insisted on traditional supply-based solutions to the issue of water shortages, the labour party demanded more sustainable alternatives. This is not to say that politics is the main driver of change; in most cases it is not. Political changes, however, can offer opportunities to gain attention for new ideas and to secure a place for new issues on the political agenda.

20.4.5 Connecting informal to formal networks: the exploitation, manipulation and creation of venues

The transitions analysed in countries as different as Indonesia, China, Mexico and the Netherlands all feature examples of actors who deliberately manipulate the composition of advisory or decision-making fora. In the China case study, the WWF managed to gain a seat on an influential national task force on river basin management (te Boekhorst et al., Chapter 6, this volume); a minister in Indonesia purposefully moved a specific department from one ministry to another in order to minimize the influence of the conservative old guard in the Ministry of Public Works (Bhat and Mollinga, Chapter 9, this volume); in Mexico, an environmental minister engineered the transfer of the national water commission from the Agriculture and Fisheries Ministry to the new Ministry of the Environment and Natural Resources (Wilder, Chapter 5, this volume); and in the Netherlands, decision-making on the Eastern Scheldt storm surge barrier saw the government deliberately instal an interdisciplinary advisory commission, thereby breaking the monopoly of the influential Rijkswaterstaat engineers (Huitema and Meijerink, Chapter 19, this volume). Those who engage in venue manipulation aim to have their own ideas or their coalition represented where policy-relevant issues are discussed and to bypass those who resist the change proposed.

Venue-shopping has also proven to be an effective strategy in several cases. The Spanish case study, for instance, shows how a coalition of parties seeking support for alternatives to traditional supply-based engineering solutions played the various EU venues successfully; this strategy increased pressure on the Spanish national government to change the planned policy programme (Font and Subirats, Chapter 15, this volume). The need to link various scales and levels of government is noted by most authors. For local initiatives to be scaled up, they need to be linked to policy arenas on a higher level, as demonstrated, for example, in the case studies from the Netherlands and Germany.

Finally, policy entrepreneurs may deliberately create new fora to bring people together, often to discuss problems, as seen for example in the CalFed case, described by Ingram and Lejano (Chapter 4), and the dialogues on water policy in Indonesia organized by international organizations and the Ford Foundation discussed by Bhat and Mollinga (Chapter 9). The Tanzania case shows that such fora, even if formally adhering to participatory principles, can be rigged to produce biases towards certain policy innovations like the privatization of water services.

20.4.6 Institutional design

Policy entrepreneurs operate within particular institutional settings, but may also try to change the rules of the game. Bureaucratic entrepreneurs in particular often have numerous opportunities to change the setting for networks and to influence the design of decision-making processes. This raises questions that include: how open or closed is the policy process? Is the policy process organized as blueprint planning or as a social learning process? Bureaucratic change agents may not always direct change deliberately in one direction, but instead purposefully design institutions that offer opportunities for learning, for instance by stimulating a variety of problem definitions and the development of a variety of policy options.

Olsson and Galaz (Chapter 17, this volume), referring to Ostrom (2005), argue that unicentric, simple, large-scale governance units do not, and cannot, have the variety of response capabilities that polycentric, multilevel governance systems can have. Awareness of which institutional settings are more conducive than others to learning helps policy entrepreneurs to change institutions purposefully in order to create a 'learning environment' that allows for variation and experimentation. Moreover, Olsson and Galaz specifically address the need for a new role for central managers, which may entail a change in management function from authoritative allocation to activation: 'instead of superimposing ready-to-use plans for ecosystem management on local contexts, the role of central authorities and agencies could hence be to legislate to enable self-organization processes' (Olsson and Galaz, Chapter 17, this volume). Ingram and Lejano, in their contribution on US water policies (Chapter 4) citing O'Leary and Bingham (2007), argue that 'cross-[ways of knowing] networks can be advanced by policy entrepreneurs through such practices as identifying the appropriate players and establishing legitimacy and transparency of policy processes', thus recognizing the usefulness, too, of careful institutional design.

From a normative point of view, we fully agree with the need for polycentric institutions allowing for variety and spaces for (social) learning. However the case studies in this volume show discrepancies between this normative standpoint and the strategies that policy entrepreneurs use in the real world. Successful entrepreneurs are good networkers, boundary-spanners or brokers who aim to create institutions for experimenting and social learning using a large arsenal of strategies. As agents of change, policy entrepreneurs use tactics designed to gain access to decision-making fora, manipulate venues, frame crises, build strategic alliances to pool resources and frame narratives to attract support. To effect change in policies and institutions, entrepreneurs need also to be able to play a political power game. Transition management is not only creative and cooperative, but also adversarial and political.

A closely related phenomenon is the greater interest often shown by advocates of policy change in institutionalizing their new policy paradigm or 'way of knowing', that is, in discursive closure (a fixation of meaning), than in creating a learning environment. Take the various examples of pilot projects, which sometimes have an experimental and testing element, but more often one of demonstration intended to sell a new approach to the public. Those in power can afford not to learn. This is why, as Baumgartner and Jones (1991) argue, newly institutionalized policy images (frames or ways of knowing) often are just as fixed as once were those of the policy that they disrupt or replace. The case studies in this volume reveal a wide gap between the normative ideal of adaptive institutions facilitating social learning processes and the empirical reality. In practice, a newly established policy often becomes just as stable – or rigid – as the old one, thus ironically creating barriers to future change.

20.4.7 Strategies to block change

The authors of this volume have focused mainly on change agents and their strategies, yet some case studies present interesting insights into players with an interest in maintaining the status quo and how they try to block change. As Lebel et al. (Chapter 8, this volume) say: 'Those who wish to be "good" after all, are not the only ones interested or skilled in transition management.' Most of the strategies discussed above have also been used to block change and maintain the status quo. Those resisting change in the cases analysed tried to disseminate their ideas, to build coalitions and to control the various decision-making fora. Narrative strategies, too, found eager users in the no-change camps (Lebel et al., Chapter 8, this volume).

As discussed above, defenders of the status quo or those trying to divert new policy can be particularly successful in the implementation stage. Narain (Chapter 7, this volume), Bhat and Mollinga (Chapter 9, this volume) and Goldin and Kibassa (Chapter 12, this volume) all refer to the work of Thomas and Grindle (1990) on policy implementation, which maintains that donor organizations have invested many resources in strengthening policy analysis and in developing models of 'good governance', but have neglected implementation. Narain (Chapter 7, this volume) describes how the street-level bureaucracy of an irrigation department in India feared the loss of opportunity to extract illicit payments from farmers for releasing water. They either withdrew support for management turnover or tried to delay the rehabilitation of projects. Large and powerful farmers who benefited from the status quo, for example, threatened policy entrepreneurs using letters and blackmail. Key stakeholders in the South African mining industry also used an armory of strategies to block policy change. As Turton (Chapter 11, this volume) says: 'it had already prepared by learning how to outmanoeuvre government attempts to regulate it'. Tactics included destroying evidence of liability and/or culpability, attacks on the credibility of scientists and provision of inadequate budgets for public participation processes. While Lebel et al.'s analysis (Chapter 8) raises questions about how to know who are the good change agents, Narain's and Turton's suggest the complementary notion of opponents ready to block with 'dirty' tactics. The case studies in this volume seem to show that proponents of policy change typically do not employ underhand strategies. Those vested in extant policy have recourse to actions made possible by the advantages of power and/or possession, such as bureaucratic delay and obstruction.

20.5 How institutions matter

Much evidence exists for the relevance of the typology of change strategies that we present in our introductory chapter on theory (Chapter 2). Although we expected that these strategies would prove relevant in some cases, we had not anticipated their applicability in so many different institutional contexts. Policy entrepreneurs around the world appear to use quite similar strategies of coalition-building, exploitation of problem or political windows, and venue manipulation and venue-shopping. This does not mean that institutional differences are not relevant to water policy transitions. The case studies show that they do, in fact, affect the potential to achieve a policy transition and for a transition to endure. Different countries clearly offer different opportunity structures to individual or collective entrepreneurs. Hughes and McKay, for example, in their case study of water transitions in South Australia (Chapter 10) conclude that:

> While strong leaders and passionate advocates are often successful in creating policy change, the range of options available to them, the pace at which they may be able to effect change and the longevity of the changes they do achieve will likely be affected by the rules and boundaries within which they operate.

While in some countries advocates of change may feel frustrated as they encounter an institutional void and a lack of opportunity to articulate ideas, those advocating change in other countries may complain about institutional complexity. Both an absence and an abundance of institutions are often said to constrain the ability of change agents to bring about a policy transition. In addition individual institutional contexts each offer a particular opportunity structure. Complex decision-making processes in which many different parties and levels of government are involved usually feature a wide range of venues where change agents may place their issues on the agenda or seek support for their ideas. As Richardson has argued elsewhere, multiple venues form an important resource for interest groups (Richardson, 2000). Becker, in his chapter on transitions in German flood management (Chapter 18), also correlates multiple venues with policy entrepreneur opportunity, arguing that the German federal state structure, precisely because it offers a large number of venues for simultaneous discussion of similar issues, acts as fertile ground for the development of new policy approaches. As Lebel et al. (Chapter 8, this volume) in a similar vein conclude, the studies overall seem to show that: 'The increase in institutional density, complexity and redundancy is not dysfunctional. It provides alternative platforms for deliberation, policy influence at multiple levels'.

More centralized and less democratic government systems, however, offer other opportunities. Once policy entrepreneurs have managed to gain access to and support from the central decision-makers, they may be quite effective in realizing policy change in such systems (see te Boekhorst et al. on China, Chapter 6, this volume). The state organization of some countries might receive criticism for various normative reasons, but in terms of change management, critiquing is far less important than knowing a particular system well and using strategies that fit within the institutional context. Turton (Chapter 11, this volume), however, argues that in most of South African history, there was no space for individual change agents. It took the end of the apartheid regime for possibilities to arise for individual policy entrepreneurs to make a difference in the management of mining water.

20.6 Conclusions and possibilities for transition management

We provide Table 20.3 as a summary of the main conclusions of this book. The patterns of continuity and change in water management show that the policy changes observed are all partial (findings 1 and 2). All the transitions exhibit remnants of former regimes or discourses, and it is clear that a policy change does not guarantee successful implementation.

Some findings of this book strongly corroborate the policy science literature discussed in Chapter 2. Findings on characteristics of policy entrepreneurs (3), on the use of advocacy and brokerage strategies (6) and on the need to anticipate on windows (9) are

Table 20.3 Findings on water policy transitions, policy entrepreneurs and change strategies: lessons learned

On patterns of change	
1	New policy ideas (paradigms, discourses or ways of knowing) do not replace the 'old' ones, rather they are placed alongside or integrated with them, hence most transitions are partial
2	After new policies have been adopted, those who have an interest in maintaining the status quo have ample opportunities to delay or frustrate policy implementation
On policy entrepreneurs	
3	Policy entrepreneurs can be found anywhere, but what they have in common is a good reputation within their respective communities, good networking skills and perseverance
4	Successful entrepreneurship often is collective entrepreneurship in which individuals play complementary roles
On strategies (and institutions)	
5	A combination of bottom-up and top-down strategies makes most transitions happen, and their relative importance depends largely on the particular institutional context or opportunity structure
6	Successful (individual or collective) entrepreneurs are able to balance advocacy and brokerage strategies
7	Successful policy entrepreneurs build networks across different ways of knowing water (different meanings)
8	Successful policy entrepreneurs use narratives to frame issues strategically and thereby justify change and attract supporters
9	Successful policy entrepreneurs anticipate windows of opportunity by developing and testing attractive policy alternatives and demonstrating their feasibility
10	Successful policy entrepreneurs employ strategies of venue manipulation and venue-shopping and/or create new venues to be able to insert new ideas into decision-making processes
11	Policy entrepreneurs try to institutionalize new ideas (discourses, images or ways of knowing) and in this way create barriers to future change
12	Successful policy entrepreneurs have a full and thorough knowledge of the institutional system they are working in and know how to use that system

in perfect alignment with Kingdon's multiple streams model (Kingdon, 1995). Finding 9 supports the argument of Olsson et al. (2006) that systems need to be prepared for change. Baumgartner and Jones's concepts of venue manipulation and venue-shopping (Baumgartner and Jones, 1991) have been very useful for understanding change agents' strategic behaviour (finding 10); the same is true of their conclusion that newly institutionalized policy (new policy monopolies) can be just as stable as those they replace. Indeed change agents usually try to institutionalize their worldview, and as a result, set barriers in place to future change (finding 11).

Other insights, such as the importance of collective entrepreneurship (finding 4), the need for context-dependent balancing between top-down and bottom-up strategies (finding 5), the usefulness of network-building across different ways of knowing water (finding 7) and the possibilities for strategic use of narratives (finding 8), were derived from the case study material more inductively, and form a useful addition to or refinement of the strategies discussed in Chapter 2.

As mentioned above, there are similarities, but also some notable differences between developing countries and the developed world when it comes to identifying key change agents and their strategies. In the developing world donor organizations are crucial change agents. They have succeeded in changing governance paradigms almost overnight in many cases, but implementation following such change tells a different story (finding 2). The observed failure of the top-down strategies employed by the coalitions between donor organizations and national governments fits very well with principal findings of other research on policy implementation (for example Pressman and Wildawsky, 1973; Thomas and Grindle, 1990).

Our final finding pinpoints the need for policy entrepreneurs to adjust their strategies to the particular institutional context in which they are operating. Different institutional contexts produce not only different institutional constraints, but also different opportunity structures. Change agents need to be able to recognize and exploit the opportunities and peculiarities of a particular institutional system (finding 12).

Using the main findings of this book, we come now to answer the final research question: 'What are the implications of these findings for those who aim to direct change?' We argue in the introductory chapter (Chapter 1) that one of the main reasons for choosing a focus on agency is that this would yield lessons about successful strategies of transition initiation and management. Was this a reasonable calculation? The in-depth case studies on the roles and strategies of policy entrepreneurs have at minimum contributed to our knowledge of the complexity of change trajectories. This knowledge absorbed, a policy entrepreneur may be able to turn a 'trained eye' on newly developing change processes, equipped to evaluate and interpret them and to recognize opportunities for change and their characteristics. Examination of policy transitions must account for their intrinsically dynamic nature. 'The experience of managing in complex adaptive systems is more similar to catching waves or looking for emergent corridors for action than pulling strings or working levers' (Westley, 2002, p. 354). A simple list of 'dos and don'ts' would not do justice to the analyses presented in this book. Nevertheless we cannot resist the temptation to formulate some general recommendations for those who seek to direct policy change:

1. Develop a thorough knowledge of the characteristics and peculiarities of the relevant institutional system. This step is necessary to be able to recognize and characterize opportunities and to assess the likely effectiveness of specific change strategies.
2. Persevere and be willing to spend resources, especially time, to see a transition process through to implementation.
3. Attract supporters to the policy alternatives and build coalitions. Strike a balance between advocacy and brokerage strategies – negotiation and compromise can be as important as communication of the new ideas.
4. Recognize the various ways of knowing water. Frame the problem and develop the narrative strategy around those that fit the institutional and social context.
5. Anticipate windows of opportunity, whether they are opened by shock events or by political changes. Be prepared to exploit these opportunities and to insert the new ideas into the political debate. Pilot projects are useful to demonstrate feasibility, but should not exclude experimentation and testing.

6. Play a political game. Issues or disasters need to be framed strategically. Fora may require manipulation and forum-shopping can be necessary in order to bypass change-blockers.

20.7 Research agenda

Although the chapter authors all focus on the key questions of this book and present in-depth case analyses, we recognize that an edited volume will always have limitations. We hope that additional effort will take the research past the limits of this work. A useful next step would be a more systematic test of one or more hypotheses on the strategies used by policy entrepreneurs. All findings, as summarized in Table 20.3, can also be read as hypotheses that need further testing and refinement in a broader range of cases. We see three research topics as particularly fruitful: the nature of coalitions and the art of coalition-building; the interaction between agency and windows of opportunity; and change strategies in the developing world. In terms of framing the research, a great deal of conceptual work needs to be done in the policy sciences.

In this chapter we have made an analytical distinction between three types of coalitions: those based on shared worldviews or meanings; those formed around shared policy preferences (strategic alliances); and those centred on resource dependencies. But how do policy entrepreneurs choose which of these types to build, and when, and using which strategies? We have seen, for example, how policy entrepreneurs may try to bridge and relate multiple ways of knowing water (meanings), which is an example of policy brokering. However they may also use narrative strategies to improve their advocacy and attract supporters. How do entrepreneurs balance advocacy and brokerage strategies? What makes them select a particular strategy or a combination? These kinds of questions go beyond an explanatory typology of strategies related to policy transition and could further guide the work of policy entrepreneurs.

Policy entrepreneurs need opportunities in the form of problem and political windows to get their new policy ideas and plans accepted and realized. We have learned that both anticipating windows and the framing or 'painting' of windows can be effective strategies. The cases show, however, that usually it is a combination or a series of windows that produce a real opening for change to occur. More research is needed on how policy entrepreneurs exploit windows of opportunity that open simultaneously or successively.

One of the striking observations in this book is the failure of implementation in most water policy transitions in the developing world. Donor agencies, such as the World Bank, face huge difficulties in implementing new governance models, such as the concept of water user associations. The burning question, then, is whether more practical alternatives can be developed. Which policies fit better with the institutional (historical and cultural) characteristics of these societies? Would it be more fruitful to build on existing institutional capacity rather than assuming the need for a new institution to take on the task of implementation? Would a leave-alone strategy be effective? Case-by-case responses to such questions might lead to more successful implementation.

Confusion and overlap are rife in regard to concepts in the policy science literature. In this book we have grouped concepts that show considerable overlap, such as the concepts of frames, narratives and ways of knowing. This rough grouping of theories and concepts works for this book, but clearly entails an interesting theoretical research agenda. How do these concepts relate to each other, how exactly do they differ, and when do these differences prove useful? This kind of conceptual research would help to continue to develop a theory on (the management of) water policy transitions.

We very much hope that the case analyses, the findings and research agenda presented in this book will stimulate others to investigate further the strategies of policy entrepreneurs in realizing water policy transitions. In this way we hope also that we might be of some help to those engaged in bringing about improvements that are so urgently needed globally in freshwater management.

References

Baumgartner, F.R. and B.D. Jones (1991), 'Agenda dynamics and policy subsystems', *Journal of Politics*, **53** (4), 1044–74.

Birkland, T.A. (1997), *After Disaster: Agenda Setting, Public Policy, and Focusing Events*, Washington, DC: Georgetown University Press.

Boin, A., P. 't Hart and A. McConnell (2009), 'Crisis exploitation: political and policy impacts of framing contests', *Journal of European Public Policy*, **16** (1), 81–106.

Kingdon, J.W. (1995), *Agendas, Alternatives and Public Policies*. 2nd edn, New York: HarperCollins.

Meijerink, S. (2005), 'Understanding policy stability and change: the interplay of advocacy coalitions and epistemic communities, windows of opportunity, and Dutch coastal flooding policy 1945–2003', *Journal of European Public Policy*, **12** (6), 1060–77.

O'Leary, R. and L. Bingham (2007), 'A manager's guide to resolving conflicts in collaborative networks', report, IBM Centre for the Business of Government.

Olsson, P., L.H. Gunderson, S.R. Carpenter, P. Ryan, L. Lebel, C. Folke and C.S. Holling (2006), 'Shooting the rapids: navigating transitions to adaptive governance of social–ecological systems', *Ecology and Society*, **11** (1), 18, www.ecologyandsociety.org/vol11/iss1/art18/.

Ostrom, E. (2005), *Understanding Institutional Diversity*, Princeton, NJ: Princeton University Press.

Pierson, P. (2000), 'Increasing returns, path dependence, and the study of politics', *American Political Science Review*, **94** (2), 251–67.

Pressman, J.L. and A. Wildawsky (1973), *Implementation. How Great Expectations in Washington are Dashed in Oakland; Or, Why it's Amazing that Federal Programs Work at All, This being a Saga of the Economic Development Administration as Told by Two Sympathetic Observers Who Seek to Build Morals on a Foundation of Ruined Hopes*, Berkeley, CA: University of California Press.

Richardson, J. (2000), 'Government, interest groups and policy change', *Political Studies*, **48** (5), 1006–25.

Sabatier, P.A. (1993), 'Policy change over a decade or more', in P.A. Sabatier and H.C. Jenkins-Smith (eds), *Policy Change and Learning: An Advocacy Coalition Approach,* Boulder, CO: Westview Press, pp. 13–39.

Thomas, J.W. and Merilee S. Grindle (1990), 'After the decision: implementing policy reforms in developing countries', *World Development*, **18** (8), 1163–81.

Westley, F. (2002), 'Adaptive management on the front lines', in L.H. Gunderson and C.S. Holling (eds), *Panarchy: Understanding Transformations in Human and Natural Systems*, Washington, DC: Island Press, pp. 333–60.

Index